Safe Handling of Foods

FOOD SCIENCE AND TECHNOLOGY

A Series of Monographs, Textbooks, and Reference Books

1. Flavor Research: Principles and Techniques, *R. Teranishi, I. Hornstein, P. Issenberg, and E. L. Wick*
2. Principles of Enzymology for the Food Sciences, *John R. Whitaker*
3. Low-Temperature Preservation of Foods and Living Matter, *Owen R. Fennema, William D. Powrie, and Elmer H. Marth*
4. Principles of Food Science
   Part I: Food Chemistry, *edited by Owen R. Fennema*
   Part II: Physical Methods of Food Preservation, *Marcus Karel, Owen R. Fennema, and Daryl B. Lund*
5. Food Emulsions, *edited by Stig E. Friberg*
6. Nutritional and Safety Aspects of Food Processing, *edited by Steven R. Tannenbaum*
7. Flavor Research: Recent Advances, *edited by R. Teranishi, Robert A. Flath, and Hiroshi Sugisawa*
8. Computer-Aided Techniques in Food Technology, *edited by Israel Saguy*
9. Handbook of Tropical Foods, *edited by Harvey T. Chan*
10. Antimicrobials in Foods, *edited by Alfred Larry Branen and P. Michael Davidson*

11. Food Constituents and Food Residues: Their Chromatographic Determination, *edited by James F. Lawrence*
12. Aspartame: Physiology and Biochemistry, *edited by Lewis D. Stegink and L. J. Filer, Jr.*
13. Handbook of Vitamins: Nutritional, Biochemical, and Clinical Aspects, *edited by Lawrence J. Machlin*
14. Starch Conversion Technology, *edited by G. M. A. van Beynum and J. A. Roels*
15. Food Chemistry: Second Edition, Revised and Expanded, *edited by Owen R. Fennema*
16. Sensory Evaluation of Food: Statistical Methods and Procedures, *Michael O'Mahony*
17. Alternative Sweeteners, *edited by Lyn O'Brien Nabors and Robert C. Gelardi*
18. Citrus Fruits and Their Products: Analysis and Technology, *S. V. Ting and Russell L. Rouseff*
19. Engineering Properties of Foods, *edited by M. A. Rao and S. S. H. Rizvi*
20. Umami: A Basic Taste, *edited by Yojiro Kawamura and Morley R. Kare*
21. Food Biotechnology, *edited by Dietrich Knorr*
22. Food Texture: Instrumental and Sensory Measurement, *edited by Howard R. Moskowitz*
23. Seafoods and Fish Oils in Human Health and Disease, *John E. Kinsella*
24. Postharvest Physiology of Vegetables, *edited by J. Weichmann*
25. Handbook of Dietary Fiber: An Applied Approach, *Mark L. Dreher*
26. Food Toxicology, Parts A and B, *Jose M. Concon*
27. Modern Carbohydrate Chemistry, *Roger W. Binkley*
28. Trace Minerals in Foods, *edited by Kenneth T. Smith*
29. Protein Quality and the Effects of Processing, *edited by R. Dixon Phillips and John W. Finley*
30. Adulteration of Fruit Juice Beverages, *edited by Steven Nagy, John A. Attaway, and Martha E. Rhodes*
31. Foodborne Bacterial Pathogens, *edited by Michael P. Doyle*
32. Legumes: Chemistry, Technology, and Human Nutrition, *edited by Ruth H. Matthews*
33. Industrialization of Indigenous Fermented Foods, *edited by Keith H. Steinkraus*
34. International Food Regulation Handbook: Policy • Science • Law, *edited by Roger D. Middlekauff and Philippe Shubik*
35. Food Additives, *edited by A. Larry Branen, P. Michael Davidson, and Seppo Salminen*
36. Safety of Irradiated Foods, *J. F. Diehl*
37. Omega-3 Fatty Acids in Health and Disease, *edited by Robert S. Lees and Marcus Karel*
38. Food Emulsions: Second Edition, Revised and Expanded, *edited by Kåre Larsson and Stig E. Friberg*
39. Seafood: Effects of Technology on Nutrition, *George M. Pigott and Barbee W. Tucker*
40. Handbook of Vitamins: Second Edition, Revised and Expanded, *edited by Lawrence J. Machlin*

41. Handbook of Cereal Science and Technology, *Klaus J. Lorenz and Karel Kulp*
42. Food Processing Operations and Scale-Up, *Kenneth J. Valentas, Leon Levine, and J. Peter Clark*
43. Fish Quality Control by Computer Vision, *edited by L. F. Pau and R. Olafsson*
44. Volatile Compounds in Foods and Beverages, *edited by Henk Maarse*
45. Instrumental Methods for Quality Assurance in Foods, *edited by Daniel Y. C. Fung and Richard F. Matthews*
46. *Listeria*, Listeriosis, and Food Safety, *Elliot T. Ryser and Elmer H. Marth*
47. Acesulfame-K, *edited by D. G. Mayer and F. H. Kemper*
48. Alternative Sweeteners: Second Edition, Revised and Expanded, *edited by Lyn O'Brien Nabors and Robert C. Gelardi*
49. Food Extrusion Science and Technology, *edited by Jozef L. Kokini, Chi-Tang Ho, and Mukund V. Karwe*
50. Surimi Technology, *edited by Tyre C. Lanier and Chong M. Lee*
51. Handbook of Food Engineering, *edited by Dennis R. Heldman and Daryl B. Lund*
52. Food Analysis by HPLC, *edited by Leo M. L. Nollet*
53. Fatty Acids in Foods and Their Health Implications, *edited by Ching Kuang Chow*
54. *Clostridium botulinum*: Ecology and Control in Foods, *edited by Andreas H. W. Hauschild and Karen L. Dodds*
55. Cereals in Breadmaking: A Molecular Colloidal Approach, *Ann-Charlotte Eliasson and Kåre Larsson*
56. Low-Calorie Foods Handbook, *edited by Aaron M. Altschul*
57. Antimicrobials in Foods: Second Edition, Revised and Expanded, *edited by P. Michael Davidson and Alfred Larry Branen*
58. Lactic Acid Bacteria, *edited by Seppo Salminen and Atte von Wright*
59. Rice Science and Technology, *edited by Wayne E. Marshall and James I. Wadsworth*
60. Food Biosensor Analysis, *edited by Gabriele Wagner and George G. Guilbault*
61. Principles of Enzymology for the Food Sciences: Second Edition, *John R. Whitaker*
62. Carbohydrate Polyesters as Fat Substitutes, *edited by Casimir C. Akoh and Barry G. Swanson*
63. Engineering Properties of Foods: Second Edition, Revised and Expanded, *edited by M. A. Rao and S. S. H. Rizvi*
64. Handbook of Brewing, *edited by William A. Hardwick*
65. Analyzing Food for Nutrition Labeling and Hazardous Contaminants, *edited by Ike J. Jeon and William G. Ikins*
66. Ingredient Interactions: Effects on Food Quality, *edited by Anilkumar G. Gaonkar*
67. Food Polysaccharides and Their Applications, *edited by Alistair M. Stephen*
68. Safety of Irradiated Foods: Second Edition, Revised and Expanded, *J. F. Diehl*
69. Nutrition Labeling Handbook, *edited by Ralph Shapiro*

70. Handbook of Fruit Science and Technology: Production, Composition, Storage, and Processing, *edited by D. K. Salunkhe and S. S. Kadam*
71. Food Antioxidants: Technological, Toxicological, and Health Perspectives, *edited by D. L. Madhavi, S. S. Deshpande, and D. K. Salunkhe*
72. Freezing Effects on Food Quality, *edited by Lester E. Jeremiah*
73. Handbook of Indigenous Fermented Foods: Second Edition, Revised and Expanded, *edited by Keith H. Steinkraus*
74. Carbohydrates in Food, *edited by Ann-Charlotte Eliasson*
75. Baked Goods Freshness: Technology, Evaluation, and Inhibition of Staling, *edited by Ronald E. Hebeda and Henry F. Zobel*
76. Food Chemistry: Third Edition, *edited by Owen R. Fennema*
77. Handbook of Food Analysis: Volumes 1 and 2, *edited by Leo M. L. Nollet*
78. Computerized Control Systems in the Food Industry, *edited by Gauri S. Mittal*
79. Techniques for Analyzing Food Aroma, *edited by Ray Marsili*
80. Food Proteins and Their Applications, *edited by Srinivasan Damodaran and Alain Paraf*
81. Food Emulsions: Third Edition, Revised and Expanded, *edited by Stig E. Friberg and Kåre Larsson*
82. Nonthermal Preservation of Foods, *Gustavo V. Barbosa-Cánovas, Usha R. Pothakamury, Enrique Palou, and Barry G. Swanson*
83. Milk and Dairy Product Technology, *Edgar Spreer*
84. Applied Dairy Microbiology, *edited by Elmer H. Marth and James L. Steele*
85. Lactic Acid Bacteria: Microbiology and Functional Aspects: Second Edition, Revised and Expanded, *edited by Seppo Salminen and Atte von Wright*
86. Handbook of Vegetable Science and Technology: Production, Composition, Storage, and Processing, *edited by D. K. Salunkhe and S. S. Kadam*
87. Polysaccharide Association Structures in Food, *edited by Reginald H. Walter*
88. Food Lipids: Chemistry, Nutrition, and Biotechnology, *edited by Casimir C. Akoh and David B. Min*
89. Spice Science and Technology, *Kenji Hirasa and Mitsuo Takemasa*
90. Dairy Technology: Principles of Milk Properties and Processes, *P. Walstra, T. J. Geurts, A. Noomen, A. Jellema, and M. A. J. S. van Boekel*
91. Coloring of Food, Drugs, and Cosmetics, *Gisbert Otterstätter*
92. *Listeria*, Listeriosis, and Food Safety: Second Edition, Revised and Expanded, *edited by Elliot T. Ryser and Elmer H. Marth*
93. Complex Carbohydrates in Foods, *edited by Susan Sungsoo Cho, Leon Prosky, and Mark Dreher*
94. Handbook of Food Preservation, *edited by M. Shafiur Rahman*
95. International Food Safety Handbook: Science, International Regulation, and Control, *edited by Kees van der Heijden, Maged Younes, Lawrence Fishbein, and Sanford Miller*
96. Fatty Acids in Foods and Their Health Implications: Second Edition, Revised and Expanded, *edited by Ching Kuang Chow*

97. Seafood Enzymes: Utilization and Influence on Postharvest Seafood Quality, *edited by Norman F. Haard and Benjamin K. Simpson*
98. Safe Handling of Foods, *edited by Jeffrey M. Farber and Ewen C. D. Todd*
99. Handbook of Cereal Science and Technology: Second Edition, Revised and Expanded, *edited by Karel Kulp and Joseph G. Ponte, Jr.*
100. Food Analysis by HPLC: Second Edition, Revised and Expanded, *edited by Leo M. L. Nollet*
101. Surimi and Surimi Seafood, *edited by Jae W. Park*

Additional Volumes in Preparation

Handbook of Water Analysis, *edited by Leo M. L. Nollet*

Handbook of Nutrition and Diet, *B. B. Desai*

Seafood and Freshwater Toxins: Pharmacology, Physiology, and Detection, *edited by Luis M. Botana*

Safe Handling of Foods

edited by

Jeffrey M. Farber
Ewen C. D. Todd

Health Canada
Ottawa, Ontario, Canada

MARCEL DEKKER

MARCEL DEKKER, INC. NEW YORK • BASEL

ISBN: 0-8247-0331-6

This book is printed on acid-free paper.

Headquarters
Marcel Dekker, Inc.
270 Madison Avenue, New York, NY 10016
tel: 212-696-9000; fax: 212-685-4540

Eastern Hemisphere Distribution
Marcel Dekker AG
Hutgasse 4, Postfach 812, CH-4001 Basel, Switzerland
tel: 41-61-261-8482; fax: 41-61-261-8896

World Wide Web
http://www.dekker.com

The publisher offers discounts on this book when ordered in bulk quantities. For more information, write to Special Sales/Professional Marketing at the headquarters address above.

Current printing (last digit):
10 9 8 7 6 5 4 3 2 1

PRINTED IN THE UNITED STATES OF AMERICA

Preface

Foodborne disease is now thought to be a major cause of gastrointestinal (GI) illness worldwide. Reported cases, however, are believed to represent only a small fraction of the people who are actually ill after eating contaminated food. Most people, in fact, do not visit a physician or go to a hospital unless the GI symptoms are severe or prolonged, by which time it may not be possible to isolate a pathogen, even if clinical specimens are sent to a laboratory. For instance, hemolytic uremic syndrome may arise from an infection of *Escherichia coli* O157 or other verotoxin-producing strains, but by the time the kidneys start failing the pathogens are often not present to isolate from stools. For *Listeria monocytogenes,* incubation periods can be longer than one month.

Case estimates have been made in several countries, such as the United States, the United Kingdom, Canada, and Australia, with numbers of cases up to 100 times those reported from lab isolation data or 350 times those reported in outbreaks. These estimates indicate that there are millions of people each year suffering from foodborne disease. The latest estimates for the United States alone is over 70 million cases per year. Although data on foodborne disease in developing countries are scarce, the actual case numbers are probably much higher than those reported in industrialized nations, because of the larger populations and fewer well-developed public health systems.

A small percentage of foodborne disease cases may go on to develop long-lasting sequelae, such as chronic diarrhea, reactive arthritis, or CNS complications, and these can result in a lower quality of life and lost earnings. Much remains to be learned about the number and types of sequelae that are associated with foodborne disease agents. In addition, up to 5000 deaths have been estimated to occur annually in the United States alone. Therefore, the impact of morbidity and mortality from foodborne disease is substantial. Expressed in monetary terms, the costs for human illness, value of lives shortened through death, lost productivity to society, constraints on the health care and public health systems, recall and destruction of food, loss of sales to the food industry, and legal settlements are in the billions of dollars. Safe handling of foods is one way to reduce the likelihood that such a disease will strike the consumer. This book discusses the production of all types of foods by the processing and foodservice industries, where the risk factors are, and what the consumer can do to prepare safe food.

The main types of problems that lead to foodborne illness result from foods that are easily contaminated, can readily support the growth of foodborne pathogens, and have limited treatment to remove or destroy the organisms. These include raw foods of animal and plant origin that are not cooked or are lightly heated. Another group of high-risk foods are ready-to-eat foods that have been frequently handled during processing and may be warmed only before consumption. People at particular risk from eating contaminated foods are travelers, those in developing countries, and those in unique or specialized occupations. In addition, certain populations, such as the very young, the elderly, the ill, and the immunocompromised are vulnerable to severe symptoms if they contract a foodborne infection—they should take precautions to avoid the more hazardous foods.

We are likely to see an increase in the incidence of foodborne illness as our population ages and we increase our leisure travel and import more exotic foods—adding to these pressures will be changes in our dietary habits, changes in food processing and packaging, the globalization of our food supply, and global warming. We have already seen an increase in the numbers of foodborne illnesses due to *Vibrio* spp. that have been associated with warmer seawater temperatures around North America as an effect of global warming. Emerging infectious disease threats will continue to be a major source of concern not only nationally but also from abroad. Because of the advent of new molecular typing methods along with advanced surveillance, the cause of many more foodborne outbreaks will be identified and organisms that we thought were only commensals will come to the fore as true or opportunistic pathogens. Antibiotic resistance will also be a major force to reckon with in the future. For example, we are now seeing an increase in the number of quinolone-resistant *Campylobacter jejuni* infections in the United States, largely due to the acquisition of resistant strains from poultry.

There are ways in which risks can be minimized through proper handling that will reduce the chances of pathogens coming in contact with foods. One should bear in mind, however, that a goal of "absence of any risk" is unattainable. In the future, mandated HACCP will be the norm for all countries across all industries, and its impact in terms of reducing the incidence of foodborne illness will be watched closely and evaluated. A variety of new processing technologies, including high-intensity light and high pressure, will be incorporated into HACCP systems. In addition, some forms of food irradiation will eventually become acceptable for finished-product pasteurization. However, these do not obviate the need for correct food handling. Many countries will require the training of all foodservice and food processing workers in safe food handling techniques, especially as these new technologies are developed and almost-sterile foods are marketed. Since such foods may allow rapid growth of pathogens if they become contaminated after opening, safe food handling will be seen as the responsibility of all parties, including the consumer, along the farm-to-fork continuum. The future for food safety trainers and educators will definitely be a challenging one as we begin the new millennium!

Jeffrey M. Farber
Ewen C. D. Todd

Contents

Contributors

Robert E. Brackett Center for Food Safety and Quality Enhancement, Department of Food Science and Technology, University of Georgia, Griffin, Georgia

Frank L. Bryan Food Safety Consultation and Training, Lithonia, Georgia

Jeffrey M. Farber Microbiology Research Division, Food Directorate, Health Canada, Ottawa, Ontario, Canada

Pina M. Fratamico Eastern Regional Research Center, Agricultural Research Service, U.S. Department of Agriculture, Wyndmoor, Pennsylvania

Colin O. Gill Lacombe Research Centre, Agriculture and Agri-Food Canada, Lacombe, Alberta, Canada

Chris Griffith Food Safety Research Group, University of Wales Institute, Cardiff, South Wales, United Kingdom

John J. Guzewich Center for Food Safety and Applied Nutrition, U.S. Food and Drug Administration, Washington, D.C.

Ashton Hughes Microbiology Evaluation Division, Health Canada, Ottawa, Ontario, Canada

Thomas Jemmi Microbiology Section, Swiss Federal Veterinary Office, Bern, Switzerland

Young-jae Kang Catering Services Department, Asiana Airlines, Inc., Seoul, Korea

Kelley P. Knight Food Research Program, Southern Crop Protection and Food Research Centre, Agriculture and Agri-Food Canada, Guelph, Ontario, Canada

Marilyn B. Lee School of Occupational and Public Health, Ryerson Polytechnic University, Toronto, Ontario, Canada

Robin C. McKellar Food Research Program, Southern Crop Protection and Food Research Centre, Agriculture and Agri-Food Canada, Guelph, Ontario, Canada

Thomas E. Rippen Sea Grant Extension Program, University of Maryland, Princess Anne, Maryland

Don Schaffner Food Risk Analysis Initiative, Rutgers University, New Brunswick, New Jersey

Michel Schmitt Department of Permits and Inspections, Swiss Federal Veterinary Office, Bern, Switzerland

Elizabeth Scott Consultant in Food and Environmental Hygiene, Newton, Massachusetts

James L. Smith Eastern Regional Research Center, Agricultural Research Service, U.S. Department of Agriculture, Wyndmoor, Pennsylvania

O. Peter Snyder, Jr. Hospitality Institute of Technology and Management, St. Paul, Minnesota

Gloria I. Swick Health Commissioner, Perry County Health Department, New Lexington, Ohio

Ewen C. D. Todd Microbiology Research Division, Food Directorate, Health Canada, Ottawa, Ontario, Canada

Donald W. Warburton Microbiology Evaluation Division, Food Directorate, Health Canada, Ottawa, Ontario, Canada

1

Safe Handling of Raw Meat and Poultry Products

Colin O. Gill
Agriculture and Agri-Food Canada, Lacombe, Alberta, Canada

I. HANDLING OF ANIMALS TO ENHANCE SAFETY

The hazards that may be controlled at the farm level are the inclusion in meat tissues of objects that may injure a consumer; the contamination of meat tissues with chemicals, pesticides, or drugs; overt disease of meat animals; and the symptomless carriage and shedding of enteric pathogens that could infect consumers.

The effective control on-farm of hazards related to meat is dependent on the identification of animals and the maintenance of records for stock on relevant matters. The extents to which animals can be identified varies with species and with the farming practices under which they are reared. Large animals that are intensively reared in monitored and/or controlled environments can be and often are individually identified (88,89). Such stock would include pigs, dairy cattle, and feedlotted beef cattle. Individual identification of small animals that are intensively reared, such as most poultry, is not economically feasible. Rather, since such animals are raised in flocks of individuals of the near same age, with all being treated as a group, identification must be by reference to the flock, not the individual. The identification of range-reared animals, such as cattle and sheep, is commonly by herd or flock and principally for the purpose of establishing ownership. Of course, farming practices in some locations may fall between the extremes of range and intensive rearing, or they may involve both types of rearing for animals at different times. For example, feedlotting of range-reared beef cattle for 3 or 4 months before their slaughter is a usual practice in several major beef-producing countries.

Ideally, the information recorded for each animal should include the times of routine injections, with note of each injection site and any needle breaking event that might have left a hazardous piece of metal in the flesh; the chemicals used and the times of treatments to control parasites or to treat infections; the times and types of any other treatments (castration, dehorning, shearing, etc.); the times of onset or occurrence, and resolution, of disease conditions, wounds, or disabilities that resolved without treatment or had not resolved by the time of slaughter; and any circumstances or incident that might suggest the possibility of exposure to toxic chemicals such as pesticides or heavy metals (114). The routine recording of that detailed information is possible for individual, large animals that are intensively reared, but for intensively reared small animals or range-reared animals, the only data that can be obtained are those on diseases in and perhaps exposure to toxic chemicals, or treatments, of flocks or herds.

The routine collection of information on the health and treatment of animals is justified only if the information is used to correct or, preferably, to prevent undesirable economic consequences for animal or meat production, or hazardous conditions of meat. Developed systems for animal production are usually operated under some disease control program with the inoculation of animals against some diseases, the routine treatment of animals to control infestations by parasites, surveillance of animals for symptoms of disease or injury, and appropriate segregation and treatment, or slaughter and disposal of diseased or injured animals. Such systems may include practices for handling feedstuffs, water, animals, and wastes in ways that help prevent diseases (65,121) and procedures for handling medicated animals to ensure that any drug residues in meat tissues when the animal is presented for slaughter are at acceptably low levels (87,113).

The value of the information collected on-farm can be greatly augmented if it is shared with, and supplemented by information obtained from, the animal slaughtering plant (6,66). Information on the health, drug residue status, and possible presence in individual animals of broken needles is useful to the packing plant for deciding on the appropriate handling and use of individual carcasses (Fig. 1). In return the plant can provide the farm with information on disease conditions, such as plural adhesion and liver defects, which are apparent only when the carcass is dressed. Moreover, plants can provide information on carcass conformation, and muscling and fat cover, that can aid the development of optimal breeding strategies and feeding regimes (82). Programs for the sharing of information between plant and farm usually involve the payment of bonuses for animals of superior quality (Fig. 2).

The matters of physical and chemical hazards associated with meat tissues, as well as overt diseases in animals, can be addressed by the maintenance of suitable records, provided there is appropriate and timely response to the information collected. However, the situation is rather different with respect to the carriage and shedding by symptomless animals of enteric pathogens. The contamination of raw

PORK PROGRAM
Pig Delivery Form

Producer Name:			
Tattoo(s) Used:			
Number of Pigs in This Shipment:			
Date:		Time Leaving Farm:	

Drug Residues:

These Pigs are free from drug residues ☐ Yes ☐ No

All drugs used are recorded in accordance with program requirements ☐ Yes ☐ No

Broken Needles:

There may be a broken needle in pig marked:______________________

Location of broken needle:____________________________________

I certify that the above information is correct:

______________________________ ______________________________

Print Name Signature

Figure 1 Information on drug residue status and possible presence of broken needles provided to a pork packing plant by pig producers participating in a pork quality program.

meat with enteric pathogens originating from animals is the hazard that poses the greatest risk to consumers (17). The organisms that may be involved in such hazardous contamination include *Salmonella,* pathogenic strains of *Escherichia coli, Campylobacter,* and *Yersinia enterocolitica* (80). All meat species may carry those organisms, and *Salmonella* are of concern with all (112), while pathogenic *E. coli, Campylobacter,* and *Yersinia* are found with particular frequency in cattle, poultry, and pigs, respectively (2,84,94).

Since the enteric organisms are frequently found in feral as well as domestic animals, the rearing of animals assuredly free of those pathogens is practicable only in laboratory circumstances. However, the numbers of those organisms in domestic animals may be controlled by various practices, such as the measures adopted in Denmark for the control of *Salmonella* in pigs (4). These comprise the production of *Salmonella*-free feeds, the testing of breeding herds and slaughtered

PORK PROGRAM

Pork Report for the Period: d/m/y to d/m/y

Producer Identification No:
Name:

Number of Head Received:
Number of Head Probed:

Carcass Bonus

Level one: Weight 88+ kg; Fat 12-18 mm; Lean 54-58 kg
Number of Head at $x each

Level two: Weight 88+ kg; Fat 12-18 mm; Lean 58+ kg
Number of Head at $2x each

Carcass Bonus $

Health Bonus

Number of Head:
Less:
Dead and condemned
Tattoo problems
Black hair
Number Eligible at $y each

Less:
Inspection demerits
Chest adhesions
Liver/Heart scars
Total Demerits at $y each

Health Bonus $

Program Bonus $ (Carcass and Health Bonuses)

Figure 2 Information for the computation of bonuses provided by a pork packing plant to pig producers participating in a pork quality program.

stock for antibodies specific to *Salmonella,* the offering of advice on control measures to those managing herds with high incidences of animals yielding antibodies to *Salmonella,* and segregated slaughter of animals from those herds until the prevalence of animals exposed to *Salmonella* is reduced to an acceptable level. Control measures include control of vermin, prevention of contamination of feed and drinking water, appropriate disposal of waste, and cleaning of rearing facilities between batches of animals (62).

A similar approach can be applied to the control of *Salmonella* in poultry, assuming that the incidence of the organism in fecal samples is determined. In addition, with poultry, the carriage of enteric pathogens by birds can be reduced by competitive exclusion (64).

The carriage of pathogens by birds involves colonization of the intestinal tract, particularly the cecum, by pathogenic bacteria. Chicks are highly susceptible to such colonization if they are exposed to the enteric organisms before the normal flora of the gut has fully developed. Once established, the normal gut flora tends to inhibit colonization of the gut, apparently because the niches required by the pathogenic species are occupied by other organisms (8). Establishment of the normal gut flora in the chick can be accelerated by feeding feces from pathogen-free birds. Although it would be desirable to establish competitive exclusion by feeding chicks cultures of known organisms, attempts to produce effective, defined cultures have so far been unsuccessful. However, the establishment of competitive exclusion by feeding suitable preparations of faeces is now a common practice in some regions (86).

With cattle there has been much interest in the possibility of controlling fecal shedding of the pathogen *Escherichia coli* 0157:H7 by manipulation of the diet for a period before slaughter. That interest derives from studies showing that volatile fatty acids in the large intestines of monogastric animals inhibit colonization by *Salmonella* (96), and other studies indicating that the numbers of *E. coli* and *Salmonella* in the rumen are affected by the concentrations of short chain length free fatty acids in the rumen fluid (83). The fatty acids are produced by the rumen flora's fermentation of the feed. Thus, when feed is withdrawn, or a readily degraded feed such as grain is replaced by one that is attacked less rapidly by bacteria (e.g., hay), the concentrations of free fatty acids tend to fall and the numbers of enteric organisms tend to rise (76). However, it is uncertain how that effect might be exploited. Feeding of animals during transport to a slaughtering plant is not practical, while slaughtering plants wish to receive animals that have been off feed for several hours, since feed withdrawal reduces the volume of paunch (first and largest stomach) contents and with that the risk of spilling paunch contents onto the carcass during evisceration. Moreover, the effects of shifting diets may not be consistent, since both increases and decreases in *E. coli* numbers as a result of switching from grain to hay have been reported (19,78).

In recent years it has become the fashion to suggest that hazard analysis–critical control Point (HACCP) systems can be developed for all stages of food production, including the rearing of animals (69). HACCP systems are a special case of control of product quality through control of the production process, where the quality to be assured is the safety of the product (30). For such a system to be effective, all possible hazardous contamination of the product during the production process must be either identified and maintained at an acceptable level, or prevented, by appropriate control of each point in a process at which a contaminant may be added or removed (15). For an HACCP system to work, the process to which it is applied must be performed consistently: when a process is inconsistent, there can be no certainty at any time that the hazards and risks associated with the process are being identified, far less controlled. It seems unlikely that any animal production process is maintained in a state sufficiently consistent for an HACCP system applied to it to be anything more than notional. In those circumstances, it would seem best not to confuse the concept of HACCP by applying it inappropriately to the partial control of hazards in inconsistent processes such as animal rearing. Instead, it would seem appropriate to refer to the containment of known on-farm hazards for foods by the application of good management practices (GMP), with the recognition that practices that are optimal for food safety may differ for similar farm products in different circumstances. Thus, GMPs should preferably be established from information that is directly relevant to each production process, rather than being assumed from general considerations.

II. SAFE HANDLING OF MEAT AT THE SLAUGHTERING PLANT

A. Inspection and HACCP Implementation

Animals presented for slaughter are subject to veterinary inspection for overt symptoms of disease or injury. The organs and carcasses of slaughtered animals are similarly inspected during the carcass dressing process, with condemnation of any overtly diseased, parasitized, or otherwise possibly hazardous organs, carcasses, or appropriate parts of carcasses that are detected. Thus, insofar as diseased or possibly hazardous, damaged tissues can be detected by inspection, they are eliminated from the meat supply (90). Inspection will, however, often involve procedures that are inappropriate for particular stock populations (68) and the condemnation of material that is not hazardous, such as bruised tissue, which is not microbiologically compromised unless it is associated with a penetrating wound (102).

Carcass and edible organs are also inspected for visible contamination by feces, ingesta, hair, or nondescript filth, with the requirement that visibly contaminated tissue be cut off and discarded. It is assumed that eliminating such tissue will serve to control the contamination of meat with enteric pathogens (115). However, despite increasingly stringent inspection of slaughtering plants and carcasses, enteric diseases associated with the consumption of meat have continued to increase (116). Consequently regulatory authorities in most countries have conceded that traditional meat inspection cannot assure the safety of meat with respect to contamination with enteric pathogens. Instead, most regulatory authorities are now encouraging or mandating the development of HACCP systems at meatpacking plants (16,118).

In the absence of any other recognized method of assessing the hygienic performances of meat plant processes, those attempting to develop HACCP systems at meatpacking plants have had recourse to traditional inspection beliefs and practices (71). Thus, it is assumed that controlling activities should be largely focused on the carcass dressing process (Fig. 3). Before an HACCP system for carcass dressing is implemented, a plant must meet regulatory specifications with respect to plant maintenance, control of vermin, and the cleaning of plant facilities and equipment. As before, the adequacy of the measures in those areas is decided by inspection. Then, the HACCP system for carcass dressing is developed on the basis of subjective judgment about which operations are likely to contaminate the meat with enteric pathogens (120). The HACCP system is therefore constructed and operated in the absence of any certain knowledge of the microbiological effects of any of the activities that are supposed to control the microbiological condition of the product. It seems self-evident that such an approach to control cannot be satisfactory.

The alternative approach is to base control of meat plant processes on appropriate microbiological data. In the context of an HACCP system, microbiological data cannot be used for process monitoring because the effort required for the collection and processing of microbiological samples severely restricts the amounts of data that can be practicably and economically collected, while there is an inevitably long delay between the collection of a sample and the enumeration of the bacteria it yields. Instead, microbiological data are used to describe the hygienic performance of a process, to objectively identify the critical control points (CCPs), to ascertain that actions aimed at improving the process do in fact improve the microbiological condition of the product, and to verify the effective operation of the HACCP system (28). Monitoring of the process and maintenance of control are obtained by establishing standard operating procedures (SOPs) at critical control points, with appropriate, routine checking procedures to assure adherence to SOPs.

B. Assessment of the Microbiological Condition of Product

The microbiological condition of product at any stage of a process can be assessed by collecting random samples from product passing through the process and calculating values for the mean and standard deviation of the $\log_{10}$ bacterial counts, on the assumption that bacteria on or in the product are log normally distributed (13). However, the microbiological conditions of product at different stages of a process or from different processes cannot be properly compared on the basis of the mean log values, as is commonly done. Handling of product often results in a more even distribution of bacteria on the product. This will lead to a reduction of the variance of the distribution of the log bacterial counts and an increase in the mean log counts when no further bacteria are added to the product (Fig. 4). Thus,

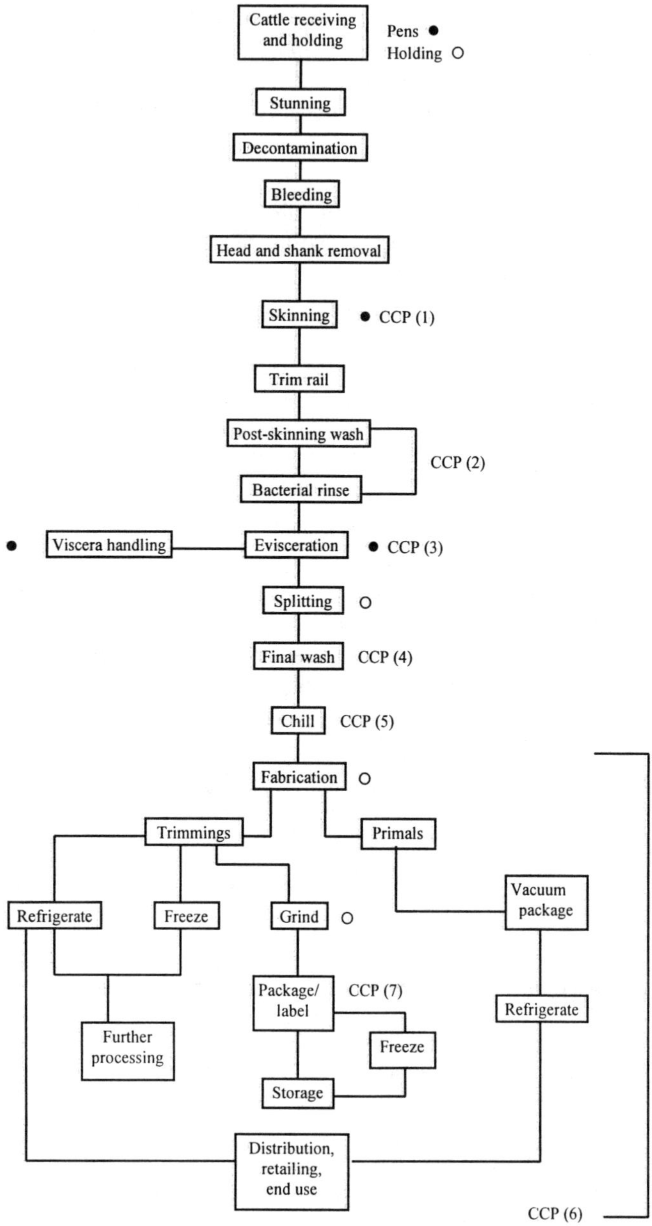

Figure 3 Generic HACCP for beef slaughter, fabrication and packaging: open points, potential sites of minor contamination; solid points, potential sites of major contamination; CCP, critical control point. (From Ref. 120.)

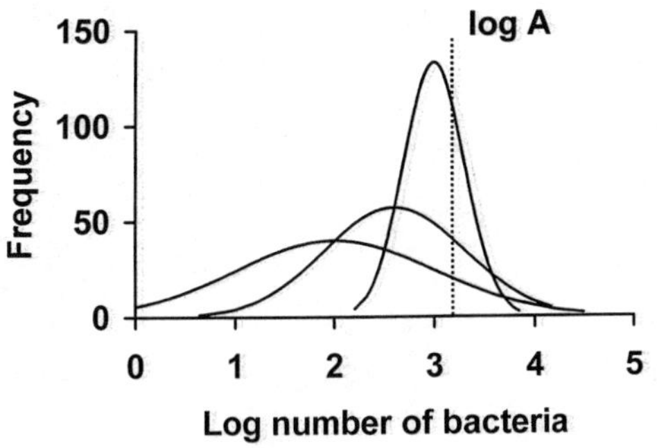

Figure 4 Three different distributions of log numbers of bacteria for the same log mean numbers (log *A*) on product.

comparison of microbiological conditions on the basis of mean log values can result in an operation or process being viewed as contaminating the product when in fact bacteria on the product were merely redistributed.

To avoid that error, it is necessary to calculate the log of the arithmetic mean (log *A*), by application of the formula: $\log A = \bar{x} + \log_n 10\,\sigma^2/2$, where $\bar{x}$ is the mean log and σ is the standard deviation of the set of log values (77). By that method, the log mean number of bacteria on the population of product items produced during or from a process can be estimated from a set of as few as 21 samples. However, in some circumstances a large portion of the samples in a set may not yield the bacteria that are being counted. Then, even though the estimation of the log mean is uncertain because of the nonnormal distribution of the log counts, some assessment of the microbiological effects of a process or operation can be made by reference to the log total numbers of bacteria recovered in each set (42).

C. Control of Carcass Dressing Processes

It is desirable that feed be withheld for a few hours before animals are presented for slaughter, since digestive tracts filled with feed are more likely than relatively empty ones to be punctured, with spillage of gut contents onto carcasses (106). Animals are usually stunned by electricity or gassed by carbon dioxide before they are stuck and bled. Stunning, sticking, and bleeding should not lead to any contamination of the meat unless the implements used are grossly contaminated (81).

The dressing of animals after bleeding varies with the species. With cattle, the stunned animal is usually shackled by one rear leg and raised to the dressing rail before it is stuck (Fig. 5). After bleeding, the head-down carcass is subjected to a series of operations to remove the skin (51). The esophagus is clipped or tied and the head is removed before the skinned carcass is eviscerated. The eviscerated

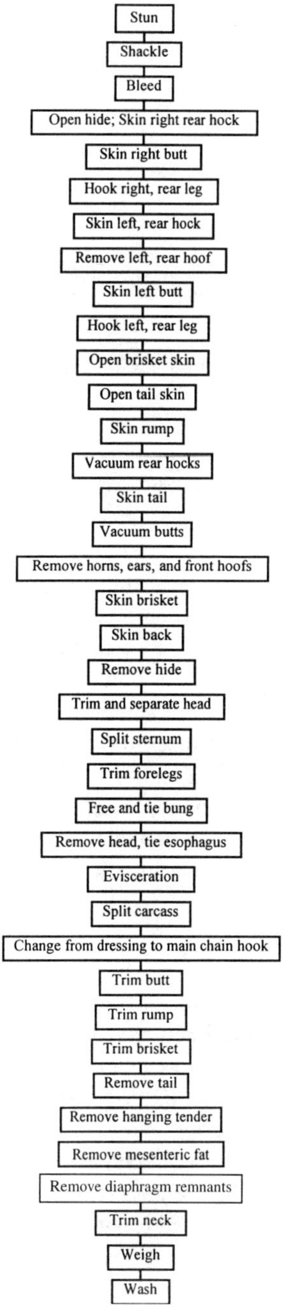

Figure 5 Operations in a beef carcass dressing process. (From Ref. 51.)

carcass is split down the backbone into two sides, which are trimmed or otherwise treated to remove visible filth and hairs before the sides are washed and dispatched to a carcass chiller.

Bacteria from the hide, and possibly from equipment, are deposited on the carcass during the skinning operations. It appears that the condition of the hide has little effect on the numbers of bacteria that are deposited on the meat because the hygienic performance of a carcass dressing process is generally consistent throughout the year, despite the varying states of the hides of cattle (Table 1). The carcasses of feedlotted cattle and culled cows emerging from the same process are of the same microbiological condition even though the hides of the former animals generally are dirtier than those of the latter (42). Direct studies indicate that the state of the hide does not generally affect the microbiological condition of the carcass (122), although visible contamination may be greater when hides are dirty, and wet hides (regardless of whether they appear to be relatively clean) are likely to cause somewhat more microbiological contamination of the meat than dry hides (7). Therefore, washing of animals before slaughter, or cleaning of carcass hides before skinning will not improve the microbiological condition of meat and may in fact increase bacterial contamination if the hides are wet when skinning is performed. It also follows that on-farm handling to maintain clean hides is not essential to assure food safety, although it may ease the efforts required for skinning and reduce the visible contamination of carcasses.

The control of contamination during beef carcass skinning involves minimizing contact, direct or indirect, between the outer surface of the hide and the meat (52). Thus, opening cuts should preferably be made by excising a strip of skin from the belly, then making all subsequent cuts by inserting a knife under the skin at an edge of the strip, and cutting from the inner surface of the skin to extend cuts as required, around the anus, along each leg and down the brisket. In subsequent stripping of the skin from the carcass, skinning operations should be implemented to avoid any rolling over of the skin such that the outside surface contacts

Table 1 Log Mean Numbers of Total Aerobes and *Escherichia coli* on Carcasses Leaving Dressing Processes for Cattle, Sheep, or Pig Carcasses

	Log mean numbers	
Species	Aerobes (no./cm^2)	*E. coli* (no./100 cm^2)
Cattle	3.62	1.74
Sheep	3.11	2.81
Pigs	3.47	1.53

Source: Data from Refs. 33, 42.

the meat, or any contact of the meat by hands or equipment that have previously contacted the outside surface of the skin.

The head is usually removed before the evisceration of the skinned beef carcass. Before the head is removed, the throat and floor of the mouth are opened and the tongue is freed from its attachment to the trachea. Since the mouth contains large numbers of bacteria, which can include enteric organisms (53), operations should be arranged so that those who handle the head do not handle any other part of the carcass. After operations on the head, the esophagus is usually clipped or banded near to the stomach to close the esophagus, and so prevent stomach contents being pushed out during the eviscerating operations. This practice is necessary because the stomach contents carry large numbers of bacteria, including enteric organisms (59).

In addition, before evisceration, the anus should be cut free and pulled from the carcass along with a length of the rectum. The anus should then be enclosed in a plastic bag secured around the rectum by means of a band. The enclosed anus can then be pushed into the body cavity for removal along with the rest of the intestines. During manipulation of the anus, contact between the carcass and both the anus and hands or equipment that have touched it should be wholly avoided, to prevent the transfer of fecal material and organisms to the meat.

The eviscerating operations should be performed with care to prevent rupture of the stomach or intestines, the opening of the esophagus, or the uncovering of the anus. After evisceration, the carcass is split and the sides then subjected to various treatments to remove visible contamination and associated bacteria. If the equipment is properly cleaned at the end of each working day, the splitting operation should not cause hazardous contamination to the meat.

The dressing of sheep carcasses may be similar to the dressing of beef carcasses, with the carcass hanging head down throughout skinning and evisceration. Since, however, the skin over the thorax is difficult to remove, it is common practice to secure the forelegs as well as the rear legs and raise the carcass to a horizontal position for skinning operations on the thorax. Further, the practice of dressing sheep carcasses as they hang by the fore legs only is increasingly followed (Fig. 6). That inverted dressing tends to reduce contamination, particularly of the rump and rear legs, since the skin can be pulled from those with only minimal cutting.

There are wide variations in the extents to which carcasses are contaminated with bacteria during different dressing processes. However, the contamination of sheep carcasses with *E. coli* may be generally greater than the contamination of beef carcasses with those organisms (33), probably because most of the sheep carcass surface must be handled during skinning operations, while large areas of the beef carcass surface need be contacted during skinning by a knife blade only, or not contacted at all. Nevertheless, the numbers of total aerobic bacteria on beef and sheep carcasses can be similar.

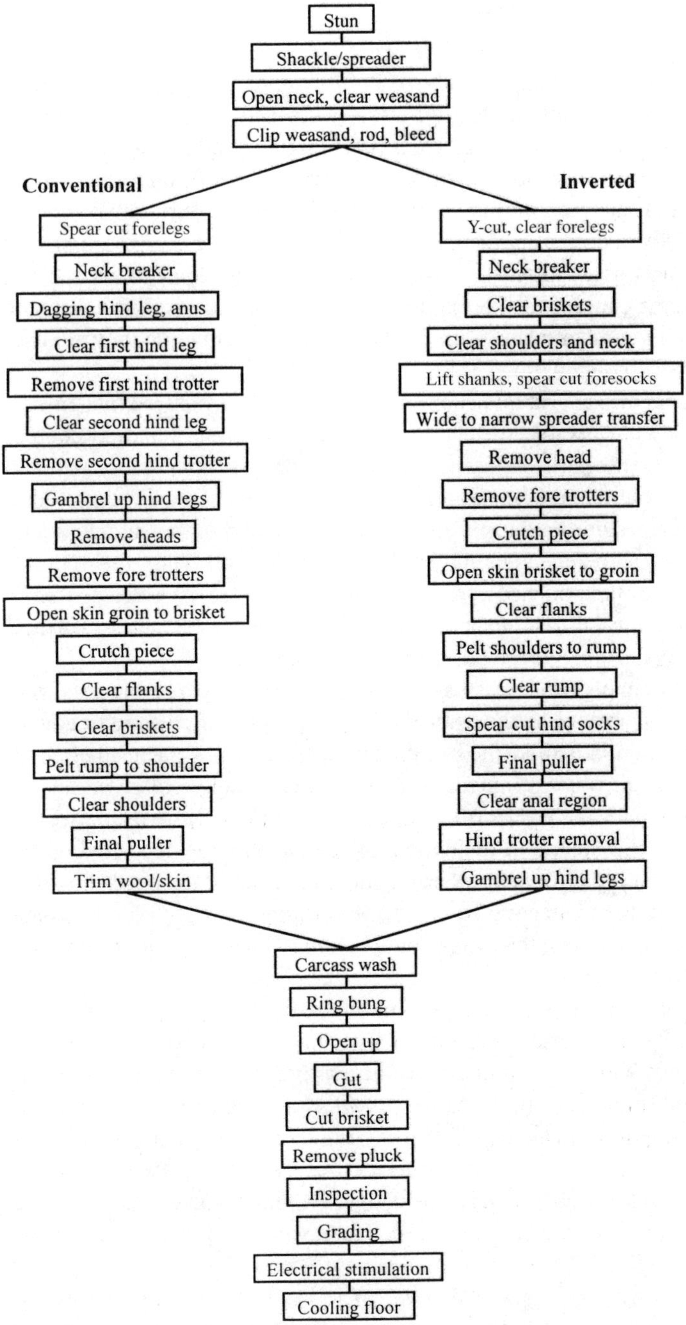

Figure 6 Operations in conventional and inverted dressing processes for sheep carcasses. (From Ref. 5.)

In contrast to beef and sheep carcasses, most pig carcasses are dressed with the skin on. Before evisceration, the pig carcass is cleaned and freed from hair by the sequential operations of scalding, dehairing, singeing, and polishing (Fig. 7). The scalding operation, in which the carcass is held in water of hotter than 60°C for several minutes, destroys all but spores and a few thermoduric organisms on the skin (110). However, during passage through the dehairer, feces, saliva, and nasal secretions are forced or washed from the carcass into cooler water (≈40°C) that circulates in the equipment. Consequently, the carcasses that emerge from the dehairer are heavily contaminated with enteric organisms, although they are usually free of visible contamination (37).

The numbers of bacteria on dehaired pig carcasses are reduced by singeing with naked flames to raise the surface temperatures to several hundred degrees Celsius. However, some parts of the carcass, such as the inside of the forelegs, will be protected from the flames. Bacteria remaining at those sites will be spread to the rest of the carcass during the polishing operation and may be augmented by bacteria that grow and persist in the polishing equipment (36). Thus, the microbiological condition of the polished carcass will be enhanced if the singeing flames are arranged and adjusted to contact as much as possible of the carcass surface, and by meticulous cleaning of the polishing equipment.

After polishing, pig carcasses may be further contaminated by bacteria from the mouth during operations on the head and by fecal organisms during freeing of the anus and evisceration. Unlike the handling of beef carcasses, the head often remains with the pig carcass until just before the carcass is split, although most of the operations on the head are performed soon after polishing. Since it appears that the mouth is the major source of the enteric organisms that are deposited on pig carcasses after polishing (48), operations on and inspection of the head should be conducted to avoid the contacting by hands or equipment of both the head and other parts of the carcass, and the head should be removed as early in the dressing process as possible.

With pig carcasses, it is usual to free the anus by means of equipment that is designed to suck the liquid contents from the large intestine and to simultaneously cut around the anus with a coring blade. As the equipment is withdrawn, the anus is pulled a few millimeters from the carcass surface. Subsequent contact between the anus and the carcass can be avoided by covering the anus via procedures similar to those used in the dressing of beef carcasses or by drawing the anus into the body cavity only after the abdomen has been opened, and placing the anus to contact only the intestines as they are removed. The contamination of pig carcasses after polishing can then be well controlled, but at the end of the dressing process, pig carcasses are likely to carry total aerobic bacteria and *E. coli* at numbers similar to those on beef carcasses, largely because of the organisms that were present on the polished carcasses (Table 1).

As with pigs, the carcasses of poultry are not skinned, but instead are scalded, then mechanically defeathered (Fig. 8). Scalding is for far shorter times

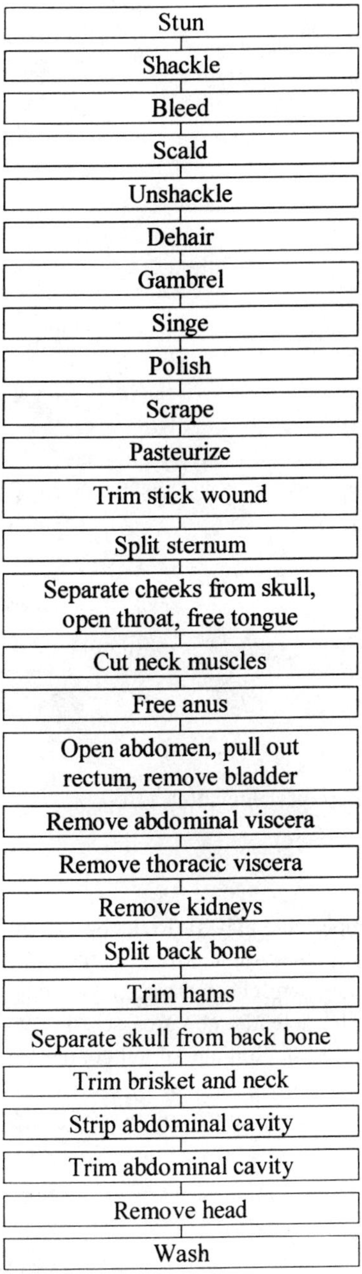

Figure 7 Operations in a pig carcass dressing process. (From Ref. 46.)

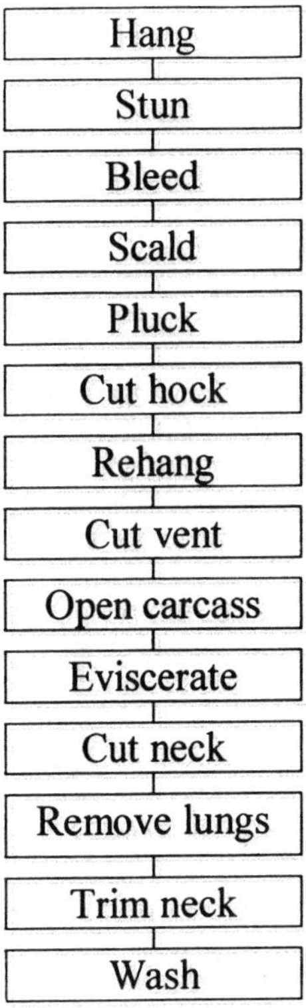

Figure 8 Operations in a poultry carcass dressing process. (From Ref. 11.)

than with pig carcasses, and often at lower temperatures (108). Reduction in the numbers of bacteria on the skins of poultry as a result of scalding are therefore limited, and the carcasses are anyway contaminated with bacteria from the equipment during the defeathering operation (85). The defeathered carcasses are mechanically eviscerated after removal of feet and heads. Spillage of gut contents is not uncommon during eviscerating operations (79). At the end of the carcass dressing process, poultry carcasses in general, will carry larger numbers of bacteria of all types than red meat carcasses.

D. Decontaminating Treatments of Carcasses

Decontaminating treatments of carcasses are of two kinds, those that are intended to remove visible contamination with incidental removal of any associated bacteria, and those that are intended to reduce the numbers of bacteria on carcasses.

Treatments of the first kind are trimming, vacuum cleaning, vacuum cleaning with simultaneous delivery of hot water or steam onto the carcass surface, and washing of carcasses or carcass sides at the end of the dressing process. The last type of treatment is usually applied to the whole carcass surface, but the other treatments are applied only to areas that are seen by an operator to be visibly contaminated. All four treatments are effective for removing visible contamination from carcasses, and in experimental circumstances all have been shown to be effective for the removal of bacteria previously inoculated onto meat (21,67). When applied during commercial carcass dressing processes, however, they generally do little to reduce the numbers of bacteria on carcasses (Table 2). This is because in the cases of the trimming and cleaning treatments, while visible contamination may indicate the presence of large numbers of bacteria, the absence of visible contamination certainly does not indicate that few bacteria are present. For example, polished pig carcasses will usually be free of visible filth but often are heavily contaminated with bacteria. Thus, the treatment of only visibly contaminated areas cannot result in any substantial reduction of the overall numbers of bacteria on carcasses.

In contrast, carcass washing can apparently in some circumstances reduce bacterial numbers (41). That in most commercial processes it does not may be due to the way in which the treatment is applied. The primary purpose of washing is to remove visible contamination. So, often with red meat carcasses, water is delivered first onto the upper end of the hanging carcass, then around the midline, and finally onto the lower end. That is effective for washing particles of filth and tissue down and finally off the carcass, but apparently not for removing bacteria.

Table 2 Effects on the Microbiological Conditions of Beef Carcasses of Treatments Applied in Commercial Processes for the Removal of Visible Contamination

		Log mean numbers	
Treatment	Stage of the treatment	Total Aerobes (log cfu/cm^2)	*E. coli* (log cfu/100 cm^2)
Vacuum cleaning	Before	3.72	2.40
	After	3.43	1.78
Hot water +vacuum cleaning	Before	4.25	4.36
	After	3.71	4.28
Trimming	Before	3.96	2.58
	After	3.72	2.43
Washing	Before	3.71	2.38
	After	3.37	2.06

Source: Data from Refs. 29, 31, 38.

However, if water is delivered onto the whole of the carcass throughout a typical washing treatment of some 20–30 seconds, bacteria also can be removed.

Washing of fecal material from poultry carcasses appears to have little effect on their microbiological condition, as concluded from reports of a lack of correlation between the presence of visible contamination and the microbiological condition of poultry carcasses (11).

Treatments of the second kind include spraying or dipping in solutions of chemicals with antimicrobial activities, such as chlorine, chlorine dioxide, organic acids, or trisodium phosphate, and pasteurizing treatments with steam or hot water.

In laboratory circumstances, treatments of meat with solutions of chemicals have been shown to be effective for reducing bacterial numbers (20,89). However, the presence of organic matter tends to protect bacteria from the effects of antimicrobial agents, while different species and strains of bacteria vary widely in their susceptibilities to those agents. For example, enteric organisms are notably resistant to inactivation by short chain length organic acids (12). In addition, treatment of carcasses by spraying tends to be incomplete because coverage of the whole surface of a carcass requires the use of uneconomically large volumes of solution. Thus, in commercial practice, treatment of carcasses with solutions of antimicrobial agents tends to give only modest or no useful reductions of bacteria numbers.

Pasteurizing treatments can be rather more effective (Table 3). Carcasses can be pasteurized with either steam or hot water. When steam is used, the carcass surface must be free of soil and detritus, and because portions of the surface covered with detritus or a film of water will not be heated by the condensing steam, it must be dry. The carcass must be exposed to steam in a sealed chamber that either has been evacuated or is raised above atmospheric pressure when filled with steam so that a uniform atmosphere of steam is formed (92). Then, the steam will condense onto the whole of the carcass surface, with rapid and uniform rise in the surface temperature. When steam is so applied to raise the surface temperature to about 100°C for 6.5 seconds, the numbers of aerobic bacteria and *E. coli* on beef carcasses are reduced by one and two orders of magnitude, respectively (38).

For pasteurizing with water to be fully effective, the water must be delivered as sheets, rather than as sprayed drops, which cool quickly to fall below the temperature required for rapid reduction of bacterial numbers (18). Carcasses pasteurized with water need not be dry before the treatment, while small particles of detritus on the carcass will not prevent effective pasteurization of all surfaces. The treatment can be applied to carcasses passing continuously through an apparatus that is open at both ends (35). When water hotter than 80°C is applied to raise the carcass surface to that temperature for 10 seconds, the numbers of total aerobes and *E. coli* on beef, pork, and lamb carcasses are reduced by two or more orders of magnitude (40,49).

Unfortunately, pasteurizing treatments for poultry carcasses are likely to be ineffective because of the porous nature of the skin, which tends to retain large

Table 3 Effects on the Microbiological Conditions of Beef and Pig Carcasses of Pasteurizing Treatments with Steam or Hot Water Applied During Commercial Processes[a]

Carcass type	Treatment medium	Stage of treatment	Total aerobes		*E. coli*	
			log *A* (log cfu/cm^2)	*N* (log cfu/25cm^2)	log *A* (log cfu/100 cm^2)	*N* (log cfu/2500 cm^2)
Beef	Steam	Before	3.36	5.23	2.20	3.84
		After	2.24	4.19	—	1.11
Beef	Water	Before	3.77	5.21	1.33	3.79
		After	1.65	3.09	—	0.00
Pig	Water	Before	2.81	4.32	—	3.33
		After	1.18	2.62	—	0.00

[a] log *A* = log mean; *N* = log of the total number recovered from 25 samples; — = insufficient bacteria-positive samples for calculation of the statistic.

Source: Data from Refs. 38, 40, 49.

amounts of water. Consequently, substantial destruction of bacteria by heating poultry carcasses requires a treatment that will cause extensive cooking of the tissues, which is commercially unacceptable (10).

E. Carcass Cooling

After dressing, carcasses must be cooled from the live body temperature to temperatures at which the growth of bacteria will be slowed or prevented. Cooling must usually be achieved without freezing the carcasses, both to allow the breaking down or packing of carcasses after cooling and because most meat is sold to consumers in the preferred, chilled state.

The reference temperature for the cooling of red meat carcasses is derived from the minimum temperature for growth of mesophilic, enteric organisms such as *E. coli* or *Salmonella*. That temperature is taken to be 7°C (77). Thus, some regulatory authorities require that red meat carcasses be cooled to a temperature of 10°C or less at the warmest, deep leg region before being removed from a carcass chiller or further processed, on the assumption that the growth of mesophilic pathogens will be acceptably slow at that temperature (116). Others stipulate a deep leg temperature of 7°C or less on the assumption that this will prevent the growth of mesophilic pathogens entirely (24).

The overnight cooling of pig and sheep carcasses to temperatures of 7°C or below is readily achieved, despite the inevitable wide variation of the rates at which carcasses within a batch cool (45). Variations in the rates of cooling are unavoidable because of large differences in the rates of flow of refrigerated air experienced by carcasses at different locations in a filled carcass chiller (73). Because beef carcasses are large, they cool relatively slowly. So, after overnight cooling it is usual for a more or less substantial portion of beef carcasses leaving a chiller to have deep leg temperatures well above 10°C (39). That circumstance is often overlooked, and the carcasses are fabricated or dispatched from the plant anyway.

Regarding the growth of bacteria on the cooling carcasses, only the surface temperatures are relevant because the deep tissues are sterile (26). While surfaces remain moist and warm, bacteria of all types can grow rapidly on the nutrient-rich environment provided by the meat (97). Growth can be controlled by either cooling or drying the surface (91). In most countries, regulations are in place to try to ensure that the surfaces of carcasses dry rapidly upon entry into a chiller. For example, there are requirements that carcasses be well spaced to allow the free flow of cool, dry air around the carcasses with consequent evaporation of water from and drying of the carcass surface. Spraying of carcasses during chilling is prohibited.

When conditions are such that surface drying occurs, the numbers of total aerobic bacteria on the carcasses tend to decrease. Decreases of *E. coli* and, presumably, similar gram-negative organisms are far larger (47). Thus, a well-controlled air cooling process can act as a treatment for decontaminating carcasses from enteric pathogens.

In North America, however, it is the usual practice to spray red meat carcasses with cold water during the first few hours of the carcass cooling process. That is done to suppress the evaporation of water from the tissues and so prevent loss of carcass weight (74). It might be supposed that such a treatment would allow the rapid proliferation of bacteria on the warm wet surfaces, but, in practice, large increases do not seem to occur (61). No explanation has been established for the failure of bacterial numbers on carcasses to greatly increase during spray chilling, but it can be suggested that the repeated washing of the carcass surface by typically spraying for about one minute at 10-min intervals removes bacteria from the carcass surface to give a small or negative increase in numbers despite rapid rates of growth.

With most spray chilling processes, the numbers of *E. coli* as well as those of total aerobic bacteria are probably little altered by the process. Large reductions in the numbers of *E. coli* but not total aerobic bacteria, as in air chilling with surface drying are, however, possible (Table 4). How such reductions are produced is not established, but they likely result because the bacteria are frozen into a film of water on the surface when the air temperature is reduced from above zero to about $-5°C$ when the spraying phase of the cooling process ends (39).

Although it is possible to operate both air and spray chilling processes in manners that will substantially reduce the numbers of enteric organisms on carcasses, any such effects obtained in current commercial practice are largely fortuitous, since few meatpacking plants have knowledge of the effects of their cooling processes on the microbiological conditions of carcasses.

Cooling of poultry carcasses may be by immersion in cold water or by air cooling in chilling tunnels. Air cooling can result in increased contamination of carcasses when contaminated water is blown from the equipment (111). Water cooling can reduce contamination somewhat, particularly if antimicrobial chemicals are maintained at appropriate concentrations in the water (97). Despite such treatments, water-chilled poultry carcasses remain heavily contaminated with bacteria, while immersion in a common bath ensures that pathogens are distributed to all carcasses (123).

The relatively small sizes of most poultry carcasses (compared with red meat carcasses), and the types of cooling process employed, allow the assured cooling of poultry carcasses to 2°C or below before any cutting or packing and storage.

F. Cutting and Packing

Meat may be dispatched from slaughtering plants as whole sides or quarters of carcasses, or as primal cuts. Trade in hanging meat (i.e., red meat in the forms of whole carcasses or carcass sides or quarters) is generally being displaced by trading in primal cuts, which may be vacuum-packed and boxed or assembled unwrapped in bulk bins, which each hold approximately 900 kg of meat.

Table 4 Effects on the Microbiological Conditions of Sheep and Beef Carcasses of Air or Spray Cooling Processes[a]

			Total aerobes		*E. coli*	
Carcass type	Treatment medium	Stage of treatment	log *A* (log cfu/cm^2)	*N* (log cfu/25cm^2)	log *A* (log cfu/100 cm^2)	*N* (log cfu/2500 cm^2)
Sheep	Air	Before	3.33	4.74	3.57	4.28
		After	2.86	4.25	1.49	2.86
Beef	Spray 1	Before	4.03	5.19	—	1.72
		After	3.58	4.91	—	1.34
Beef	Spray 2	Before	3.12	4.55	2.01	3.76
		After	2.48	3.76	—	1.15

[a]log *A* = log mean; *N* = log of the total number recovered from 25 samples; — = insufficient bacteria-positive samples for the calculation of the statistic.

Source: Refs. 39, 47.

The breaking down of carcasses to primal cuts should be performed in rooms maintained at an air temperature of 10°C or below to meet regulatory requirements (109). In practice, temperature gradients exist in most cutting facilities, and temperatures that are too low are uncomfortable for workers. Thus, air temperatures a few degrees above 10°C are not uncommon in cutting rooms. Temperature alone then gives only limited and uncertain control over the growth of bacteria that were initially present on the meat. Consequently, meat should not be held in cutting rooms for extended periods, but should be dispatched to a storage chiller or freezer as quickly as possible after the preparation and packaging of cuts.

The growth of bacteria on meat in cutting rooms can be of concern when product is slowly assembled there during a working day. Of far greater concern, however, is the possible contamination of meat by bacteria that grow on detritus that persists from day to day in improperly cleaned equipment.

The equipment used in the fabrication of carcasses includes personal items, such as steel mesh gloves, knives, scabbards, and meat hooks; movable tools, such as reciprocating saws and powered knives; and large, fixed items such as band saws and conveyors. Elementary sanitary considerations and regulations require that all equipment used with meat be thoroughly cleaned at the end of each working day (58). The adequacy of cleaning is assessed each day, by regulatory as well as plant personnel, by visual inspection and often by microbiological sampling of cleaned work surfaces (57). It is then assumed by most in the meat industry that proper cleaning of fabricating equipment is routinely assured.

In fact it is not. The cleaning of personal equipment is left to the discretion of the individual workers. Consequently, cleaning is highly varied and often inadequate. For example, cleaned mesh gloves are usually found to harbor large populations of bacteria that can include *E. coli* or other organisms indicative of the presence of pathogens (34). Clearly, specialized facilities for cleaning and decontaminating personal equipment are required at all locations where raw meat is fabricated, with the institution of procedures that ensure that all items of personal equipment receive an adequate treatment at the end of each working day.

Of even greater concern is the persistence of bacteria-bearing detritus in obscured or inaccessible parts of fixed items of equipment. The meat industry has never developed specifications for equipment to assure its cleanability (25). Moreover, equipment is often modified after installation, to accommodate the peculiarities of usage at individual plants. Thus, areas where detritus may persist unobserved can exist within equipment. Product can be extensively contaminated by small quantities of such detritus, which may carry bacteria at numbers of 10^9 colony-forming units per gram (cfu/g). Persisting sources of microbial contamination can be detected by determining the microbiological conditions of product entering and leaving a fabrication process (32). Unambiguous increases in the numbers of total aerobes or of indicator organisms on product show that contaminants are being added during the process (Table 5).

The source or sources of the contamination may be identified by determin-

Table 5 Effects on the Microbiological Condition of Meat of the Carcass Breaking Processes at Four Beef Packing Plants[a]

		Total aerobes		*E. coli*	
Plant	Stage of the process	Log *A* (log cfu/cm^2)	*N* (log cfu/25 cm^2)	Log *A* (log cfu/100cm^2)	*N* (log cfu/2500 cm^2)
A	Before	2.55	3.86	—	n.d.
	After	2.90	4.38	3.96	4.97
B	Before	2.56	4.02	—	2.05
	After	3.24	4.50	3.56	4.38
C	Before	3.42	4.95	—	1.57
	After	4.36	4.94	1.78	2.74
D	Before	3.44	4.68	—	1.59
	After	3.33	4.65	—	1.86

[a]log *A* = log mean; *N* = log of the total number recovered from 25 samples; — = insufficient bacteria-positive samples for calculation of the statistic; n.d. = none detected.
Source: Data from Ref. 32.

ing the point or points in the process at which the product is contaminated, and then examining the associated equipment to identify the location of contaminating material and to develop cleaning procedures to assure its routine removal. Examination of product for several types of organism (e.g., total aerobes, coliforms, *E. coli,* aeromonads) may be necessary for assurance that no source of contamination exists in a fabrication process, although increases in any one would indicate the existence of a source of contamination (34).

Although much poultry is sold in carcass form, increasing amounts are being fabricated to portions. Because poultry carcasses are usually heavily contaminated with a wide variety of bacteria, the addition during carcass fabrication of more bacteria of types already present on the carcass may be difficult to discern. The possibility of detecting inadequately cleaned equipment used for poultry carcass processing by enumeration of indicator organisms on the product has not been examined.

III. SAFE HANDLING OF MEAT DURING STORAGE AND DISTRIBUTION

When product has been packaged, it will be protected from extraneous contamination. Bulk-packed product should also be protected from extraneous contamination by providing lids for bulk containers or, as is common, loading the product into a plastic bag within each container and clipping the mouth of the bag when the container is filled. Although much meat is carried without wrapping, even hanging carcasses, sides, or quarters may be wrapped for transport. Extraneous contamination is then of limited concern during storage at and transport of product from slaughtering facilities. Instead, the major concern must be the growth of pathogenic organisms on the product.

Concern about the microbiological condition of meat tends to focus on the mesophilic pathogens, which do not grow at chiller temperatures. However, some, cold-tolerant pathogens can grow on meat at 0°C or below (Table 6). Thus, chiller temperatures can restrict, but not entirely prevent, all growth of pathogens (93). However, freezing does prevent the growth of all pathogens.

The rates at which bacteria grow increase rapidly at temperatures above their minima for growth (Fig. 9). Therefore, it is important that raw, chilled meat be maintained at the lowest temperature that is practicable, for even relatively short periods of exposure to higher temperatures can result in large increases in the numbers of pathogenic organisms.

The temperatures of concern for bacterial growth are those of the product, not the operating temperatures of refrigeration equipment. That might seem obvious, but in practice many implicitly assume that product placed in a refrigerated facility immediately assumes the gauge temperature set for the facility. Of course,

Table 6 Minimum Temperatures for Growth of Some Pathogenic Bacteria Associated with Foods

Organism	Temperature (°C)
Campylobacter jejuni	32
Clostridium botulinum group III (types C, D)	15
Clostridium perfringens	15
Clostridium botulinum group I (types A, B, F)	10
Clostridium botulinum group IV (type G)	10
Shigella	10
Salmonella	8
Escherichia coli	7
Bacillus cereus	5
Staphylococcus aureus	5
Vibrio parahaemolyticus	5
Clostridium botulinum group II (types B, E, F)	3
Aeromonas hydrophila	2
Listeria monocytogenes	0
Yersinia enterocolitica	−2

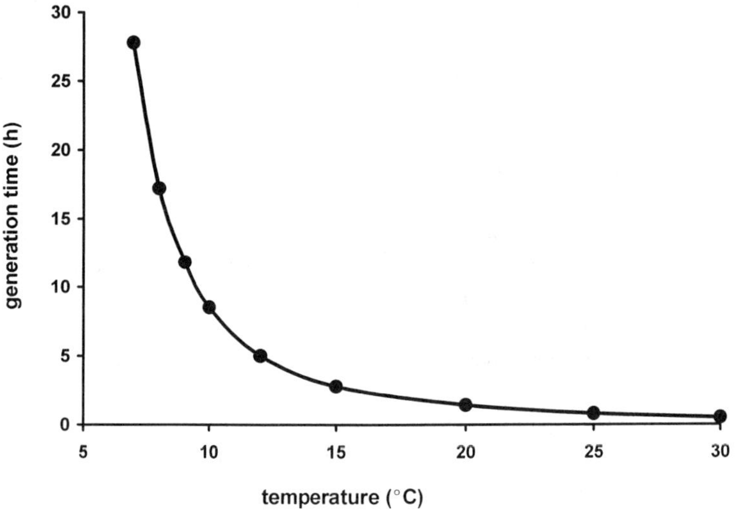

Figure 9 Effect of temperature on the generation time of *Escherichia coli.* (Data from Ref. 98.)

that is not so. Instead, product at a temperature higher than that of the facility air will cool at a rate determined by the difference in temperature between the product and the air, the rate of flow of the air over the product surface, and the size of the product unit (72).

Since newly fabricated product may well be at temperatures toward the higher end of, or even above the chill temperature range of 7 to −2°C, storage chillers for newly packaged product should be capable of rapidly reducing the temperature of product to an optimal temperature of −1.5 ± 0.5°C. Unfortunately, newly packaged meat is often transferred to chillers that are designed for storage, rather than cooling (44). That is, they are designed to maintain, not reduce product temperatures, with weak airflows and stacking arrangements that shield much of the product from any flow of air. Moreover, such chillers are often operated at temperatures of 2 ± 2°C, which will slow the rate of cooling compared with the optimal temperature. Even if conditions were such that the product actually cooled to the air temperatures, the storage life at 2°C would be only half that at the optimal temperature, while all cold-tolerant pathogens would be able to grow (95).

With bulk containers, the large size of the units precludes rapid cooling even when air temperatures are low and airflow rates are high. It is then a common practice to cool the product in bulk containers by adding carbon dioxide snow as the containers are filled (54). However, the amounts of carbon dioxide snow added are usually crudely estimated, so the cooling of product achieved is often varied and uncertain. A procedure for estimating the mean temperature of the product being loaded, and from that determining the amount of CO_2 snow to add to obtain a product temperature approaching -1.5°C, would be desirable.

Refrigerated transport facilities such as road trailers, railcars, or sea containers are designed to circulate cold air around the inner walls of the unit, and so prevent the contained product from being warmed by heat from the outside (107). Thus, like storage chillers, they can maintain product temperature, but they will usually cool product only slowly. Despite that, it is common practice to load on to transport facilities chilled product well above the optimum temperature, not only for immediate dispatch but also for storage for sometimes lengthy periods.

Cooling of product will, of course, be aided if the stacking allows some circulation of air through the stack, and will tend to be faster in railcars or sea containers than in road trailers, since former means of transport have more powerful refrigeration units than the latter. Thus, hanging meat in rail cars may cool to the car air temperatures within 2 or 3 days, but block-stacked boxed beef in road trailers may take weeks to cool throughout to the trailer air temperature (55). It is then evident that the proper practice should be to load transport facilities only with product that has been cooled to near the optimal storage temperature. That is at present the norm only with poultry, because of the reliable cooling of poultry carcasses and the short storage life for the product if storage temperatures are allowed to rise.

IV. SAFE HANDLING OF MEAT AT THE RETAIL LEVEL

Most retail outlets that sell meat operate a facility on the premises for fabricating primal cuts of red meats to retail-ready forms. However, increasing quantities of retail-packaged meat are being prepared at central cutting and packing facilities (70). Many such facilities are operated by meat processing companies that draw product from slaughtering plants as it is available and required; but some are further processing facilities of individual slaughtering plants, while preparation of retail-ready products is common at poultry slaughtering plants. In addition, grinding, which is traditionally done at the time of retail pack preparation, is increasingly performed at slaughtering plants, with packaging of the meat in portions sufficient for the filling of several retail packs at a retail outlet.

In handling meat at the retail level actions should be taken to ensure (a) that it meat is received in an acceptable condition and properly stored when it arrives; (b) that contamination of the meat during preparation of retail-ready product is avoided; and (c) that product is maintained at an adequately low temperature during display.

Criteria for the acceptance of product for retail preparation rarely amount to more than ascertaining by inspection that packagings are not damaged and that the product is not overtly spoiled, as evidenced by discoloration or off-odors, and by spot-checking that product temperatures are not above 5°C. In practice, failure of the last criterion will rarely lead to rejection if the product is needed to meet current demand. Thus, inspection for acceptance will generally do little for safety except to exclude product that has been grossly and obviously abused before delivery. The acceptable state of arriving products becomes more certain if consignments are accompanied by documentation that establishes the times of storage and transport and the temperatures experienced during these periods. On delivery, meat should be moved without delay to a storage chiller, for product may warm rapidly if left on an unrefrigerated dock, particularly if exposed to direct sunlight.

At retail facilities, storage refrigeration is usually of a walk-in type operated at 2 ± 2°C. Operation at lower temperatures is preferable, while operation at higher temperatures is likely to be hazardous (43). Product that is relatively warm on delivery should be placed to allow good circulation of cold air with consequent cooling. Product that is at or below the facility temperature on delivery can be block-stacked in areas of low air flow to maintain the lower temperatures.

The entrances to walk-in refrigerators must be closed except during the entry or exit of personnel and product. Otherwise, the common practice of leaving the entrance open while product is moved in and out during the working day can result in temperatures of air and product in the facility rising above the chill temperature range.

The prevention of contamination during the preparation of retail-ready product is largely a matter of ensuring that all equipment is properly cleaned between working periods. In order of importance for controlling bacterial contamination by cleaning are (a) the removal of detritus and adherent organic material, (b) drying, and (c) treatment of equipment with sanitizing agents. The physical removal of organic matter, with the assistance of detergents for emulsifying fats and lifting protein films, removes the sources of nutrients for bacteria as well as the organisms themselves. Since all microorganisms require water for growth, drying of the equipment will then prevent any subsequent bacterial growth and ensure the inactivation of a large fraction of any remaining organisms, particularly gram-negative species (105).

If equipment can be freed of organic matter and well dried, the use of sanitizing agents will not greatly enhance safety. However, if drying is difficult or uncertain, sanitizing agents should be used to control the growth of bacteria in water remaining in or on equipment (22). Sanitizing agents are also needed when scarring, pitting, or other damage to surfaces prevents the thorough removal of organic material (23). Thus, the use of sanitizing agents is appropriate in most circumstances, but as a supplement to, not a substitute for cleaning and, preferably, drying.

Cleaning procedures should be documented and adhered to, with particular concern to ensure that no item of equipment is neglected, and that each item is treated appropriately to remove all organic matter even from areas that might be difficult to access. Equipment should be replaced when surfaces become worn or damaged, if they cannot be polished or resurfaced, since damaged surfaces can never be cleaned as effectively as undamaged ones.

If red meats and poultry are prepared in the same facility, their preparation must be strictly separated, because poultry will commonly carry larger numbers of pathogens than red meats while, unlike poultry, some red meats are traditionally cooked under conditions that do not assure the destruction of all bacteria. In fact, some red meat dishes are prepared without cooking of the meat. Thus, after the preparation of poultry, a facility must be thoroughly cleaned before it is used for the preparation of any red meat. A usual practice is to prepare poultry after completion of the preparation of red meat on any day when both are prepared.

Although facilities for cutting retail-ready product are cooled during working periods to a nominal temperature of 10°C or below, temperatures are in practice often higher, particularly if there are a number of frequently used entrances. Certainly, product should be moved expeditiously into and out of the cutting facility to minimize the warming of product. Product for cutting should be removed from the storage chiller in small batches as required. Finished product should be moved to display or returned to the storage chiller. It should not be allowed to accumulate in the cutting facility no matter how convenient that might be to the display area. Ground product is likely to be warmed by the grinding operation and by the inclusion of trimmings from primal cuts that have warmed to the temperature of the cutting facility. It is then good practice to cool product for grinding by addition of carbon dioxide snow, if that is possible, or to transfer the retail-packed product to a chiller where it will cool before display.

Any hazardous warming of product during the preparation of retail-ready packs will likely be compounded by the inadequate control of temperatures in the retail display case. As with refrigerated facilities for storage and transport, display cases are designed to maintain rather than reduce product temperatures. With display cases that are open, a curtain of cold air is usually blown across the opening to maintain the temperature within (72). That curtain can be and is often disrupted by the stacking of product above the load line, by the placement of advertising material within the case, or by drafts of warm air that blow into the case. Moreover, with cases of all types, the distribution of refrigerated air can be unbalanced, so that large, persisting temperature gradients can occur in some display cases (Table 7). Moreover, the refrigeration equipment must undergo periodic defrosting, usually at 12-hour intervals, when nonrefrigerated air is blown into the case and air temperatures rise above the chiller range, sometimes for an hour or more (60).

Because of the highly uncertain control of product temperature in the retail case, retail display continues to be the weakest link in the cold chain for all chilled foods (9). The inherent limitations on the control of product temperature during display can be minimized by ensuring (a) that product is cooled to near the optimum temperature before it is loaded to display; (b) that cases are never overloaded or allowed to contain objects for advertising or other purposed that will disrupt flows of air; (c) that the refrigeration is regularly checked for maintenance of set air temperatures and the balanced distribution of air; and (d) that cases are effectively shielded from currents of warm air that may blow into them; also helpful is

Table 7 The Effects of Position in a Retail Display Case on the Surface Temperatures of Steaks During a 48-hour Period

Distance from back of case (cm)	Distance from load line (cm)	Steak surface temperature (°C)		
		Maximum	Minimum	Mean
8	– 17	5.3	0	1.7
	– 13	7.5	1.0	2.7
	0	8.0	1.3	3.3
	+4	8.3	2.8	4.5
40	– 17	8.0	4.3	6.1
	– 13	10.0	4.3	6.9
	0	10.0	4.0	6.7
	+4	11.8	7.5	10.0
75	– 17	7.3	4.0	5.6
	– 13	9.0	1.3	6.8
	0	9.5	5.8	7.2
	+4	10.3	6.0	8.3

Source: Data from Ref. 60.

the practice of moving product to a storage chiller between display periods, or at least covering the opening of display cases overnight. Temperature control can also be improved by the use of fluorescent rather than incandescent lighting, since the latter tends to heat product surfaces by radiation.

Whatever practices are adopted, some loss of control over product temperatures is likely during display. There should then be an established procedure for withdrawing and discarding product after a maximum period of display, which is decided for each display case on the basis of the temperature control that can be assured in the worst circumstances. Product that is withdrawn from display will be compromised to some extent by bacterial growth (50). The common practice of grinding and reoffering for sale meat that has been withdrawn from display because of discoloration or other signs of deterioration is therefore hazardous and should not be countenanced (Table 8).

V. HANDLING OF RAW MEATS IN THE RESTAURANT OR HOME

Meat purchased for the home or for smaller restaurants will usually be transported without refrigeration. The interiors of vehicles can warm rapidly in the hotter seasons, particularly if the vehicle is exposed to heating by direct sunlight, and a pas-

Table 8 Microbiological Conditions of Hamburger Patties from Five Retail Outlets[a]

		Log mean numbers (cfu/g)	
Outlet	Chilled or frozen	Total aerobes	*E. coli*
A	Frozen	5.44	1.87
B	Frozen	8.54	1.91
C	Frozen	4.66	0.71
C	Chilled	8.54	2.17
D	Chilled	7.59	2.65
E	Chilled	4.82	1.38

[a]Preparation of frozen patties from product abused during display or abuse of chilled patties during display was indicated at outlets B, C, and D.
Source: Data from Ref. 56.

senger compartment will likely be warm whatever the season. The time between the purchase of meat and replacing it under refrigeration should be minimized by, for example, purchasing meat last rather than early in a series of stops to purchase items. If possible, meat should be placed in an insulated container for transport.

It is important to recognize, when chilled meat is stored in a domestic refrigerator, that control of the air temperature is often uncertain, with a substantial fraction of refrigerators apparently being consistently operated at temperatures above the chill temperature range (75). Determination of the actual operating temperature of a domestic refrigerator, with adjustment of the temperature if necessary, can then be recommended. Even so, temperatures are likely to rise during the day if the refrigerator is opened frequently. It can then be suggested that chilled raw meat not be stored in a domestic refrigerator for more than 2 days.

Indeed, it is a common practice for consumers to freeze meat that is purchased in the chilled state. When that is done, it is desirable that the meat be wrapped as individual portions before it is frozen. That will avoid a subsequent need to thaw more meat than is required on a particular occasion, so that the excess product may be separated and refrozen, and will allow more rapid thawing of portions that are separated rather than frozen together in a block.

Care should be exercised in thawing meat. Thawing at room temperatures is undesirable because rapid bacterial growth at surfaces and in exudate may well be initiated before the center of a meat mass has thawed. Thawing in a refrigerator is often recommended, but that can take an impractically long time because of the small difference in temperature between the product and the air, and weak or no flow of air over the product (103). Consequently, with large pieces of meat or poultry carcasses, thawing in a cool area, at temperature of about 10°C, such as a flagged pantry or basement cold room, may be the only option. However, if practicable, rapid thawing by microwaving or, if in a watertight wrapping, immersion in cold water would generally be the safest ways of thawing.

Raw meat should be prepared on a plastic cutting board used for that purpose only. Wood should not be used, for not only will it tend to become deeply scored, but it will also absorb water and swell, and so be difficult to clean (101). After preparation of the meat, the cutting board and instruments should be thoroughly cleaned and allowed to drain and dry. Fixed surfaces likely to have been contacted by the meat or exudate should be cleaned of visible contamination, and possibly wiped over or sprayed with a sanitizing agent (3). Cloths used for cleaning after the cutting of meat should be discarded immediately afterward for washing or disposal, since large populations of bacteria can readily develop in damp cleaning cloths (104).

As elsewhere, poultry should be prepared separately from other raw meats, and the preparation of all other foods should be separated from the preparation of meat. In particular, great care should be exercised to keep raw meat from any direct or indirect contact with foods that are to be eaten without any further cooking (14).

Meat dishes should be cooked to expose all parts that may be contaminated with bacteria to time/temperature treatments that will be adequate for the destruction of all vegetative organisms (117). Adequate cooking may then involve heating for short periods to relatively high temperatures, or for long periods at relatively low temperatures (Table 9). Meat may be cooked rare if the portion is a single mass of tissue, because the deep tissues are sterile. Consequently, cooking such a portion to a depth of a few millimeters will destroy all the bacteria on the product. However, if a portion has been mechanically tenderized, restructured, or rolled, or has separated while raw along junctions between muscles, the portion must be cooked throughout to assure its safety (100). Any ground or otherwise comminuted meat must likewise be cooked thoroughly because bacteria will be distributed throughout such product. Thorough cooking must extend also to any stuffings or fillings used with meats, since these will be contaminated with exudate from the raw meat. For the dish to be assuredly safe, the bacteria associated with the exudate must be inactivated (1).

VI. CONCLUSIONS

For meat to be a safe food, all who rear food animals or handle the product must exercise due care within the processes for which they are responsible. This is implicit in the HACCP approach to food safety. Thus leaving a hazard for the next person in the production and distribution chain to deal with is becoming decreasingly viable as an economic strategy. Even so, there remains much uncertainty and

Table 9 Times for Maintenance of the Stipulated Temperature at the Center of a Product Unit for Adequately Safe Cooking of Meat Pieces

Temperature (°C)	Time (min)
55	97
56	62
57	37
58	32
59	19
60	12
61	8
62	5
63	<1

Source: Data from Ref. 119.

misunderstanding of how health risks associated with meat might best be controlled. While in some areas research is evidently required, in others only the wider dissemination and application of established knowledge is needed to enhance meat safety. It is to be hoped that at least the latter areas can be effectively addressed in the near future, to allay in part the ongoing concerns about the safety of the meat supply.

REFERENCES

1. Ahmed, N. M., D. E. Conner, and D. L. Huffman. 1995. Heat resistance of *Escherichia coli* O157:H7 in meat and poultry as affected by product composition. J. Food Sci. 60:606–610.
2. Anderson, J. K., R. Sorenson, and M. Glensberg. 1991. Aspects of the epidemiology of *Yersinia enterocolitica:* A review. Int. J. Food Microbiol. 13:231–238.
3. Assanta, M. A., D. Roy, and K. Machika. 1996. Efficiency of disinfecting agents to destroy *Listeria monocytogenes, Yersinia enterocolitica* and *Staphylococcus aureus.* Dairy Food Environ. Sanit. 16:426–430.
4. Bager, F., H. D. Emborg, L. L. Sorenson, C. Halgaard, and P. T. Jensen. 1995. Control of *Salmonella* in Danish pork. Fleischwirtsch. Int. 4:27–28.
5. Bell, R. G., and S. C. Hathaway. 1996. The hygienic efficiency of conventional and inverted lamb dressing systems. J. Appl. Bacteriol. 81:225–234.
6. Berends, B. R., J. M. A. Snijders, and J. G. Van Logtestijn. 1993. Efficacy of current EC meat inspection procedures and some proposed revisions with respect to microbiological safety: A critical review. Vet. Rec. 133:411–415.

7. Biss, M. E., and S. C. Hathaway. 1996. Effect of pre-slaughter washing of lambs on the microbiological and visible contamination of the carcass. Vet. Rec. 128:82–86.
8. Blankenship, L. C., J. S. Bailey, N. A. Cox, N. J. Stern, R. Brewer, and O. Williams. 1993. Two-step mucosal competitive exclusion flora treatment to diminish salmonellae in commercial broiler chickens. Poultry Sci. 72:1667–1672.
9. Bogh-Sorenson, L., and P. Olson. 1990. The chill chain, in T. R. Gormley (ed.), Chilled Foods, the State of the Art. Elsevier Applied Science, London, pp. 245–267.
10. Bolder, N. M. 1997. Decontamination of meat and poultry carcasses. Trends Food Sci. Technol. 8:221–227.
11. Bolder, N. M. 1998. The microbiology of the slaughter and processing of meat and poultry, A. R. Davies and R. G. Board (eds.), The Microbiology of Meat and Poultry. Blackie Academic, London, pp. 158–173.
12. Brackett, R. E., Y.-Y. Hao, and M. P. Doyle. 1994. Ineffectiveness of hot acid sprays to decontaminate *Escherichia coli* O157:H7 on beef. J. Food Prot. 57:198–203.
13. Brown, M. H., and A. C. Baird-Parker. 1982. The microbiological examination of meat, in M. H. Brown (ed.), Meat Microbiology. Applied Science Publishing, London, pp. 423–520.
14. Bryan, F. L. 1988. Risks of practices, procedures and processes that lead to outbreaks of foodborne disease. J. Food Prot. 51:663–673.
15. Bryan, F. L. 1990. Application of HACCP to ready-to-eat chilled foods. Food Technol. 44:70–77.
16. Canadian Food Inspection Agency. 1994. HACCP generic model: Beef slaughter. CFIA, Ottawa, Ontario, Canada.
17. Carosella, J. 1988. Control of microbiological contamination remains priority. Nat. Provision. 179:36–49.
18. Davey, K. R., and M. G. Smith. 1989. A laboratory evaluation of a novel hot water cabinet for the decontamination of sides of beef. Int. J. Food Sci. Technol. 24:305–316.
19. Diez-Gonzalez, F., T. R. Callaway, M. G. Kizoulis, and J. B. Russel. 1998. Grain feeding and the dissemination of acid-resistant *Escherichia coli* from cattle. Science 281:1666–1668.
20. Dorsa, W. J., C. N. Cutter, and G. R. Siragusa 1997. Effects of acetic acid, lactic acid and TSP on the microflora of refrigerated beef carcass surface tissue inoculated with *Escherichia coli* O157:H7, *Listeria innocua* and *Clostridium sporogenes.* J. Food Prot. 60:619–624.
21. Dorsa, W. J., C. N. Cutter, G. R. Siragusa, and M. Koohararaie. 1996. Microbial decontamination of beef and sheep carcasses by steam, hot water spray washes, and a steam-vacuum sanitizer. J. Food Prot. 127–135.
22. Dunsmore, D. G., A. Twomey. W. G. Whittlestone, and H. W. Morgan. 1981. Design and performance of systems for cleaning product-contact surfaces of food equipment: A review. J. Food Prot. 44:220–240.
23. Edelmeyer, H. 1982. Cleaning and disinfection. Inst. Meat Bull. 116:20–26.
24. European Economic Community Council Directive: Commission Regulation No. 2226/78. Sept. 25, 1978.
25. European Hygienic Equipment Design Group. 1992. A method for assessing the in-place cleanability of food-processing equipment. Trends Food Sci. Technol. 3:325–328.

26. Gill, C. O. 1979. Intrinsic bacteria in meat. J. Appl. Bacteriol. 47:367–378.
27. Gill, C. O. 1986. The control of microbial spoilage in fresh meats, in A. M. Pearson and T. R. Dutson (eds.), Advances in Meat Research, Vol. 2. AVI Publishing, Westpoint, CT, pp. 49–88.
28. Gill, C. O. 1995. Current and emerging approaches to assuring the hygienic condition of red meats. Can. J. Anim. Sci. 75:1–13.
29. Gill, C. O. 1997. Microbial effects of trimming, vacuum cleaning or vacuum-hot-water cleaning of beef or sheep carcasses in commercial dressing processes. Technical Bulletin No. 1997-4E, Agriculture and Agri-Food Canada Research Centre, Lacombe.
30. Gill, C. O. 1998. Microbiological contamination of meat during slaughtering and butchering of cattle, sheep and pigs, in A. R. Davies and R. G. Board (eds.), The Microbiology of Meat and Poultry. Blackie Academic, London, pp. 118–157.
31. Gill, C. O., M. Badoni, and T. Jones. 1996. Hygienic effects of trimming and washing operations in a beef carcass dressing process. J. Food Prot. 59:666–669.
32. Gill, C. O., M. Badoni, and J. C. McGinnis. 1999. Assessment of the adequacy of cleaning of equipment used for breaking beef carcasses. Int. J. Food Microbiol. 46:1–8.
33. Gill, C. O., and L. M. Baker. 1998. Assessment of the hygienic performance of a sheep carcass dressing process. J. Food Prot. 61:329–333.
34. Gill, C. O., L. P. Baker, and T. Jones. 1999. Identification of inadequately cleaned equipment used in a sheep carcass breaking process. J. Food Prot. 62:637–643.
35. Gill, C. O., D. Bedard, and T. Jones. 1997. The decontaminating performance of a commercial apparatus for pasteurizing polished pig carcasses. Food Microbiol. 14:71–79.
36. Gill, C. O., and J. Bryant. 1992. The contamination of pork with spoilage bacteria during commercial dressing, chilling and cutting of pig carcasses. Int. J. Food Microbiol. 16:51–62.
37. Gill, C. O., and J. Bryant. 1993. The presence of *Escherichia coli, Salmonella* and *Campylobacter* in pig carcass dehairing equipment. Food Microbiol. 10:337–344.
38. Gill, C. O., and J. Bryant. 1997. Decontamination of carcasses by vacuum–hot water cleaning and steam pasteurizing during routine operations at a beef packing plant. Meat Sci. 47:267–276.
39. Gill, C. O., and J. Bryant. 1997. Assessment of the hygienic performances of two beef carcass cooling processes from product temperature history data on enumeration of bacteria on carcass surfaces. Food Microbiol. 14:593–602.
40. Gill, C. O., J. Bryant, and D. Bedard. 1999. The effects of hot water pasteurizing treatments on the appearances and microbiological conditions of beef carcass sides. Food Microbiol. 16:281–289.
41. Gill, C. O., J. Bryant, and J. C. McGinnis. 1999. Microbial effects of carcass washing operations at three beef packing plants. Fleischwirtsch. Int. (in press).
42. Gill, C. O., B. Deslandes, K. Rahn, A. Houde, and J. Bryant. 1998. Evaluation of the hygienic performances of the processes for beef carcass dressing at 10 packing plants. J. Appl. Microbiol. 84:1050–1058.
43. Gill, C. O., M. Friske, A. K. W. Tong, and J. C. McGinnis. 1995. Assessment of the hygienic characteristics of a process for the distribution of processed meats, and of storage conditions at retail outlets. Food Res. Int. 28:131–138.

44. Gill, C. O., and S. D. M. Jones. 1992. Efficiency of a commercial process for the storage and distribution of vacuum packaged beef. J. Food Prot. 55:880–887.
45. Gill, C. O., and T. Jones. 1992. Assessment of the hygienic efficiencies of two commercial processes for cooling pig carcasses. Food Microbiol. 9:335–343.
46. Gill, C. O., and T. Jones. 1997. Assessment of the hygienic characteristics of a process for dressing pasteurized pig carcasses. Food Microbiol. 14:81–91.
47. Gill, C. O., and T. Jones. 1997. Assessment of the hygienic performances of an air-cooling process for lamb carcasses and a spray cooling process for pig carcasses. Int. J. Food Microbiol. 38:85–93.
48. Gill, C. O., and T. Jones. 1998. Control of the contamination of pig carcasses with *Escherichia coli* from their mouths. Int. J. Food Microbiol. 44:43–48.
49. Gill, C. O., T. Jones, and M. Badoni. 1999. The effects of hot water pasteurizing treatments on the microbiological conditions and appearances of pig and sheep carcasses. Food Res. Int. 31:273–278.
50. Gill, C. O., and J. C. McGinnis. 1993. Changes in the microflora on commercial beef trimmings during their collection, distribution and preparation for retail sale as ground beef. Int. J. Food Microbiol. 18:321–332.
51. Gill, C. O., J. C. McGinnis, and M. Badoni. 1996. Assessment of the hygienic characteristics of a beef carcass dressing process. J. Food Prot. 59:136–140.
52. Gill, C. O., J. C. McGinnis, and J. Bryant. 1998. Microbial contamination of meat during the skinning of beef carcass hindquarters at three slaughtering plants. Int. J. Food Microbiol. 42:175–184.
53. Gill, C. O., J. C. McGinnis, and T. Jones. 1999. Assessment of the microbiological conditions of tails, tongues and head meats at two beef packing plants. J. Food Prot. 62:674–677.
54. Gill, C. O., J. C. McGinnis, K, Rahn, and A. Houde. 1996. Control of product temperatures during the storage and transport of bulk containers of manufacturing beef. Food Res. Int. 7:647–651.
55. Gill, C. O., and D. M. Phillips. 1993. The efficiency of storage during distant continental transportation of beef sides and quarters. Food Res. Int. 26:239–245.
56. Gill, C. O., K. Rahn, K. Sloan, and L. M. McMullen. 1997. Assessment of the hygienic performances of hamburger patty production processes. Int. J. Food Microbiol. 36:171–178.
57. Goodfellow, S. J. 1995. Implementation of the HACCP program by meat and poultry slaughterers, in A. M. Pearson and T. R. Dutson (eds.), HACCP in Meat, Poultry and Fish Processing. Blackie Academic, London, pp. 58–71.
58. Gould, W. A. 1994. The implications of current good manufacturing practices (CGMP's) and food plant sanitation, in CGMP's/Food Plant Sanitation. CTI Publications, Baltimore, pp. 1–6.
59. Grau, F. H. 1987. Prevention of microbial contamination in the export beef abattoir, in J. M. Smulders (ed.), Elimination of Pathogenic Organisms from Meat and Poultry. Elsevier, Amsterdam, pp. 221–23.
60. Greer, G. G., C. O. Gill, and B. D. Dilts. 1994. Evaluation of the bacteriological consequences of the temperature regimes experienced by fresh chilled meat during retail display. Food Res. Int. 27:371–377.
61. Greer, G. G., S. D. M. Jones, B. D. Dilts, and W. M. Robertson. 1990. Effects of

spray chilling on the quality, bacteriology and case life of aged carcasses and vacuum packaged beef. Inst. Food Sci. Technol. J. 23:82–86.
62. Grossklaus, D. 1987. The future role of the veterinarian in the control of zoonosis. Vet. Q. 9:321–331.
63. Gunvig, M., and L. Bogh-Sorensen. 1990. Time–temperature distribution of meat, in P. Zuethen, J. C. Cheftel, C. Erikson, T. R. Gormley, P. Linko, and K. Paulus (eds.), Chill Foods: The Revolution in Freshness. Elsevier Applied Science, London, pp 3-244–3-247.
64. Hakkinen, M. 1997. Control of enteropathogens in poultry production by competitive exclusion. Proceedings of the World Congress on Food Hygiene, Hague, August 1997, pp. K9–K13.
65. Hancock, D. D., D. H. Rice, D. E. Herriott, T. E. Besser, E. D. Ebel, and L. V. Carpenter. 1997. Effects of farm manure-handling practices on *Escherichia coli* O157:H7 prevalence in cattle. J. Food Prot. 60:363–366.
66. Harbers, A. H., J. F. Smeets, and J. M. A. Snijders. 1991. Predictability of post mortem abnormalities in slaughter pigs as an aid to meat inspection. Vet. Q. 13:74–80.
67. Hardin, M. D., G. R. Acuff, L. M. Lucia, J. S. Osman, and J. W. Savell. 1995. Comparison of methods for decontamination from beef carcass surfaces. J. Food Prot. 58:358–374.
68. Hathaway, S. C., and M. S. Richards. 1993. Determination of the performance attributes of post-mortem meat inspection procedures. Prevent. Vet. Med. 16:119–131.
69. Heidelbaugh, N. D. 1992. Recommendations from the AVMA workshop on the safety of foods of animal origin. J. Am. Vet. Med. Assoc. 202:201–204.
70. Hood, D. E., and G. C. Mead. 1993. Modified atmosphere storage of fresh meat and poultry, in R. T. Parry (ed.), Principles and Applications of Modified Atmosphere Packaging of Foods. Blackie Academic, London, pp. 269–298.
71. Hudson, W. R., G. C. Mead, and M. H. Hinton. 1996. Relevance of abattoir hygiene assessment to microbial contamination of British beef carcasses. Vet. Rec. 139: 587–589.
72. James, S. 1996. The chill chain "from carcass to consumer." Meat Sci. Suppl. 43: S203–S216.
73. James, J. S., and C. Bailey. 1989. Process design data for beef chilling. Int. J. Refrig. 12:42–49.
74. James, J. S., and C. Bailey. 1990. Chilling of beef carcasses, in T. R. Gormley (ed.), Chilled Foods, The State of the Art. Elsevier Applied Science, London, pp. 159–181.
75. James, S., and J. Evans. 1990. Temperatures in the retail and domestic chill chain, in P. Zeuthen, J. C. Cheftel, C. Ericksson, T. R. Gormley, P. Linko, and K. Paulus (eds.), Chilled Foods: The Revolution in Freshness. Elsevier Applied Science, London, pp. 3,273–278.
76. Jordan, D., and S. A. McEwen. 1998. Effect of duration of fasting and a short-term high-roughage ration on the concentration of *Escherichia coli* biotype 1 in cattle feces. J. Food Prot. 61:531–534.
77. Kilsby, D. C., and M. E. Pugh. 1981. The relevance of the distribution of microorganisms within batches of food to the control of microbiological hazards from foods. J. Appl. Bacteriol. 51:345–354.

78. Kudva, I. T., C. W. Hunt, C. J. Williams, U. M. Mance, and C. J. Houde. 1997. Evaluation of dietary influences on *Escherichia coli* O157:H7 shedding by sheep. Appl. Environ. Microbiol. 63:3878–3886.
79. Lillard, H. S. 1990. The impact of commercial processing procedures on the bacterial contamination and cross-contamination of broiler carcasses. J. Food Prot. 53: 202–204.
80. Mackey, B. 1990. The incidence of food poisoning bacteria on red meat and poultry in the United Kingdom. Food Sci. Technol. Today 3:246–250.
81. Mackey, B. M., and C. M. Derrick. 1979. Contamination of the deep tissues of carcasses by bacteria on the slaughtering instruments or in the gut. J. Appl. Bacteriol. 46:355–366.
82. Madsen, K. B., C. Hagdrup, U. Thrane, K. B. Rasmussen, and W. K. Jensen. 1991. Production management and process control. Proceedings of the 37th International Congress on Meat Science Technology, Kulmbach, Germany, September 1991, pp. 1019–1033.
83. Mattila, T., A. J. Frost, and D. O'Boyle. 1988. The growth of *Salmonella* in rumen fluid from cattle at slaughter. Epidemiol. Infect. 101:337–345.
84. Marks, S., and T. Roberts. 1993. *E. coli* O157:H7 ranks as the fourth most costly foodborne disease. Food Rev. 16:51–59.
85. Mead, G. C., C. Gibson, and D. B. Tinker. 1995. A model system for the study of microbial colonization in poultry defeathering machines. Lett. Appl. Microbiol. 20: 134–136.
86. Mead, G. C., M. J. Scott, T. J. Humphrey, and K. McAlpine. 1996. Observations on the control of *Campylobacter jejuni* infection of poultry by "competitive exclusion." Avian Pathol. 25:69–79.
87. Mitchell, J. M., M. W. Griffiths, S. A. McEwen, W. B. McNab, and A. J. Yee. 1998. Antimicrobial drug residues in milk and meat: Causes, concerns, prevalence, regulations, tests and test performance. J. Food Prot. 61:742–756.
88. Montanari, D. 1996. The role of animal identification systems in controlling the spread of pathogens, in J. J. Sheridan, R. L. Buchannan, and T. J. Montville (eds.), HACCP: An Integrated Approach to Assuring the Microbiological Safety of Meat and Poultry. Food and Nutrition Press, Trumbull, CT, pp. 23–34.
89. Morrison, G. J., and G. H. Fleet. 1985. Reduction of *Salmonella* on chicken carcasses by immersion treatments. J. Food Prot. 48:939–943.
90. Murray, G. 1986. Ante-mortem and post-mortem meat inspection: An Australian Inspection Service perspective. Aust. Vet. J. 63:211–215.
91. Nottingham, P. M. 1982. Microbiology of carcass meats, in M. H. Brown (ed.), Meat Microbiology. Applied Science Publishers, London, pp. 13–66.
92. Nutsch, A. L., R. K. Phebus, M. J. Riemann, D. E. Schafer, J. E. Boyer Jr., R. C. Wilson, J. D. Leising, and C. L. Kastner. 1997. Evaluation of a steam pasteurizing process in a commercial beef processing facility. J. Food Prot. 60:485–492.
93. Palumbo, S. A. 1986. Is refrigeration enough to restrain foodborne pathogens? J. Food Prot. 49:1003–1009.
94. Par, R. W. A., P. L. Griffiths, and G. S. Moreno. 1991. Sources and survival of campylobacters: Relevance to enteritis and the food industry. J. Appl. Bacteriol. Symp. Suppl. 70:97S–106S.

95. Pooni, G. S., and G. C. Mead. 1984. Prospective use of temperature function integration for predicting the shelf-life of non-frozen poultry–meat products. Food Microbiol. 1:67–87.
96. Prohaszka, L., B. M. Jayarao, A. Fabian, and S. Kovacs. 1990. The role of intestinal volatile fatty acids in the *Salmonella* shedding of pigs. Zentrabl. Vet. Med. B 37: 570–574.
97. Rathgeber, B. M., and A. L. Waldroup. 1995. Antibacterial activity of a sodium acid pyrophosphate product in chilled water against selected bacteria on broiler carcasses. J. Food Prot. 58:530–534.
98. Reichel, M. P., D. M. Phillips, R. Jones, and C. O. Gill. 1991. Assessment of the hygiene adequacy of a commercial hot boning process for beef by a temperature function integration technique. Int. J. Food Microbiol. 14:27–42.
99. Robinson, R. 1995. Electronic identification of animals and carcasses, in S. D. Morgan Jones (ed.), Quality and Grading of Carcasses of Meat Animals. CRC Press, Boca Baton, FL, pp. 201–213.
100. Rocelle, M. R. S., L. R. Beauchat, and M. P. Doyle. 1998. Thermal inactivation of *Escherichia coli* O157:H7 isolated from ground beef and bovine feces, and suitability of media for enumeration. J. Food Prot. 61:285–289.
101. Rodel, W., H. Hechelmann, and J. Dresel. 1996. The hygiene aspects of wooden and plastic cutting boards. Fleischwirtsch. Int. 1996 (2):16–21.
102. Rogers, S. A., N. W. Hollywood, and G. E. Mitchell. 1992. The microbiological and technological properties of bruised beef. Meat Sci. 32:437–447.
103. Rosset, R. 1982. Chilling, freezing and thawing, in M. H. Brown (ed.), Meat Microbiology. Applied Science Publishers, London, pp. 265–318.
104. Rusin, P., P. Orosz-Coughlin, and C. Gerba. 1998. Reduction of faecal coliform, coliform and heterotrophic plate count bacteria in the household kitchen and bathroom by disinfection with hypochlorite cleaners. J. Appl. Microbiol. 85:819–828.
105. Schmidt, U. 1983. Cleaning and disinfection of slaughterhouses and meat processing factories, Fleischwirtschaften 63:1188–1191.
106. Schoonderwoerd, M. 1997. Main factors responsible for visible pork carcass contamination. Proceedings of the World Congress on Food Hygiene, The Hague, August 1997, p. 67.
107. Scrine, G. R. 1985. Refrigerated vehicles—What next? in Long Distance Refrigerated Transport: Land and Sea. International Institute of Refrigeration, Paris, pp. 17–22.
108. Slavik, M. F., J. W. Kim, and J. T. Walker. 1995. Reduction of *Salmonella* and *Campylobacter* on chicken carcasses by changing scalding temperature. J. Food Prot. 58:689–691.
109. Smith, M. G. 1985. The generation time, lag time and minimum temperature of growth of coliform organisms on meat, and the implications for codes of practice in abattoirs. J. Hyg. Cambridge 94:289–300.
110. Sorquist, S., and M. L. Danielsson-Tham. 1986. Bacterial contamination of the scalding water during the vat scalding of pigs. Fleischwirtschaft 66:1745–1748.
111. Stephan, F., and K. Fehlhaber. 1995. Geflugelfleischgewinnig: Untersuchung zur Hygiene des Luft-Spruch-Kuhlverfahrens. Fleischwirtschaft 74:870–873.
112. Tauxe, R. V. 1991. *Salmonella:* A postmodern pathogen. J. Food Prot. 54:563–568.

113. Telling, G. M. 1990. Control of veterinary drug residues in food. Proceedings of the EuroResidue Conference, Noordwijkerhout, The Netherlands, May 1990, p. 19.
114. Troutt, H. F., J. Gillespie, and B. I. Osburn. 1995. Implementation of HACCP program on farms and ranches, in A. M. Pearson and T. R. Dutson (eds.), HACCP in Meat, Poultry and Fish Processing. Blackie Academic, London, pp. 36–57.
115. U.S. Department of Agriculture, Food Safety and Inspection Service. 1993. Beef carcass trimming versus washing study. Fiscal year 1993. Fed. Regist. 58:33925–33932.
116. U.S. Department of Agriculture, Food Safety and Inspection Service. 1995. Pathogen reduction: Hazard analysis and critical control point system; proposed role. Fed. Regist. 60:6774–6889.
117. U.S. Department of Agriculture, Food Safety and Inspection Service. 1996. Performance standards for the production of certain meat and poultry products. Fed. Regist. 61:19564–19582.
118. U.S. Department of Agriculture, Food Safety and Inspection Service. 1996. Pathogen reduction; hazard analysis and critical control point system; final rule. Fed. Regist. 61:38805–38989.
119. U.S. Department of Agriculture, Food Safety and Inspection Service. 1998. Requirements for the production of cooked beef, roast beef, and cooked corned beef. Code of Federal Regulations, Title 9, Vol. 2, part 318, sec. 318.17.
120. U.S. Department of Agriculture, National Advisory Committee on Microbiological Criteria for Foods. 1993. Generic HACCP for raw beef. Food Microbiol. 10:449–488.
121. Urlings, H. A. P., P. G. H. Bijkers, and J. M. A. Snijders. 1996. The control of feed and water as a source of pathogens for food animals, related to veterinary public health, in M. H. Hinton and C. Rowlings (eds.), Factors Affecting the Microbial Quality of Meat. University of Bristol Press, Bristol, U.K., pp. 1–8.
122. Van Donkergoed, J., K. W. F. Jericho, H. Grogan, and B. Thorlakson. 1997. Preslaughter hide status of cattle and the microbiology of carcasses. J. Food Prot. 60: 1502–1508.
123. Waldroup, A. L. 1996. Contamination of raw poultry with pathogens. World Poult. Sci. J. 52:7–25.

2
Safe Handling of Dairy and Egg Products

Robin C. McKellar and Kelley P. Knight
Agriculture and Agri-Food Canada, Guelph, Ontario, Canada

I. INTRODUCTION

Dairy products are considered to be relatively safe foods, accounting for 1–3% of reported outbreaks in the United States (70). The majority of outbreaks from dairy products can be attributed to postprocessing contamination (31,38,70), since the efficacy of pasteurization processes has been well established. Thermal processing of milk has been routinely performed since the latter part of the nineteenth century, when it was introduced to combat such diseases as tuberculosis, diphtheria, and typhoid.

Reports of illness attributed to cheese consumption, dating back to the nineteenth century (138), eventually led to the introduction in the 1940s of regulations requiring pasteurization of cheesemilk, or holding of cheese for 60 days prior to sale (70). In recent years, however, the emergence of pathogens such as *Listeria monocytogenes, Escherichia coli* O157:H7, and *Salmonella,* which are considered high-risk for dairy products, has raised concerns about the validity of regulations governing dairy products.

Cheese and other dairy products are considered to be ready to eat, requiring no additional processing before consumption. Consumers are relatively experienced in the handling of dairy products, and are less likely to subject them to temperature abuse than meat and poultry products. Changes in milk handling practices over the last 30 years, the advent of new processing technologies, and the need for the dairy industry to deal with outdated equipment suggest the need for strict guidelines for the handling of dairy products.

Eggs are an important food commodity throughout the world, especially as a protein source (122). Historically, the egg has been thought of as a "clean" food owing to its protective outer shell, inner membrane, and internal antimicrobials (68). Recent findings have disputed this belief. It has been observed that *Salmonella enteritidis* can contaminate an egg prior to laying. In addition, the shell may come in contact with numerous microorganisms during the laying process or, afterward, from the environment. Under poor storage conditions, bacteria may enter the egg, grow, and cause spoilage and/or illness. There has also been a trend to commercial size egg production flocks and hatcheries, largely as a result of the automation of operations (123). As the industry undergoes such changes, it is imperative that the handling of eggs and egg products be done correctly to assure the safety of eggs and egg products.

Until the 1940s, processed egg products consisted mainly of frozen whole eggs and separated whites and yolks. World War II brought on a resurgence of interest in egg products, especially dried whole eggs. Since then, the industry has expanded to include other markets such as egg products for institutional use (122). In the past, the majority of egg-breaking operations and processing plants were not inspected. A need was identified for improved information and direction on the processing of egg products. The pasteurization of eggs was developed mainly to address the control of salmonellae (137). In 1971 the U.S. Department of Agriculture (USDA) followed up with mandatory inspection of egg-breaking operations.

The Centers for Disease Control and Prevention's (CDC) most recent survey of foodborne illness outbreaks, for 1988–1992, cited *S. enteritidis* as a leading cause of foodborne illness and death, with the number of outbreaks increasing over the 5-year period (12). Most of these outbreaks were attributed to eating undercooked, infected eggs. *S. enteritidis* also caused more deaths than any other pathogen, with the majority of deaths affecting nursing home residents. Improper holding temperatures and poor personal hygiene of food handlers were cited.

This chapter outlines some of the common handling practices that may be applied to dairy and egg products and assesses the extent to which the various sectors of the "farm-to-fork" continuum share a common responsibility to ensure the safety of foods.

II. DAIRY PRODUCTS

A. Regulations

Dairy products are regulated by virtually every country in the world. Initiated by the International Dairy Federation (IDF), the Codex Alimentarius Commission was founded in 1962 to implement the Joint FAO/WHO Food Standards Program (120). The Joint FAO/WHO Committee of Government Experts on the Code of Principles concerning Milk and Milk Products, established in 1958 by IDF, became the Codex Committee on Milk and Milk Products (CCMMP). This group has provided a code of principles for the handling of milk and milk products (67,120). IDF has also published a bulletin outlining recommendations for the hygienic manufacture of milk and milk-based products (66).

In the United States, the Pasteurized Milk Ordinance covers standards and sanitation requirements for the milking farm, and for the processing and shipping of fluid milk (132). Canadian regulations for dairy products may be accessed from the Dairy Plant Inspection Manual (48), published by Agriculture and Agri-Food Canada, and more recently from the Canadian Food Inspection Agency (CFIA) Dairy Products Regulations (21). In Canada, fluid milk is covered by provincial regulations. For instance, in Ontario these are published by the Ontario Ministry for Agriculture, Food and Rural Affairs (OMAFRA) as the Milk Act Regulations (90).

B. Incidence

The incidence of foodborne pathogens in milk and other dairy products has been extensively studied, and several comprehensive reviews on individual microorganisms have appeared in recent years. It is beyond the scope of this chapter to compile all data on foodborne pathogens in milk and dairy products, but a brief summary of each is included. The reader is encouraged to consult cited reviews, as well as appropriate original papers.

1. Salmonella

Salmonella, probably the best-known causative agent of human foodborne illness, has been extensively studied and reviewed (30,38,70,100,106,141). *Salmonella typhimurium* is the most common serovar causing human foodborne illness. The serovars *S. typhimurium, S. enteritidis, S. infantis,* and *S. heidelberg* were found to be the most prevalent in a survey of 109 countries. *Salmonella* is found in the intestinal tract of animals and humans, in animal feed and food products, and products made with them. Foods harboring *Salmonella* include dairy products such as raw and pasteurized milk and cheese. Incidence of up to 23% has been noted in raw milk samples. The growth range for *Salmonella* is from 5°C (41°F) to 45.6°C (114°F), although the lower limit for growth depends greatly on the physiological state of the cells and the food product. The optimal pH for growth is 6.5 to 7.5, but growth can occur at as low as pH 4.0. *Salmonella* does not survive below pH 4.0. The growth of *Salmonella* is generally inhibited by 3–4% salt. Strains other than *S. senftenberg* have no particularly high resistance to heating and are destroyed by pasteurization. *Salmonella* is known to survive in Cheddar and other types of cheese; however, the cooking step involved in the production of mozzarella and cottage cheese inactivates this pathogen.

2. Listeria monocytogenes

L. monocytogenes is one of the pathogens of most concern with respect to dairy products, and its presence and behavior have been well documented (26,41,42,59, 78,92,101,106,107,111,114,141). As a psychrotroph, *L. monocytogenes* is capable of growth over a wide temperature range: 1.7°C (35°F) to 45°C (113°F). This pathogen has been isolated from a wide variety of environmental sources including farm animals, mud, silage, dairy barns, and dairy processing plants. *L. monocytogenes* has also been shed by cows affected by listeric mastitis. In processing plants, it is found on floors and in drains, around coolers, in air handling units, and in cooling systems. *L. monocytogenes* has also been isolated from raw and pasteurized milk, ice cream, yogurt, and many types of cheese, in particular soft cheeses. Incidence of this pathogen in raw milk has been reported as high as 45%; however, incidences of less than 10% are more common. In soft cheeses, incidences of 10% are common, with up to 10^7 cfu/g found in some samples. *L. monocytogenes* is found in approximately 1.2% of Canadian dairy products.

L. monocytogenes is considered to be quite heat resistant, but it is destroyed by high temperature, short time (HTST) pasteurization at 72°C for 16 seconds. It is also very resistant to acid conditions, being reported to grow as low as pH 4.4, and to survive at pH 3.3. Acid tolerance may increase at lower temperatures. *L. monocytogenes* is also resistant to high levels of salt. For these reasons, *L. monocytogenes* survives well in cheeses, up to 180 days in some cases. Survival and growth of this pathogen in soft cheese is enhanced by virtue of the higher pH, prompting the development of a risk assessment model that determined the probability of acquiring listeriosis in Canada from soft and semisoft cheese consumption (42).

3. Escherichia coli

Escherichia coli is considered a member of the normal facultative flora of the intestine; however, there are four categories of *E. coli* that cause diarrhea (34,35, 37,70,105,106,141): (a) enterotoxigenic (ETEC), a major cause of travelers' and infant diarrhea in developing countries, (b) enteroinvasive (EIEC), cause of diarrhea and illness similar to shigellosis, (c) enteropathogenic (EPEC), causing infant diarrhea, and (d) enterohemorrhagic (EHEC), the cause of hemorrhagic colitis and hemolytic uremic syndrome. The principal member of this last group, the serotype O157:H7, was first identified as causing a clinical illness in 1982. Temperature range for O157:H7 is between 2.5°C (36.5°F) and 45°C (114°F), although it grows poorly at 44–45°C, and growth on food is rarely if ever seen below 8–10°C. It grows within a pH range of 4.4–9.0, and at a minimum a_w of 0.95. This pathogen is quite acid resistant, surviving in fermented foods as low as pH 3.7; it has no special heat resistance, however, and is destroyed by pasteurization. Dairy cattle have been identified as a reservoir for O157:H7, and this serotype is found primarily in calves. Thus, this pathogen has been found in raw milk and in various soft and semisoft cheeses as well as yogurt. In cheeses such as brick, Camembert, and Colby, *E. coli* survived the manufacturing process and was inactivated during curing.

4. Campylobacter

Campylobacter jejuni became an important cause of foodborne diarrheal illness during the 1980s (19,70,104,106,123). Its growth range is 30°C (86°F) to 45°C (113°F). It is associated with foods of animal origin contaminated by feces and has been associated with raw milk. *C. jejuni* does not grow well in foods, since it requires a low oxygen concentration and does not compete well with common food microflora. It does not survive pasteurization and has limited resistance to drying, storage at room temperature, and acidic conditions; thus heavy contamination of a food is necessary for illness to occur. *C. jejuni* was shown not to survive in Cheddar cheese after storage for 30–60 days.

5. Yersinia

Yersinia enterocolitica was identified as a foodborne pathogen in the 1970s (26,70,102,106,110,117). *Y. enterocolitica* is a psychrotroph, growing between 0°C (32°F) and 44°C (111.2°F). It grows well over the pH range of 4.6–9.0, is sensitive to heat [50°C (122°F)] and to salt concentrations above 7%. It is widely distributed in the environment and has been isolated from many animals; however, most isolates are avirulent for humans. It has been isolated from raw, pasteurized, and chocolate milk, and has caused foodborne illness with these as vehicles. The incidence in raw milk can approach 48%. *Y. enterocolitica* survives up to 8 weeks in Colby cheese, but dies off during maturation of Cheddar cheese. It can survive for long periods in water and frozen foods, and has survived on the outside of milk cartons.

6. Staphylococcus aureus

Foodborne illness resulting from ingestion of *Staphylococcus aureus* enterotoxin has been recognized for many years (15,32,70,99,118). *S. aureus* is salt tolerant, and grows at a wide range of pH values (4.5–7.0). The temperature range for growth is 6.5°C (44°F) to 50°C (122°F); however, toxin production is best at 21°C (70°F) to 37°C (98.7°F). The pathogen is killed by heating. So, if food poisoning does occur, it is due to postprocessing contamination. *S. aureus* is found in the human respiratory tract, and the primary mode of food contamination with this pathogen is from handling of food by an infected person. It is also found in raw milk as a result of shedding by mastitic cows. Up to 50% of raw milk samples can be positive for *S. aureus,* with as many as 4.2% of the isolates producing toxin. The pathogen is thus found in dairy products such as chocolate milk, cheese curds, cheeses such as Cheddar, Colby, and Swiss, and in cream-filled pastries. It can also be found in milk powder.

7. Bacillus cereus

Bacillus cereus, which is widespread in the environment, is a common contaminant of raw foods (36,74,103,117,118). It can be found in raw milk as a result of contamination from milking sheds and from mastitic cows. It has been isolated from pasteurized milk, milk powder, Cheddar cheese, and ice cream. It produces heat-resistant spores and thus survives normal food processing, and can become a problem if products are temperature-abused. Because of the heat resistance of spores, up to 48% of milk samples treated at ultrahigh temperatures have been found to be positive for *B. cereus.* Typical environmental strains grow at 10°C (50°F) to 50°C (122°F). Environmental strains are generally not able to grow below 10°C; however, 17% of food isolates tested grew at 7°C or below (36). *B. cereus* strains can produce heat-labile diarrheal toxins and heat-stable emetic toxins. The pH range for growth is 4.3–9.0, with a minimum a_w of 0.95.

C. Outbreaks

Outbreaks and incidences of foodborne illness attributed to dairy products have been extensive, and many reviews have summarized these events (7,13–15,30, 34,35,37,38,41,59,70,74,78,92,110,123,128,129,141). From these report important trends can be highlighted. Outbreaks have been mainly attributed to cheese and raw milk, and the high profile cases usually have resulted from infection with *L. monocytogenes* or *Salmonella.* Some of the best-known cases include the 1983 outbreak of *L. monocytogenes* in pasteurized milk in Massachusetts, the 1984 outbreak of *Salmonella* in Cheddar cheese in Canada, the 1985 outbreak of *Salmonella* in pasteurized milk in Illinois, the 1985 outbreak of *L. monocytogenes* in Mexican-style soft cheese in California, and the 1996 outbreak of *E. coli* O157:H7 from contaminated pasteurized milk in Scotland.

An analysis of these trends emphasizes practices that appear to lead to outbreaks in dairy products. Practices such as consumption of raw milk, improper pasteurization, contamination of pasteurized milk with raw milk, and the use of unpasteurized milk for cheese making may increase the hazard of foodborne illness in dairy products. Soft cheeses have been implicated more than any other type of cheese in foodborne illness. In addition, the ability of *L. monocytogenes* to proliferate in soft cheese is of particular concern. The concern is exacerbated when one considers that the sector of the population most at risk (infants, hospital patients, pregnant women, the elderly, and the immunocompromised) are likely to consume dairy products as a good source of nutrients.

These potential hazards have stimulated recommendations, including that from the Institute of Food Science and Technology (IFST) to closely monitor the presence of *L. monocytogenes* in soft cheese (6). Because of the considerable hazard posed by *L. monocytogenes* in soft cheese, Health Canada has classified soft cheeses as category 1, with an action level for *L. monocytogenes* of > 0 cfu/50 g, with immediate action being a class I recall to retail level (42). In a more general sense, it has been recommended that all cheese be made from pasteurized milk (7,31).

D. Farm

Table 1 gives the U.S. standards applied to farms for the handling of raw milk (132).

A number of publications have summarized the various good farm handling practices, which are described in brief here (28,29,37,48,66,70,90,108,132). Some of these target specifically the control of mastitis or foodborne pathogens such as *L. monocytogenes* or *E. coli* O157:H7; however, the various practices are applicable to control of all contaminating microorganisms.

1. Be sure that muddy, dusty conditions are controlled, especially in cow yards and around milk houses.
2. Keep milking equipment clean and properly stored to protect it from dust and dirt.
3. Sanitize all milk contact surfaces just prior to use.
4. Keep cows as clean as possible.
5. Use a sanitizer to prep cows, and make sure udders are dry before milkers are attached.
6. Keep all birds and other farm animals out of milking barns and away from milking equipment. Establish good insect control.
7. Feed only high quality materials and be certain fermented feeds are properly cured.

Table 1 U.S. Regulations as Applied to Farms for the Handling of Raw Milk

Specification	Regulation
Temperature	Cooled to 7°C (45°F) or less within 2 hours after milking, provided the blend temperature after the first and subsequent milkings does not exceed 50°F (10°C).
Bacterial limits	Individual producer milk not to exceed 100,000/mL prior to commingling with other producers' milk. Not to exceed 300,000/mL as commingled milk prior to pasteurization.
Drugs	No positive results on drug residue detection methods as referenced in Section 6 of Ref. 132.
Somatic cell count	Individual producer milk: not to exceed 750,000/mL.

Source: Ref. 132.

8. Check all silage for proper pH (<5), especially silage stored in bulk silos or trenches.
9. Keep the milk house and bulk truck loading dock clean and free of mud and animal waste.

In addition to reducing environmental contamination, good raw milk management practices can help in reducing the microbiological load. Milk cooling on the farm must be well controlled, especially since pickup times have been increased (i.e., longer times between milk pickups) as a cost-saving measure (70). Some areas of concern include malfunctioning cooling units, electric current failure, incorrectly adjusted thermostats, and mixing of properly cooled and improperly cooled milk (70). Another strategy to improve raw milk quality can include on-farm thermization (63–65°C for 15–20 s) (70) to reduce counts of psychrotrophic bacteria.

When applying hazard analysis–critical control point (HACCP) or good manufacturing practices (GMP) to the handling and transportation of raw milk, it is important not to restrict the view to a single element, rather the "cow to consumer" concept should be considered (94). An incidence of 4.7% for *Salmonella* in milk trucked to dairies (38) emphasizes the need for the good hygienic collection of raw milk (7) to reduce the incidence of contamination (124). For example, dairy cattle have been implicated as a reservoir for *E. coli* O157:H7 (37). As a result, milk tankers and drivers contaminated with fecal material from the farm can carry the pathogen back to the dairy plant (37). Recent work with *Listeria* ribotypes common to both dairy processing and farm environments clearly implicates the farm as a source of dairy plant contamination (8). Thus, the following guidelines might be utilized to prevent entry of pathogens into the plant environment (28):

1. Isolate receiving area and everyone associated with it from the processing and packaging areas of the plant.
2. Have no direct openings into the processing area.
3. Prohibit anyone having contact with receiving (haulers, lab personnel) from going into, passing through, or working in the processing area.
4. Assure that no raw product comes in contact with the floor anywhere around processing and packaging equipment.
5. Isolate raw milk tanks.
6. Keep receiving room walls and floors clean.
7. Cleaned and sanitize drains daily.
8. Limit the on-farm contact of bulk haulers to the milk house area.
9. Avoid reworking of returned product; this practice is dangerous because *L. monocytogenes* and other contaminants may be present.
10. Do not use sponges and rags.

E. Dairy/Cheese Plant

Milk provides an ideal environment for the survival and growth of a wide variety of foodborne pathogens. The incidence of human pathogens such as *Mycobacterium tuberculosis* in raw milk prompted early attempts to heat milk to improve the safety. Indeed, prior to the introduction of HACCP, the dairy industry utilized these concepts to eradicate *M. tuberculosis* from milk supplies (112). Thus, heating of milk is considered to be one of the most important mechanisms for reducing the risk of foodborne illness in milk and milk products.

Westhoff (140) wrote an excellent review on the historical development of pasteurization. Heating of milk to improve keeping qualities was first initiated in the late nineteenth century, with the first commercial milk pasteurizer being built in 1882. In the early part of the twentieth century, continuous HTST treatment was developed as an alternative to batch heating. Considerable effort was expended to develop time–temperature combinations to ensure the destruction of *M. tuberculosis,* and standardized processing conditions were established. HTST systems were rapidly developed in the 1930s and 1940s, and by the 1950s, ultrahigh temperature (UHT) processing was being developed. Table 2 gives the temperature requirements for HTST milk in Canada (48); conditions described for the United States are similar (132).

Research on pasteurization for cheese making began in the early part of the twentieth century (69). The primary objective was to enhance the quality of the product; however, safety was also an important factor. There was, and continues to exist, considerable controversy over the effect of pasteurization on the quality of cheese. In general, it is believed that unpasteurized milk produces cheese of higher quality (69), and it was considerable time before pasteurization of cheese milk was accepted as a standard practice in most countries.

Table 2 Pasteurization Standards for HTST Milk in Canada

Milk product	Holding time (s)	Temperature (°C)
<10% milk fat	16	72
	1.0	89
	0.5	90
	0.1	94
	0.05	96
	0.01	100
>10% milk fat	16	75
Frozen dairy mixes	25	80
Eggnog	16	83

Source: Ref. 48.

It is clear that consumption of raw milk presents a considerable hazard and should be avoided (69); however, it is not clear that full pasteurization of cheese milk is required for consumer safety. Recommendations were made that cheese produced in the United States be (a) made from pasteurized milk, (b) pasteurized, or (c) cured for a minimum of 60 days (69). There exists some doubt, however, that the 60-day hold is effective (69), which has led to the recommendation that all milk for cheese making be pasteurized (7,31). Alternatively, it has been proposed that milk destined for cheese making be thermized or heat-treated at 64.4°C (148°F) for 16 seconds to reduce the potential hazard due to foodborne pathogens (69).

The thermal stability of several foodborne pathogens such as *L. monocytogenes, Salmonella* spp., and *E. coli* O157:H7 has been extensively studied in milk and milk products, and the results reviewed (26,34,38,41,59,70,92). The consensus is that correct pasteurization will ensure that these pathogens are eliminated from milk. A hazard assessment of *L. monocytogenes* in the processing of bovine milk revealed that the probability was less than 2 in 100 that one cell occurs in every 2 gallons of milk processed at 71.7°C for 15 seconds (93). Considerable research has also been conducted on the survival of these pathogens in a wide variety of cheeses, and this research has been extensively reviewed (26,34,38,41,59,92,108). It is generally felt that these pathogens survive readily, and even grow, in most cheeses, especially soft-ripened cheeses. Thus, pasteurization is considered to be the most effective means of ensuring the safety of milk and dairy products.

Appropriate handling practices for dairy and cheese plants are detailed in Canadian (48,90) and U.S. (132) regulations, as well as in numerous articles and reviews (25,28,37,38,57,66,70,92,97,108,112,124,135,137). The following subsections summarize the various practices.

1. Sources of Contamination

The primary source of pathogens such as *L. monocytogenes* is raw materials coming into the plant. Pathogens may also be found in unchlorinated water supplies, clean-in-place (CIP) lines, hollow equipment legs and framework, refrigeration units, conveyors, ingredient container surfaces, and heavily etched stainless steel surfaces. Floors and floor drains are also a common source, and drains should not be located close to filling and packaging equipment. Proper location, construction, and cleaning of floor drains is essential. Items that absorb water (e.g., sponges) should not be used, since they might harbor pathogens, as porous surfaces such as wood are known to do. Aerosols from high pressure hoses, unshielded pumps, or condensates can also serve as sources.

2. Design

Initial design of processing establishments is one of the most crucial aspects of maintaining good product quality. Serious flaws in design can lead to contamination of product. Pasteurizer designs have not changed since the 1930s, and addition of extra tanks and pipelines to older systems can result in the development of biofilms that are hard to clean and might harbor pathogens. To alleviate some of these concerns, several key items should be kept in mind: raw and processed product must be well isolated; floors must be designed for good drainage and cleaning, and pipelines and equipment for easy disassembling and cleaning.

3. Environment

Coolers and freezers should be kept free of condensate, and floors, walls, and ceilings should be cleaned frequently. Belt systems and conveyors must be cleaned and sanitized regularly. Clean surfaces also increase the efficacy of sanitizers. Pooling of milk or water in ducts or grouting should be eliminated, and product splashing should be avoided, as should aerosols. Use of hot water during processing can lead to condensation.

4. Pasteurization

Pasteurization may be carried out by vat or HTST systems. It is common for small operations to utilize batch systems when the amount of milk processed is too small for HTST. Some of the problems with the vat system can include improper equipment design, absence of proper outlet valves and airspace thermometers, and improperly operated airspace heaters. To ensure pasteurization, the airspace temperature must be at least 2.8°C higher than the product. Valves and connections must be properly designed to prevent pockets of cold milk, and foam must be minimized during filling, heating, and holding.

All pasteurization equipment must be properly designed, installed, and operated. Improperly functioning pasteurizers have been linked to outbreaks of foodborne pathogens. Common problems with HTST systems include stress cracks or pinholes in heat exchanger plates, leaking gaskets, improper flow diversion valves, and inadequate cleaning and sanitizing. Regular testing should be carried out on holding times, thermometers, flow diversion valves, and flow controls. Positive pressure should be maintained between the product and heating medium as well as the product and cooling medium. Sweetwater and glycol should be kept at a lower pressure than the milk. Plates should be inspected regularly for cracks or holes, and the cooling water monitored for foodborne pathogens. Maintenance of accurate records and chart recordings is recommended. Problems can also arise when dairy ingredients such as ice cream mix are pasteurized and transported to another location.

5. Clarifiers and Separators

Bactofugation, or the removal of bacteria by high speed centrifuge, can be effective in removing from raw milk spore-forming bacteria such as *Bacillus* spp. and eukaryotic cells such as leukocytes. This technique is routinely used in the Netherlands prior to the manufacture of Gouda cheese. Waste material from clarifiers must be removed from the plant as soon as possible to avoid contamination. Both clarifiers and separators should be cleaned by hand.

6. Traffic Patterns

It is very important to isolate the raw and processed sections of the plant to prevent carryover from the home or farm, machine shop, or raw milk area. Special emphasis in training employees in avoiding the spread of pathogens within the plant environment from outside the plant is needed. The use of foot baths should be encouraged, and such stations should be monitored routinely for proper disinfectant strength and cleanliness. Access to the processing area must be restricted, and the movements of pallets and forklifts from raw milk, case wash, or dock into the processing or packaging areas should be controlled.

7. Sanitation

Cleaning and sanitizing regimens must be reviewed regularly. The correct sanitizers should be used at the appropriate strengths and contact times. Table 3 lists the common sanitizers used.

Charts and records should be kept, and the data verified regularly. Sanitizers must be applied only to clean contact surfaces, since organic material will reduce the efficacy of sanitizers. Casers, cappers, stackers, undersides of equipment and parts tables, walls, ceilings, and floors should be sanitized regularly. Floors should be flooded with 300 ppm of quaternary ammonium sanitizers, and exterior surfaces should be fogged with 800–1000 ppm of quaternary ammonium sanitizers.

8. Pipelines

Pipelines are used to transfer raw and pasteurized product from the various areas throughout the factory. Numerous violations, including cross-connections be-

Table 3 Strengths of Common Sanitizers

Sanitizer type	Strength (ppm)
Chlorine	100
Quaternary ammonium compounds	100
Acidic ionics	200
Iodophors	25

tween raw and pasteurized milk lines and between CIP and product lines, have been disclosed by inspections of U.S. plants. It is essential to prevent bypass of raw milk around the pasteurizer; otherwise postprocessing contamination might occur. Up-to-date blueprints of product flow must be kept, and unwanted piping, dead ends, illegal cross-connections, or unauthorized changes should be eliminated. It is essential to eliminate trapping of washing or sanitizing solutions, and to ensure free draining of all lines. CIP lines should be self-draining or self-evacuating.

9. Filling and Packaging

Postpasteurization contamination usually occurs during filling and packaging, as a result of contact with surfaces. Product may be contaminated from mandrels, drip shields, prefilling coding equipment, overhead shielding, conveyor belts, extruder heads, chain rollers, supports, and lubricants. It is essential to sanitize all exposed surfaces carefully both during and after production runs. Extruder heads, which are particularly prone to contamination, should be sanitized frequently during filling. It is important to reduce product handling as much as possible. Manual handling, filling, and capping should be avoided.

10. Air Quality

Heating, ventilating, and air conditioning systems must be well designed and periodically cleaned. Drip pans and drain lines should be checked regularly. Protective shielding must be installed to reduce contamination of the product from airborne microbes. Where product is exposed, a positive pressure must be maintained, and air from the raw milk area must be minimized or excluded. Outside air should be filtered, and direct flow of air on the product should be avoided. Air filters should be designed to reduce particulate matter, condensate, and microorganisms. Freezers must be easily cleanable and designed to reduce contamination. Sanitary check valves must be inspected to prevent product backup, and air and blow lines must be manually cleaned. Airflow should be from processed area to raw.

11. Cross-contamination or Postprocessing Contamination

Cross-contamination or postprocessing contamination can be a serious problem in dairy or cheese plants. To limit this problem, sweetwater and glycol cooling systems must be monitored regularly for the presence of pathogens, as should tanks, jacketed vessels, and cooling plates. Cracks in storage tanks, leaking valves, improper welds, and irregular surfaces can lead to sanitation problems. The use of brushes (rather than sponges or cloths) and impervious materials is recommended. Temperature abuse of the product should also be avoided. Employee cleanliness is an important factor in limiting cross-contamination. Street clothes should not be allowed in the plant, and work clothes should be properly laundered by the plant.

Hands and tools must be carefully washed and sanitized, especially when a person is moving from the raw to processed areas. Regular microbiological examination of areas of concern will assist in establishing a high level of cleanliness.

12. Rework

Reprocessing of product returned to the factory is a very high risk operation. Some of the potential problems include failure to repasteurize returned product, pumping of returned product through pasteurized lines without repasteurizing, reuse of outdated product, and reuse of temperature-abused product or product from leaking containers. All returned product should be treated as raw milk and processed at temperatures above the required minimum; it should not be used as grade A milk. External carton contamination should be carefully avoided, and reclaimed product should be isolated from all other plant operations.

F. Food Service

In his foodborne disease summary for Canada for 1982, Todd (128) estimated that food service establishments accounted for 38.1% of all mishandled food incidents, and 75.7% of all cases involving mishandling of foods. Problems in hotels and restaurants accounted for more than half of the incidents associated with food service mishandling, but only 21.1% of the cases (128). These findings were confirmed in a 10-year summary (1975–1984) also conducted in Canada (129). Costs (in $0.5) from outbreaks from food service establishments have ranged from $16,690 to over $1 million (127). Other workers have estimated that the food service industry is responsible for more than half of the food poisoning outbreaks of known cause in western countries, including Australia (13,88). It has also been estimated that 85% of Americans consume at least one meal outside the home every 2 weeks (106), and that by the year 2000, one-third to one-half of all meals will be consumed outside the home. Poor personal hygiene on the part of food service workers can account for 25–40% of foodborne illness (116). This is of particular concern when we consider that the food service industry provides for several high risk populations most affected by foodborne illness (106,119). Clearly, the food service industry has a great responsibility to improve the safety of food offered to the consumer.

Dairy products such as milk, cream, yogurts, and soft cheese are defined as chilled foods: uncured, potentially hazardous foods with a water activity of 0.93 or above, and a pH of 4.6 or above (43,98). Potentially hazardous foods are those that can support the rapid and progressive growth of infectious or toxigenic microorganisms (98). There are three categories of chilled foods: those handled close to freezing; those handled just above freezing (0–5°C); and those that require temperature control but not to the same degree (0–8°C) (43). Dairy products, other than ice cream are considered under category 2.

Specific examples of the handling of dairy products in food service operations are difficult to find. Munce (88) has described an HACCP plan for an egg/dairy uncooked dessert. Recommendations for the handling of dairy products are limited to the use of pasteurized dairy foods in clean sealed containers. Bryan (18) noted that critical control points (CCPs) for dairy products exist in dairy plants, but not at the food service level. Milk and dairy products should be pasteurized in plants that (a) implement an HACCP system, (b) monitor the pasteurization process, and (c) verify control by testing finished products to ensure effectiveness of pasteurization, and absence of postpasteurization contamination.

Much of the published material relative to food service operations is concerned with commodities of greater hazard than dairy products. In general, the following guidelines can be applied to chilled foods (18):

1. Cool foods rapidly [e.g., so that they do not remain within the temperature range of 21–49°C (70–120°F) for longer than 2–3 h (2 h if feasible)].
2. Continually cool these foods so that their temperatures fall to 7°C (45°F) in another 4–6 hours.
3. During cooling, store foods in depths not exceeding 7.5 cm (3 in.), unless some other measures are used to cool foods rapidly.
4. During cooling, do not stack pans on top of each other.
5. During cooling, keep food containers at least 5 cm (2 in.) apart.
6. Never store foods in deep containers.
7. Maintain refrigerator air temperatures below 4°C (40°F).

G. Home

There is little information on the specific handling of dairy products by the consumer; indeed dairy products have enjoyed a good record in the home. This is linked to the experience consumers have had with dairy products. Some general guidelines have been published in scientific papers, and several Internet sites provide consumer-oriented handling information. The material summarized below was abstracted from a variety of such sources (38,40,49,58,62,94,125,126).

Clearly, the most critical aspect of handling foods in the home is storage temperature. Since pathogens grow in the "danger zone" of 4°C (40°F) to 60°C (140°F), it is essential that home refrigerators be kept at 4°C or less. Temperatures should be verified with a thermometer.

The other important aspect of food handling is personal cleanliness. Hands should be washed thoroughly with soap and hot water before and after handling dairy products, particularly cheeses. Counters, utensils, and equipment should also be washed before and after use. Cutting boards, of particular importance because they are often used for both raw and cooked food, should be washed and sanitized with chlorine before cheese is cut on them. Food should be served on clean plates, and leftovers should be refrigerated promptly, within 2 hours, if possible.

When purchasing dairy products, buy only containers that are cold. Dairy products should be purchased last, and refrigerated immediately on return home. Check "use by" dates and select items that will stay fresher longer. Soft cheeses, which can support the growth of *L. monocytogenes,* are of particular concern. They should be used promptly after purchase and not stored for extended periods.

H. HACCP Examples

General handling procedures with the emphasis on dairy products have been discussed. To bring a more scientific focus to correct handling procedures, the concept of HACCP has been introduced to the food industry (80,81). The HACCP system is described in detail in other chapters.

Cullor (29) examined the hazards associated with the spread of bovine mastitis in a dairy farm. He identified the maternity pen, the sick pen, the free stalls and pastures, and the milking parlor as locations where mastitis pathogens can affect the lactating cow. The maternity pen was identified as a CCP, and means to control pathogens were developed. These included appropriate teat dip, appropriate medical management of clinical cases, segregation of infected cows, and control of the environment (i.e., adequate drainage; manure removal; clean, dry bedding).

Goff (57) discussed an HACCP system for ice cream plants. He emphasized that (a) plant flow diagrams are important to identify problem areas, (b) raw materials entering the plant must adhere to strict specifications, and (c) colors and flavorings must be closely controlled. Pasteurization was identified as a CCP because it is the only biological control point at which pathogens are destroyed. Freezing of the product was also identified as a CCP, since freezing does not kill pathogens that may reside in the product as a result of improper pasteurization or postprocessing contamination. Proper, rapid freezing is important, as well as sanitation of the ingredient feeder, a major source of contamination. It was recommended that reruns not be added back to the flavor tank, that filler heads be kept clean, and that manual capping be performed with great care.

A recent example of an HACCP plan for bulk Cheddar cheese was reported by Bernard and Sreum (16). This was a hypothetical plan for a small (<50 employees) privately owned cheese manufacturer preparing bulk Cheddar cheese for food service and food processing. Pasteurization was identified as a CCP. To ensure proper pasteurization, the process must be constantly monitored for effective seals and timing pump, and the holding tube exit temperature monitored by chart recorder. Proper operation of the diversion valve is also essential, and this should be tested at start-up. Milling (cutting of the curd) was also identified as a CCP, since there is the risk of *S. aureus* growth if acid development is inadequate. The presence of *S. aureus* and pH are monitored, and product is diverted for further thermal processing if necessary.

Generic HACCP models have also been developed by the National Dairy Council of Canada for partially skimmed milk, aged Cheddar cheese, and ice cream novelties. These identify raw cream/milk storage as CCPs owing to the potential growth of pathogens and the possibility for toxin production. Cream and milk should be stored below 5°C, with a maximum storage period of 72 hours. Education of the producer and transporter were suggested as ways of controlling hazards.

III. EGG PRODUCTS

A. Regulations

Both production and sale of shell and processed eggs are regulated. In Canada, shell eggs are governed by the Canada Agriculture Products Act—Egg Regulations (22). This act regulates the registration and suspension of egg stations, the conditions applying to egg stations, inspection and certification, packaging and labeling, grading, and trade. Processed eggs are regulated by the Canada Agriculture Products Act—Processed Egg Regulations (23). This act regulates similar aspects of the processing industry as in the Egg Regulations. In the United States, foods are governed by the Food Code (46). Processed eggs are specifically addressed in the Egg Pasteurization Manual (136), which outlines processing conditions for specific egg products, and provides guidelines for the procedures that govern packaging, labeling, sanitation, facilities, equipment, plant layout, and operations. Processed eggs also fall under the 1970 USDA Egg Products Inspection Act. In 1995 the USDA's Food Safety and Inspection Service took over the inspection of egg products (51).

B. Incidence

Eggs consist of a shell (9.5%), albumen (63%), and a yolk (27.5%) with total solids of the albumen, yolk and whole egg being 12, 52, and 24%, respectively (96). Eggs also have been reported to impart protective effects to microorganisms. This has been attributed to the yolk portion, which is high in fat and solids that are able to bind organic materials such as chlorine that might otherwise be bactericidal. Eggs are also high in nutrients, neutral in pH, and acceptable in terms of water content. All these conditions lend themselves to microbial growth.

1. Salmonella enteritidis

The organism of most concern in eggs is *Salmonella* and, in particular *Salmonella enteritidis*. *S. enteritidis* is of concern owing to its mode of infection. There is evidence that the organism is passed through the reproductive tissue, as opposed to causing contamination after laying. Kim et al. (71) inoculated eggs with six different concentrations of *S. enteritidis* (10^1, 10^2, 10^3, 10^4, 10^5, 10^6 cfu/mL) and incubated at five temperatures (4, 10, 16, 21, and 27°C) for 30 days. The growth rate of *S. enteritidis* increased with temperature regardless of the initial inoculum. At the end of the incubation period, increases in *S. enteritidis* were observed, even at 4°C. Growth can occur within 2–3 days when inoculated eggs are exposed to abuse temperatures such as room temperature (109). A risk assessment carried out in 1990 (86) on *S. enteritidis* and eggs reported about 1 in 200 eggs from infected hens will be contaminated; from an epidemic area, about 1 in 10,000 to 14,000 eggs were contaminated. It was concluded that 0.9% of eggs may be eaten in dishes requiring no cooking; however the number of eggs eaten undercooked is unknown. Canadian commercial egg-producing flocks were surveyed to estimate the prevalence of *S. enteritidis* and other *Salmonella* spp. (95). Samples taken from feces, eggbelts, and feed showed 52.9% of randomly selected flocks to be contaminated with 35 different *Salmonella* serovars. The most common serovar was *S. heidelberg* with a 20% incidence; *S. enteritidis* was isolated from 2.7% of the flocks tested. More recently, the American Egg Board (4) reported that in areas where outbreaks have occurred, tested flocks have shown an average of 2–3 infected eggs out of 10,000. A liberal estimate country-wide is 1 out of 20,000 eggs (4). The likelihood of finding an infected egg is about 0.005%.

2. Listeria monocytogenes

Another organism that has become of concern in egg and egg products in recent years is *Listeria monocytogenes*. *L. monocytogenes* is able to survive in refrigerated raw egg and to grow well in cooked eggs (115). *Listeria* spp., including *L. monocytogenes,* have been found at low levels in 2 of 42 commercially broken raw liquid eggs (75). In another study, Moore and Madden (85) found 37.8% of 173 samples of blended whole egg from the in-line filters of an egg pasteurization plant to be positive for *L. monocytogenes*. However, 500 daily samples of pasteurized product were found to be negative for *Listeria* spp. *L. monocytogenes* has been reported to survive in frozen liquid egg products and powdered egg products for up to 180 days (17). The organism can grow in liquid whole egg (LWE) or LWE that has been salted (5% NaCl), particularly when there is temperature abuse (39,44). Depending on initial contamination levels, *L. monocytogenes* can survive the current minimum pasteurization requirements (10,44,45,91). Reductions of 1.7 to 4.4 log cycles of viable *L. monocytogenes* result from this treatment in LWE products (10,44). In LWE with 10% NaCl, pasteurization resulted in a 0.2 to 0.6 $\log_{10}$ reduction of *L. monocytogenes* (10).

3. Other Pathogens

Although *S. enteritidis* is the main foodborne pathogen associated with eggs, with some attention to *L. monocytogenes* in recent years, there is a potential for other pathogenic infections to occur. *Campylobacter jejuni* (33) and *Yersinia enterocolitica* (89) are associated with poultry, and thus not surprisingly may be linked to the safety of eggs, since the possibility exists for eggshells to become contaminated. *Staphylococcus aureus* may also be present, through postprocessing contamination from infected food handlers.

C. Outbreaks

Reviews and publications outline foodborne outbreaks involving eggs and egg products (24,82,87,121,127,128,130,131,139). A majority of outbreaks occur in the food service sector and are caused by improper food handling techniques. Risk factors include the pooling of large numbers of eggs that are permitted to stand unrefrigerated and the undercooking of mishandled eggs.

One outbreak in a Maryland restaurant involving scrambled eggs is of particular interest because it had seven serious breaks in food handling rules (77): as many as 1800 eggs were pooled; the mixture was left at room temperature for up to 6 hours; additional eggs were added to the raw egg mixture; leftover eggs were used the next day; cooks were urged not to overcook the eggs, which may encourage undercooking; freshly cooked eggs were combined with allready cooked eggs; and holding temperatures were sometimes reduced to prevent drying out of the eggs. Morris (86) estimated the risk of eating eggs at the implicated restaurant to be more than 2000-fold greater than the risk to the person who prepares eggs individually and eats them promptly.

There have also been outbreaks of staphylococcal infection. In 1983 approximately 300 children in California became ill with staphylococcal food poisoning from intact boiled, dyed Easter eggs (79). It was discovered that the hands of a food handler bore lesions of a staphylococcal infection, and the eggs had been left unrefrigerated for 3–5 days. It was further observed experimentally that heated eggs can absorb up to 2 mL of contaminated water through an intact shell. In 1985 lasagna was the vehicle for an outbreak in which inadequately pasteurized liquid egg was used in the production of pasta. Prolonged holding times of unrefrigerated raw pasta dough allowed for the growth of *S. aureus* and the production of toxin.

D. Farm

Farm practices greatly influence productivity, and the farm is the starting point for the control of contamination of the laying facility, the laying flocks, and eggs. In 1990, when *Salmonella* outbreaks were affecting the European and American egg markets, the Canadian Egg Marketing Agency in conjunction with Agriculture and Agri-Food Canada developed a voluntary program of recommended codes of farm management practices (1), which are summarized in the subsections that follow. The program is based on an HACPP approach for controlling risks in food production.

1. Biosecurity Areas

On the farm, biosecurity zones such as the laying house, the egg collection room, and the cooler, should be well marked and the distinctions adhered to. Visitors to these restricted areas should be kept to a minimum. To prevent the introduction of contamination, protective outer clothing such as coveralls, boots, hairnets, and gloves are to be worn by all persons entering these zones. Special attention and care is required by persons having had recent contact with other poultry or livestock, to avoid the spread of disease-causing agents. It is strongly recommended that only one kind of poultry production unit be located on a given farm. If this is not feasible, extra care must be taken, especially with personal hygiene and the changing of protective clothing from one area to another. Regulations governing the personal hygiene of those, entering restricted biosecurity zones should be strictly enforced, especially those mandating hand washing.

2. Facility

The facility itself must be hygienic. Manure, dirt, or cobwebs should not be allowed to accumulate. The optimal time to clean and sanitize facilities is right after depopulation. Any repairs to the facility should be made prior to cleaning. Cleaning procedures consists of a pressure wet-wash, drying, disinfecting, and airing of floors, walls, light fixtures, vents, exhausts, and equipment. The disinfectant should be an approved agent with a pest control product number. Drinking cups and nipples should be individually scrubbed. A chlorine solution should be left in contact with water lines for 24 hours, after which the lines should be rinsed with fresh water. The cleaning process takes a minimum of 7 days.

Maintenance operations include keeping all work areas, as well as cages and feed troughs, clean. The speed, volume, and direction of supplied air should minimize the presence of gases, dust, odors, and bacteria in the air to ensure a healthy flock environment.

One should start with a disease-free flock, purchasing chicks or pullets only from suppliers who have a *Salmonella* control program in place. Flocks should arrive in clean and sanitized containers, and all chicks should be of the same age group. The first precaution makes complete cleaning and disinfecting possible; in addition, multiage flocks attract rodents.

A major source of contamination is from rodents and insects. All breeding areas for these pests should be eliminated and all building openings screened. There should be no stagnant water within 60 m of buildings, and at least 4.5 m immediately surrounding buildings should be kept free of debris and long grass. Garbage should be stored in a sealed area separate from production areas. Spilled feed, excrement, or waste should not be allowed to build up. Dead or sickly birds are to be removed immediately.

Another major source of contamination is feed and water supplies. Processed, (i.e., pelleted or heat-treated) feed should be purchased from suppliers with a *Salmonella* control program in place. Feed should be transported in a way that minimizes contamination (e.g., stored in closed containers).

3. Egg Handling

Hygiene is extremely important because eggs are food. Eggs should be collected as soon as possible after laying (i.e., at least twice a day) and placed in coolers. Great care should be taken to avoid contamination and to minimize breakage. In cleaning the egg collection system, special care is to be given to eggbelts, which have been found to be contaminated with salmonellae more often than fecal samples (95). Cracked or manure-soiled eggs should be removed and handled separately. Eggs should be stored in clean, sanitized carts, and not placed on the floor. Washing of eggs is to be carried out by registered graders only. All workers should wash their hands before and after the collection of eggs.

Storage conditions should be strictly adhered to, with recommended temperatures being between 10 and 13°C; lower temperatures can cause condensation, leading to mold growth, whereas higher temperatures may lead to bacterial growth. Humidity should be set at 75–85%. Thorough records of all sanitation and handling practices should be kept.

E. Wash Station/Breaking Plant

1. General

All plants should adhere to rodent control and good manufacturing practices set down by the appropriate jurisdiction. The summary that follows is based on Agriculture and Agri-Food Canada's sanitation requirements (1).

Use only suppliers who follow the Code of Farm Management Practice. All biosecurity restrictions should be followed. Without exception, product flow protocol should be observed. Sanitation control should be checked through microbial analysis such as environmental swabs of equipment, plant, and washwater, and the cleaning and sanitizing of all carts, trays, pallets, and equipment should be ensured.

The storage temperature of eggs should not exceed 13°C (as set by Agriculture and Agri-Food Canada). Proper inventory control should be exercised—first in, first out. All persons within the plant should wear clean clothing and footwear and change these garments as necessary. In addition, employees should use proper hand-washing procedures following contact with ungraded or broken eggs and dirty items, and before and after going to the washroom or lunchroom. Facilities are to be provided for these purposes. Staff training is imperative for any sanitation program to be effective and successful. All staff should report any illness or incidents that may put eggs at risk. All immediate areas surrounding the building(s) must be clean and free of debris. An HACCP program is recommended.

2. Shell Eggs

Graded and ungraded eggs should be stored separately to avoid cross-contamination. Shell eggs undergo a washing process designed to improve the overall appearance of the egg and to remove surface dirt such as feces and soil, as well as bacteria that could contaminate eggs during the breaking process (83). Most egg-washing systems are automated and use recycled water. Eggs are placed on a conveyor belt and passed through the washer. Rollers, brushes, and sprayed water clean the eggs.

Washwater should be of the appropriate temperature, pH, and chlorine levels at all times and should be changed every 2–4 hours. A temperature of 43 ± 3°C and a pH of 10 or more are recommended. Agriculture and Agri-Food Canada developed these guidelines to eliminate pathogens and reduce the overall bacterial load of shell eggs (47). Washwater should not contain more than 10^5 of cfu of aerobic bacteria per milliliter. A high grade approved egg-washing alkaline detergent, should be used according to the manufacturer's recommended dosage, with the alkalinity being monitored throughout the washing process.

A relationship has been established between washwater quality and total aerobic bacterial counts on washed eggs. As the washing process occurs, egg solids, manure, dirt, and bacteria from the eggs end up in the recycled water and build up over time (60,61). The accumulation of egg solids and dirt in the washwater will lower the pH to sublethal levels (11,73,84). Holley and Proulx (65) reported that a temperature of 42°C and pH above 10 were required to prevent the survival of *Salmonella* spp. in egg washwater. Bartlett et al. (11) found that temperature and pH alone were insufficient to control the bacterial load in washwater. The addition of chlorine to washwater was found to be to be beneficial. For temperature of 40°C or more and pH values of 10 or more, a minimum of 0.45 mg/L of available chlorine should be maintained in the washwater. The monitoring of turbidity (%T) of samples was also found to be beneficial in monitoring washwater quality. Leclair et al. (76) reported that the addition of chlorine (20 μg/mL) to washwater at moderate temperature (42°C) and pH (10.5) at the beginning of a run (i.e., while egg solids were not present) was effective in controlling *S. typhi-*

murium and *L. monocytogenes.* When eggs solids were present in the washwater, increases in temperature and pH to at least 47.4°C and ≥10.8, respectively, were required for the control of these microorganisms.

Once washed, eggs should be rinsed in potable water. A final hypochlorite rinse spray of 50 mg/L residual (active) chlorine is often used. Spray nozzles should not be plugged because plugging leads to reduced and variable spray flows.

Eggs are dried using a high velocity down draft of filtered air. Intake air filters should be cleaned and replaced regularly. Eggs are then oiled with a mineral oil, which acts as a sealant against the transfer of moisture and gases through the shell. The eggs next undergo a candling process, during which eggs with cracked or weak shells are removed, as well as eggs that are excessively contaminated with dirt, fecal matter, or blood that had not been removed at the farm. These eggs should be washed at the end of each shift just before dumping the washwater tank. Finally, shell eggs are packaged and stored at 4 to 7°C before shipping.

3. Processed Eggs

Egg products are defined as eggs that have been removed from their shells for processing (i.e., breaking, filtering, mixing, stabilizing, blending, pasteurizing, cooling, freezing, drying, and packaging). Egg products include whole eggs, whites, yolks, and various blends, with or without the addition of non-egg ingredients such as salt and sugar. Egg products are preferred because they add convenience, are relatively easy to handle and store, and offer an enhanced degree of food safety (51). Egg products are utilized as an ingredient in many foods such as baked goods and salad dressings, by bakeries, candymakers, food service institutions and, increasingly, consumers.

Processed eggs are pasteurized and made available as either liquid (LWE, yolks alone or whites alone) or dried products. Pasteurization of eggs mainly addresses the elimination of *Salmonella* spp. Pasteurization causes no significant changes in the color, flavor, or nutritive value of eggs. The functional properties of eggs are detrimentally affected by heating. The margin between effective kill and irreversible functional changes is small; thus much care must be taken by the processor.

The initial numbers of bacteria should be low, since low numbers take longer to reach dangerous levels than larger numbers. Also the degree of protection achieved is greater with smaller numbers. Thus sanitary measures are important. Eggs used for processing must have edible interiors to begin with.

Most commercial pasteurizers are HTST units. Specific guidelines for the installation and operation of HTST liquid egg pasteurizers deal with adherence to building guidelines, inspections, and validations (136).

Processing is based on the ability of elevated temperatures to destroy bacteria. In thermal bacteriology, the term D-value is used to describe the amount of time needed to reduce a microbial population by 90% or 1 log cycle. Pasteuriza-

Table 4 Processing Conditions Required for Egg Products

Egg product	Temperature °C	Temperature °F	Holding time (min)
Whole eggs	60	140	3.5
Salted whole egg (2–12%)	63	146	3.5
Plain yolks	61	142	3.5
Salted yolk (2–12%)	63	146	3.5
Sugared yolk (2–12%)	63	146	3.5

Source: Data from Refs. 23, 136.

tion standards are based on a 9-D effect (kill) of *Salmonella* (136). Table 4 summarizes the time–temperature combinations for the HTST pasteurization of some egg products (23,136). Pasteurization standards for other egg products and alternative time–temperature combinations can be found the regulations cited.

Once an egg product has been pasteurized, it is important to ensure that recontamination does not occur. Pasteurization is ineffectual if the proper precautions are not adhered to in order to keep the processed egg product "clean." Packaging under proper conditions, product cooling, and appropriate storage of the product will reduce the likelihood of recontamination. Guidelines are available (23,46,136).

F. Food Service

The U.S. Centers for Disease Control and Prevention have reported that the majority of outbreaks involving eggs in the United States have occurred within the retail segment of the food industry. This has been attributed to the number of eggs involved, the number of people involved in their preparation, high employee turnover, and possible inability of employees to read or understand instructions (4). The mishandling of the food was cited as the main cause of most outbreaks (4, 12,46,55).

Food safety is especially important in the food service sector because of the increased exposure to high risk groups, including immunocompromised people, alcoholics, diabetics, transplant recipients, AIDS and cancer patients, very young infants, steroid users, and patients with chronic renal disease and iron storage disorders (40).

A number of publications and resources geared specifically toward food service outline safe food handling practices for egg and egg products in this sector (2,5,9,20,27,51,55,64,72). The recommended food handling practices are summarized next.

It is recommended that food service institutions not serve raw eggs or foods containing raw eggs. Instead, recipes are to be altered to allow the substitution of pasteurized egg products in, for example, Caesar salad dressing, hollandaise sauce, ice cream, eggnog, mayonnaise, key lime pie, and mousse.

Although pasteurization destroys bacteria such as *Salmonella,* products must be handled carefully to avoid recontamination. Use only products that have been certified and inspected. Receive product only in tightly sealed containers that have been stored properly (refrigerated or frozen). To thaw frozen egg products put containers in the refrigerator or under running cold water. Dried eggs should be kept in a dry cool place before and after opening; reconstituted mixtures should be refrigerated in tightly closed, clean containers. Remove only what is required, and do not return any unused product to the original container; instead place in a clean new container, seal it tightly, and store it appropriately.

Frozen egg products can be stored for up to 1 year at 0°C (32°F) or below, but they should not be refrozen after thawing. Thawing should be done in the refrigerator or under running cold water, not on a counter. Use by the "best before" date and store appropriately. Liquid eggs should be stored at 4°C (40°F) or below for up to 7 days without exceeding 3 days after opening. Do not freeze open containers. Dried egg products are to be stored in a dry cool place, and once opened, should be stored in the refrigerator. Reconstituted egg products should be used. USDA Commodity Dried Egg Mix should be stored at less than 10°C (50°F) in the refrigerator on the day of preparation and used within 7–10 days of opening. Reconstitute only what is needed, then use immediately, or refrigerate and use within 1 hour.

The practice of pooling eggs represents a major area of risk. Pooled eggs are eggs that are broken together, such that one contaminated egg can contaminate the whole batch. Preferably, pasteurized egg products should be used for recipes requiring pooled eggs. Pooled eggs should not be left at room temperature for more than 2 hours.

Temperature control is critical. Eggs should be cooked until both yolk and white are firm. Egg dishes should be served [holding temperature of 60°C (140°F)] immediately after cooking or refrigerated (4°C) (40°F) at once by placing in shallow containers. Scrambled eggs should be cooked in small batches of 3 L maximum until there is no visible liquid egg. New batches should not be added to older batches in the steam table. In addition, do not reuse utensils or containers that held raw egg mixture, even for another raw egg mixture. Wash and sanitize between uses.

Also imperative is the observation of good sanitation practices of the food service kitchen. All efforts should be made to create a clean environment and to avoid contamination of the product. The Canadian Restaurant and Food Services Association has prepared a Sanitation Code for food service operations. It is Canada's code of safe food handling practices, which incorporates provincial and federal health regulations for the food service sector. The association also offers aid for the development of HACCP programs by food service establishments.

The Canadian regulations specify that when shell eggs are used, one should purchase only Canada grade A eggs, which are free of cracks and dirty shells. Eggs that have been stored under refrigeration should be refrigerated upon delivery, and used before the "best before" date. In addition one should buy eggs in small quantities, frequently (i.e., buy only enough for 1–2 weeks).

G. Home

Eggs are an important source of many essential nutrients and should be a part of our daily diet. Consumers can prevent the contamination or spread of bacteria by following safe food handling and preparation practices within the home. The following guidelines will aid the consumer (4,9,20,27,49–56,63,72,125,133,134).

1. Purchase of Eggs

When purchasing eggs, check the "best before" date on the carton. With proper storage and handling, eggs retain their freshness and quality for up to 3 weeks after they have been packaged. Beyond this, eggs are best used for baking, scrambling, or hard-cooking rather than frying or poaching. Buy only the amount of eggs you expect use prior to the "best before" date. Buy only uncracked grade A or AA eggs that are clean (i.e., have no dirt on the shell) and have been held at appropriate refrigeration temperatures, and never buy eggs that have been sitting at room temperature.

When running errands, visit the grocery store last, so that perishables are not left in the car for extended periods of time, especially in warmer climates. While in the grocery store, leave the perishables aisles, including eggs and egg products, until the end.

2. Storage of Eggs

Once home, store eggs in the refrigerator in their original carton, which protects the eggs from absorbing the flavor and odors of other foods. Do not store eggs on the door in the tray which is built in on many refrigerators. The door area tends not to be cold enough. It is a good idea to purchase an inexpensive thermometer to monitor the temperature of your refrigerator. Refrigerator temperatures should be 4°C(40°F). In addition, never wash eggs before storing them, for this removes the protective coating. Washing also encourages bacterial penetration through the egg shell.

If any eggs become cracked, break them open into a clean container and use within a few days. Whites and yolks will keep for up to 4 days if placed in the refrigerator in airtight containers. Whole eggs can be frozen for up to 4 months. Beat the whole egg, place the mixture in a freezer container, and seal tightly. Whites can be frozen as is without special precautions, but yolks will thicken or gel when frozen. To avoid this, beat in either 0.5 mL (1/8 tsp) salt or 7 mL ($1\frac{1}{2}$ tsp) sugar or corn syrup per 63 mL (1/4 cup) of yolk. Freeze in small quantities and thaw only what is required. Thaw the frozen eggs in the refrigerator. Storage times are summarized in Table 5 (20).

3. Cooking with Eggs

Prior to using eggs, the cook's hands, as well as all utensils, equipment, and work areas, should be washed with hot soapy water. All utensils and equipment should be washed between uses even if reuse will entail another raw egg mixture. Household bleach at 30–45 mL (1–$1\frac{1}{2}$ tsp) in 4 L (1 gallon) of water can be used to sanitize kitchen equipment. Eggs should not be left at room temperature for more than 2 hours.

The consumption of raw or undercooked eggs is not recommended. For recipes that traditionally contain raw eggs, such as eggnog, Caesar salad dressing, ice cream, and hollandaise sauce, one should substitute pasteurized eggs or egg substitutes. Alternatively, a cooked egg base could be utilized by heating the raw eggs with liquid or other ingredients until the mixture reaches 60°C (140°F) for 3, 5 minutes or reaches an end point of 71°C (160°F). Eggs should be cooked so that both whites and yolks are firm. Fried eggs should be cooked for 2–3 minutes at 121°C (250°F) on each side or in a covered pan for 4 minutes. Scrambled eggs are to be cooked for 1 minute at 121°C (250°F) or until firm throughout. Poach eggs in boiling water for 5 minutes, or boil in the shell for 7 minutes. For those who insist on soft yolk eggs, cook or microwave until the egg white is completely firm and the yolk is thickened but not hard. Sunnyside eggs are to be cooked uncovered for 7 minutes at 121°C (250°F) or covered for 4 minutes at the same temperature. When preparing quiches, casseroles, or any egg dishes, use an inexpensive, easy-to-read thermometer to check internal temperatures or ensure that a knife inserted in the

Table 5 Home Storage Conditions for Eggs

	Storage times	
Eggs	Refrigerator	Freezer
Fresh raw	3 weeks	4 months
Fresh yolk and white	2–4 days	4 months
Hard cooked, yolk	1 week	Not recommended
Hard cooked, whole	1 week	Not recommended

Source: Ref. 20.

center comes out clean. Persons in high risk groups should avoid lightly cooked egg-containing foods such as French toast.

For an indoor party or get-together, serve hot dishes from the stove and cold dishes from the refrigerator or from containers placed in ice. For picnics or outdoor parties, pack cold egg dishes with ice or commercial coolant in an insulated cooler or bag.

Egg dishes should be served immediately after cooking, or refrigerated and served within 3–4 days. To refrigerate dishes made ahead or leftovers, place in shallow containers in the refrigerator. For longer periods of storage, dishes may be frozen.

H. HACCP Examples

Ali and Spencer (3) conducted a hazard analysis of 6 food preparation sites and 16 school cafeterias for the production of lunch foods including egg sandwiches. Several hazards were identified. Fillings and sandwiches were prepared several hours prior to consumption. Foods were transported and held at temperature of 17–40°C, which allows for the growth of bacteria and germination of spores. There was also much handling of the sandwiches by food service personnel. To control these hazards, foods should not be prepared long in advance of serving, appropriate holding temperatures are to be maintained, and personnel must take extra care while handling food to avoid contaminating the finished product.

Bryan (18) has described HACCP systems for retail food and restaurant operations. Identified hazards included the type of food, preparation methods, storage and display temperatures, and the time between preparation and consumption. Cooking, chilling, and using foods from safe sources and/or reheating were identified as CCPs for cook/hold hot systems, cook/chill and cook/freeze systems, and assemble/serve systems, respectively. Bryan also described the monitoring, action, and verification components of a HACCP program. A specific application of HACCP was given for the production of potato salad with eggs. The cooking of eggs was identified as a CCP, since this step can eliminate bacteria. Other CCPs were formulation, mixing, cooling, and cold storage—steps that offer some assurance of control.

I. Where to Get Information

There are many resources for information on the safe handling of egg and egg products for both the food service sector and the home consumer alike. (See also Chapter 18). Any of the following organizations or regulatory agencies have a wealth of information.

Canadian Egg Marketing Agency
Place de Ville

320 Queen St, Suite 1900
Ottawa, ON
Canada K1R 5A3
Tel: (613) 238-2514
Fax: (613) 238-1967
Web page: *www.canadaegg.ca*
e-mail: *info@CanadaEGG.ca*

Health Protection Branch
Health Canada
Publications
Ottawa, ON
Canada K1A 0K9
Fax: (613) 952-7266
Web page: *www.hc_sc.gc.cc/food-aliment*

American Egg Board
1460 Renaissance Drive
Park Ridge, IL 60068
Web page: *www.aeb.org/safety/egg-handling.html*

U.S. Department of Agriculture
Food Safety and Inspection Service
Washington, DC 20250
Consumer hot line: 1-800-535-4555
Media and educators contact: USDA (202) 720-5604
FSIS (202) 720-7410
Web page: *www.usda.gov/fsis/eggprod.htm*

The National Food Safety Database
USDA/FDA Foodborne Illness Education Information Center
National Agricultural Library/Food and Nutrition Information Center
10301 Baltimore Blvd.
Rm. 304
Beltsville, MD 20705-2351
Tel: (301) 504-5719
Fax: (301) 504-6409
Web page: *www.foodsafety.org/*

The International Dairy Federation
41 Square Vegote, 1030 Brussels, Belgium
Tel: +322 733 9888
Fax: +322 733 0413
Web page: *www.fil-idf.org*
e-mail: *info@fil-idf.org*

Codex Alimentarius Commission
Secretariat of the Joint FOA/WHO Food Standards Programme
Food and Agriculture Organization of the United Nations
Viale delle Terme di Caracalla
00100 Rome, Italy
Tel: +39(6)5705.1
Fax: +39(6)5705.3152/5705.4593
Web page: *www.fao.org/waicent/faoinfo/economic/esn/codex/codex.htm*
e-mail: *CODEX@FAO.Org*

REFERENCES

1. Anonymous. 1990. Start Clean—Stay Clean, 2nd Ed. Canadian Egg Marketing Agency, Ottawa, ON, Canada.
2. Albrecht, J. A., S. S. Sumner and A. Henneman. 1992. Food safety in child care facilities. Dairy Food Sanit. 12:740–743.
3. Ali, A. A. and N. J. Spencer. 1996. Hazard analysis and critical control point evaluation of school food programs in Bahrain. J. Food Prot. 59:282–286.
4. American Egg Board. 1997. Salmonella and egg safety. *www.aeb.org/safety/index.html#safety*
5. Anonymous. 1988. Managing a sanitary and safe foodservice operation. Dairy Food Sanit. 8:410–411.
6. Anonymous. 1995. IFST: Current hot topics—*Listeria monocytogenes* in cheese. *www.ifst.org/hottop2.htm*
7. Anonymous. 1997. IFST: Current hot topics—Food safety and cheese. *www.ifst.org/hottop15.htm*
8. Arimi, S. M., E. T. Ryser, T. J. Pritchard, and C. W. Donnelly. 1997. Diversity of *Listeria* ribotypes recovered from dairy cattle, silage, and dairy processing environments. J. Food Prot. 60:811–816.
9. Baird, P. 1989. Using eggs safely. Restaurant Bus. 88:82–85.
10. Bartlett, F. M., and A. E. Hawke. 1995. Heat resistance of *Listeria monocytogenes* Scott A and HAL 957E1 in various liquid egg products. J. Food Prot. 58:1211–1214.
11. Bartlett, F. M., J. M. Laird, C. L. Addison, and R. C. McKellar. 1993. The analysis of egg wash water for the rapid assessment of microbiological quality. Poult. Sci. 72:1584–1591.
12. Bean, N. H., J. S. Goulding, C. Lao, and F. J. Angulo. 1996. Surveillance for foodborne-disease outbreaks—United States, 1988–1992. CDC Surveillance Summaries, MMWR 45 (No. SS-5):1–68.
13. Bean, N. H., and P. M. Griffin. 1990. Foodborne disease outbreaks in the United States, 1973–1987: Pathogens, vehicles, and trends. J. Food Prot. 53:804–817.
14. Bean, N. H., P. M. Griffin, J. S. Goulding, and C. B. Ivey. 1990. Foodborne disease outbreaks, 5-year summary, 1983–1987. J. Food Prot. 53:711–728.
15. Bergdoll, M. S. 1989. *Staphylococcus aureus,* in M. P. Doyle (ed.), Foodborne Bacterial Pathogens. Marcel Dekker, New York, pp. 464–523.

16. Bernard, D., and W. Sveum. 1994. Industry perspectives on *Listeria monocytogenes* in foods: Manufacturing and processing. Dairy Food Environ. Sanit. 14:140–143.
17. Brackett, R. E., and L. R. Beuchat. 1991. Survival of *Listeria monocytogenes* in whole egg and egg yolk powders and in liquid whole eggs. Food Microbiol. 8:331–337.
18. Bryan, F. L. 1990. Hazard analysis critical control point (HACCP) systems for retail food and restaurant operations. J. Food Prot. 53:978–983.
19. Butzler, J. P., and J. Oosterom. 1991. *Campylobacter:* Pathogenicity and significance in foods. Int. J. Food Microbiol. 12:1–8.
20. Canadian Egg Marketing Agency. 1997. Food handling: How to buy, use and store fresh eggs.
www.canadaegg.ca/english/recipes/handle.html
21. Canadian Food Inspection Agency. 1997. Dairy products regulations.
www.cfia-acia.agr.ca/english/actsregs/dairy/home.html
22. Canadian Food Inspection Agency. 1997. Egg regulations.
www.cfia-acia.agr.ca/english/actsregs/eggs/home.html
23. Canadian Food Inspection Agency. 1997. Processed egg regulations.
www.cfia-acia.agr.ca/english/actsregs/eggproc/home.html
24. Centers for Disease Control. 1990. Update: *Salmonella enteritidis* infections and shell eggs—United States. MMWR 39:909–912.
25. Chambers, J. V. 1987. Securing the "safety net" in the dairy industry. Dairy Food Environ. Sanit. 7:340–343.
26. Champagne, C. P., R. R. Laing, D. Roy, A. A. Mafu, and M. W. Griffiths. 1994. Psychrotrophs in dairy products: Their effects and their control. Crit. Rev. Food Sci. Nutr. 34:1–30.
27. C. Williamson and the National Food Safety Database. 1997. Egg handling handbook.
www.foodsafety.org/fs/fs067.htm
28. Coleman, W. W. 1986. Controlling *Listeria* hysteria in your plant. Dairy Food Sanit. 6:555–557.
29. Cullor, J. S. 1995. Implementing the HACCP program on your clients' dairies. Vet. Med. 90:290, 292–295.
30. D' Aoust, J.-Y. 1989. *Salmonella,* in M. P. Doyle (ed.), Foodborne Bacterial Pathogens. Marcel Dekker, New York, pp. 328–445.
31. D'Aoust, J. Y. 1989. Manufacture of dairy products from unpasteurized milk: A safety assessment. J. Food Prot. 52:906–914.
32. DeLuca, G., F. Zanetti, and S. Stampi. 1997. *Staphylococcus aureus* in dairy products in the Bologna area. Int. J. Food Microbiol. 35:267–270.
33. Doyle, M. P. 1984. Association of *Campylobacter jejuni* with laying hens and eggs. Appl. Environ. Microbiol. 47:533–536.
34. Doyle, M. P. 1991. *Escherichia coli* O157:H7 and its significance in foods. Int. J. Food Microbiol. 12:289–301.
35. Doyle, M. P., and V. V. Padhye. 1989. *Escherichia coli,* in M. P. Doyle (ed.), Foodborne Bacterial Pathogens. Marcel Dekker, New York, pp. 236–281.
36. Dufrenne, J., P. Soentoro, S. Tatini, T. Day, and S. Notermans. 1994. Characteristics of *Bacillus cereus* related to safe food production. Int. J. Food Microbiol. 23:94–109.

37. Duncan, S. E., and C. R. Hackney. 1994. Relevance of *Escherichia coli* O157:H7 to the dairy industry. Dairy Food Environ. Sanit. 14:656–660.
38. El-Gazzar, F. E., and E. H. Marth. 1992. Salmonellae, salmonellosis, and dairy foods: A review. J. Dairy Sci. 75:2327–2343.
39. Erickson, J. P., and P. Jenkins. 1992. Behavior of psychrotropic pathogens *Listeria monocytogenes, Yersinia enterocolitica,* and *Aeromonas hydrophila* in commercially pasteurized eggs held at 2, 6.7 and 12.8 degrees C. J. Food Prot. 55:5–12.
40. Farber, J. M., and A. Hughes. 1995. General guidelines for the safe handling of foods. Dairy Food Environ. Sanit. 15:70–78.
41. Farber, J. M., and P. I. Peterkin. 1991. *Listeria monocytogenes,* a food-borne pathogen. Microbiol. Rev. 55:476–511.
42. Farber, J. M., W. H. Ross, and J. Harwig. 1996. Health risk assessment of *Listeria monocytogenes* in Canada. Int. J. Food Microbiol. 30:145–156.
43. Farquhar, J., and H. W. Symons. 1992. Chilled food handling and merchandising: A code of recommended practices endorsed by many bodies. Dairy Food Environ. Sanit. 12:210–213.
44. Foegeding, P. M., and S. B. Leasor. 1990. Heat resistance and growth of *Listeria monocytogenes* in liquid whole egg. J. Food Prot. 53:9–14.
45. Foegeding, P. M., and N. W. Stanley. 1990. *Listeria monocytogenes* F5069 thermal death times in liquid whole egg. J. Food Prot. 53:6–8, 25.
46. Food and Drug Administration. 1997. 1997 Food Code. *vm.cfsan./fda./gov/~dms/fc-toc.html*
47. Food Production and Inspection Branch. 1983. Mannual of Operation for Effective and Efficient Washing of Eggs. Food Production and Inspection Branch, Agriculture Canada, Ottawa, ON, Canada.
48. Food Production and Inspection Branch. 1992. Dairy Plant Inspection Manual. Agriculture and Agri-Food Canada, Guelph, ON, Canada.
49. Food Safety Inspection Service. 1996. Update: Food safety in the kitchen: A HACCP approach. *www.fsis.usda.gov/OA/pubs/haccpkit.htm*
50. Food Safety Inspection Service. 1997. Update: Focus on: Cutting board safety. *www.fsis.usda.gov/OA/pubs/cutboard.htm*
51. Food Safety Inspection Service. 1997. Update: Focus on: Egg products. *www.fsis.usda.gov/OA/pubs/eggprod.htm*
52. Food Safety Inspection Service. 1997. Update: Hotline offers advice on safe handling of meats, eggs for springtime festivities. *www.fsis.usda.gov/OA/news/spring97.htm*
53. Food Safety Inspection Service. 1997. Update: It's party time—But let's make sure the food is safe. *www.fsis.usda.gov/OA/pubs/party.htm*
54. Food Safety Inspection Service. 1996. Update: Meat and poultry hotline offers advice on holiday egg recipes. *www.fsis.usda.gov/OA/news/xmaseggs.htm*
55. Food Safety Inspection Service. 1997. Update: The food safety educator. *www.fsis.usda.gov/OA/educator/educator.htm*

56. Food Safety Inspection Service. 1996. Update: USDA offers advice for safe handling of shell eggs.
www.fsis.usda.gov/OA/news/shellegg.htm
57. Goff, H. D. 1988. Hazard analysis and critical control point identification in ice cream plants. Dairy Food Environ. Sanit. 8:131–135.
58. Griffith, C. J., and D. Worsfold. 1994. Application of HACCP to food preparation practices in domestic kitchens. Food Control 5:200–205.
59. Griffiths, M. W. 1989. *Listeria monocytogenes:* Its importance in the dairy industry. J. Sci. Food. Agric. 47:133–158.
60. Hamm, D., G. K. Searcy, and A. J. Mercuri. 1974. A study of the waste wash water from egg washing machines. Poult. Sci. 53:191–197.
61. Harris, C. E., and W. A. Moats. 1975. Recovery of egg solids from washwaters from egg-grading and -breaking plants. Poultry Sci. 54:1518–1523.
62. Health Canada: Food Quality and Safety Programs. 1997. Education on safe food handling.
www.hc-sc.gc.ca/main/hc/web/datahpsb/npu/agr6.htm
63. Health Protection Branch, Health Canada. 1994. Handling eggs safely at home. Issues. March 4.
64. Health Protection Branch, Health Canada. 1994. Safe handling of eggs by food service institutions. Issues. March 4.
65. Holley, R. A., and M. Proulx. 1986. Use of egg washwater pH to prevent survival of *Salmonella* at moderate temperatures. Poult. Sci. 65:922–928.
66. International Dairy Federation. 1994. Recommendations for the hygienic manufacture of milk and milk based products. Bulletin No. 292. International Dairy Federation, Brussels, Belgium.
67. International Dairy Federation. 1997. International Dairy Federation: Codex Committee on Milk and Milk Products.
www.fil-idf.org/codex.htm
68. Jay J. M. 1992. Modern Food Microbiology, 4th ed., Van Nostrand Reinhold, New York.
69. Johnson, E. A., J. H. Nelson, and M. Johnson. 1990. Microbiological safety of cheese made from heat-treated milk. Part I. Executive summary, introduction and history. J. Food Prot. 53:441–452.
70. Johnson, E. A., J. H. Nelson, and M. Johnson. 1990. Microbiological safety of cheese made from heat-treated milk. Part II. Microbiology. J. Food Prot. 53:519–540.
71. Kim, C. J., D. A. Emery, H. Rinke, K. V. Nagaraja, and D. A. Halvorson. 1989. Effect of time and temperature on growth of *Salmonella enteritidis* in experimentally inoculated eggs. Avian Dis. 33:735–742.
72. Kinderlerer, J. L. 1994. *Salmonella* in eggs. BNF Nutr. Bull. 19:11–18.
73. Kinner, J. A., and W. A. Moats. 1981. Effects of temperature, pH, and detergent on survival of bacteria associated with shell eggs. Poult. Sci. 60:761–767.
74. Kramer, J. M., and R. J. Gilbert. 1989. *Bacillus cereus* and other *Bacillus* species, in M. P. Doyle (ed.), Foodborne Bacterial Pathogens. Marcel Dekker, New York, pp. 22–70.

75. Leasor, S. B., and P. M. Foegeding. 1989. *Listeria* species in commercially broken raw liquid whole egg. J. Food Prot. 52:777–780.
76. Leclair, K., H. Heggart, M. Oggel, F. M. Bartlett, and R. C. McKellar. 1994. Modelling the inactivation of *Listeria monocytogenes* and *Salmonella typhimurium* in artificial egg washwater. Food Microbiol. 11:345–353.
77. Lin, F.-Y. C., J. G. Morris, D. Trump, D. Tilghman, P. K. Wood, N. Jackman, E. Israel, and J. P. Libonati. 1988. Investigation of an outbreak of *Salmonella enteritidis* gastroenteritis associated with consumption of eggs in a restaurant chain in Maryland. Am. J. Epidemiol. 128:839–844.
78. Lovett, J. 1989. *Listeria monocytogenes,* in M. P. Doyle (ed.), Foodborne Bacterial Pathogens. Marcel Dekker, New York, pp. 284–310.
79. Merrill, A. G., S. B. Werner, R. G. Bryant, D. Fredson, and K. Kelly. 1984. Staphylococcal food poisoning associated with an Easter egg hunt. J. Am. Med. Assoc. 252:1019–1022.
80. Microbiological and Food Safety Committee of the National Food Processors Association. 1992. HACCP and total quality management—Winning concepts for the 90's: A review. J. Food Prot. 55:459–462.
81. Microbiological and Food Safety Committee of the National Food Processors Association. 1993. Implementation of HACCP in a food processing plant. J. Food Prot. 56:548–554.
82. Mishu, B., J. Kochler, L. A. Lee, D. Rodrique, F. H. Brenner, P. Blake, and R. V. Tauxe. 1994. Outbreaks of *Salmonella enteritidis* infections in the United States, 1985–1991. J. Infect. Dis. 169:547–552.
83. Moats, W. A. 1978. Egg washing—A review. J. Food Prot. 41:912–925.
84. Moats, W. A. 1981. Antimicrobial activity of compounds containing active chlorine and iodine in the presence of egg solids. Poult. Sci. 60:1834–1839.
85. Moore, J., and R. H. Madden. 1993. Detection and incidence of *Listeria* species in blended raw egg. J. Food Prot. 56:652–654, 660.
86. Morris, G. K. 1990. *Salmonella enteritidis* and eggs: Assessment of risk. Dairy Food Environ. Sanit. 10:279–281.
87. Morse, D. L., G. S. Birkhead, J. Guardino, S. F. Kondracki, and J. J. Guzewich. 1994. Outbreak and sporadic egg-associated cases of *Salmonella enteritidis:* New York's experience. Am. J. Public Health 84:859–860.
88. Munce, B. A. 1984. Hazard analysis critical control points and the food service industry. Food Technol. Aust. 36:214–217.
89. Norberg, P. 1981. Enteropathogenic bacteria in frozen chicken. Appl. Environ. Microbiol. 37:32–34.
90. Ontario Ministry of Agriculture, Food, and Rural Affairs. 1997. Milk Act Regulations. Regulation No. 761: Milk and Milk Products.
www.gov.on.ca:80/OMAFRA/english/livestock/dairy/legal/reg761_pg1.htm
91. Palumbo, M. S., S. M. Beers, S. Bhaduri, and S. A. Palumbo. 1995. Thermal resistance of *Salmonella* spp. and *Listeria monocytogenes* in liquid egg yolk and egg yolk products. J. Food Prot. 58:960–966.
92. Pearson, L. J., and E. H. Marth. 1990. *Listeria monocytogenes*—Threat to a safe food supply: A review. J. Dairy Sci. 73:912–928.

93. Peeler, J. T., and V. K. Bunning. 1994. Hazard assessment of *Listeria monocytogenes* in the processing of bovine milk. J. Food Prot. 57:689–697.
94. Peta, C., and K. Kailasapathy. 1995. HACCP—Its role in dairy factories and the tangible benefits gained through its implementation. Aust. J. Dairy Technol. 50: 74–78.
95. Poppe, C., R. J. Irwin, C. M. Forsberg, R. A. Clarke, and J. Oggel. 1991. The prevalance of *Salmonella enteritidis* and other *Salmonella* spp. among Canadian registered commercial layer flocks. Epidemiol. Infect. 106:259–270.
96. Powrie, W. D., and S. Nakai. 1986. The chemistry of eggs and egg products, in W. J. Stadelman and O.J. Cotterill (eds.), Egg Science and Technology, 3rd ed. AVI Publishing, Westport, CT, pp. 97–140.
97. Rampling, A., C. E. D. Taylor, and R. E. Warren. 1987. Safety of pasteurised milk. Lancet 2:1209.
98. Reed, G. H. J. 1993. Safe handling of potentially hazardous foods (PHF)—A checklist. Dairy Food Environ. Sanit. 13:208–209.
99. Reed, G. H. 1993. Foodborne illness. 1. Staphylococcal ("Staph") food poisoning. Dairy Food Environ. Sanit. 13:642–642.
100. Reed, G. H. 1993. Foodborne illness. 2. Salmonellosis. Dairy Food Environ. Sanit. 13:706–706.
101. Reed, G. H. 1994. Foodborne illness. 10. *Listeria monocytogenes.* Dairy Food Environ. Sanit. 14:482–483.
102. Reed, G. H. 1994. Foodborne illness. 11. Yersiniosis. Dairy Food Environ. Sanit. 14:536.
103. Reed, G. H. 1994. Foodborne illness. 4. *Bacillus cereus* gastroenteritis. Dairy Food Environ. Sanit. 14:87.
104. Reed, G. H. 1994. Foodborne illness. 5. Foodborne campylobacteriosis. Dairy Food Environ. Sanit. 14:161–162.
105. Reed, G. H. 1994. Foodborne illness. 8. *Escherichia coli.* Dairy Food Environ. Sanit. 14:329–330.
106. Ryser, E. T., and E. H. Marth. 1989. "New" food-borne pathogens of public health significance. J. Am. Diet. Assoc. 89:948–956.
107. Ryser, E. T., and E. H. Marth. 1991. *Listeria,* Listeriosis and Food Safety. Marcel Dekker, New York.
108. Ryser, E. T., and E. H. Marth. 1991. Incidence and control of *Listeria* in food-processing facilities, in E. T. Ryser and E. H. Marth (eds.), *Listeria,* Listeriosis, and Food Safety. Marcel Dekker, New York, pp. 531–575.
109. Saeed, A. M., and C. W. Koons. 1993. Growth and heat resistance of *Salmonella enteritidis* in refrigerated and abused eggs. J. Food Prot. 56:927–931.
110. Schiemann, D. A. 1989. *Yersinia enterocolitica* and *Yersinia pseudotuberculosis,* in M. P. Doyle (ed.), Foodborne Bacterial Pathogens. Marcel Dekker, New York, pp. 602–672.
111. Schlech, W. F. 1996. Overview of listeriosis. Food Control 7:183–186.
112. Shapton, N. 1988. Hazard analysis applied to control of pathogens in the dairy industry. J. Soc. Dairy Technol. 41:62–63.
113. Shapton, N. 1989. Starters—A suitable case for HACCP? Dairy Ind. Int. 54:25–29.

114. Shelef, L. A. 1989. Listeriosis and its transmission by food. Prog. Food. Nutr. Sci. 13:363–382.
115. Sionkowski, P. J., and L. A. Shelef. 1990. Viability of *Listeria monocytogenes* strain *Brie*-1 in the avian egg. J. Food Prot. 53:15–17, 25.
116. Snyder, O. P. J. 1992. HACCP—An industry food safety self-control program. Part VI. Dairy Food Environ. Sanit. 12:362–365.
117. Snyder, O. P., and D. M. Poland. 1990. America's "safe" food. Dairy Food Environ. Sanit. 10:719–724.
118. Snyder, O. P., and D. M. Poland. 1991. America's "safe" food. Part 2. Dairy Food Environ. Sanit. 11:14–20.
119. Snyder, O. P. J. 1992. HACCP—An industry food safety self-control program. Part IV. Dairy Food Environ. Sanit. 12:230–232.
120. Spomer, D. R. 1997. Codex milk product standards. Dairy Food Environ. Sanit. 17:281–283.
121. St. Louis, M. E., D. L. Morse, M. E. Potter, T. M. DeMelfi, J. J. Guzewich, R. V. Tauxe, and P. A. Blake. 1988. The emergence of grade A eggs as a major source of *Salmonella enteritidis* infections: New implications for the control of salmonellosis. J. Amr. Med. Assoc. 259:2103–2107.
122. Stadelman. W. J. 1982. The egg industry, in W. J. Stadelman and O. J. Cotterill (eds.), Egg Science and Technology, 3rd ed. AVI Publishing, Westport, CT, pp. 1–10.
123. Stern, N. J., and S. U. Kazmi. 1989. *Campylobacter jejuni,* in M. P. Doyle (ed.), Foodborne Bacterial Pathogens. Marcel Dekker, New York, pp. 71–110.
124. Surak, J. G., and S. F. Barefoot. 1987. Control of *Listeria* in the dairy plant. Vet. Hum. Toxicol. 29:247–249.
125. The National Food Safety Database. 1997. Guide to safe food handling. *www.foodsafety.org/fs/fs002.htm*
126. The National Food Safety Database. 1997. Keep food safe: Prepare and cook with care. *www.foodsafety.org/ky/ky002.htm*
127. Todd, E. C. D. 1985. Economic loss from foodborne disease outbreaks associated with foodservice establishments. J. Food Prot. 48:169–180.
128. Todd, E. C. D. 1988. Foodborne and waterborne disease in Canada—1982 annual summary. J. Food Prot. 51:56–65.
129. Todd, E. C. D. 1992. Foodborne disease in Canada: A 10-year summary from 1975 to 1984. J. Food Prot. 55:123–132.
130. Todd, E. C. D. 1996. Risk assessment of use of cracked eggs in Canada. Int. J. Food Microbiol. 30:125–143.
131. Todd, E. C. D. 1996. Worldwide surveillance of foodborne disease: The need to improve. J. Food Prot. 59:82–92.
132. U.S. Department of Health and Human Services. 1995. Pasteurized Milk Ordinance, Public Health Service/Food and Drug Administration Publication No. 229.
133. U.S. Food and Drug Administration. 1997. Can your kitchen pass the food safety test? *www.fda.gov/fdac/features/895 kitchen.html*
134. U.S. Food and Drug Administration. 1997. Handling eggs safely at home. *vm.cfsan.fda.gov:80/~dms/eggs.html*

135. U.S. Food and Drug Administration, Milk Industry Foundation and International Ice Cream Association. 1988. Recommended guidelines for controlling environmental contamination in dairy plants. Dairy Food Environ. Sanit. 8:52–56.
136. U.S. Department of Agriculture. 1969. Egg Pasteurization Manual ARS 94-48. Poultry Laboratory, Agricultural Research Services, U.S. Department of Agriculture, Albany, CA.
137. Vasavada, P. C., and M. A. Cousin. 1993. Dairy microbiology and safety, in Y. H. Hui (ed.), Dairy Science and Technology Handbook, Vol. 2., Product Manufacturing. VCH Publishers, New York, pp. 301–426.
138. Vaughan, A. C. 1984. Poisonous or sick cheese. Public Health Papers and Reports. Am. Public Health Assoc. 10:241
139. Vugia, D. J., B. Mishu, M. Smith, D. R. Tavris, F. W. Hickman Brenner, and R. V. Tauxe. 1993. *Salmonella enteritidis* outbreak in a restaurant chain: The continuing challenges of prevention. Epidemiol. Infect. 110:49–61.
140. Westhoff, D. C. 1978. Heating milk for microbial destruction: A historical outline and update. J. Food Prot. 41:122–130.
141. Zottola, E. A., and L. B. Smith. 1991. Pathogens in cheese. Food Microbiol. 8: 171–182.

3
Safe Handling of Fruits and Vegetables

Robert E. Brackett
University of Georgia, Griffin, Georgia

I. INTRODUCTION

In recent years, consumers have been demanding and purchasing more foods that they consider healthful. Fruits and vegetables have always been viewed as healthy, but recently their importance in the diet has been especially emphasized. Consequently, consumers are purchasing more of these products (14). Although canned and frozen products are still important, sales of fresh fruits and vegetables have enjoyed a dramatic increase in domestic production and in international trade. Not coincidentally, the consideration of fruits and vegetables as a source of foodborne illness, once a rarity, has become a major concern for government regulatory agencies. This chapter addresses the aspects of production, processing, and distribution of produce that are important to food safety. Ways to minimize risk are also discussed.

II. SAFE FOOD HANDLING ON THE FARM

Although food safety is often considered an issue to be dealt with by food processors and food service personnel, steps to assure food safety must begin earlier in the distribution chain. Indeed, the most efficient way to maximize food safety is to use what is known as a "systems approach" (8). That is, all aspects of the production, processing, and distribution chain, including preparation and consumption, must be considered as one integrated system rather than as separate parts.

A. History of Land

Many people assume that growth, cultivation, and harvesting of crops are the initial points in the food system. However, several factors that come into play before crops are even planted can affect the ultimate microbiological safety of foods. The sanitary quality of the soil in which vegetables are grown can directly influence these products. For example, croplands on which farm animals have recently been grazing are much more likely to be contaminated with organisms of fecal origin. Moreover, some foodborne pathogens can survive for months or years in soil. For example, Watkins and Sleath (31) demonstrated that *Salmonella* and *Listeria monocytogenes* could survive for months in sewage sludge applied to agricultural soils. Outbreaks of *Escherichia coli* O157:H7 (4,2) and *Cryptosporidium* (2) in apple cider have also led to the realization that tree-borne fruits could become contaminated with fecal contaminants. While one would normally not think of tree-borne fruit as being highly at risk to acquire fecal contamination, it was suspected that the problem occurred when farms used apples that had been lying on the ground (so-called drops) making cider. Sources of the fecal contamination could be either domestic or wild animals (27).

Another factor that could affect the sanitary quality of croplands is a history of flooding. This situation can become especially problematic when floodwaters cover areas on which farm animals have been grazed or confined. In these cases, surface waters can become polluted with animal waste and carry the contaminants downstream, where they may also flood over croplands. Major flooding has also caused rivers to cover sewage treatment plants, whereupon the rivers have been heavily contaminated with human, municipal, and industrial wastes. Once microorganisms have been deposited on croplands, it may take months or years for them to be eliminated. Crops produced during that time could be contaminated with the residual organisms.

B. Soil

Once present in soil, pathogens may either grow or survive long enough to contaminate crops. However, the likelihood of an organism surviving will differ depending on soil type as well as how the fields were cultivated. Weis and Seeliger (33) surveyed soils in Germany for the presence of *L. monocytogenes* and found it in 12.2% of the cultivated soils examined. In contrast, they found that 44% of uncultivated soils contained the bacterium. Based on these results, the authors speculated that this organism not only has the ability to survive in agricultural environments, but that soil is a normal ecosystem for it. More recently, Dowe et al. (12) obtained remarkably similar results after surveying Canadian soils. These investigators hypothesized that a reported preference by *L. monocytogenes* for the plant rhizosphere may be one reason for its higher prevalence in uncultivated than cultivated soil. Dowe et al. (12) also observed that when *L. monocytogenes* was present, populations were usually low (4–90 organisms per gram of soil). Hence, any produce items that did become contaminated during production would likely contain populations so low that they would be difficult to detect during routine sampling.

Several authors have noted the influence of soil type on survival of *L. monocytogenes*. In the early 1960s, Welshimer (34) reported that populations of *L. monocytogenes* steadily decreased in clay soils, ultimately becoming undetectable after about 6 months of storage. Similar experiments by Van Renterghem et al. (30) revealed that *L. monocytogenes* populations likewise decreased in sandy loam soil after 6 weeks and became undetectable after about 2 months. In contrast, Dowe et al. (12) reported that *L. monocytogenes* initially grew in clay and sandy loam, and sandy soils, but slowly decreased thereafter in clay loam and sandy soils. However, the behavior differed depending on the initial inoculum. High populations of *L. monocytogenes* (ca. 10^5–10^6 cfu/g) remained essentially stable, with populations in sandy and sandy loam soils decreasing slightly. In contrast, populations initially increased to about 10^5 cfu/g after 15 days at ambient (25–30°C) temperatures when low populations (10^2 cfu/g) were introduced into all types of soil. However, populations then decreased about 1 log in sandy and sandy loam soils. The authors suggested that the relative moisture-bearing capacity of the soils may have influenced survival.

C. Equipment

Little has been published on the role of farm equipment in the transfer of foodborne pathogens. Nevertheless, common sense and experience with plant pathogens suggest that the sanitary condition of such equipment is important. Brodie (9) observed that plows and harrows were responsible for spreading the nematode *Globodera rostochiensis*. Based on his results, it is very conceivable that virtually any organism could be transferred by agricultural tools from soil on a farm to other locations.

One obvious example of a likely candidate for source of contamination is the manure spreader. This common piece of farm equipment is routinely used to haul solid animal waste to fields, where the waste is chopped and distributed on fields. The manure spreader can contribute to reducing the safety of fruits and vegetables in several ways. First, the practice of using raw animal waste increases the presence of human pathogens in the soils and plants on which the waste is deposited. Second, manure spreaders are rarely if ever properly cleaned and so serve as a source of contamination and attract insects, such as flies, which can also spread contamination.

D. Cleaning and Sanitation

Even equipment not intended for use with raw manure or soil can influence the sanitary quality of fruits and vegetables. Harvesting equipment that has direct contact with fruits and vegetables would be one example of such equipment. In past years, farmers rarely or never cleaned wagons or field conveyors used in the harvest of produce. However, the increased concern over potential transfer of foodborne pathogens has prompted producers to begin cleaning and sanitizing such equipment.

E. Farm Workers

Most produce has been grown and harvested with heavy reliance on human labor. Because many, if not most, foodborne illnesses are transmitted by humans, the workers involved in farming can have an important impact on the microbial safety of the products they handle. It is therefore of the utmost importance for agricultural workers to adhere to proper sanitary procedures.

Actual production fields are usually quite distant from primary farm buildings that might house toilet facilities. This situation makes it difficult or impossible for farm workers working in the field to use these toilet facilities. On many farms, workers typically use secluded areas such as irrigation ditches, tall crops, or other hidden areas to relieve themselves. Clearly, such a practice is entirely unacceptable as far as sanitary practices are concerned. Since under these circumstances it is also impossible for agricultural workers to wash their hands, pathogens may be transferred to tools or produce.

The simplest means to encourage or prevent the use of production areas as open toilets is to place toilets in strategic locations to which workers have relatively easy access. However, on some large farms it may be financially prohibitive to station toilets in every field, particularly since human activity occurs only sporadically in any particular field. Some farmers have dealt with this problem by using portable toilets that can be moved to serve fields in which work is being conducted. In fact, many local or regional regulatory agencies specify minimum ratios of toilets for number of workers. Regardless of the system being used, it is also important to provide hand-washing stations as part of the toilet facility.

Although providing toilet and hand-washing facilities in the close proximity to workers can do much to improve sanitation, it is no guarantee that the facilities will in fact be used as intended. Appreciation for proper sanitation will depend on the country, social position, and background of workers. In some cultures and social strata, toilet paper and indoor or chemical toilets are luxuries to which the workers are unaccustomed. Moreover, the habit of hand washing after using the toilet is absent even in developed countries, but is especially a problem in certain population groups. Hence, despite the availability of sanitary facilities, their use must be emphasized, and it may still be necessary to instruct some workers on the proper use of the facilities.

One of the primary steps producers can take to ensure sanitary handling of their products is to properly train workers. Of greatest importance is impressing upon workers the importance of adhering to rules and procedures related to sanitation. Moreover, it is very valuable in the long run to also provide reasons for the implementation of the procedures and to explain how they help ensure the safety of the product. However, managers and supervisors should also stress that compliance with sanitation procedures is nonnegotiable regardless of how insignificant these procedures may appear to be.

Even if sanitary facilities are provided and workers trained, it may still be desirable to somehow monitor workers' use of these facilities. It would be impractical as well as degrading to continually observe workers' sanitary habits. Nevertheless, less personal means of monitoring personal sanitary habits do exist. For example, one might wish to place hand-washing stations in open and obvious locations. In this case, workers would be more likely to feel compelled to wash their hands than if the stations were in more private locations. Other, more technical means such as computerized badges that record hand washing are also being proposed.

F. Water

As mentioned in Sec. II. A, water can have a profound influence on the microbiological quality of fresh produce. Fruits and vegetables receive the water they need to grow by three general means. Natural precipitation is the least expensive and best source of water. Unless atmospheric conditions have somehow become contaminated with external materials, such as particulates arising from dust storms or air pollution, rain and snow are normally quite clean and sanitary. However, natural precipitation is in many growing regions quite unpredictable and unreliable, and in other semiarid to arid regions it is insufficient to support production. Consequently, there is often a need to rely on the second source of water, irrigation.

Sources of irrigation waters differ depending on the climate, the geology of the production areas, and the relative amount of water required to sustain a particular crop. Some regions have an abundance of nearby surface waters such as rivers, lakes, or ponds from which to draw water. In other cases, surface waters may not be close or convenient but may still be the main source for irrigation water. In this case, surface waters that are drawn from reservoirs or impoundments may need to be transported to the production areas via viaducts or pipes. Regardless of the source, surface waters are usually the irrigation source most likely to be contaminated with pathogenic microorganisms and toxic chemicals. As mentioned earlier, surface waters can become contaminated via various routes.

Some municipalities still employ outdated or poorly designed sewage treatment plants that allow untreated or undertreated municipal waste to be dumped into rivers. Similarly, streams flowing through agricultural regions in which animals are raised are often subject to runoff from feedlots or barnyards. In either case, fields and crops irrigated with such water can become contaminated with any pathogens that are present.

The third source of irrigation water is from underground aquifers, commonly referred to as groundwater. In general, groundwaters are likely to be of higher microbiological quality than surface water because they are less likely to be exposed to animal or human waste, and the originating water is usually filtered in its migration through soil layers and rock to the aquifiers. However, many people incorrectly assume that groundwaters are always potable. In fact, groundwaters can be contaminated with pathogenic microorganisms. Groundwaters become contaminated when raw sewage from septic tanks or holding ponds leaches through soils into underground aquifers. Moreover, the source of contamination often is located some distance from the well. Consequently, all ground source waters should be tested for the presence of fecal contaminants before being used for irrigation of vegetable and fruit crops. The type of soil or rock influences the likelihood of water movement.

Municipalities can also provide waters that could be potentially used for irrigation, although this can bc quite expensive. In this case, the original source of the water can be either treated surface water or groundwater (wells). Some people assume that municipal sources of irrigation waters will always be free of contamination because they are treated. However, although municipal water supplies are usually of very high microbiological quality, mistakes and accidents can still allow harmful microorganisms to survive. Indeed, one of the largest outbreaks of waterborne illness in U.S. history was traced to a contaminated municipal water supply. In this outbreak, an estimated 403,000 people in Milwaukee became ill from consuming the contaminated municipal water. The outbreak was ultimately blamed on flaws in the coagulation–filtration system during treatment, which allowed the waterborne parasite *Cryptosporidium* to contaminate drinking water (20). The chlorination system used for the water was ineffective against the *Cryptosporidium.*

Finally, it is possible to use wastewater for irrigating crops. Use of this source of irrigation has been prevalent in many parts of the world, but especially in arid regions, where water supplies are limited or local economic conditions make the use of other sources of water too expensive (24). Wastewater, as well as other forms of irrigation, is usually applied to the fields by one of three basic techniques. The most well known is spray application, in which waters are piped to the area requiring the irrigation and then sprayed over the land. Overland flow involves a controlled flooding of entire fields with a shallow layer of water. Finally, ridge and furrow-irrigation also involves flooding fields but differs from overland flow in that waters are allowed to flow in furrows located between rows of crops.

All these methods involve direct contact of both soils and crops by the irrigation waters. In the case of wastewater, these waters can be heavily contaminated with microorganisms, including human pathogens. Moreover, in arid climates these methods also suffer from the disadvantage of allowing significant wastage of waters through evaporation into the atmosphere. Hence, some (24) have suggested that a fourth method, known as drip irrigation, might be a better alternative. Drip irrigation involves the strategic application of water directly to the root areas of plants via conduits placed either above or below the surface of the soil. Drip irrigation has the advantages of requiring less water and being able to direct the application specifically to the plants with little or no loss from evaporation.

Sources of wastewater can range from processing waters in food processing plants to raw sewage. Hence, the microbiological quality of wastewater can vary dramatically. As one might expect, the use of wastewater on vegetable crops can introduce significant microbiological as well as chemical hazards. Kirkham (18) briefly categorized the nature of hazards associated with using wastewater on vegetable crops as physical, chemical, and biological. Physical problems were those affecting agronomic characteristics of the soils (e.g., clogging of soils by suspended solids, reduction in soil aeration) but otherwise not adversely affecting food safety. Likewise, chemical problems also included issues affecting production such as accumulation of salts or overfertilization. However, chemical problems also include issues of potential health concern such as accumulation of heavy metals. Potential biological problems associated with use of wastewater are directly tied to the deposition of human pathogens on both soils and food crops. However, Kirkham (18) points out that human illness has not been demonstrated to result from systemic uptake of pathogens by crops irrigated with properly treated wastewater.

Relatively little has been published on the hazards involved in the use of wastewater to irrigate fruits and vegetables. However, sufficient information exists to suggest that the use of raw wastewater for irrigation is risky. Mutlak et al. (22) likewise cautioned against the use of "sewage farming," which relies on untreated municipal sewage for irrigation. In Baghdad, Iraq, these authors observed viable salmonellae in sewage as well in surface and subsurface soils irrigated with raw sewage. Moreover, these organisms remained recoverable for at least a week. Van Donsel and Larkin (29) likewise noted that *Mycobacterium bovis* could be recovered from garden soil, radishes, and lettuce irrigated with raw sewage effluent inoculated with the organisms. They also found that the D-value (90% reduction time) for *M. bovis* in the soil and on radishes was 11 and 6 days, respectively. Although the authors were unable to determine D-values on lettuce, they detected the organism on lettuce for up to 35 days after irrigation. In a companion study, where viruses were inoculated into wastewater before application onto vegetables, Larkin et al. also were able to detect poliovirus (19).

Potential problems associated with spray irrigation have also been noted by authors not directly concerned with hazards associated with the safety of fruits and vegetables. For example, Katzenelson and Teltch (17) investigated the dispersal of aerosolized pathogens generated by spray irrigation of wastewater as it relates to the spread of human disease by contaminated air. They found that enteric bacteria could be isolated from the air at a distance of 350 m and *Salmonella* 60 m downwind of the source of irrigation. Katzenelson and Teltch's suspicion that spray irrigation of wastewater could have an adverse effect on human health was confirmed by Fattal et al. (13). The latter authors determined that children in the areas surrounding farms suffered twice as much enteric disease when wastewater was used for irrigation than when it was not used. Although the thrust of this chapter does not deal with airborne illness, these examples clearly demonstrate that human pathogens present in wastewater can be dispersed significant distances and then deposited on fruits and vegetables.

Although one would predict that overland flow and ridge and furrow irrigation with wastewater would be less likely to spread hazardous organisms, the large-scale exposure of soil and plants to the contaminated waters is still undesirable. Conversely, one would also predict that drip irrigation would cause the minimum amount of contamination to soils and crops. Israel is particularly motivated to use recycled wastewater because the amount of affordable fresh water for irrigation is very limited. Consequently, research in Israel on drip irrigation has been conducted to determine what risks this method entails. Sadovski et al. (24) evaluated the contribution of drip irrigation with sewage effluent on the microbiological quality of soil and vegetables. They found that when cucumbers and eggplant were exposed to sewage effluent by drip irrigation, fecal coliform counts were 38-fold higher than for the same vegetables irrigated with potable water. However, the investigators also discovered that there were 10-fold fewer fecal coliform organisms when the drip was delivered subsoil instead of to the soil surface. Similarly, covering the soil with plastic sheets reduced fecal coliform contamination by nearly 13-fold. Hence, the authors suggested that subsurface delivery on covered soils be considered a preferable alternative to other forms of wastewater irrigation. In addition, they noted that vegetables that were irrigated with wastewater up to the point of flowering and irrigated with good quality water thereafter did not contain significantly more fecal coliform organisms than vegetables irrigated entirely with pure water.

Sadovski et al. (24) further investigated the potential for using drip irrigation to deliver wastewater and determined the dissemination of traceable microorganisms (*E. coli* and poliovirus) in soil and on vegetables. Again, they found that drip irrigation under plastic sheet covers or buried 10 cm under the soil surface significantly reduced contamination of cucumbers. However, they also found that irrigation conduits and soil irrigated by the wastewater remained contaminated for at least 8 and 18 days, respectively.

G. Fertilizers

All plants require nutrients for growth. In most cases, farmers will provide these nutrients to soil in the form of fertilizer. As in the case of water, fertilizer can be provided to the plant in at least three ways. The most common means used by modern farmers is to fertilize crops with commercial chemical fertilizers, specifically blends of nitrogen, phosphorus, and potassium. Although this type of fertilizer has fallen in to some disfavor, especially among environmentalists and natural food proponents, it is least risky from a microbiological perspective. The most common means of delivery is with the use of tractor-pulled sprayers or subsoil injectors. However, it is also possible to include fertilizers and other agricultural chemicals in irrigation spray systems, a technique known as chemigation.

The next least risky means of fertilizing crops is to use composted or treated waste or manure. This type of fertilizer involves treating animal waste or municipal sewage sufficiently to completely kill all harmful bacteria. Treatment methods usually entail composting, heating, or traditional sewage treatment. Compost treatment time is a function of outdoor temperature. Regardless of method, an essential requirement is to either eliminate any pathogenic microorganisms.

Finally, the most risky form of fertilizer is raw animal waste or untreated sewage. As discussed above, such materials are commonly contaminated with a variety of viruses, parasites, and bacteria (6). Although the use of raw manure or sewage is uncommon in commercial farming operations in developed countries, this is not always true for small farms and for agricultural enterprises in developing countries (22,24). In such cases, it is sometimes easier or more economical to utilize readily available sources of nutrients, such as animal manure. A typical example is the practice of some dairy farmers of distributing cattle manure on pastures. This act could conceivably constitute a potential hazard if produce production areas are adjacent to or downstream from the treated areas.

Although previously not considered to be a microbiological concern, agricultural chemicals could also pose a hazard if not prepared in a sanitary manner. Many agricultural chemicals such as pesticides or nematocides are applied directly to the fruits and vegetables. Hence, application could pose a microbiological hazard if poor quality water is used to prepare the chemicals. Indeed, it has been suggested that several outbreaks arising from the parasite *Cyclospora cayetanensis* on raspberries might have been caused by contaminated surface water used to spray raspberries with fungicide before harvest (27).

H. HACCP

The hazard analysis–critical control points (HACCP) strategy has been promoted as a means by which safe foods can be produced while minimizing the need for expensive and often inefficient end-product testing. A complete discussion of HACCP is beyond the scope of this chapter. However, the concept is briefly reviewed. Willocx et al. (35) published a detailed discussion of the use of HACCP to ensure the safety of minimally processed endives.

In general, HACCP identifies points (critical control points or CCP) in the production and processing chain at which failure to maintain control of the process could lead to foodborne illness. For example, irrigation could be considered to be a CCP in the production of fruits and vegetables because it is a point at which it is essential to control contamination by irrigation waters.

HACCP has been almost universally accepted in the food processing industry, and some have suggested the implementation of HACCP in farming. However, this concept is new to production agriculture, and many aspects of production do not lend themselves to the HACCP system as well as in the food processing industry. In particular, many if not most steps in food production are virtually uncontrollable. That is, it is difficult and impractical to maintain sanitation, temperature control, and other essentials of typical good manufacturing practices. For example, harvested or processed foods can be held in sanitizable containers or areas that can be kept free from outside contamination and at a proper temperature. In contrast, preharvest fruits and vegetables are grown outside with complete exposure to the elements, blowing soil, wild animals and birds, and uncontrollable fluctuations in temperature. Consequently, any HACCP system for production agriculture would have to be designed with these constraints in mind. Nevertheless, it is possible for a modified HACCP plan to be developed for many products that would identify controllable points and means for control. Although the modified system might not meet the strict criteria for HACCP, it would make great strides in reducing the risk of foodborne illness arising at production. Currently, a HACCP plan is not considered to be mandatory for incoming raw products and ingredients.

III. SAFE FOOD HANDLING AT THE PROCESSING LEVEL

In general, the food processing industry has had significant experience and has made great strides in maintaining the safety of foods. The primary goal of food processing is food preservation, and most of the technologies that most efficiently preserve foods are aimed at eliminating or controlling the growth of microorganisms. Consequently, control of pathogenic microorganisms is usually inherent in the technologies of preserving foods. Although food processors deal with postharvest rather than preharvest problems, they also face many of the same issues with regard to food safety. However, the food safety problems faced by food processors will differ slightly depending on the commodity. Unlike most other foods such as meats or dairy products, fresh fruits and vegetables are alive. That is, fresh produce continues to respire and senesce even after harvesting. Therefore, processors of fresh produce must deal with a dynamic product whose characteristics change continuously as the product continues to mature. This makes the job of predicting and dealing with microbial issues that much more challenging.

A. Water

The food industry uses large quantities of water. The uses include obvious steps such as the primary washing of fruits and vegetables as well as procedures such as hydrochilling. In many facilities, however, flume waters that move products within the plant likely account for the largest use of water.

Processors usually have fewer choices in the source of their water than do producers. Most industrialized nations have promulgated regulations requiring the use of potable water that must meet minimum standards. As with irrigation, waters used in food processing come from several sources. In rural areas and some cities, fruit and vegetable processors rely on ground-water. Food processing plants will sometimes find it more economical to drill a private well than to purchase water from municipalities. Often the food processing company will likewise build its own chlorination or treatment plant to bring the available water to at least minimum requirements.

Because food processors do use so much water, many try to recycle processing water. Processing wastewater is less likely to contain high concentrations of pathogenic microorganisms than would municipal wastes, but it can contain high concentrations of other microorganisms. In addition, recycled water may contain pathogens acquired from contaminated produce. Hence, processors must take extra precautions and steps to be sure that the recycled water is not a source of contamination for newly received items.

B. Effects of Sanitizers

Sanitizers are often added to the water used for processing fruits and vegetables. This practice has led to the common but mistaken assumption that the sanitizers are used to treat the fruit or vegetable. In fact, sanitizers are usually added to maintain the microbiological quality of the water and, by extension, to maintain the quality of the products contacted by the water. However, the potential for using sanitizers to reduce or eliminate pathogens on produce has not escaped the attention of processors. The use of sanitizers to disinfect fresh produce has been a high priority in fruit and vegetable research in recent years. However, the results of such studies have not been altogether promising. Most have found that sanitizers are of little or no benefit for disinfecting produce. For example, Brackett (7) observed that 200 μg/mL (ppm) of chlorine quickly eliminated up to 10^6 cfu/mL of *L. monocytogenes* from water. In contrast, the same concentration reduced populations on brussels sprouts by only 2 log units. Moreover, water containing no chlorine reduced populations by 1 log.

More recently, Zhang and Farber (36) evaluated the antilisterial effect of a variety of disinfectants in fresh-cut vegetables. Their results were similar to those of Brackett (7) in that treatment with chlorine reduced populations of *L. monocytogenes* by 2 log units or less. However, other treatments were also unimpressive (Table 1).

In addition to *L. monocytogenes,* other bacteria present on fresh produce appear to survive treatment with sanitizers. Zhuang et al. (37) and Wei et al. (32) both demonstrated that chlorine dips [200 μg/mL (ppm)] could reduce populations of

Table 1 Reductions in Population of *L. monocytogenes* in Lettuce by Various Disinfection Treatments

Treatment	Reduction in *L. monocytogenes* populations (log cfu/g)
Sodium hypochlorite, 200 μg/L	<1.5 log
Chlorine dioxide, 5 μg/L	ca. 1 log
Salmide®, 200 μg/L	<1 log
Lactic acid, 1%	ca. 0.5 log
Acetic acid, 1%	<0.5 log
Trisodium phosphate, 2%	No decrease
Tergitol, 0.1%	<0.5 log
NaOCl, 0.1% + 100 μg/L	ca. 0.5 log

Source: Adapted from Ref. 36.

Salmonella montevideo on the surface of tomatoes (1.2 logs on the surface; 0.7 log in the core), but could not be relied on to eliminate the pathogen. Wei et al. (32) observed that salmonellae located in cracks and stem scars were even more resistant to chlorine than cells located on unbroken surface of tomatoes. It also appears that other sanitizing chemicals are of limited value in eliminating salmonellae from minimally processed produce.

Beuchat and Ryu (6) compared treatment of alfalfa sprouts and cantaloupe cubes with chlorine, hydrogen peroxide, and 70% ethanol. Their results showed that both the type of product and the sanitizer tested influenced the efficacy of treatment. In the case of alfalfa sprouts, results were similar to those observed for *L. monocytogenes* (7,36,37) in that 200 μg/mL of chlorine was only slightly more effective at reducing populations of *Salmonella* than was water alone. Although higher concentrations did increase elimination of *Salmonella,* 2000 μg/mL was required to reduce populations of *Salmonella* to undetectable levels (10^1–10^2 cells/g seed initially present). In the case of cantaloupe cubes, chlorine concentrations as high as 2000 μg/mL reduced populations of *Salmonella* by only about 1 log. Similarly, although hydrogen peroxide and 70% ethanol effected substantial (2–5 logs) reductions of *Salmonella* spp. on alfalfa sprouts, these treatments had minimal effects on this bacterium on cantaloupe. Undoubtedly, current and future research will continue to look for, and possibly discover, new and more effective disinfection techniques. However, it also appears likely that the use of sanitizers and other chemicals to completely eliminate pathogens from fresh produce is an unrealistic goal. Consequently, producers will still need to pay attention to good manufacturing practices and HACCP to account for the majority of their food safety efforts.

C. Facilities

The physical design of packing sheds, processing plants, and related facilities is an important consideration in assuring the microbiological safety of fruits and vegetables. Because fresh fruits and vegetables are raw commodities, they are inherently subject to potential contamination. Hence, they are also a potential source of contamination in the plant. Once present, pathogenic bacteria can become established in a plant and can serve as a source of cross-contamination for other products that pass through. Although no design or precautions can completely eliminate this possibility, proper design and maintenance of the facilities can minimize potential hazards.

The specific design requirements depend on the type of facility in question. For example, requirements for packing sheds that primarily deal with whole fresh products and bulk shipments are less rigorous than those for processing plants in which cut, ready-to-eat products are produced. Although a detailed discussion of plant design is beyond the scope of this chapter, several basic design requirements are discussed. For a more detailed discussion of the role of plant design in sanitation, the reader is directed to the book by Troller (28).

In general, sanitary design of plants can fall into one of several main foci. The first involves plant location and layout. Food processing plant location can have a direct influence on the ease and success of sanitation. For example, locating food processing facilities too close to obvious sources of contamination (landfills, feedlots, wildlife refuges, etc.) increases the chance that birds, rodents, or even human activity will transfer contaminants to the plant. Similarly, it is important to maintain separation of areas directly involved in product handling from those that are not. Nonproduct areas, such as offices or sales outlets, are more likely to experience human foot traffic from unknown outside locations. Consequently, nonproduct areas should have their own entrances.

The overall design of product handling areas is also important. Incoming, unprocessed products should be separated from finished products as much as possible. This separation may be accomplished as simply as by locating unfinished and finished products at opposite ends of processing areas or, at the other end of the spectrum, by setting up a dedicated room for each product. Likewise, it is important to avoid human traffic as well as movement of equipment (e.g., forklifts) between the two areas. Maintaining appropriate separation in these situations minimizes chances of contamination of finished from unprocessed products.

Air handling is an often-ignored factor that can influence contamination in the plant. Intake vents should be located to prevent the drawing of dust and debris from roofs or exhaust into the product area. Likewise, air handling systems should be designed so that air currents flow generally opposite to product flow. That is, the air stream should be flowing from finished product to newly received products rather than drawing potentially contaminated air from receiving areas into finished product areas.

The actual construction of the building can either enhance or reduce the sanitation in an operation. Materials used in product areas should be nonporous and cleanable, and walls, windows, and floors should be constructed with exclusion of animals and insects as a goal. It is also important that the facilities be kept on a regular maintenance schedule to assure that sanitary design is not defeated by nonexistent or poor maintenance.

Finally, facilities for personal hygiene of workers are of utmost importance. Ideally, toilets and locker rooms should be located in easily accessible locations but separate from production areas. Fixtures such as foot-pedal-operated faucets and flushing mechanisms allow for less contamination of hands and more effective hand washing. In some plants, locker rooms with showers enable workers to change from street clothes into approved sanitary garments. However, as mentioned earlier, management must take extra efforts to encourage and motivate workers to properly use the facilities.

D. Equipment

The sanitary condition of fruit and vegetable processing equipment can have a dramatic influence on the microbiological quality of the product. As early as the 1960s, contaminated equipment surfaces were identified as a major cause of high microbial populations in vegetables (26). It was reported even earlier that visual cleanliness is not necessarily a guarantee of the sanitary condition of food processing equipment (16). Hence, it is important that fruit and vegetable processing equipment be maintained properly. Yet despite the importance of proper maintenance of all types of equipment, the neglect of some types of equipment is especially critical with respect to food safety.

In general, one can categorize fruit and vegetable processing equipment into one of several broad categories: equipment for product movement, equipment used to cut product, food preservation apparatus and associated equipment, and food packaging equipment (Table 2). Past research has indicated that equipment from the first two categories is most often responsible for contaminating products during processing. For example, Splittsoesser et al. (26) identified inspection belts as a primary site of contamination during the processing of peas and corn. They noted that older belts, which are often cracked or pitted, were more likely to be a source of microorganisms than were newer belts. In addition, they found that cleaning and sanitizing older belts was less likely to eliminate the problem, possibly because microorganisms were hidden in cracks and crevices.

Equipment that in some way cuts fruits and vegetables can also adversely affect microbiological quality. Splittsoesser et al. (25) found that significant recontamination often occurred when French-style green beans were sliced after being blanched, a process that normally reduces populations of microorganisms. The influence of cutting operations on microbiological quality was also observed by Garg et al. (15). They observed a 10- to 100-fold increase in microbiological

Table 2 Examples of Equipment Important in Fruit and Vegetable Processing

Class of equipment	Examples
Product movement	Flumes
	Belts
	Hoppers
	Bin and crates
	Crates
Product cutting	Peelers
	Slicers
	Corers
Product preservation	Blanchers
	Retorts
	Freezing tunnels
Product packaging	Film shrinkers
	Vacuum chambers
	Film wrappers

populations after shredding or slicing of several vegetables (Fig. 1). Although the authors did not provide reasons for the population increases, it is likely that the equipment harbored microorganisms and served to inoculate any product that was sliced or shredded.

Cutting equipment is particularly troublesome because it is constantly becoming soiled with juices and other organic matter during the process of cutting or slicing. These materials can support the growth of microorganisms, hence potentially can accumulate biofilms of microorganisms that could serve to inoculate products. However, cutting can also sometimes reduce populations. For example, Garg et al. (15) also found that peeling carrots (Fig. 1) reduced microbial populations by 100-fold. In this case, the peeling process likely eliminated soil-borne organisms present on the surface of the unpeeled carrots. However, it is also possible that the slicing releases antimicrobial compounds (5).

Although equipment used in preservation processes such as canning is unlikely to contaminate fruits and vegetables, it is still important to keep this equipment well maintained and in sound working order. In particular, any errors leading to improper thermal processes could lead to spoilage, and perhaps even serious risk of foodborne illnesses, such as botulism (see Chapter 11).

E. Personnel

As with production workers, personnel involved with processing can dramatically affect the microbiological safety of fruits and vegetables. However, processors

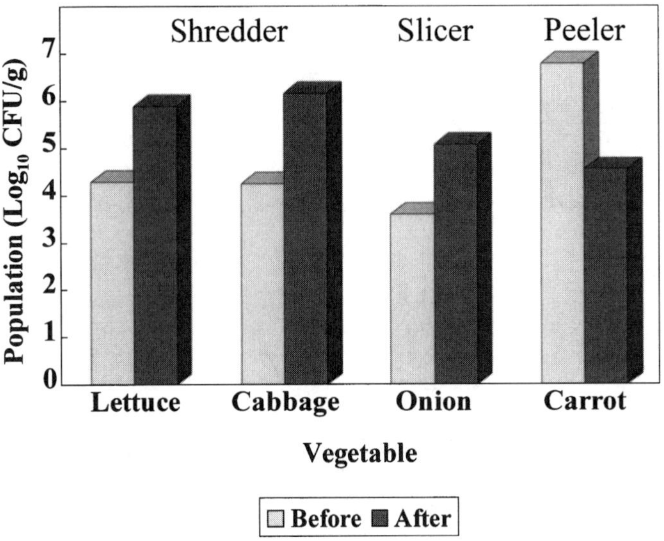

Figure 1 Effects of various processing steps on populations of microorganisms in vegetables. Counts given are for before and after the individual process mentioned at the top of each set of bar graphs. (From Ref. 15).

have the advantage of operating in a more controlled environment. Nevertheless, some of the same concerns and strategies surrounding proper personal hygiene apply to workers in both categories.

Two of the most important factors in assuring that personnel do not contribute to microbial safety problems are proper training and adequate toilet and hand-washing facilities. Workers must be informed of sanitation policy implemented by the processor and, if necessary, instructed in the importance of proper personal hygiene and sanitary practices. As with farm workers, one must not assume that all workers are familiar with even the most basic of sanitary practices.

If one is to have a properly educated workforce, it is also important for supervisors to receive continuing education that will enable them to instruct line workers on a day-to-day basis. Many universities and private consulting firms offer "train the trainer" short courses aimed at providing supervisors with the necessary tools for instructing their workers.

Workers should be instructed in two major areas. The first, as already mentioned, are the basics of personal hygiene and their role in food safety. Second, workers should be instructed in the aspects of their jobs that contribute to the production of a clean and sanitary product. For example, many workers and even supervisors fail to realize the importance of cleaning food contact surfaces before sanitizing. Consequently, excessive amounts of sanitizer are often used in place of proper cleaning in the hope that the sanitizer will eliminate all the microorganisms. Unfortunately, this practice not only does not work but may damage equipment. In addition, workers should be instructed to identify situations that could lead to potential food safety problems and encouraged to notify supervisors or management of those situations.

F. Transportation

Although not normally considered to be a part of processing, transportation to and from handling and processing facilities is an important operation that affects both the quality and the safety of fruits and vegetables. Practices employed by the transporters can affect safety in several ways. First, failure to maintain products at refrigeration temperatures can allow microorganisms to grow. Transportation to the processing facility is often done using open or unrefrigerated trucks that may expose freshly harvested products to hot summer temperatures. The specific impact of this exposure depends on the length of time and the temperatures involved. Obviously, if products can be transported quickly and at cool temperatures, safety is less likely to be compromised.

In general, conditions to which products are exposed are better controlled in postprocessing transportation. Processors typically use refrigerated trucks to transport whole or fresh-cut fruits and vegetables. In theory, refrigerated trucks keep the products sufficiently chilled to minimize growth of microorganisms and maintain quality. In practice, however, several factors can allow products to become exposed to less than ideal conditions, even in refrigerated trucks. First, it is important that refrigeration units be maintained in optimal working condition. Hot conditions experienced while traveling through desert areas or during summer can put extra stress on refrigeration units. The units are less likely to fail, however, if kept in good condition. Second, the manner in which products are packed into trucks can have a dramatic effect on product temperature. When trucks are overfilled with product, airspace may be insufficient, thereby inhibiting circulation and return of chilled air. This type of loading results in hot spots in various sections of trailers. Second, products closest to the walls, ceilings, and floors of trucks are likely to be subjected to greater radiant heating from the outside even when trucks are properly loaded. This effect is especially likely when trucks are overfilled and leave little or no airspace between the product and these walls, ceilings, and floor. Therefore, it is important that fruits and vegetables be loaded to allow free circulation of air and to minimize exposure to external surfaces. In addition, drivers may turn off refrigeration units at rest stops.

A second general way in which transportation can affect safety is by cross-contamination. In particular, the practice of "backhauling" is especially risky.

Backhauling refers to the transporting of one type of cargo in one direction and another on the return or subsequent trips. For example, a farmer might haul hogs or cattle on one day and use the same truck to haul fresh produce on another. The risk of cross-contamination from such a practice is obvious, particularly if the truck was not cleaned and sanitized between trips. Backhauling is more common among small farm operators and in developing countries, where it is cost that forces transporters to the same use trucks for multiple purposes. However, backhauling has also been practiced by large, well-financed trucking companies in modern, developed countries. Unfortunately, backhauling led to a major outbreak of disease in the United States in 1994: ice cream premix was transported in tanker trailers that had previously carried non pasteurized liquid egg, and over 200,000 cases of salmonellosis resulted (15a).

Because of the risks inherent in backhauling, most large produce processors either maintain their own fleet of trucks, providing them with more control over the product, or specify temperatures and loading plans. In addition, most specifically forbid such practices when contracting transportation.

G. HACCP

As mentioned earlier in this chapter, the well-known HACCP concept is used by most food processors, including fruit and vegetable processors. Most of the critical control points are similar whether the product is to be sold as fresh-cut, frozen, or canned. For example, Willocx et al. (35) listed some typical critical control points for processing of endive that are similar to those of other vegetables (Table 1). Of those listed, storage, preparation, washing, cutting, and transporting procedures are important critical control points that are almost universal in the produce industry. However, other critical control points differ slightly and are specific to final products. Adhering to the established reporting protocol would be an example of a specific critical control point in the production of canned vegetables. New critical control points will undoubtedly arise as technology changes.

IV. SAFE FOOD HANDLING AT THE FOOD SERVICE LEVEL

It is estimated that many consumers in industrialized nations purchase at least one meal per day away from home. Consequently, it should not be surprising that a large proportion of foodborne illness cases arise from such foods. Although "food service" traditionally refers to food served in restaurants and institutions, retail food outlets have begun marketing new products known as "meal solutions." This new class represents a blend of traditional food service, grocery foods, and home-cooked meals. Some are fully cooked and packaged meals produced at retail food outlets and designed to be taken off-site for consumption. Others are partially cooked and require reheating. As meal solutions become more popular with consumers, they too will likely reveal some new food safety issues.

A. Personnel

The food service industry is one of the most labor-intensive and service-oriented sectors of the food industry. People dictate virtually every aspect of the business from preparation through sales. Unlike production or processing, however, many workers in food service are primarily part-time employees who view their jobs as temporary. This situation leads to several common weaknesses in the food service industry that make it more difficult to ensure the microbiological safety of the foods prepared. First, many food service workers receive little or no training in food sanitation. Managers are often hesitant to invest too much time or effort into training these individuals because they do not expect the employees to be with the establishment for any prolonged period of time. Second, food service workers are often poorly paid and receive minimal or no benefits. This lack of pay and benefits leads to a lack of loyalty on the part of the workers and prompts them to come to work even when sick so that they lose no pay. In addition, language is a problem for some food service workers. Hence, it is important to change the mind-set of food service managers as well as employees before meaningful progress on ensuring food safety can be accomplished. Both parties must recognize the value of training and education in sanitation and the importance of personal health and hygiene to avoid foodborne illness in patrons.

B. Equipment

In general, three types of equipment are found in food service establishments. These are devices used in the preparation, heating, and storage of foods. With regard to fruits and vegetables, preparation and storage equipment has the greatest impact on microbiological safety.

Equipment used for the preparation of fruits and vegetables primarily consists of cutting equipment (knives, choppers, etc.), mixing containers, and preparation surfaces, such as cutting boards. As one would expect, the main hazard associated with such equipment is the potential for cross-contamination, especially if the equipment used for raw meats and poultry also is used for fresh fruits and vegetables. It is vitally important that common use equipment be thoroughly cleaned and sanitized after each use to prevent cross-contamination. The choice of specific brands or models of equipment can be important in this regard. For example, there has been debate recently on whether wood or synthetic (plastic, nylon, etc.) cutting boards are easier to sanitize or better resist survival and growth of microorganisms, thus affecting food safety (10,21). The consensus to date seems to indicate that attention to cleaning and sanitizing is more important than the material per se. Rather than continuing to debate issues of materials used and care in cleaning and sanitizing, a better solution is for food service establishments to designate completely separate utensils for meats and poultry, raw produce, and cooked foods. In the case of major pieces of equipment such as mixers, food service managers should consider using equipment approved by the National Sanitation Foundation (NSF) or similar agencies to ensure that the equipment has been designed for cleanability and sanitizing.

Most storage equipment used for fruits and vegetables in foodservice establishments is primarily for refrigeration. An obvious factor in the safety of such equipment is temperature control. Refrigerators or display cases should be properly maintained and kept as cold as possible without damaging the food. It is also a good idea, and in many cases the law, for refrigerators to have a working thermometer in plain view. Many people often forget that refrigerators also are food contact surfaces that can transfer pathogenic microorganisms. Indeed, at least one report indicated that strains of *L. monocytogenes,* associated with listeriosis, were present in patients' home refrigerators (23). To prevent the establishment of contamination of pathogens in refrigerators, these pieces of equipment should be placed on a regular cleaning and sanitation schedule. In addition, food service establishments should consider having separate refrigerators for raw meats and poultry, raw produce, and cooked foods, as was suggested for other types of equipment.

C. Preparation Practices

The relative proximity of different types of foods in the preparation and serving areas can also influence microbiological safety. In many facilities such as small restaurants, it is not unusual for raw meats and poultry to be cut, mixed, and prepared close to or in the same workspace as raw produce. This situation is obviously less than optimal and could lead to contamination of the produce by meat microorganisms. A safer alternative would be to have separate preparation areas for these types of food. At the very least, these areas require extra care in cleaning and sanitation.

Self-serve buffet lines also present a unique combination of food safety hazards. Such serving lines often have cooked meats and poultry, raw produce, and condiments placed close together or adjacent to each other in the display cases. In such serving situations, it is common for various products to spill over in different containers, creating a risk of cross-contamination. Although spillage is not always entirely preventable in buffet lines, the potential risk can be minimized by keeping incompatible foods far apart and by frequent cleaning of the display case. "Topping off" can also lead to microbiological hazards associated with fruits and vegetables in food service. This practice involves adding new fresh product to partially full containers of previously served product. Topping off without periodically discarding the original contents of containers is a bad practice. It allows older product to develop higher populations of microorganisms, which then contaminate fresh product. Conscientious establishments always exchange older containers and their contents for new clean containers with fresh contents.

V. SAFE FOOD HANDLING IN THE HOME

Safe fruit and vegetable handling in the home shares some of the same issues with food service. However, there are also some very important differences between food safety in the home and all other situations discussed thus far. That is, the people involved in and responsible for food safety in the home are probably the least educated in food safety, hence are most likely to make mistakes leading to foodborne illness. In addition, unlike business enterprises, food safety practices in the home are rarely, if ever, subject to governmental regulation. Hence, the home is probably a major source of foodborne disease outbreaks, although only large outbreaks are likely to come to the attention of health agencies.

A. Prewashing

Many of the same comments provided in connection with foodservice also apply to the home. Fresh unprocessed produce has been subjected to minimal cleaning steps and should be considered to be contaminated with pathogens. Although there is currently nothing consumers can do to completely remove pathogens, they can take steps to reduce populations. For example, Adams et al. (1) found that simply removing the outermost layers of whole lettuce reduced populations of microorganisms by 1 log. Similarly, they (1) and others (7) have observed that washing reduces populations by a similar amount. In general, packaged fresh-cut fruits and vegetables are subjected to a more rigorous washing procedure than are whole products, and some manufacturers even advertise that washing of their product is unnecessary. However, rinsing such products immediately before use does not adversely affect their microbiological safety and may help to remove additional microorganisms. A number of new washing aids are being developed specifically to disinfect fresh fruits and vegetables. These products can include food-grade surfactants, acidulants, and other ingredients that enhance the removal of microorganisms as well as pesticide residues. Regulatory agencies have been slow to approve or endorse these products because published information on their efficacy is lacking. However, some may offer significant advantages over washing with water alone.

B. Utensils

As with foodservice, the potential for cross-contamination to and from fresh produce is a threat in the home. The sources of cross-contamination are varied but usually originate with raw meats and poultry. Public awareness regarding the importance of proper handling of raw meats and poultry has increased in recent years. However, many consumers are still unaware that microorganisms from these products can be transferred to fresh produce if common cooking utensils or cutting surfaces are used. Similarly, the importance of separating various foods in the refrigerator is also important. Ready-to-eat foods such as fresh produce should be stored in such a way that they will not become contaminated by juices exuded by meats and poultry. Some useful practices include not storing fresh fruits and vegetables directly beneath shelves on which raw meat are stored. It is also helpful to keep fresh produce in closed containers, and to have separate cutting boards for each type of food.

C. Storage and Handling

The form in which fresh produce is purchased and stored has changed in recent years. Fresh produce has traditionally been sold as whole, uncut products. Such products are quite perishable and, depending on the specific fruit or vegetable, can spoil relatively soon after purchase. However, the development of new technologies such as barrier films and modified atmosphere packaging have dramatically extended the useful shelf life of many fresh fruits and vegetables. In addition, these technologies have enabled processors to market many new types of products, such as packaged, fresh-cut fruits and vegetables mixes. Because fresh-cut fruits and vegetables are so convenient and generally of high quality, they are becoming increasingly popular and account for most of the increased sales of fruits and vegetables observed in recent years (14).

Convenient, packaged fresh-cut fruits and vegetables require to be handled and stored differently from whole products. The visual quality of packaged produce can be a poor indication of microbiological safety. In general, the relatively short shelf life of whole products generally causes them to become spoiled before high populations of pathogenic bacteria can accumulate. In contrast, modern packaging technologies maintain the sensory quality of the many fresh products long enough for pathogens to grow to high populations without adversely affecting sensory quality (3). Hence, the extended shelf life may allow more time for pathogens to grow and potentially create a hazardous situation, even when the products are properly refrigerated. Consumers can avoid this situation by using the product as soon as possible after purchase, despite how good it might look during the whole shelf life of the product. Likewise, products that have been stored for extended times are best discarded, even if they do not appear to be spoiled.

Another important difference between whole and packaged fresh produce is that packaged produce can be more susceptible to growth of pathogenic bacteria than is the whole product. Both whole and fresh-cut products continue to respire during storage, resulting in the consumption of oxygen and the release of carbon dioxide. This metabolic activity increases with higher storage temperatures. In unpackaged products, the CO_2 is quickly diluted into the atmosphere, and atmospheric O_2 always surrounds the plant tissue. In contrast, the free flow of gases is retarded in packaged products, resulting in a reduction of O_2 and an increase in CO_2

concentrations in the package. If the CO_2 concentration becomes high enough and the O_2 concentration becomes low enough, it is possible for *Clostridium botulinum* to grow and produce toxin (11). Proper refrigeration of packaged products will enhance safety by (a) minimizing the respiration rate of the product, thereby minimizing hazardous concentrations of CO_2, and (b) preventing or slowing the growth of *C. botulinum* and other pathogens. Consumers need to be educated on the proper handling of new types of produce product, and they should be instructed to always discard products that have been subjected to abuse temperatures. In the future, indicators may be included in packages that warn consumers of products that have been poorly handled.

VI. CONCLUSIONS

Fruits and vegetables have been considered to be among the safest of foods. However, they are not without some risks. Controlling and minimizing the risk entails paying careful attention to how these products are handled not only after arriving into the marketplace, but also during production and harvesting. As new technologies change the way fruits and vegetables are grown, harvested, processed, and marketed, farmers, processors, and consumers will likewise have to change the precautions needed to maintain safety. Moreover, food scientists and microbiologists will likely need to continue research on maintaining the safety of fruits and vegetables as more is learned about new and emerging pathogens.

REFERENCES

1. Adams, M. R., A. D. Hartley, and L. J. Cox. 1989. Factors affecting the efficacy of washing procedures used in the production of prepared salads. Food Microbiol. 6: 69–77.
2. Anonymous. 1997. Outbreaks of *Escherichia coli* O157:H7 infection and cryptosporidiosis associated with drinking unpasteurized apple cider—Connecticut and New York, October 1996. MMWR 46(1):1–8
3. Berrang, M. E., R. E. Brackett, and L. R. Beuchat. 1990. Microbial, color and textural qualities of fresh asparagus, broccoli, and cauliflower stored under controlled atmospheres. J. Food Prot. 53:391–395.
4. Besser, R. E., S. M., J. T. Weber, M. P. Doyle, T. J. Barrett, J. G. Wells, and P. M. Griffin. 1993. An outbreak of diarrhea and hemolytic uremic syndrome from *Escherichia coli* O157:H7 in fresh-pressed apple cider. JAMA 269:2217–2220.
5. Beuchat, L. R., and R. E. Brackett. 1990. Inhibitory effects of raw carrots on *Listeria monocytogenes.* Appl. Environ. Microbiol. 56:1734–1742.
6. Beuchat, L. R., and J.-H. Ryu. 1997. Produce handling and processing practices. Emerg. Infect. Dis. 3:459–465.

7. Brackett, R. E. 1987. Antimicrobial effect of chlorine on *Listeria monocytogenes.* J. Food Prot. 50:999–1003.
8. Brackett, R. E. 1994. Microbiological spoilage and pathogens in minimally processed fruits and vegetables, in R. C. Wiley (ed.), Minimally Processed Refrigerated (MPR) Fruits and Vegetables. Van Nostrand Reinhold, New York, pp. 269–312.
9. Brodie, B. B. 1993. Probability of *Globodera rostochiensis* spread on equipment and potato tubers. J. Nematol. 24:291–296.
10. Carpenter, B. 1997. Sanitary quality of meat chopping board surfaces; a bibliographic study. Food Microbiol. 14:31–37.
11. Conner, D. E., V. N. Scott, and D. T. Bernard. 1989. Potential *Clostridium botulinum* hazards associated with extended shelf-life refrigerated foods: A review. J. Food Saf. 10:131–153.
12. Dowe, M. J., E. D. Jackson, J. G. Mori, and C. R. Bell. 1997. *Listeria monocytogenes* survival in soil and incidence in agricultural soils. J. Food Prot. 60:1201–1207.
13. Fattal, B., Y. Wax, M. Davies, and H. I. Shuval. 1986. Health risks associated with wastewater irrigation: An epidemiological study. Am. J. Public Health. 76:977–979.
14. Food Marketing Institute. 1996. Trends in the United States: Consumer attitudes and the supermarket. FMI, Washington, DC.
15. Garg, N., J. J. Churey, and D. F. Splittstoesser. 1990. Effect of processing conditions on the microflora of fresh-cut vegetables. J. Food Prot. 53:701–703.
15a. Hennessey, T. W., C. W. Hedberg, L. Slutsker, K. E. White, J. M. Besser-Wiek, M. E. Moen, J. Feldman, W. W. Coleman, L. M. Edmonson, K. L. MacDonald, M. T. Osterholm, and the Investigation Team. A national outbreak of *Salmonella enteritidis* infections from ice cream. N. Engl. J. Med. 334:1281–1286.
16. Hucker, G. J., R. F. Brooks, and A. J. Emery. 1952. The source of bacteria in processing and their significance in frozen vegetables. Food Technol. 6:147–155.
17. Katzenelson, E., and B. Teltch. 1976. Dispersion of enteric bacteria by spray irrigation. J. Water Pollut. Control Fed. 48:710–716.
18. Kirkham, M. B. 1986. Problems of using wastewater on vegetable crops. HortScience 21:24–27.
19. Larkin, E. P., J. T. Tierney, and R. Sullivan. 1976. Persistence of virus on sewage-irrigated vegetables. J. Environ. Eng. Div. ASCE 102:29–35.
20. Mac Kenzie, W. R., N. J. Hoxie, M. E. Proctor, M. S. Gradus, K. A. Blair, D. E. Peterson, J. J. Kazmierczak, D. G. Addiss, K. R. Fox, J. B. Rose, and J. P. Davis. 1994. A massive outbreak in Milwaukee of *Cryptosporidium* infection transmitted through the public water supply. N. Engl. J. Med. 331:161–167.
21. Miller, A. J., T. Brown, and J. E. Call. 1996. Comparison of wooden and polyethylene cutting boards: Potential for the attachment and removal of bacteria from ground beef. J. Food Prot. 59:854–858.
22. Mutlak, S. M., Y. A. Hamdi, M. A. Nour, N. Bakal, M. Al-Gassal, and N. Ayar. 1980. Sewage farming in Iraq: A potential hazard for pollution. J. Environ. Qual. 9: 677–680.
23. Pinner, R. W., A. Schuchat, B. Swaminathan, P. S. Hayes, K. A. Deaver, R. E. Weaver, B. D. Plikaytis, M. Reeves, C. V. Broome, and J. D. Wenger. 1992. Role of foods in sporadic listeriosis. II. Microbiologic and epidemiologic investigation. JAMA 267: 2046–2050.

24. Sadovski, A. Y., B. Fattal, and D. Goldberg. 1978. Microbial contamination of vegetables irrigated with sewage effluent by the drip method. J. Food Prot. 41:336–340.
25. Splittsoesser, D. F., M. Groll, D. L. Downing, and J. Kaminski. 1977. Viable counts versus incidence of machinery mold (*Geotrichum*) on processed fruits and vegetables. J. Food Prot. 6:402–405.
26. Splittsoesser, D. F., W. P. Wettergreen, and C. S. Pederson. 1961. Control of microorganisms during preparation of vegetables for freezing. II. Peas and corn. Food Technol. 15:332–334.
27. Tauxe, R. V. 1997. Emerging foodborne diseases: An evolving public health challenge. Emerg. Infect. Dis. 3:425–434.
28. Troller, J. A. 1983. Sanitation in Food Processing. Academic Press. New York, pp. 21–41.
29. Van Donsel, D. J., and E. P. Larkin. 1977. Persistence of *Mycobacterium bovis* BCG in soil and on vegetables spray-irrigated with sewage effluent and sludge. J. Food Prot. 40:160–163.
30. Van Renterghem, B., F. Huysman, R. Rygole, and W. Verstraete. 1991. Detection and prevalence of *Listeria monocytogenes* in the agricultural ecosystem. J. Appl. Bacteriol. 80:316–320.
31. Watkins, J., and K. P. Sleath.. 1981. Isolation and enumeration of *Listeria monocytogenes* from sewage, sewage sludge and river water. J. Appl. Bacteriol. 50:1–9.
32. Wei, C. I., T. S. Huang, J. M. Kim, W. F. Lin, M. L. Tamplin, and J. A. Bartz. 1995. Growth and survival of *Salmonella montevideo* on tomatoes and disinfection with chlorinated water. J. Food Prot. 58:829–836.
33. Weis, J., and P. R. Seeliger. 1975. Incidence of *Listeria monocytogenes* in nature. Appl. Microbiol. 30:29–32.
34. Welshimer, H. J. 1960. Survival of *Listeria monocytogenes* in soil. J. Bacteriol. 80:316–320.
35. Willocx, F., P. Tobback, and M. Hendrickx. 1994. Microbial safety assurance of minimally processed vegetables by implementation of the hazard analysis critical control point (HACCP) system. Acta Aliment. 23:221–238.
36. Zhang, S., and J. M. Farber. 1996. The effects of various disinfectants against *Listeria monocytogenes* on fresh-cut vegetables. Food Microbiol. 13:311–321.
37. Zhuang, R.-Y., L. R. Beuchat, and F. J. Angulo. 1995. Fate of *Salmonella montevideo* on and in raw tomatoes as affected by temperature and treatment with chlorine. J. Food Prot. 61:2127–2131.

4

Safe Handling of Seafood

Thomas Jemmi and Michel Schmitt
Swiss Federal Veterinary Office, Bern, Switzerland

Thomas E. Rippen
University of Maryland, Princess Anne, Maryland

I. INTRODUCTION

Fish and shellfish are very popular and nutritious food, and in some countries they are the most important protein source. Traditionally, the consumption of seafood was very localized, occurring close to the areas of harvesting. Developments in food technology, particularly freezing, chilling, canning, and other stabilizing techniques, have allowed seafood products to become less perishable. Therefore, they are now common items of international trade and are consumed in areas where they used to be almost unknown. The increase of aquaculture has also contributed to this. Nevertheless, fish and shellfish are still mostly harvested from the sea and are subject to problems related to availability and quality. Fishery stocks often fluctuate widely from year to year, but generally they are decreasing. Environmental and weather conditions also greatly influence their periodic availability. Seafood quality is affected by harvesting methods and onboard handling and stowage procedures. Products harvested in the wild are "subject to a variety of both natural and man-made hazards" (26), which is a concern for seafood safety. Aquacultured products are associated with other potential hazards, including water pollutants, drug residues, and agriculture chemicals used near farm ponds. These concerns that lead to human illness associated with the consumption of fish and shellfish are addressed in depth in this chapter.

In the 1980s there was a dramatic increase in seafood-borne diseases (130). This increase was mainly due to the rising consumption of these products, especially in countries or regions where seafood was not traditional. Therefore, knowledge of how to handle seafood safely often was poor or even nonexistent. Moreover, such wholesome and easily digestible food often was consumed by people with underlying diseases or general health problems. In the 1990s, this increase in seafood-borne diseases seems to have slowed down.

Hazards in seafood, including *Salmonella* and enteric viruses, have been known for decades. However, some new pathogens and safety problems emerged. Centralization of food production and the widespread use of refrigeration in industrialized countries may have contributed to the emergence of some psychrotrophic pathogens as *Listeria monocytogenes* or *Aeromonas hydrophila* (130). In addition, as aquaculture grew in importance, some formerly unknown residue problems also arose.

Some definitions are essential to avoid misunderstandings. In particular, the word *fish* is used not only as a synonym for seafood but to specify finfish. The expressions used in this chapter are defined as follows:

Seafood: generic term including all animal aquatic life and their products, but excluding birds and mammals

Fish: fresh and saltwater finfish belonging to the zoological classes of Pisces, Elasmobranchii, and Cyclostoma

Shellfish: aquatic animals belonging to the zoological class of bivalves (e.g., oysters, scallops, clams, mussels)

Cephalopods: aquatic animals belonging to decapodes and octopodes (e.g., cuttlefish, squid, octopus)

Crustaceans: aquatic animals belonging to the zoological class of Crustacea, characterized by a chitinous exoskeleton (e.g., lobsters, crabs, shrimps, prawns, crayfish)

II. UTILIZATION AND PROPERTIES OF SEAFOOD

A. Utilization of Seafood

In the last decades, as more fish were caught, the amount sold frozen increased, while the percentage of fish used or sold as fresh, salted, smoked, or marinated went down to approximately 50% (Table 1). Both the way fish are consumed, and the preferred types of fish and fish product are largely influenced by traditional and/or ethnic background.

B. Properties of Seafood

1. Main Components of Seafood

Water, proteins, and fat are the main components of all fish tissues, and indeed fish is consumed principally as a source of protein that is high in nutritional quality in

Table 1 Utilization of Worldwide Catch of Fish

	1950	1960	1970	1980	1990
Total catch, million metric tons	21.1	40.0	70.0	72.2	92.2
Used as fresh fish, %	45.0	41.8	27.8	20.6	22.7
Used as frozen fish, %	5.2	8.8	13.8	22.6	24.9
Used as salt-, smoked, marinated fish, %	26.1	18.7	11.6	15.7	11.5
Used for human consumption, %	85.8	78.5	62.1	73.9	72.2
Fishmeal and oil	9.5	19.0	36.5	25.1	26.2
Other products	4.7	2.5	1.4	1.0	1.5

Source: Ref. 71.

terms of digestibility and amino acid composition (171). Other components, like mineral salts, carbohydrates, and vitamins, usually are present only in a low proportion but may be very important in a nutritional or technical point of view. All components are subject to variations including fish species, sexual maturity, age, nutritional condition, and environmental factors. The composition of fish muscle is shown in Table 2 (186). Lipid content in muscle varies greatly primarily because fish keep energetic reserves in the form of fat in muscle or in other organs (especially the liver), according to their species. In terms of food safety, a large number of contaminants (e.g., organochlorine compounds) are lipophilic and the degree of contamination is closely related to the lipid content (see Sec. III.A.5.). Examples of the composition of some shellfish tissues are listed in Table 3 (71).

2. Properties of Fish Muscle

In comparison to mammalian muscle, fish muscle contains only small amounts of connective tissue: 1–3% of total protein versus 10–15% in mammals (186). In addition and in contrast to warm-blooded animals, the muscles and groups of mus-

Table 2 Composition (%) of Fish Muscle

	Water	Protein	Lipid	Ash
Average	74.8	19	5	1.2
Range	28–90	6–28	0.2–64	0.4–1.5
Ratio of high to low values	3.2	4.7	320	3.8

Table 3 Examples of the Composition of Shellfish Tissues

	Water (%)	Protein (%)	Lipid (%)	Ash (%)
Shrimp	78.40	18.60	1.44	1.33
Crayfish	83.10	15.00	0.47	1.26
Cuttlefish	81.00	16.10	0.90	1.00
Mussel	83.20	9.84	1.34	1.70

cles are wrapped by connective tissue only weakly or not at all. This results in easy penetration of microorganisms throughout the whole fish.

Furthermore, fish muscle is high in nonprotein, nitrogen-containing compounds. As shown in Table 4, the amounts vary with the type of fish and are also influenced by age. These small molecules are dissolved in the tissue fluids and seem to play an important role during bacterial spoilage (168). Nonprotein nitrogen sources perform an important physiological function for fish, serving as osmoregulators. Compounds such as trimethylamine oxide assist the tissues by providing osmotic balance with the surrounding environment. Free amino acids are often present in certain fish species, including many in the family Scombridae (tuna and mackerel). Under certain conditions, histidine is converted to histamine that may induce allergy-like symptoms in humans shortly after the fish are consumed. This illness, often referred to as scombroid poisoning, is described in detail in Sec. III.A. Fish muscle also contains proteases, which, in many species, break down protein quickly after death. Since fish are cold-blooded (poikilothermic), these enzymes are active over a wider range of temperatures than those present in warm-blooded animals. As a consequence, enzymatic softening of the tissues occurs even at refrigeration temperatures. The peptides and other small molecular products resulting from these catabolic processes are directly usable as a protein source by spoilage bacteria, contributing to the short shelf life of much

Table 4 Nonprotein Nitrogen (NPN) Material in Fish

	Cod (mg %)	Herring (mg %)	Haddock (mg %)	Elasmobranch (mg %)
Total NPN	419.2	437.0	380.0	—
Bases	257.6	269.9	239.4	—
Creatine	163.0	182.7	—	—
Urea	1.9	4.0	14.5	2000.0
Trimethylamine oxide	~95.0	~40.0	~70.0	~275.0

seafood. The mix of native and bacterial enzymes also produces amine by-products, some of which (putrescine and cadaverine) may potentiate the toxicity of histamine (180).

C. Classes of Seafood Products

According to the type and degree of preparation and/or preservation, seafood products can be classified as follows:

Live seafood: animals that are kept alive after harvesting, then offered for sale, bought, and in some cases even consumed alive.

Raw seafood products: harvested seafood, which is only gutted, filleted, peeled, or shucked. These products are preserved only by chilling on ice, or in some cases are not preserved at all. Whole fish or any minimally processed raw form is included in this category.

Frozen seafood: from the point of view of food hygiene, it seems to be useful to divide this group of fish and fish products into two subcategories.

Frozen seafood: raw seafood preserved by freezing.

Frozen seafood products: a large number of different further or highly processed products. In all cases, their specific ingredients, their way of production, and/or the way they must be prepared at home is more characteristic than the fact that they are frozen.

Dried products: the aim is to reduce moisture, generally to a content not greater than 15–20%. The drying process can be carried out under natural (suspending over fire, exposing to the sun) as well as under artificial (drying tunnels) conditions.

Salted products: preservation is achieved by the inhibiting action on microorganisms and enzymes of high salt concentration and the dehydration that accompanies the process. Either of two methods is commonly followed: brine salting or dry-salting. The former method retains moisture from the fish to form brine liquor, and the fish are stored in brine. The latter method allows moisture to drain away from the fish during the salt penetration phase, which is followed by a drying step as described for dried products.

Enzyme-processed products: products that achieve a ready-to-eat form, mainly or exclusively as a result of the action of proteolytic enzymes, without any cooking or other heating step (e.g., cured anchovies and hydrolyzed fish sauces).

Thermally processed products: products that undergo cooking or another heating process to be preserved and/or to achieve their typical taste and texture (e.g., fish sausages, fish rissoles, hot smoked products, grilled fish, or any other preparation that exposes the product to a higher temperature as 60–65°C). Process schedules may be established by performing heat penetration studies and calculating the lethality of accumulated heat exposure on pathogens and spoilage organisms.

Sous-vide products: have become very popular because they allow the consistency of centralized food preparation but with improved eating qualities. Flavors blend and intensify without yield losses. The objective is quality enhancement, not pathogen destruction. Seafood or prepared seafood dishes are also produced using this technology. Raw material is portioned into plastic bags or containers and vacuum packed, and then cooked or steamed to reach a core temperature of at least 63°C (i.e., generally lower than pasteurization temperatures). The product is subjected to shock cooling to rapidly reach a core temperature below 10°C and is then stored at a temperature between 0 and 3°C (200). Therefore shelf life depends uniquely on the refrigeration process, the only safety control measure. Otherwise surviving pathogens may encounter nearly ideal growth conditions.

Canned fish: the preservative action of the canning process depends on the heat inactivation of microorganisms and intrinsic enzymes and the protection of atmospheric oxygen. As a result of the high temperature during the production, the canned product picks up the character of a thermally processed product. Commercial sterility as applied to canned seafood is similar to that for other low acid canned foods. A process that imparts a minimum of 12 decimal reductions for heat-resistant strains of *Clostridium botulinum* (spore-forming bacteria responsible for botulism) is required. When left in the original containers, these products have a long shelf life at room temperature (3).

Smoked fish: produced from chilled or frozen fish, these foods are primarily brined and then exposed to freshly produced wood smoke. In traditional smoked fish, preservation is achieved through a combination of dehydration, salt and, to a lesser extent, phenols and other compounds found in wood smoke. In most countries, contemporary versions are processed for their sensory qualities and lack significant shelf life extension owing to higher moisture and lower salt contents. These products require refrigeration.

Imitation seafoods: these products consist of highly processed fish proteins that are structured, colored, and flavored to look and taste like another product (e.g., crab-flavored Surimi made of myofibrillar proteins of fish).

Fishmeal: most fishmeal is used in animal feeds. It is produced from the solids remaining after low value fish have been cooked and pressed to release oil and water. When produced under strict hygienic standards, fishmeal can be used in human nutrition, especially as an additive in protein-deficient parts of the world.

Fish oil: produced out of fatty raw material and partially or totally hydrogenated, it is used in fish canneries, in margarine production, and as an additive in the food industry.

III. FOODBORNE ILLNESS DUE TO SEAFOOD CONSUMPTION

A. Possible Hazards

Foodborne hazards are generally classified as biological, chemical, or physical. Biological hazards include bacteria, viruses, parasites, and toxins.

There are over 200 possible hazards known to cause foodborne disease. An overview of the seafood-borne hazards is listed in Table 5. The most important hazards are listed in the subsections that follow.

1. Bacteria

a. Clostridium botulinum Type E

Hazard: botulinum toxin formed by *Clostridium botulinum* type E.

Distribution: worldwide, in an environment close to water (coast, sediments).

Disease: botulism (64).

Transmission: ingestion of food contaminated with preformed botulinum toxin.

Associated food: fish and fish products (64), home-fermented fish products (126).

Resistance of hazard organism (40,64):

Growth temperature: minimum 3°C; maximum 45°C.

Thermal inactivation: D100°C: < 0.1 minute; D82°C: 0.2–1.0 minute.

pH: minimum 5.0; maximum 8.5.

Water activity a_w: minimum 0.97.

Food additives: nitrite has antibacterial activity.

Smoke: the use of natural or liquid smoke has inhibitory effect against *C. botulinum* in fish.

Prevention:

Refrigeration of fish (e.g., icing).

Adherence to good manufacturing practice (GMP) to control growth of *C. botulinum* (64): combination of thermoprocessing, salt content, and refrigeration.

b. Listeria monocytogenes

Hazard organism: Listeria (L.) monocytogenes.

Distribution: worldwide.

Diseases: listeriosis: meningoencephalitis, septicemia, neonatal listeriosis; mainly in persons with underlying conditions (72).

Transmission: ingestion of contaminated food.

Associated food: primarily milk, meat, vegetables, and derived products;

Table 5 Possible Seafood-Borne Hazards and Their Origins[a]

Bacteria	*Aeromonas* spp.	A, B, C	*Clostridium botulinum* type E	A	*Clostridium perfringens*	C
	Escherichia coli	B, C	*Listeria monocytogenes*	B, C	*Plesiomonas shigelloides*	A, B, C
	Salmonella typhi	B, C	*Salmonella* spp.	B, C	*Shigella* spp.	C
	Staphylococcus aureus	C	*Vibrio cholerae*	B	*Vibrio parahaemolyticus*	A
	Vibrio vulnificus	A				
Viruses	Hepatitis A	B, C	Norwalk-like viruses and the small round viruses	B	Rotavirus	B, C
Parasites	*Anisakis simplex*	A	*Capillaria philippinensis*	A	*Clinostomum complanatum*	A
	Clonorchis sinensis	A	*Diphyllobothrium* spp.	A	*Dioctophyme renale*	A
	Echinochasmus perfoliatus	A	*Echinostoma* spp.	A	*Gnathostoma* spp.	A
	Heterophydae	A	*Isoparorchis hypselobagri*	A	*Metorchis conjunctus*	A
	Nanophyetes salmincola	A	*Opisthorchis* spp.	A	*Paragonimus* spp.	A, C
	Pseudoterranova decipiens	A				
Toxins	Amnesic shellfish poisoning (ASP): domoic acid	A	Ciguatera poisoning: ciguatoxins	A	Clupeotoxism	A
	Diarrhetic shellfish poisoning (DSP): okadaic acid and derivates, pectinotoxins, and yessotoxins	A	Gempylid poisoning	A	Ichthyoallyeinotoxism (hallucinogenic fish)	A
	Ichthyotoxism or fish roe poisoning	A	Neurologic shellfish poisoning (NSP): brevitoxins	A	Paralytic shellfish poisoning (PSP): saxitoxin, neosaxitoxin, and gonyautoxins	A
	Poisonous marine animals (e.g., whelks, turban shells, tropical reef crabs)	A	Scombroid fish poisoning: histamine and other biologically active amines	A, C	Tetrodon or puffer poisoning: tetrodotoxin	A
Residues	Antibiotics/Antiparasitics	B	Heavy metals	B	Polychlorinated biphenyls	B
	Radioactivity	B	Sulfites	C	Nitrosamines	C
Other hazards	Hard foreign objects	C	Seafood allergy	A		

[a]A, agent naturally present in the aquatic environment; B, pollution of the aquatic environment, including veterinarian therapy in aquaculture; C, contamination during processing and food handling.

seafood may be a vehicle in sporadic listeriosis cases. However, there are no reports on epidemic outbreaks due to consumption of fish products (72,113,117).

Resistance of hazard organism (40):

Growth temperature: minimum 0°C; maximum 44°C.

pH: minimum 4.5; maximum 8.0.

Water activity a_w: minimum 0.92.

Prevention:

Attitude of plant personnel: good hygienic practices during processing; regular cleaning and disinfecting of working installations, tools, clothes, hands, etc.; elimination of processing waste.

GMP.

Shorter keeping time; limitation of "best before" dates; rapid consumption.

Supplying the consumer with objective information on the consequences of temperature abuse and long storage periods.

c. *Salmonella* spp.

Hazard organism: Salmonella spp.

Distribution: worldwide.

Diseases: salmonellosis (gastroenteritis), typhoid fever (201).

Transmission: ingestion of contaminated food.

Associated food: all kinds of food.

Resistance of hazard organism (40:)

Growth temperature: minimum 6°C; maximum 46°C.

pH: minimum 3.8; maximum 9.0.

Water activity a_w: minimum 0.95.

Prevention:

GMP.

Thorough cooking of food.

Avoidance of cross-contamination in the household.

d. *Shigella* spp.

Hazard organism: Shigella spp.

Distribution: worldwide, epidemics very common in Central America, Southeast Asia, India, and Africa.

Disease: shigellosis, dysentery (140).

Transmission: fecal–oral: consumption of contaminated water and food; low infectious dose (3).

Associated food: all kinds of food; water (140).

Resistance of hazard organism (40):

Survival in food for 50 days at room temperature.

Survival for 10 days in acidic food.

Prevention:

Maintenance of good personal hygiene.
Sanitation of water supplies.
Good hygienic practices to avoid contamination.
Health education of food handlers.

e. Staphylococcus aureus

Hazard: enterotoxins formed by *Staphylococcus aureus.*

Distribution: worldwide.

Diseases: staphylococcal food poisoning: vomiting, nausea, sometimes diarrhea (139,149).

Transmission: ingestion of contaminated food; frequently via postprocessing contamination with subsequent improper storage at high temperatures (> 15°C).

Associated food: all kinds of food, contaminated by food handlers (139).

Resistance of hazard organism (40):

Growth temperature: minimum 7°C; maximum 48°C.
pH: minimum 4.3; maximum 9.0.
Water activity a_w: minimum 0.86.
Enterotoxin production: minimum temperature 14°C, maximum 45°C; a_w minimum 0.92, heat resistance high (139).

Prevention:

Keep susceptible food refrigerated at all stages except when being served and being prepared.
Quick cooling of the entire mass.
GMP.
Strict personal hygiene.

f. Vibrio cholerae

Hazard organism: Vibrio cholerae.

Distribution: sporadic cases worldwide, epidemics mainly in Asia, Africa, and South America.

Disease: cholera: profuse, watery diarrhea, without fever (134).

Transmission: fecal–oral; consumption of contaminated water and food (29, 134).

Associated food: water and raw seafood.

Resistance of hazard organism (40):

Growth temperature: maximum 43°C; minimum 10°C.
pH: minimum 6.0; maximum 11.0.
Water activity a_w: minimum 0.97.
Phosphates increase survival of *V. cholerae* at low temperatures (192).

Prevention:

Sanitation of water supplies (e.g., chlorination).
Maintenance of good personal hygiene.
Vaccination.
Restriction in fishing.
Good hygienic practices to avoid contamination.
Not eating raw or partially cooked seafood.
Thorough cooking and holding temperatures.
Prevention of postprocessing contamination.

g. Vibrio parahaemolyticus

Hazard organism: Vibrio parahaemolyticus. Most isolates obtained from seafood and marine environment are avirulent (184).

Distribution: estuarine and coastal waters, worldwide (25).

Diseases: mild diarrhea (48), sometimes septicemia in persons with underlying chronic diseases (98).

Transmission: ingestion of raw or recontaminated food.

Associated food: seafood, in particular crustaceans and mollusks (25).

Resistance of hazard organism (25,40):

Growth temperature: minimum 5°C; maximum 44°C, numbers decline when held at 10°C or below.

pH: minimum 4.8; maximum 9.0.

Water activity a_w: minimum 0.93.

Food additives: potassium sorbate, sodium benzoate, monolaurin, and monocaprin are effective inhibitors (48).

Prevention:

Harvesting restricted to approved sources; summer harvesting of shellfish restricted (4).
Maintenance of the refrigeration from harvesting to consumption.
Good hygienic practices to avoid contamination.
Not eating raw or partially cooked shellfish.
Thorough cooking and holding temperatures.
Prevention of postprocessing contamination.

h. Vibrio vulnificus

Hazard organism: Vibrio vulnificus.

Distribution: worldwide, in warmer coastal waters.

Diseases: wound infections, septicemia (154). Victims often have underlying chronic diseases, especially liver diseases (110).

Transmission: ingestion of contaminated or undercooked raw seafood, especially oysters (46); contamination of skin lesions with seawater and/or shellfish.

Associated food: raw seafood, especially oysters.
Resistance of hazard organism (40):
Growth temperature: minimum 8°C; maximum 43°C.
pH: minimum 5.0; maximum 10.0.
Water activity a_w: minimum 0.94.
Prevention:
Harvesting restricted to approved sources; summer harvesting of shellfish restricted (4)
Raw or partially cooked shellfish should not be eated especially by persons with underlying diseases.
Thorough cooking and holding temperatures.
Prevention of postprocessing contamination.
Education and information of people with underlying diseases regarding risk.

2. Foodborne Viruses

Most of the viruses that may cause foodborne illness originate from the human gastrointestinal tract. Transmission is mainly directly by a fecal–oral route, with food or water occasionally interposed as a vehicle (51). Viruses do not multiply in food but are generally highly infective. Since even very small numbers of viruses may cause disease in man, effective sanitary installations are necessary to avoid infection cycles. In developed countries, even although sanitary systems are in general well established, sewage is contaminated with pathogenic viruses. Untreated sewage may be discharged and into waters from which fish and shellfish are harvested. Even if sewage is treated, the reduction in contamination may be insufficient (95). In developing countries, money is often lacking to introduce good and functional sanitary installations. Sometimes even such basic hygiene tools as clean water and soap are missing.

a. Hepatitis A

Hazard organism: hepatitis A virus.
Distribution: worldwide.
Disease: hepatitis, usually with jaundice (55).
Transmission: fecal–oral: direct contact with infected persons or consumption of contaminated water and food.
Associated food: shellfish (53,70).
Resistance of hazard organism:
Quite stable in the environment (86): resistant in dried feces (141); acid-fast (165).

Heat sensitivity may not be totally eliminated by pasteurization (55); total inactivation in shellfish at 90°C during 1.5 minutes (145); in vitro inactivation at 80°C after 5 seconds (157).

Prevention:

Immunoprophylaxis (189).
Restriction in harvesting and marketing of shellfish.
Relaying or depuration of shellfish.
Thorough cooking of shellfish.
Not eating raw or partially cooked shellfish.
Public education.

b. Norwalk Virus and Other Small Round Viruses

Hazard organism: Norwalk virus and unclassified small round structured viruses (20).

Distribution: worldwide.

Disease: mild gastroenteritis.

Transmission: fecal–oral: direct contact with infected persons or consumption of contaminated water and food (66).

Associated food: shellfish (20,92,174), food contaminated by infected food handlers or chefs (116).

Resistance of hazard organism:

Acid-stable (65).
Heat sensitivity may not be totally eliminated by pasteurization (20).

Prevention:

Restriction in harvesting and marketing of shellfish.
Relaying or depuration of shellfish.
Thorough cooking of shellfish.
Not eating raw or partially cooked shellfish.
Public education.

3. Parasites

a. Diphyllobothriosis

Hazard organism: Diphyllobothrium spp. (82,144,147,191), tapeworms.

Distribution:

Diphyllobothrium latum: central and northern Europe, Siberia, Manchuria, Japan, and North America (144).

Diphyllobothrium pacificum: Peru, Chile, and Japan (56,191).

Disease: diphyllobothriosis of the small intestine. Although the infection usually produces few symptoms, a parasite-born vitamin B_{12} deficiency may cause intestinal obstruction or megaloblastic anemia. Furthermore, toxic substances excreted by the parasite may affect the nervous system.

Transmission: consumption of raw, undercooked, or inadequately prepared fish.

Associated food: many fish species; the most important are as follows (191):

Esox lucius, Lota lota, Perca fluviatilis, and *Acerina cernua* (Eurasia)

Esox lucius, Stizostedion vitreum, Stizostedion canadense griseum, Lota maculosa, Perca flavescens, Oncorhynchus (O.) spp. (North America)

O. masu, O. gorbuscha, O. keta, and *O. nerka* (Japan)

Resistance of hazard organism: freezing at −10°C for 1–3 days should kill the plerocercoids in fish (56). Parasites are destroyed when held 60°C for at least 1 minute (130).

Prevention:

Avoidance of consumption of raw or undercooked fish.

Freezing fish at −20°C for at least 24 hours.

b. Opisthorchiosis

Hazard organisms and distribution (82):

Opisthorchis tenuicollis: southern, central, and eastern Europe, Turkey, southern part of former Soviet Union, Vietnam, India, Japan, Puerto Rico, and possibly other Caribbean Islands (191).

Opisthorchis viverrini: China, Laos, Vietnam, Thailand (56,108).

Clonorchis sinensis: Korea, China, Japan, Taiwan, Vietnam (191).

Metorchis conjunctus: North America.

Diseases: opisthorchiosis; mainly cholangitis or hepatitis. When a high number of parasites is present, the bile duct epithelium shows papillomatous hyperplasia, which finally can turn into a certain form of cancer, cholangiocarcinoma.

Transmission: consumption of raw or inadequately processed freshwater fish.

Associated food: several species of freshwater fish (56,191,195).

Resistance of hazard organism: thorough cooking or freezing will kill the parasites.

Prevention:

Thorough cooking of fish.

Not eating raw or partially cooked shellfish.

Public education.

c. Paragonimiasis (Lung Flukes)

Hazard organism: Paragonimus spp., primarily *Paragonimus westermani* (82).

Distribution: Africa, southern and eastern Asia, western Pacific, South and North America (56,108).

Disease: pulmonary disease is the major problem.

Transmission: man becomes infected by eating raw or undercooked freshwater crabs, crayfish, or their juices (56). Cross-contamination during processing is possible.

Associated food: freshwater crabs of the genera *Erlocheir, Potamon,* and *Sundathelphusa,* and crayfish of the genus *Cambaroides* in Africa and Latin America (56).

Resistance of hazard organism: cooking of crustaceans at 55°C for 5 minutes as well as thorough freezing will kill Metacercariae (56).

Prevention:

Not eating raw or undercooked crabs or crayfish.

Not eating freshly salted or pickled crabs or crayfish.

Care in preparation of crustaceans.

Not marinating food in crayfish juices.

d. Anisakiasis

Hazard organism: anisakine nematodes (147,191): *Anisakis simplex, Pseudoterranova decipiens* (also known as *Terranova decipiens, Porrocaecum decipiens, Phocaema decipiens*) (88). Larvae are frequently found coiled in capsules on the surface of viscera; the majority occur in the body cavity of fish (147). After death of the host, nematodes tend to escape from their capsules and penetrate into the skeletal muscle, especially into the hypaxial parts surrounding the body cavity. This observation was repeatedly made in fatty species like herring and mackerel, but not in nonfatty species, such as Gadidae (169).

Distribution: worldwide, with prevalence varying in different parts of the world (191).

Diseases:

The larvae, released in the stomach by digestion of the fish tissue, penetrate the gastric or intestinal mucosa, which leads to hemorrhages, abscesses, and necrosis. Worms may also enter the peritoneal cavity, where they may migrate to other organs. First clinical symptoms of a noninvasive anisakiasis (epigastric pain, nausea, vomiting) may occur within 4–6 hours after ingestion (147). Symptoms may decrease within a few days, but abdominal pain with intermittent nausea and vomiting can persist for weeks. The symptoms of an invasive anisakiasis start within 7 days after ingestion. Severe abdominal pain, diarrhea, nausea, and possibly fever can be observed (136,191); see also overview (56).

Anisakis simplex can also induce anaphylactic reactions (23). The allergen is not destroyed by cooking or freezing.

Transmission: Man becomes infected by eating fish containing live larvae of *Anisakis simplex* or *Pseudoterranova decipiens.*

Associated food: large variety of fish species (e.g., Gadidae or Clupeidae).

Resistance of hazard organism:

Freezing to −30°C kills larvae within 5 minutes (94). Freezing to −20°C normally kills the larvae within 24 hours (56) but they may survive this temperature for 4–5 days (Nygard, 1967, cited in Ref. 147). After 16 hours, freezing to −30°C and subsequent storage at −12°C for one week, all *Anisakis* were dead in commercial 100 lb blocks (147).

Cooking for 5 minutes at 60°C or a shorter time at higher temperatures kills the larvae (58).

Dry salt kills the infectious agent within 10 minutes. In a solution of 50 g NaCl/L, *Anisakis* can survive for several weeks (146).

4% of *Anisakis* survive 60 days in herring kept in a solution of 4% acetic acid and 6% salt (Houwing, 1969, cited in Ref. 147).

Prevention:

Cooking fish and squid at 60°C or above.

Freezing fish at −20°C for 7 days.

Eviscerating fish immediately and properly after catch.

Cutting off hypaxial parts of the fillet of certain fish species known to be highly infected (e.g., Alaska pollock or blue whiting) may diminish the risk but is no guarantee that the fish is free from *Anisakis.*

4. Toxins

The toxins described in the subsections that follow have a very high resistance. They are heat-stable (97), acid-fast (3) substances, unaltered by any processing (42,100,130).

a. Amnesic Shellfish Poisoning

Hazard: domoic acid produced by toxic diatoms (*Pseudonitzschia* or *Nitzschia*) (138,158); dangerous dose 2 mg/100 g of meat (3).

Distribution: Pseudonitzschia or *Nitzschia* is found in many coastal regions, although "toxigenic strains" of *P. multiseries* were only found in Canada (138) and the United States (193).

Disease: amnesic shellfish poisoning (ASP): amnesia (47,181).

Transmission: ingestion of toxic shellfish.

Associated food: shellfish, mainly mussels.

Prevention:

Restriction in harvesting and marketing of shellfish.

Relaying or depuration of shellfish.

Screening for domoic acid (112).
Public education.

b. *Ciguatera*

Hazard: ciguatoxin, maitotoxin, and derivates (129,199) produced by toxic dinoflagellates (*Gambierdiscus toxicus, Prorocentrum concavum*).

Distribution: tropics or temperate zone between 35°N and 34°S latitude.

Disease: ciguatera fish poisoning: 2–6 (for 12) hours after ingestion, victims experience neurological symptoms, diarrhea, vomiting, and cardiac troubles. Reversal of hot and cold sensation. Mortality, due to cardiovascular shock (42,199), is low.

Transmission: ingestion of toxin by eating fish.

Associated food: ciguatoxic fish, mainly tropical reef fish (100).

Prevention:

In endemic regions, caution about eating large predacious reef fish (e.g., barracuda, snapper) (100).
Not eating viscera of tropical reef fish.
Eating only small amounts of fish in endemic regions.
Avoidance of large carnivorous fish in endemic regions (toxin concentrations increase higher along the food chain).
Screening of ciguatoxic fish by means of a stick test (111).
Public education, especially for tourists and sports fishers (e.g., by means of warning signs or advisories on hazards of particular species) (5).

c. *Diarrhetic Shellfish Poisoning*

Hazard: okadaic acid and derivatives, pectinotoxins, and yessotoxins produced by toxic dinoflagellates (*Dinophysis* spp. and *Prorocentrum* spp.) (198).

Distribution: Japan, Europe, and Canada.

Disease: diarrhetic shellfish poisoning (DSP): mild diarrhea after 0.5–4 hours after ingestion, lasting for 3 days.

Transmission: ingestion of toxic shellfish (83).

Associated food: shellfish, mainly mussels, oysters, and scallops.

Prevention:

Restriction in harvesting and marketing of shellfish.
Relaying or depuration of shellfish.
Screening methods (43,54,135).
Public education.

d. *Neurotoxic Shellfish Poisoning*

Hazard: brevitoxins produced by a toxic dinoflagellate (*Gymnodinium breve,* also known as *Ptychodiscus brevis*) (97).

Distribution: toxigenic algae are found only in the Gulf of Mexico, but *Gymnodinium breve* exists in Japan, Spain, and even in the eastern Mediterranean Sea (179).

Disease: neurotoxic shellfish poisoning (NSP): paraesthesia, inversion of the hot–cold sensation, nausea. Mild course, mortality not reported (97).

Transmission: ingestion of toxic shellfish.

Associated food: shellfish, mainly clams.

Prevention:

Restriction in harvesting and marketing of shellfish.
Relaying or depuration of shellfish.
Public education.

e. Paralytic Shellfish Poisoning

Hazard: saxitoxin, neosaxitoxin, and gonyautoxins produced by toxic dinoflagellates (*Alexandrinium* spp. and *Gymnodinium* spp.) (194); 1 mg provokes sickness, 2 mg is lethal (3).

Distribution: temperate zone in latitudes greater than 30°.

Disease: paralytic shellfish poisoning (PSP): paralytic symptoms, sometimes severe, appear 30 minutes after ingestion (17).

Transmission: ingestion of toxic shellfish (83,87).

Associated food: shellfish (mainly mussels, clams, cockles, and scallops) and lobster (185).

Prevention:

Restriction in harvesting and marketing of shellfish.
Relaying or depuration of shellfish.
Regular monitoring in affected zones (8).
Public education.
Control of end products: regulatory limits for PSP toxins in food (3).

f. Scombroid Poisoning

Hazard: histamine, formed by bacterial decarboxylation of free histidine (100,132). The minimum toxic dose varies with individual (3).

Distribution: worldwide.

Disease: scombroid poisoning: symptoms resembling those associated with allergic reactions (100,180).

Transmission: ingestion of histamine-containing fish.

Associated food: fish of the family Scombridae (tuna, mackerel, bonito, etc.) and nonscombroid fish with a high amount of free histidine in flesh (sardines, anchovies, herring, etc.) (81,180); processed scombroid fish (e.g., smoked mackerel, preserved anchovies, canned tuna) (100,178,187). More recent outbreaks of scombroid poisoning have resulted from the consumption of canned products.

Prevention:

Immediate refrigeration of fish (e.g., icing).

Maintenance of the refrigeration from harvesting to consumption (temperature < 5°C).

Good hygienic practices to avoid contamination.

GMP.

Avoidance of long storage times at home.

Control of end products: regulatory limits for histamine in food (178).

g. Tetrodon or Pufferfish Poisoning

Hazard: tetrodotoxin, produced by fish of the order Tetraodontiformes (e.g., globefish, porcupinefish, or mola) (100,197); 1 mg provokes sickness, 2 mg is lethal (3).

Distribution: worldwide.

Diseases: tetrodon (Puffer) poisoning: paraesthesia, respiratory distress, extense muscular paralysis, fatality rate more than 50% (100).

Transmission: ingestion of flesh, viscera, or skin of toxic tetraodontiform fish.

Associated food: toxic fish, "fugu" (100,117), rarely mollusks (197).

Prevention:

Not eating tetraodontiform fish.

Artificial cultivation of "toxin-free" fish.

Allow only specially trained cooks to prepare these fish.

5. Residues

Chemical hazards are generally not very important. But they may be underestimated with respect to their mainly chronic toxicity and their negative effects on reproductivity.

The spread industrialization has led to increasing production of chemical waste, and every substance produced by man (including metabolites) sooner or later will find its way to the environment, including to some extent the rivers and seas (125). Lakes, rivers, or coastal embayments are especially affected. Concern is primarily focused on these waters rather than on the open ocean. A good example is the Northern Sea, an excellent fishing ground, which is suffering from direct and indirect (through contaminated rivers) disposal of waste (124,125). Therefore regulatory limits were imposed for many pollutants, in particular for polychlorinated biphenyls (34,124) and heavy metals (125,163,166).

Fish can easily accumulate toxic chemical residues, sometimes to levels that may affect the health of humans who eat such products. As a result of this bioconcentration effect, older (bigger) fish are more frequently contaminated and have higher levels than younger animals. Carnivorous fish like sharks or swordfish, being at the top of the food chain, also accumulate environmental pollutants. Therefore, as a preventive measure, especially carnivorous and old fish, must be monitored regularly.

a. Polychlorinated Biphenyls

Hazard: polychlorinated biphenyls (PCBs).

Distribution: worldwide. Fish may be useful bioindicators of the presence of PCBs in aquatic ecosystems. These substances have an extremely long half-life in the environment and are also very slowly degraded by most organisms. The environmental load may be very high in some rivers.

Disease: PCBs are carcinogenic (34); in addition, they have a negative impact on fertility, pregnancy, and the development of babies (79).

Transmission: ingestion of contaminated food.

Associated food: fish, especially large predator fish ("bioconcentration effect") (124) and shellfish.

Resistance of hazard: very stable; extremely long half-life in the environment.

Prevention:

Recognition of contaminated regions and waters.

Constant monitoring in contaminated regions.

Not eating large fish (3).

b. Toxic Heavy Metals

Hazard: mercury, lead, cadmium, arsenic, selenium, and copper (166). Cadmium intake due to seafood consumption is only 6% of the overall dietary exposure (3). For lead, the corresponding percentage is 12.5%, whereas 50% of all mercury consumption has its origins in seafood. Acceptable daily intake as follows (3):

Arsenic and selenium: uncertain

Cadmium: 51–72 mg

Lead: 429 mg

Mercury: 0.23 mg

Distribution: worldwide, emitted by industrial activities, especially from burning wastes (mercury).

Diseases: acute intoxications and chronic, long-term effects, as follows:

Mercury: neurotoxic in its methylated form: paraesthesia, visual and hearing problems, incoordination, muscle weakness, coma, and death (102). Most countries have introduced legal limits in an attempt to control the problem, usually 1 mg/kg.

Lead: neurotoxic; anemia, proteinuria (3).

Cadmium: osteoporosis (122); ill effects on kidney.

Arsenic: carcinogenic, teratogenic; gastroenteritis, nephritis, neuropathies (41).
Selenium: edema, hepatitis, infertility; teratogenic (103).
Transmission: ingestion of contaminated food.
Associated food: aquatic animals, especially large predator fish ("bioconcentration effect").
Resistance of hazard: very stable.
Prevention:
Recognition of contaminated regions and waters.
Constant monitoring in contaminated regions.
Not eating large fish (3).

c. *Veterinary Drugs*
Hazard: antibiotics (e.g., chloramphenicol) and antiparasitics (131).
Distribution: aquaculture production.
Diseases:
Development of resistance, allergic reactions, or aplastic anemia due to chloramphenicol.
A problem with abundant or uncontrolled administration of antibiotics is the antibiotic resistance that may develop (60). Although resistance declines after treatment, pathogens naturally present in the aquatic environment (e.g., *Plesiomonas shigelloides, Aeromonas hydrophila,* or *V. vulnificus*) may become resistant to some antibiotics, which renders eventual medical treatment more difficult.
Associated food: fish from aquacultures.
Prevention:
Good veterinary practice.
Sound aquaculture systems: no overcrowding.
Regular monitoring.

d. *Food Additives and Processing Aids.* Nitrosamines, hazardous due to their carcinogenic effects, are an undesirable consequence of nitrite use. They are formed as a chemical reaction between secondary amines and nitrites. However, the formation is minimized commercially by following proper cooking schedules and by the use of approved reducing agents in the cure solution. In the United States, nitrite has been authorized in the processing of smoked fish to avoid botulism (3). In other countries (e.g., Switzerland), the use of nitrite for the production of smoked fish is technologically not necessary, hence is not allowed.

Sulfites are used to prevent blackspot in shrimp (melanosis in crustaceans). These preservatives are safe to the general public when applied in concentrations accepted in international trade, but highly allergenic to susceptible individuals (e.g., steroid-dependent asthmatics), even at low concentration. Alternatives would be other less toxic substances, such as 4-hexylresorcinol, although these were less effective (143). In practice, sulfite often is used in abundance (21).

Processors must recognize that shellfish, other seafoods, certain ingredients, additives, and food colors can cause severe allergic reactions in some people. Since susceptible individuals are usually aware of their sensitivity, proper labeling is essential. Even slight contamination of a product with an undisclosed allergen is potentially life-threatening.

6. Other Hazards

a. Hard Foreign Objects

Hazard: hard foreign objects (90).

Distribution: worldwide.

Diseases: injuries of teeth and the digestive or respiratory tract, fatal obstruction of the airway.

Associated food: every type of processed food (e.g., fish scales in canned tuna) (90).

Prevention:

GMP.

Detection systems: x-ray sorter, metal detectors.

b. Seafood Allergy. Seafood is known to be one of the most frequent causes of adverse food reactions in hypersensitive individuals (44,57,109,128). If we exclude the above-described toxicological effects, these adverse reactions are based mainly on food allergies. The exact prevalence of reactivity is not known, but Daul et al. (57) estimate that up to 250,000 individuals in North America may be at risk, for an approximate rate of 0.5–1 per 1000. Because of anaphylactic reactions, symptoms occur rapidly (within 2 h of exposure) and include urticaria, asthma, gastrointestinal disorders, and, in very severe cases, shock (57). The seafoods most frequently causing symptoms are shrimp, squid, lobster, and codfish (44,57,109). Persons allergic to fish may be clinically sensitive to more than one fish species (109). Avoidance is the only measure to control food allergies. Seafood is generally easy to avoid, but the allergens may also be hidden in prepared dishes, sauces, or soups.

B. Foodborne Diseases Due to Seafood: Epidemiological Aspects

Information on causative agents of foodborne diseases has increased considerably in the last decades. Nevertheless, direct comparisons from country to country must be judged very cautiously. Notifications of foodborne diseases and outbreaks to a central health agency, epidemiological investigations on foodborne outbreaks, as well as sentinel and special epidemiological studies, are very important tools to determine a realistic level of morbidity of a foodborne disease (183). Unfortunately, such advisories are far from being complete. Thus it is difficult to postulate an objective worldwide epidemiological status of foodborne diseases due to seafood. At present, it is estimated that in developed countries about one in 10 persons suffers a foodborne illness each year (67). Fish and shellfish represent important vehicles.

1. North America

In 1980 Bryan (35) reported that seafood was implicated in approximately 11% of the outbreaks of U.S. foodborne diseases during 1970–1978. Todd et al. (184) found the same percentage in a later study.

In a survey covering the years 1977 to 1984, Bryan (36) stated that seafood products were vehicles for 24.8% of all reported outbreaks; 59% of these were due to fish (scombroid poisoning, ciguatera poisoning, and botulism), 33% to shellfish (diarrhea probably due to Norwalk agent, hepatitis A, *V. parahaemolyticus*), 6% to crustaceans (*V. parahaemolyticus*), and 2% to sea mammals (botulism).

Todd (182) reported that from 1983 to 1987, 1.9% (U.S.) and 2.8% (Canada) of all outbreaks were due to shellfish consumption, compared with 4.8 and 4.4% for fish, respectively. Interestingly, Canada reports relatively few viral outbreaks. A possible explanation is that shellfish are aquacultured in regions where there is relatively little sewage pollution. Relatively few outbreaks are due to *Vibrio* spp., but there seems to be always a direct link with consumption of raw seafood (62).

Parasitic infections are concentrated in certain ethnic groups that consume raw or undercooked fish (5).

2. Europe

According to the WHO Surveillance Programme for Control of Foodborne Infections and Intoxications in Europe (14), seafood was the vehicle in 397 outbreaks or 4.7% of all reported outbreaks between 1990 and 1992. The main hazards were scombroid poison and *Salmonella,* as well as Norwalk agent, hepatitis A virus, *S. aureus, C. botulinum, C. perfringens, V. parahaemolyticus,* DSP, and ciguatoxin. Many outbreaks, however, were of unknown etiology.

Surveillance of foodborne diseases in the 1980s and 1990s in the Netherlands showed that fish and shellfish were vehicles in 4% of all reported outbreaks and 10% of the reported single cases (153). Similar data were found in France (137).

3. Asia and Africa

Seafood-borne diseases represent 70% of foodborne outbreaks in Japan (184). The most important foodborne pathogen is *V. parahaemolyticus,* but fish poisoning

(PSP, DSP, pufferfish poisoning) is also a major concern (183). In China and Vietnam outbreaks of hepatitis A (99) and *V. vulnificus* (183) have been described; *V. parahaemolyticus, V. cholerae,* and *S. aureus* are also important. In the Middle East as well as in Africa, very few specific data are available linking seafood to foodborne illness (183), but *V. cholerae* is most important.

4. Australia, New Zealand, and Oceania

Seafood-borne diseases represent 20% of foodborne outbreaks in Australia (184). Norwalk-like viruses have been reported in Australia for many years, along with the occasional *Vibrio parahaemolyticus* and other vibrio infections. In Oceania fish poisoning is very important. Although data were not specified, ciguatera, scombroid poison, and PSP seem to be the most frequent cause of illness (183).

5. Caribbean Region; Central and South America

Fish toxins are very important in the southern half of the western hemisphere. Ciguatera is a major concern in the Caribbean, but PSP and scombroid poisoning also occur frequently (183). In 1991 *V. cholerae* O1 was reintroduced to Peru in the early 1990s and led to outbreaks (80,176).

C. Factors Contributing to the Occurrence of Foodborne Illness

The same contributory factors lead to foodborne illness everywhere (37). Relatively few factors are responsible for the majority of the problems. This instantiation of the Pareto principle (172) can lead to disease control with relatively few measures. It is therefore quite astonishing that morbidity from foodborne illness has not diminished over the last years, especially in developed countries. Some of the past efforts to cause changes in the behavior of persons who harvest, process, handle, serve, supervise, or—last but not least—eat food, have apparently failed, because outbreaks still occur (37). The continued occurrence of foodborne illness is due in part, however, to relatively recent genetic adaptations of pathogens and to global trade. The increased incidence of illness associated with psychrotrophic pathogens, as well as highly virulent bacteria capable of infecting at low doses, has confounded traditional control strategies. A development that is expected to reduce seafood illnesses worldwide in the future is the recent adoption of programs based on HACCP (hazard analysis–critical control point) systems for hazard identification and control. The most important contributory factors to outbreaks can be seen in Table 6.

Table 6 Contributory Factors to Foodborne Outbreaks

Contributory factor	*C. botulinum*	*C. perfringens*	*L. monocytogenes*	*Salmonella*	*Shigella*	*S. aureus*	*Vibrio* spp.	Viruses	Parasites	Toxins	Scombroid poisoning
Improper cooling		✓		✓	✓	✓	✓				✓
Lapse of 12 or more hours between preparing and eating	✓	✓	✓	✓		✓	✓				
Colonized person handled implicated food				✓	✓	✓		✓			
Incorporating contaminated raw food				✓			✓	✓	✓	✓	
Inadequate cooking, heat processing and reheating/Improper hot holding	✓	✓	✓	✓	✓	✓	✓	✓	✓		
Obtaining food from unsafe sources							✓	✓	✓	✓	
Cross-contamination		✓	✓	✓		✓	✓				
Improper fermentation				✓							
Improper cleaning of equipment		✓		✓		✓	✓				

Source: Ref. 36.

Intoxications due to natural toxins are mostly species-associated and are often a regional problem. People at high risk are consumers of raw molluscan shellfish, sports anglers who eat their catch, inhabitants of tropical islands, and consumers of fresh/frozen mahi-mahi, tuna, and bluefish (5).

D. Risk Analysis

Hazard analysis is the link between epidemiology and microbiology (39). It provides information to initiate control actions and to implement preventive measures. Moreover, the reduction to a minimum of biological and chemical hazards in food is technically feasible (15). Risk assessment techniques must be applied to determine the significance of hazards and to evaluate risk management strategies like HACCP (39,114). The utilization of risk assessment techniques to control adverse health effects will be an essential component in international food trade of the future. At present, there are only qualitative or semiquantitative approaches to risk assessment for biological agents. According to Ahmed (3), epidemiological data are not sufficient to perform quantitative risk assessment of risks due to the consumption of fish and shellfish. Nightingale (152) estimated that chicken was 200 times more likely to cause illness than finfish. Raw or undercooked shellfish, however, was 100 times more likely to cause illness than chicken.

The basic principles and applications can be found elsewhere (3,15) but are briefly outlined below.

1. Risk Assessment

a. Hazard Identification. Hazards in seafood can be classified into three groups:

Cause illness in healthy adults: *C. botulinum, C. perfringens, Salmonella, Shigella, S. aureus, V. cholerae, V. parahaemolyticus,* hepatitis A virus, Norwalk-like viruses, *Diphyllobothrium, Anisakis,* toxins, heavy metals, PCBs.

Do not cause illness in healthy adults, but are dangerous to susceptible people (immunocompromised individuals, children, elderly people, pregnant women, etc.): *L. monocytogenes, V. vulnificus.*

Uncertain pathogenicity: *Aeromonas hydrophila, Plesiomonas shigelloides.*

b. Dose–Response Assessment. Viral agents and some bacterial agents like *Shigella* have a very low minimal infectious dose. Most of the bacteria require a very high dose which, however, often varies considerably owing to the highly variable host susceptibility and virulence of the pathogen and to "intrinsic factors" from the food (e.g., antagonistic or protective effects).

For chemical hazards, like heavy metals, differentiation between acute, short-term, and chronic long-term exposures must be taken into consideration.

c. Exposure Assessment. The risk of exposure depends on many variables including type of food, geographical location, harvesting conditions, and processing hygiene. This makes such risks very difficult to generalize.

d. Risk Characterization. Epidemiological data indicate that raw and undercooked shellfish carry a higher risk than other seafood. Viral diseases occur more frequently than bacterial illnesses, but risks from parasitic diseases have lesser public health significance in seafood.

2. Risk Management

Control of seafood-borne diseases is achieved by excluding the agents from food, controlling eventual growth, and/or destroying the agents (3).

The possible control measures and detailed risk management procedures for fish and shellfish are outlined in Secs. IV–VII.

IV. SAFE HANDLING AT HARVESTING

An important source of hazards associated with the safety of fish and shellfish lies in the environment, and controls should be instituted at harvesting or at capture. Hazards associated with shellfish-growing waters (bacteria, viruses, marine biotoxins: see Sec. III.A) are good examples.

A. Preharvest Food Safety

Preharvest food safety is an increasingly important and essential step toward the safe production of food. The global trade of food makes a worldwide and interdisciplinary perspective essential. The basic change of control systems from finished product to production implies a new paradigm and will require new approaches in teaching and communication. Producers, including aquaculture farmers, will have to collect great amounts of data on factors influencing farm practices. These factors not only are directly linked to the classical food safety approach but include environmental, sociological, and economical issues, as well as animal welfare.

The welfare of fish and to a lesser extent shellfish has become an important issue in the public mind in developed countries. The concern is that animals poorly reared or harvested because of profit-oriented husbandry practices (e.g., overcrowding, inadequate medication) are not only politically unacceptable but lead to a decrease in product quality as a result of, say, poor health status or stress. Therefore, there are good reasons for maintaining sound and well-controlled aquaculture systems (9). Mass production animal husbandry practices require a strict medical regime to ensure the optimal health and welfare of the animals (131). This is also true for aquaculture procedures in which multiple fish species (e.g., salmon, catfish) are fattened. As well, these animals are susceptible to infections and must be carefully managed to produce high quality products at a reasonable price; such management includes veterinary treatment. When drugs are used for prophylactic and therapeutic means, there can always be residues in the animal products (see Sec. III.A.5.c). Drug residues that exceed a defined level are a result of misuse, illegal use, or negligence with respect to the withdrawal times (although metabolism of drugs in fish, shellfish, and crustaceans as poikilothermic animals is largely influenced by water temperature). To guarantee safe food, safe acceptable levels must be established on the basis of toxicological studies. Furthermore an increasing number of consumers, mainly in Western Europe, are ceasing to regard drug use (especially the prophylactic use) as acceptable.

The consumption of seafood, especially shrimp, is on a constant increase in the United States, as well as in Europe, and aquaculture has become increasingly important (156), but there are consequences that may lead to increased hazards (183). These include the use of wider new harvest areas in possible polluted water or water with risk of harmful algal blooms; distribution of products, with increased sales volume; the use of human and animal feces as nutrients; and the use of antibiotics to prevent diseases in fish and shellfish.

Frequently fish farms are located close to shellfish beds, with the result that antibiotic-treated effluents from fish farms may contaminate shellfish (32).

B. Harvesting of Fish

Fish harvesting includes selection of fishing grounds, the catching itself, as well as gutting, chilling, and packaging onboard vessels, either at sea or upon return to port.

1. Catching

Harvesting methods are mostly based on traditional practice, stock management regulations, and catch efficiency. Seafood safety is seldom a primary consideration. Several aspects of harvesting procedures may increase the animals' stress, muscle exhaustion, enzymatic degradation, or physical damage. However, a few safety issues also should be addressed. Catching gear that may kill fish long before they are brought onboard for chilled stowage can set the stage for histamine development and subsequent scombroid poisoning. Gill-netted mackerel and long-lined tuna, mahi-mahi, and marlin are examples. Gill nets and long lines must be worked at regular intervals to minimize the potential hazard. Conversely, procedures that harvest and chill susceptible species quickly offer effective control strategies. One example of this is the pumping of sardines into refrigerated brine tanks.

2. Gutting

Gutting must be performed in accordance with good hygienic practices to limit spread of contamination. The time between catching and evisceration should be as short as possible. Bacteria and parasites from the guts can contaminate the abdominal cavity and the meat surface. However, improper gutting is far more likely to result in poor quality fish than in unsafe seafood. Gutted fish should be protected from contamination with digestive juices. Knives used for gutting should be cleaned and disinfected before use; aprons should be worn and cutting boards used.

3. Chilling

Chilling of fish is a critical control point in an HACCP system (114). It is the only way to prevent histamine formation to toxic levels (see Sec. III.A.4.f). According to the size of the fish, there must be immediate icing (baitboats), storage in refrigerated seawater at approximately −1°C, or direct freezing. The aim is to reduce the temperature to 3°C or below as fast as possible (114).

Chilling of raw molluscan shellfish will also prevent the growth of such bacterial pathogens as *V. parahaemolyticus* (see Sec. III.A.1).

4. Packaging

Onboard packaging of fish, where practiced, usually amounts to boxing in ice. Finished, packaged product is likely to consist of fiberboard cartons of plate-frozen fillets, processed on large factory boats. This represents a small sector of the industry and does not present a significant food safety hazard. Vacuum and modified-atmosphere packaging for fresh distribution provide conditions that may favor anaerobic pathogens but are rarely, if ever, used on fishing vessels.

C. Harvesting of Shellfish

Shellfish of commercial importance include oysters, mussels, cockles, clams, and scallops. Normally they are harvested in coastal and estuarine waters and are therefore likely to be exposed to fecal contamination. Oysters and mussels are often grown in aquaculture, and are transported live in the shell, with or without refrigeration.

Molluscan shellfish are filter feeders. They pump water over their gills and trap and remove plankton and other small food particles. In the process, they periodically accumulate biotoxins associated with toxic algal blooms or bacterial and viral pathogens present in the water column. Most nations with a coastline have programs for the monitoring of shellfish-growing areas for such hazards. The effectiveness of these programs varies by region and by hazard.

According to Desenclos (61), key issues for the improvement of shellfish-borne disease prevention include a better knowledge of marine biology, the limitation of coastal water pollution, improved surveillance, the development of more sensitive indicators, and the responsiveness of the industry. Halliday et al. (99) proposed the following control measures:

Regulations regarding residential effluent drainage into catching areas
Supervision of drainage from fishing boats
Constant testing of water and shellfish from the catching area for coliform and other bacteria (virus testing is time-consuming and expensive)
Closing of contaminated catching areas
Removal of contaminated shellfish from the market
Depuration, (but epidemics still may occur, since pathogens survive for a long time in contaminated shellfish)
Steaming of clams until they are open

1. Fecal Contamination of Water and Shellfish

Most shellfish are grown in coastal regions, often in lagoons and close to estuaries, where conditions are better for survival and growth of the animals, as well as for efficient harvesting. Thus there is exposure to pollution originating from rivers, estuaries, etc. The degree of contamination depends on several factors:

Quality of the water of rivers and effluents
Distance between polluting entrance and place where shellfish is grown
Dilution, dispersion, and sedimentation of microorganisms
Resistance of microorganisms in the aquatic environment, depending on quantity of organic material, light intensity, and sedimentation

Outbreaks of disease may be caused by fecal contamination of shellfish. Education of oyster harvesters and enforcement of regulations regarding waste disposals by oyster harvesting boats may prevent such outbreaks (123).

Oysters and mussels may become contaminated by sewage dumped overboard by recreational and commercial boaters (16,80). To prevent this fecal contamination, boaters must not be permitted to dump sewage overboard, or beds used for harvesting must be limited to those in pollution-free waters. Shellfish may be contaminated not only with local sewage but also by wastewater pumped from ships in harbor.

In the United States, a federal and state cooperative program addresses concerns relating to shellfish-growing waters: the National Shellfish Sanitation Program (NSSP). This program is coordinated jointly by industry and government. Regulatory agencies (shellfish control authorities) for participating states, including all shellfish shipping states, meet each year to review and update strategies to improve shellfish safety. In most growing regions, coastal waters are sampled and the shoreline periodically surveyed to identify and correct potential sources of pollution. Waters that may at times be unsafe for harvesting are closed to commercial fishing. Harvesters tag shellfish to identify harvest location, and records are maintained throughout distribution. This control strategy has been recognized for many years but was not uniformly practiced until recently. The tagging requirements of the NSSP were formalized as required documentation for HACCP compliance in regulations implemented by Food and Drug Administration (FDA) in December 1997.

2. Agents Naturally Present in the Aquatic Environment

Among the vibrios, *V. parahaemolyticus* and *V. vulnificus* are pathogens of significant concern in shellfish. These microorganisms are naturally present in the aquatic environment and are not effectively managed by controlling sewage sources. Although exceptionally fast growing at elevated temperatures (short generation times), these bacteria fortunately are sensitive to cool temperatures. They are most effectively controlled by limiting exposure to summer temperatures and by timely refrigeration.

3. Traceability of Shellfish

To guarantee the effectiveness of back-tracing contaminated lots, all shell stock should be identified by tags with date of harvesting, place of harvesting, type and quantity of the merchandise, and company or vessel by which they were harvested (66). Forward tracing, however, is also very important, with respect to the rapid prevention of future cases. To have all the needed information quickly available, participants in international shellfish trade must establish a total quality management program featuring the following:

Responsibility of the producers and processors of shellfish (self-control).
Permanent records kept of each lot sold.
Traceability guaranteed (information for the consumer as well as national and international health authorities). In Europe at least the dispatch center must be indicated, with the trend going toward labeling of the waters of origin. In North America, this is considered to be excessive regulation, if shellfish safety programs are properly implemented.
Thorough recall of contaminated shellfish.

4. Restriction in Harvesting and Marketing

The Council of the European Communities has listed some basic principles (10). Production areas are classified into three zones from which bivalve mollusks can be collected (a) for direct consumption, (b) only after purification, and (c) for consumption only after relaying over a long period. Any change in demarcation or temporary or definitive closure must be announced to the competent authority. A control system including periodic monitoring is required.

In the United States, requirements for holding or depurating shellfish are referenced in the Guide for the Control of Molluscan Shellfish (19). Specific provisions are provided for moving harvested molluscan shellfish to other sites or holding facilities. The U.S. model ordinance recognizes that shellfish may be held live temporarily in floats or tanks for the purposes of storage, reducing grit and sand, or increasing salt content to improve flavor. This practice is defined as wet storage. Reducing pathogen loads is not a valid rationale for wet storage. The shellfish must come from approved growing waters or from a certified depuration facility.

Depuration is specifically intended to reduce the number of pathogenic organisms that may be present in shellfish harvested from moderately polluted (restricted) waters to levels that make the shellfish acceptable for consumption without further processing. Heavily polluted shellfish cannot be legally depurated. Depuration facilities require an extensive review of design and procedures before being certified to operate. This is unusual for the U.S. FDA, which normally does not perform preapprovals for the seafood industry.

5. Depuration and Relaying of Shellfish

Purification can be performed either in a natural clean environment or in a controlled environment: after washing, bivalves are put in basins with constant renewal of seawater, and are kept either fresh, or in disinfected closed circles.

Relaying areas must be approved, and boundaries clearly identified by buoys or other fixed means. According to the prescriptions of the European Union (10), relaying must last until the bacteriological standards have been met (< 300 fecal coliform organisms, or < 230 *E. coli* per 100 g mollusk flesh; no *Salmonella* detectable in 25 g).

To check the microbiological, chemical, and toxicological quality of relaying and purification areas, monitoring must be performed periodically.

6. Control

Outbreaks of viral gastroenteritis are very frequent and underestimated. There is an urgent need for improved indicators of viral contamination, inasmuch as coliform count and *E. coli* counts that meet national or international standards often are insufficient to indicate the presence of viruses (45,61). Bosch et al. (27), who detected rotaviruses and hepatitis A virus in 56 and 40%, respectively, of mussel samples meeting the EU criteria, concluded that all shellfish should be purified before consumption.

Because monitoring is done in the absence of reliable indicators, it is very difficult to predict whether an oyster bed or a shellfish aquaculture facility can deliver safe food. Perhaps techniques such as bacteriophage assay or direct detection by means of the polymerase chain reaction might serve as reliable indicator tests for viruses (22,49,91,127).

D. Harvesting of Crustaceans

Important commercialized crustacean species include crabs, lobsters, shrimp, and prawns. Crabs and lobsters are generally trapped and transported live, while shrimp and prawns are caught by trawlers, immediately iced, and transported to the processing plants. Shrimps and prawns are harvested from coastal waters and are therefore exposed to a variety of microbial or chemical contaminants (see Sec. III.A).

Hemolymph drawn from live crabs often contains low level bacterial infections including human pathogens, which may relate to the animal's nonvascular circulatory system (188). Although shellfish control programs rarely apply to bacterial or viral pathogens in Crustacea, crabs should be processed with this potential hazard in mind, especially those harvested from waters of questionable microbiological quality.

E. Transport from Harvest to Processing or Foodservice Level

Proper temperature control during transport of refrigerated seafood is very significant in assuring the safety of certain products. Although unlikely to contribute to the safety of most fish, elevated temperatures during transport are quite commonly responsible for histamine development in susceptible species, such as tuna, mackerel, mahi-mahi, and sardines. Such environments also constitute a very significant potential source of high pathogen loads on molluscan shellfish and crustaceans. Most noteworthy here are the pathogenic vibrios such as *V. parahemolyticus* and *V. vulnificus* (19).

Temperatures should be maintained at or below 2°C in transit whenever feasible to minimize growth of psychrotrophic pathogens (*L. monocytogenes, A. hydrophila, P. shigelloides,* and *C. botulinum* type E). The maximum time from harvest to refrigeration must not exceed 36 hours in colder months and 20 hours in the summer (75). Any packaging, that creates an anaerobic environment, including sealed bulk containers, could lead to outgrowth of *C. botulinum* if subjected to temperature abuse. *C. botulinum* type E (85,105) has been implicated several times in seafood-related illness (5,64).

At the transportation level, mild temperature abuse (4–10°C) is very common and may allow the growth of either psychrotrophic bacterial pathogens or those associated with somewhat warmer conditions, such as *S. aureus, Salmonella* spp., and *Vibrio* spp. Chilled products should be loaded rapidly and transported under active cooling or on ice.

F. Recreational and Sports Fishing

According to Ahmed (6), recreational and subsistence fishing is largely ignored in health and safety monitoring. Closures of recreational harvest areas are often forgotten and may lead to foodborne illness.

1. Shellfish

Especially in tropical and subtropical waters, live shellfish should be taken only from approved regions. Advice and maps are available for most U.S. recreational fishing grounds, and to some extent in Europe. Where such information is not available, local or regional offices of public health or local professionals should be contacted for advisories.

After harvesting, shellfish must be immediately refrigerated. Direct storage on ice may kill shellfish and is therefore not recommended. Indirect icing by means of a "box-in-a-box" system is probably the best way of storing live shellfish. The conventional recommendation is to discard all shellfish that are dead when received for processing. The concern here relates to the unknown temperature history of such products.

2. Fish

Ciguatera toxin (ciguatoxin) can occasionally be found in certain tropical reef fish. Since it cannot be detected organoleptically and is unable to be destroyed by heating or freezing, other prevention methods must be applied. Moreover, the list of possible ciguatoxic fish is very long, although some of the larger carnivores (e.g., amberjack, barracuda, snappers) have a more frequent association with the toxin. Also, the lack of a uniform nomenclature of fish species and the variety of local names make reliable identification of implicated species difficult. Sports fishers should use advice from local authorities, as well as information developed through their own investigation to acquire awareness of potentially ciguatoxic fish and other potential dangers in the areas they fish in. If particularly large tropical reef fish are sought in such areas, it may be wiser just to take a photograph and avoid consumption. Other toxic fish like pufferfish should also be avoided. Local knowledge should become global! Basic knowledge on all toxic fish must be communicated to sports and recreational fishers, as well as to tourist offices, local organizers, and other appropriate bodies.

Similar precautions are necessary to avoid the intake of fish and shellfish highly contaminated with chemicals. Health advisories are periodically issued, based on monitoring analyses and risk assessment studies.

Immediate killing and gutting of captured fish is recommended, not only to avoid unnecessary suffering, but also to improve the quality and safety of the resulting food. Fish can also be held live in holding tanks (live wells) until slaughter. The water in such tanks must be changed frequently to prevent the accumulation of metabolites and a drop in oxygen levels, which will kill the fish. Dead fish decompose quickly in warm water and become unsuitable for consumption. Gut the fish with a clean knife, leave no viscera or organs in the cavity, and wash the gutted fish with clean water. While gutting, prevent contact of the edible flesh with digestive juices or other visceral contents. This is primarily a quality concern, although natural toxins are occasionally associated with the viscera of certain fish and shellfish (e.g., pufferfish worldwide, anchovies from the U.S. Pacific coast, lobsters and crabs) (76). If in doubt about the inherent safety of a species harvested from a particular area, consult local health officials.

Immediate refrigeration or freezing of slaughtered fish is highly recommended. In particular, potentially scombrotoxic fish like tuna, marlin, and mackerel must be iced rapidly. When these species are left on a warm deck or beach for some time, a high amount of histamine is likely to be produced, which can cause serious illness in consumers of the cooked flesh. Before a fishing trip, determine how much ice is likely to be needed. Ice can easily be discarded, but to throw away inadequately chilled fish would be a pity—and to keep fish with the potential to cause illness would be worse. After unloading, cross-contamination can be avoided by discarding all ice, and washing and disinfecting containers.

V. SAFE HANDLING AT PROCESSING

Postharvest control (i.e., the control of processing) is a very important factor in maintaining seafood safety. Although there is no greater risk than for other processed foods for pathogen contamination, processing of seafood has become more and more automatic. A good example is the construction of an apparatus that processes bivalve mollusks to recover the edible portion (2). Nevertheless, there is still a good part of seafood processing that requires handling. The risks of contamination, therefore, can become a function of the number of processing steps.

Any viable organism in food or water could be a pathogen. Therefore, industry, government, and consumers should strive to secure sources of food and water having the lowest bacterial content possible. Batch testing to certify for the absence of a specific rare pathogen that can cause serious human disease with a very low inoculum can never provide confidence that such a pathogen is not present. Similarly, in view of the very high frequency with which some organisms are found in food, testing for foodborne bacteria would seem to offer nothing over less specific tests designed to detect and enumerate either total bacteria or fecal coliform organisms. Since it is currently not feasible to eliminate pathogens from raw food, other safe and practical approaches must be chosen. Food producers, processors, the food industry, the public, and health authorities must collaborate intensively, by means of such measures as the introduction and implementation of HACCP systems (38,40,114). At present, the use of HACCP measures, coupled with education and constant training of food handlers and processors, appears to be the best "broad spectrum" approach to food and water bacterial safety.

In the United States the safety regulations regarding seafood are already based on the principles of HACCP (13,26,106). Beginning in late 1997, seafood processors were required to conduct a hazard analysis for each product to determine the need for an HACCP plan, then to implement HACCP accordingly. In the European Union, HACCP-type processing is also required. In addition, the Codex Alimentarius Commission has also adopted HACCP for seafood processing. It is inevitable, therefore, that an increasing number of countries will start to introduce this system to maintain their exports. The steps that must be followed in developing an HACCP plan include the following (74):

1. Describe the food.
   a. Specify the animal species used [e.g., rainbow trout (*Oncorhynchus mykiss*)].
   b. Describe the finished product (e.g., hot-smoked rainbow trout, filleted).
   c. Describe the eventual packaging method (e.g., vacuum-packaged).
2. Describe distribution and storage (e.g., distributed chilled in trucks, stored under refrigeration).
3. Describe the intended use (e.g., to be eaten without further cooking).
4. Describe the processing of the product, preferably by means of a flow diagram.

5. Identify the potential hazards related to species and to processing.
6. Identify the critical control points (CCPs).
7. Set the critical limits.
8. Establish monitoring practices.
9. Establish corrective action procedures.
10. Establish record-keeping procedures.
11. Establish verification procedures.

Some examples of HACCP plans are mentioned later (Sec. V. C.) in connection with our discussion of the hazards associated with seafood products. Hazards due to non seafood ingredients are covered in other chapters of this book.

A. Hygiene and Good Manufacturing Practice

Good manufacturing practice and good hygienic practice are the basic principles of safe food handling (18,24).

Microorganisms normally present on the hands, which are not considered to be a threat in food processing, are more important as contributors to food spoilage, especially in precooked products (e.g., cooked and peeled shrimp). Transient microorganisms picked up from other sources (nasal cavity, feces, the environment, etc.) are far more important. A basic principle is therefore effective and proper washing and cleaning of the hands.

In seafood processing plants, treatment of effluent water is very important to be sure that one is not maintaining a contaminated environment (120,155).

B. Processing Parameters

To reduce or eliminate hazards during processing of food, such technological parameters as heat, cold, acidity, and drying are important. Technological parameters for the corresponding hazards were presented earlier (Sec. III.A). In the last decades other treatments and techniques such as irradiation and the use of bacteriocins have been introduced.

1. Temperature

a. Heat. Heat is an effective method of destroying viable microorganisms (see Sec. III.A). A heating step must be controlled within the scope of HACCP if it is implemented for pathogen control or otherwise alters the safety of the product. Undercooking may allow the survival of pathogens, leading to several unintentional but potentially hazardous conditions: (a) direct contamination of a ready-to-eat product with pathogens, (b) elimination of other less heat-resistant microflora which, if present, may suppress pathogen growth or lead to spoilage prior to significant pathogen growth, and (c) survival of pathogens with increased heat resistance to any subsequent cooking or reheating step. It is also possible for a sublethal heating step to trigger the germination of bacterial spores, producing vegetative cells capable of releasing toxin. Subsequent heating will destroy some such toxins (e.g., botulinum toxin), but others (e.g., *B. cereus* enterotoxin) may be resistant. In some cases, a cooking or heating step presents no increased risk, even if sublethal to pathogens. Examples include a blanching step to inactivate enzymes or a par-fry operation to set breading on products to be cooked by the user. To inhibit the development of pathogenic microorganisms in bivalve mollusks and marine gastropods, European Community legislation has approved heat treatment, where either a sterilization is achieved or the core temperature reaches at least 90°C for 90 seconds (11).

b. Cold. A refrigeration chain without any interruption will control the hazard of scombrotoxin (histamine) formation in fish. It is very important that proper icing be performed onboard, during transport, and in the processing plants. Incoming fish may be screened via sensory examination, and temperature control of the fish and the plant may be established.

Temperatures of 0–2°C will also control growth of bacterial pathogens (see Sec. III.A.1), even the psychrotrophs. Therefore, control of storage temperature is essential, especially to assure the absence of botulinum toxin (84). This hazard, however, occurs only in low oxygen environments (e.g., hermetically packaged products) and is not a concern in frozen products. Harrison et al. (105) inoculated packaged blue crab meat with *C. botulinum* and found that at 4 and 10°C the numbers of bacteria increased, but no botulinum toxin was detected. Garren et al. (85) made the same experiment with rainbow trout, and found toxin after 6 days' storage at 10°C. The fish, however, was noticeably spoiled. At a storage temperature of 4°C, no toxin was detected.

Modified atmosphere packaging of fish extends its shelf life, inhibiting the growth of psychrotrophic aerobic gram-negative bacteria—the main spoilage flora. Pathogenic psychrotrophic bacteria, however, may grow, especially when temperature abuse occurs (59).

Freezing is a processing technique that can be used for killing parasites in fish (see Sec. III.A.3). It is of particular importance for products such as Sashimi, in which parasites are not killed by any other processing steps.

2. Acidity

A low pH contributes to bacteriological and parasitological safety of processed food (see Secs. III.A.1 and III.A.3) but has no significant effect on other hazards. Possible survival of parasites and bacteria (with subsequent growth of bacteria) must be controlled by ensuring proper acidity, eventually in combination with other measures.

Addition of lactic acid or other organic acids to fish fillets may extend shelf life (121) and has also been used as a control measure. Although marinades may be bactericidal depending on their pH and the nature of the acid, their use is not considered a safe method to render mussels free from *L. monocytogenes* and other bacterial pathogens (31).

3. Drying

Products with a water activity of 0.85 or lower do not represent a risk with regard to the growth or toxin production of bacterial pathogens, including *S. aureus* (see Sec. III.A.1). The drying process should, therefore, be targeted to control this pathogen.

4. Trimming of Fish

Trimming away the belly flaps of possibly infected fish is an effective method of reducing the number of parasites (see Sec. III.A.3). However, this technique does not eliminate the hazard and must be supported by other means such as cooking or freezing.

Trimming away fatty tissue (belly flaps and, in some instances, skin, lateral line, and dorsal midline flesh) has been shown to lower pesticide/chemical contaminant loads but not heavy metal content of fatty fish from polluted waters (162) This information is used primarily in health advisories for recreational fishermen. Commercially harvested fish that may be problematic are generally kept off the market by regulation. In areas of the world where monitoring programs are inadequate, chemical contaminants must be seriously considered in any seafood product hazard analysis.

5. Irradiation

Gamma irradiation may be used as a food preservation method. The food is exposed to ionizing irradiation to inactivate microorganisms, especially bacterial foodborne pathogens. Organizations such as the World Health Organization and the U.S. Food and Drug Administration recommend irradiation as a valid and safe technique for improving the safety of food (7,119). Acceptance by consumers, however, seems to be quite low and has become a political and a psychological problem (161).

Gamma irradiation is a cost-effective preservation method also in seafood and fish (89,148). Bacterial pathogens as *Salmonella, Vibrio* spp., or *L. monocytogenes* will be reduced in number (30,107), as well as the spoilage flora (1), and the irradiation allows an extended shelf life for these products.

Irradiation can never be a substitute for proper sanitation; but in combination with good manufacturing practices, it contributes to a lower risk of bacterial foodborne illnesses (148). The effectiveness of this process in enhancing microbiological food safety needs to be communicated to the public and the national health authorities (33).

6. Bacteriocins

Many studies have been aimed at the control of specific foodborne pathogens by means of antimicrobial peptides or bacteriocins (101,104,150,173,190). Some bacteriocins are already used in novel food preservation applications. Muriana (151) stated that bacterocins may contribute an additional barrier in the "hurdle" concept of food safety. However, proteolytic enzymes present will eliminate their action.

C. Examples

1. Smoked Fish

a. Cold Smoked Fish. Huss et al. (113) listed the safety hazards of cold smoked salmon and stated that processing, if done according to GMP, will prevent the growth of *C. botulinum* type E, *Salmonella, S. aureus,* and pathogenic *Vibrio* spp. However, cold smoking and chill storage will not control the growth of *L. monocytogenes* (69,96). So, these products cannot conform to the "zero-tolerance" guideline or regulation on the books in some countries. The best way to control this pathogen consists of GMP in combination with appropriate shelf life and low storage temperature. Huss et al. (113) propose a shelf life of 3 weeks for smoked fish stored at 5°C or below.

Cold smoked salmon may become quite a risky product in the event of temperature abuse, since there is no subsequent control step, such as heating (113).

b. Hot Smoked Fish. The temperature of heating/smoking process is critical. A pasteurization effect must be achieved to eliminate vegetative pathogenic bacteria. The temperature also should be sufficient to damage spores, especially *C. botulinum,* to render them more susceptible to inhibition by salt (77). A core temperature of 63°C for 30 minutes would have this effect. The kiln should be loaded in a uniform manner, with the fish having approximately the same size and weight, and the internal temperature must be checked at the thickest portion of the largest fish in the kiln.

Water activity is also a critical parameter. Nitrite, if permitted, allows a lower salt level. Brining or dry salting will inhibit botulinum toxin formation. A minimum of 2.5% water phase salt in the fish muscle must be achieved for air-packaged smoked fish. For vacuum-packaged smoked fish (or for fish packaged under modified atmosphere), the percentage should not be lower than 3.5%. In addition, the product must be stored at temperatures not exceeding 3.3°C.

Control measures during production and storage of hot smoked fish include the following (118):

Plant design featuring smooth and easily washable walls, floors, and working surfaces; strict separation of raw fish and finished products.
Maintenance of the refrigerating chain during the entire manufacturing process.
Attitude of plant personnel (good hygienic practices during processing; regular cleaning and disinfecting of working installations, tools, clothes, hands, etc.); elimination of processing waste.
Strict adherence to GMP.
Elaboration of an HACCP-program.
Regular cleaning and disinfecting of all surfaces that come in contact with smoked fish.
Regular cleaning and disinfecting of the slicing machine.
Immediate refrigeration of the finished products to temperatures below 4°C.

2. Cooked and Peeled Shrimp

Shrimp is a relatively safe food item (6). The biggest problem is postprocessing contamination. Shrimps and prawns have been implicated in foodborne outbreaks caused by *S. aureus, Salmonella, Shigella,* and *V. parahaemolyticus* (35).

After harvesting, and removal of the heads, shrimp and prawns should be washed in potable water. Then they are cooked in boiling water or steamed in kettles. The core temperature should reach at least 75°C to pasteurize the product. After cooking, a rapid cooling to 10°C or less should be achieved within 4 hours. Cross-contamination must be prevented by the following measures: separate locations for handling raw and cooked shrimp, control of personnel movement, and personal hygiene (hand washing after handling raw shrimp).

The very delicate processing step of peeling is done by hand in most of the world. The inherent slowness of the process, in combination with elevated ambient temperatures, make it evident that there may be a high risk of contamination (114). A rapid cooling after peeling to 5°C or below is essential. Subsequently, the crustaceans should be frozen as soon as possible, either individually or in blocks. A temperature of −18°C should be achieved.

3. Surimi

Surimi, used as a crab substitute in seafood salads or other dishes, is manufactured mainly from Alaskan pollock (*Theragra chalcogramma*) and also contains sugars and emulsifiers. It is heated to provide a long shelf life. Since, however, such products represent an excellent growth medium, it is very important to avoid cross-contamination (6). Unfortunately, such products, when contaminated with pathogens, do not develop rapid spoilage with adverse organoleptic effects. To control bacterial growth, these products should be refrigerated to temperatures below 4°C (115).

VI. SAFE HANDLING AT THE FOODSERVICE LEVEL

Foods prepared in foodservice and retail establishments are responsible for most outbreaks of foodborne disease (36). Indeed, the hazards associated with seafood at the foodservice level, mainly bacteria and viruses (see Secs. III.A.1 and III.A.2) are often underestimated. It is felt that high risk individuals should be better informed about the potential risks.

Traditionally, in inland regions where seafood was considered to be highly unsafe, these products were not often found in local markets. This attitude has changed remarkably, mainly as a result of the globalization of the markets as well as increasing traveling and tourism. Also HACCP must be introduced at the foodservice level.

A. Sources of Hazards at the Foodservice Level

1. Contaminated Raw Material

a. Fish. Thorough cooking of fish and shellfish avoids most of the hazards (see Sec. III.A). Nevertheless, specialty items that are eaten raw, like sushi or sashimi, are gaining increasing popularity in Europe and in the United States. Eating raw fish is a hazardous practice, since the food may contain bacteria, viruses, or parasites. Because of the increasing awareness of different eating habits, the public is being more exposed to products from potentially hazardous fish species and foods that are eaten raw. A good example is the case report of an outbreak among 19 people who ate sashimi prepared from the white sucker (*Catostomus commersoni*) (133).

If raw fish absolutely must be on the menu, be sure it has been recently caught, handled hygienically, and held frozen to $-20°C$ for at least 24 hours or preferably for 7 days. By subjecting raw fish to such long-term freezing, one can avoid parasitic infections.

b. Shellfish. Raw shellfish, especially raw oysters, pose a risk to the consumer. Even when depuration is performed, outbreaks are still reported (49). There is no way to guarantee shellfish free from pathogens. Hot sauces like Tabasco® do not eliminate or significantly reduce the number of *V. vulnificus* (175). "Hot-pot" cooking (dipping shellfish into hot water) should be vigorously discouraged (50).

The preparation and serving of oysters and raw shellfish in foodservice operations entails the following steps and considerations (114):

Raw shellfish must be purchased from approved sources, which are regularly inspected or have no history of virological, bacteriological, or toxicological contamination. However, there is no absolute assurance that shellfish so qualified are free from hazards.

On receipt, shellfish must be cooled rapidly to near 0°C. Monitoring includes temperature checks and observation of stock rotation.

Proper cleaning of the shell and of the knives during shucking, in combination with good hygienic practices, will minimize the risk of contamination of shellfish by preventing the transfer of microorganisms from the outside of the shell to the meat.

Shellfish that are not immediately served should be stored on ice.

Serving of shellfish on the half-shell does not increase the contamination risk appreciably.

The foodservice facility should provide a "health warning" for those who want to eat raw oysters (e.g., a note in the menu: "Oysters are known to be associated with gastroenteritis, even when they are optimally handled and prepared") (49).

A large gastroenteritis outbreak demonstrated, however, that oysters cooked to the point at which oyster eaters consider them done or even overdone are still able to transmit virus particles (142). This observation emphasizes that the best way to prevent disease is to follow up avoidance of initial contamination by thorough cooking, measured with a thermometer.

2. Faulty Hygienic Practices

Contamination by infected food handlers must be avoided. Mainly viral but also bacterial hazards are of concern. The most effective measures are as follows:

People who are handling food should wash their hands frequently with soap and water, especially after using toilets.

A minimum of hand–food contact should be allowed.

Wherever possible, those who manipulate food should wear gloves, and discard them frequently.

Food contact surfaces should be cleaned and disinfected after contact with raw food.

3. Time–Temperature Relation

Especially at the foodservice level, the time–temperature relationship for holding and storing food plays a major role, hence is the most important preventive tool in avoiding bacterial hazards. In several outbreaks, the following contributing factors were identified: food prepared several hours before serving; improper cooling, hot holding, and heating; inadequate cooking; refuse of foods; and inadequate reheating (130,149).

The time that seafood is kept in the "danger zone" (generally between 0°C and 55°C) must be as short as possible (170). In retailed seafood, the safest practice is to store seafood on ice or keep it hot at temperatures of 60°C or higher.

Shorter keeping times and limitations of the "best before" dates will also enhance microbiological safety.

B. Health Education of Food Handlers

Constant teaching and training programs for food handlers regarding safe food handling should be encouraged, although the basic principles should already be understood from studies in school.

VII. SAFE HANDLING AT HOME

Poor food hygiene practices at home are responsible for much foodborne illnesses. Approximately 20% of the foodborne diseases occurring in Europe and in North America are acquired at home (14,114). The contributing factors are basically the same as elsewhere (Table 6). So it is very important to insist on safe handling at home and to pass the necessary information and knowledge to the consumers in an objective and easy understandable way.

Food handling at home is a matter of responsibility within the family (114). No law can dictate how food is handled, prepared, stored, or consumed by private individuals. Only by objective information and specific recommendations can people be encouraged to understand and practice good hygiene and safe handling of food. Nevertheless, it does not excuse the consumer from using common sense and taking repeated precautions while preparing and cooking potentially contaminated products.

Consumers in developed countries are exposed to a mass of information in various forms (press articles, information from radio and television, leaflets, videos, etc.). Scott (167) investigated the impact of this information on the British consumer and stated that many people had a complete lack of understanding of the nature of bacteria or viruses. In addition, raw food is generally not recognized as a source of microorganisms. Therefore, the basic instructions are not understood. It was concluded that it is not possible for consumers to grasp the complexities of safe food handling without understanding the basic principles. Food hygiene and safe food handling should be a part of the school curriculum. It is not enough to show the consumer a simple list of "do's and don'ts" (167).

Safe food handling is often compromised, however, because of situations such as lack of water, food shortage, or poverty. The normal principles of food hygiene are then disregarded, as simple survival is put before safety. Additionally, transmission of traditional knowledge on food handling from generation to generation often is disrupted as a result of global society changes, urbanization, loss of family integrity, etc., and even small changes in traditional practices can result in hazardous products.

A. General Advice

Generally at home, microbial hazards exist on the purchased food or may occur through storage, and preparation; thus these steps must be critically checked.

Consumers should be educated in factors that most significantly impact on food safety. Quality factors, such as indicators of fish freshness, are normally much less important than are identified product-specific hazards and the use of proper handling and holding conditions. Consumers should be informed that ready-to-eat products (hot smoked fish, raw oysters) are associated with certain risks and require careful handling. Cooked seafood coming into contact with drips from raw seafood is a common source of contamination. A national consumer education campaign in the United States is based on four food safety messages. (a) *clean:* wash hands and surfaces often, (b) *separate:* don't cross-contaminate, (c) *cook:* cook to proper temperatures, and (d) *chill:* refrigerate promptly (160).

1. Purchase and Storage of Seafood

In food processing and foodservice the maintenance of the refrigeration chain for susceptible products is a magnificent tool for providing safe food. Many consumers, however, fail to respect these rules, especially when shopping. Some good, sound basic seafood handling recommendations are as follows:

Buy only fresh or frozen fish. Quality of fresh whole fish can easily be checked by three criteria: (a) eyes must be clear, not sunken; (b) skin must not be covered with turbid slime; and (c) gills should be bright red. In general, all three criteria must be met. However, there are some exceptions for particular species; for example, pike perch (*Stizostedion lucioperca*) has turbid eyes by nature.

Shellfish in the shell must be alive when purchased.

Buy fish and seafood only from approved sources. In shellfish-harvesting regions, people should be warned about hazards associated with harvesting shellfish from polluted or closed areas, and consumption of shellfish from unknown sources should be discouraged (36).

Buy refrigerated and frozen food last during grocery shopping.

Take food straight home and put it into the freezer or refrigerator. Do not leave food in the car.

Keep fresh fish cold, preferably in the coldest part of the refrigerator (under the freezer, near the compressor).

Store processed fish (smoked fish, pickled fish, surimi, and canned fish after opening) in the refrigerator. Be aware of short keeping times, and eat such fish soon after purchase.

Check the temperature of the refrigerator and freezer.

Avoid cross-contamination: do not unwrap fresh fish until it is to be cooked or served; keep cooked and raw seafood separate.

These easy measures will prevent or at least slow bacterial growth. Temperature abuse in scombroid or scombroidlike fish represents a serious illness threat, since it may lead to a rapid increase in histamine levels and result in scombroid poisoning.

The emergence of psychrotrophic bacterial species such as *C. botulinum* type E, *L. monocytogenes,* and *A. hydrophila* (see Sec. III. A.1) led Schmidt-Lorenz to ask whether the storage of foods in refrigeration is still sufficiently safe (164). Products that may be stored for longer than 10 days at 5°C (e.g., pasteurized products) are critical. It is safer to store these products between 0 and 2°C. When the lag phases are over and psychrotrophic pathogens start to develop, most seafood products are already spoiled. Generally the storage of products in a refrigerator is safe, but often the temperature in household refrigerators exceeds 5°C. Many consumers are not aware of the importance of maintaining an appropriate refrigerator temperature and have no idea what temperature their home refrigerator is set at. The consequences of temperature abuse and long storage periods must be communicated (117).

2. Preparation

a. Basic Kitchen Hygiene.

Wash hands with hot water and soap before preparing food and after using the restroom.
Replace kitchen towels, sponges, etc. often.
Keep raw fish and shellfish and their juices away from other food.
Do not use the same knives, cutting boards, dishes, etc. for handling cooked and raw fish until such items have been thoroughly washed.
To avoid bacterial growth, thaw and marinate fish in the refrigerator.
When cooking ahead, divide large portions into small ones to ensure rapid and safe cooling. The same applies for large amounts of leftovers.

b. Thorough Cooking. Wherever possible, it is best to inactivate enteric pathogens. Sometimes this is not done (e.g., with oysters) because the taste might be lost or spoiled. Steaming or immersion of shellfish in boiling water is safe only when the heat penetrates into the shellfish, as opposed to reaching only the surface. Research is inadequate for accurately predicting virus particle destruction at the higher temperatures frequently encountered when cooking (no D or z values). However, based on a thermal inactivation study of oysters artificially contaminated with hepatitis A, cooking shellfish just until they open may be insufficient to destroy the virus (159). Other investigators have associated the ingestion of lightly cooked molluscan shellfish with hepatitis A infection (28,63, 145,150). However, boiling shellfish for 1 minute was found to be sufficient to inactivate the virus (68). Also Parry and Mortimer (157) reported a rapid inactivation of hepatitis A virus at 80°C. The FDA (73) has issued a brochure directed to consumers with the following helpful advice on how to cook oysters and other shellfish:

Boil for 3–5 minutes after shells open. Do not cook too many oysters at one time.

Shellfish that remain open even when tapped or put into water should not be eaten. Live shellfish are of better quality and less hazardous than dead shellfish, which spoil quickly.

Steam live oysters 4–9 minutes in a steamer that is already steaming.

It seems that microwave cooking is safe, although there are still some problems with the occurrence of so-called cold spots. Often higher temperatures are attained below the surface than on the surface as a result of evaporative cooling. Since surface contamination is frequently the greatest source of pathogens, this surface cooling effect may lead to pathogen survival. Simply covering the product to retain moisture readily controls this risk factor. Shrimp inoculated with *L. monocytogenes* that were cooked in a microwave oven mainly turned out to be free of this pathogen (93).

c. Homemade Fermented Fish Products (Native Foods). Homemade fermented fish products may lead to a hazard even without vacuum packaging or a similar form of air removal because they are normally produced at home by traditional methods that are controlled poorly or not at all (5). Fermentation conditions may create an anaerobic environment directly, often at elevated temperatures favorable to several types of *C. botulinum.* Subsequent vacuum packing and storage can lead to the enhanced growth of *C. botulinum* (64).

Fermented foods are expected to possess an extended shelf life, which provides sufficient time for pathogen growth, occasionally without the development of objectionable odors normally associated with putrefied food. The *C. botulinum* hazard is avoided in commercially produced products through standardized use of salt and/or acid in the formulation and by proper sanitation and temperature controls.

B. Dietary Advice

1. Raw Seafood

The consumption of raw muscle foods offers numerous biological hazards associated with bacteria, viruses, and parasites. The risks associated with consuming raw red meat and poultry are substantial. Similarly, it should not be surprising that raw seafoods, such as oysters and clams on the half-shell, raw fish dishes (e.g., surimi and sashimi), and lightly marinated seafoods, (e.g., seviche) pose hazards that must be considered. Many species of fish are potential sources of parasites capable of human infestation. These can be killed by first freezing for 7 days at −20°C or for 15 hours at −35°C (78).

The avoidance of raw seafood is therefore the best preventive measure. In addition, shellfish from uncertified beaches should not be eaten (87).

2. Parties, Barbecues, and Picnics

Parties and barbecues represent an elevated risk of contracting a foodborne illness. Basic hygienic rules, as described previously, are typically disregarded. As these events happen mainly in the summertime with high ambient temperatures, the risks of bacterial growth increase. The same preventive measures that apply for preparing food in the home kitchen must be respected. The most important factors are time and temperature.

All foods should be kept either hot or cold and served as quickly as possible. Large (or stacked) containers of food cook and cool much more slowly than those typically used for single-family meal preparation. The use of commercial cooking and refrigeration equipment may be required for serving large functions.

3. Consumers at Risk

The percentage of immunocompromised individuals is increasing, especially in industrialized nations, as a result of the use of immunosuppressive drug therapies, acquired immunodeficiencies such as AIDS, underlying diseases (e.g., diabetes mellitus, liver diseases), and extended life expectancy. These people are much more susceptible to many infections than apparently healthy people. A good example is *V. vulnificus:* people with an underlying liver disease have an increased risk of infection from this microorganism, and the risk of death is almost 200 times greater than for people without an underlying disease (73). Therefore, individuals with underlying diseases are at risk and should not eat raw or partially cooked shellfish or fish.

People who are allergic to seafood should carefully read labels or ask what ingredients are used in a specific dish.

In addition, high risk consumers should not eat raw or cooked food with their hands when sharing a meal with other persons. Diseases like shigellosis or viral hepatitis may be transmitted under such conditions.

VIII. CONCLUSIONS

Little is known of foodborne disease on a worldwide basis (183). In many countries essential information on incidents and causes of foodborne disease is lacking or incomplete. Centralized information is necessary. The use of the Internet may help bring food hygiene principles to a wide variety of people, especially those in developing countries. The rapid spread of Internet technology is having a significant impact on health care development, management, and practice. Rapid and complete information about foodborne diseases that are due to the consumption of fish and shellfish is very important. Also sentinel or case-control studies may contribute to a worldwide information system. However, consistent and coordinated action by policy makers and food suppliers is necessary to achieve all of this.

Seafood consumption is increasing worldwide, especially in countries with no direct access to the sea. A growing dependency on aquaculture can be observed. Safe seafood therefore has become an important public health concern. Control and monitoring is rendered more difficult by the different standards in international trade as well as the variety of new, sometimes unconventional seafood products (48).

Epidemiological studies in sentinel sites are very important. Unfortunately, in developing countries money is lacking, and the budgets for such studies are more difficult to obtain because of shrinking resources.

The Third World Congress on Foodborne Infections and Intoxications (12) drew the following very important conclusions for reducing hazards in fish and other seafood:

Surveillance programs need strengthening in developing countries, for example, by means of the promotion of epidemiological tools like sentinel studies.
Legislation of the European Community has an impact on other countries: in general, better legislation is needed, especially in developing countries.
Increases in travel and tourism underline the need for information exchange.
HACCP should be the reference method built into systems of food safety.
Teaching and training programs should be encouraged, beginning in school. Mass media must be involved.
Animal hygiene should be included in food safety.

Ahmed (5) stated the following conclusions:

Most risks due to seafood consumption originate in the environment and cannot be identified or controlled by a sensory inspection system. The best control must be performed at the origin (i.e., harvesting).
Most seafood-associated illnesses are reported from consumers of raw bivalve molluscan shellfish.
Valid indicators for human pathogen contamination of growing waters must be developed.
Adequate and proper treatment and disposal of sewage should be implemented worldwide.
Attention should be paid to new products, such as “sous vide” and the use of controlled atmosphere packaging.

The key to safe seafood handling is the continuing education and training of all personnel involved in production, foodservice, and inspection. The understanding of the principles and strategies in food hygiene should be the same in all sectors of the food industry. In fact, industry and regulatory personnel should be trained together, even although this is "an enormous job" (52) and demands complete national and international collaboration. A great challenge for the future!

REFERENCES

1. Abu-Tarboush, H. M., H. A. Al-Kahtani, M. Atia, A. A. Abou-Arab, A. S. Bajaber, and M. A. El-Mojadiddi. 1996. Irradiation and postirradiation storage at 2 ± 2°C of *Tilapia (Tilapia nilotica × T. aurea)* and Spanish mackerel (*Scomberomorus commerson*): Sensory and microbial assessment. J. Food Prot. 59:1041–1048.
2. Adcock, J. T. 1992. Shellfish processing. PCT-International-Patent-Application, WO 92/22212 A1.
3. Ahmed, F. E. (ed.). 1991. Seafood Safety. National Academy Press, Washington, DC.
4. Ahmed, F. E. 1992. Programs of safety surveillance and control of fishery products. Regul. Toxicol. Pharmacol. 15:14–31.
5. Ahmed, F. E. 1992. Assessing and managing risk due to consumption of seafood contaminated with microorganisms, parasites, and natural toxins in the United States. Int. J. Food Sci. Technol. 27:243–260.
6. Ahmed, F. E. 1992. Safety of seafoods in the USA, in Quality Assurance in the Fish Industry. Ministry of Fisheries, Denmark, pp. 283–292.
7. Anonymous. 1988. Food irradiation: A technique for preserving and improving the safety of food. WHO, Geneva.
8. Anonymous. 1991. Food safety. Paralytic shellfish poisoning. Wkly. Epidemiol. Rec. 66:185–187.
9. Anonymous. 1991. The role of the veterinarian in fish farming and aquaculture. Vet. Rec. 129:124–125.
10. Anonymous. 1991. Council directive of 15 July 1991 laying down the health conditions for the production and the placing on the market of live bivalve molluscs (91/492/EEC). Off. J. Eur. Community L 268:1–14.
11. Anonymous. 1993. Commission decision of 11 December 1992 approving certain treatments to inhibit the development of pathogenic micro-organisms in bivalve molluscs and marine gastropods (93/25/EEC). Off. J. Eur. Community L 16:22–23.
12. Anonymous. 1993. Final Report, Third World Congress on Foodborne Infections and Intoxications, 1992. Newsletter 37. Institute of Veterinary Medicine, Berlin.
13. Anonymous. 1994. FDA proposes seafood safety controls (news). J. Am. Vet. Med. Assoc. 204:841–842.
14. Anonymous. 1995. WHO Surveillance Programme for Control of Foodborne Infections and Intoxications in Europe. Sixth Report, 1990–1992. Federal Institute for Health Protection of Consumers and Veterinary Medicine, Berlin.

15. Anonymous. 1995. Application of Risk Analysis to Food Standards Issues: Report of the Joint FAO/WHO Expert Consultation. WHO, Geneva.
16. Anonymous. 1995. Multistate outbreak of viral gastroenteritis associated with consumption of oysters—Apalachicola Bay, Florida, December 1994–January 1995. MMWR 44:37–39.
17. Anonymous. 1996. Paralytic shellfish poisoning. Int. Food Hyg. 7:17.
18. Anonymous. 1996. Codex Alimentarius: FAO/WHO Food Standards. FAO, Rome.
19. Anonymous. 1997. Guide for the Control of Molluscan Shellfish, 406 pp. National Shellfish Sanitation Program, Interstate Shellfish Sanitation Conference, U.S. Department of Health and Human Services, Public Health Service, Food and Drug Administration. Government Printing Office, Washington, DC.
20. Appleton, H. 1994. Norwalk virus and the small round viruses causing foodborne gastroenteritis, in Y. H. Hui, J. R. Gorham, K. D. Murrell, and D. O. Cliver (eds.), Foodborne Disease Handbook, Vol. 2. Marcel Dekker, New York, pp. 57–79.
21. Armentia-Alvarez, A., C. Garcia-Moreno, and M. J. Peña-Egido. 1994. Residual levels of sulfite in raw and boiled frozen shrimp: variability, distribution and losses. J. Food Prot. 57:66–69.
22. Atmar, R. L., F. H. Neill, C. M. Woodley, R. Manger, G. Shay Fout, W. Burkhardt, L. Leja, E. R. McGovern, F. le Guyader, T. G. Metcalf, and M. K. Estes. 1995. Collaborative evaluation of a method for the detection of Norwalk virus in shellfish tissues by PCR. Appl. Environ. Microbiol. 62:254–258.
23. Audicana, L., M. T. Audicana, L. Fernandez de Corres, and M. W. Kennedy. 1997. Cooking and freezing may not protect against allergenic reactions to ingested *Anisakis* simplex antigens in humans. Vet. Rec. 140:235.
24. Ayulo, A. M., R. A. Machado, and V. M. Scussel. 1994. Enterotoxigenic *Escherichia coli* and *Staphylococcus aureus* in fish and seafood from the southern region of Brazil. Int. J. Food Microbiol. 24:171–178.
25. Beuchat, L. R.. 1982. *Vibrio parahaemolyticus:* Public health significance. Food Technol. 36:80–88.
26. Billy, T. J. 1995. Seafood safety: A new era. J. Assoc. Food Drug Off. 59:65–68.
27. Bosch, A., F. Xavier Abad, R. Gajardo, and R. M. Pinto. 1994. Should shellfish be purified before public consumption? Lancet 344:1024–1025.
28. Bostock, A. D., P. Mepham, and S. Phillips. 1979. Hepatitis A infection associated with the consumption of mussels. J. Infect. 1:171–177.
29. Boyce, T. G., E. D. Mintz, K. D. Greene, J. G. Wells, J. C. Hockin, D. Morgan, and R. V. Tauxe. 1995. *Vibrio cholerae* O139 Bengal infections among tourists to Southeast Asia: An intercontinental foodborne outbreak. J. Infect. Dis. 172:1401–1404.
30. Brandao Areal, H., R. Charbonneau, and C. Thibault. 1995. Effect of ionization on *Listeria monocytogenes* in contaminated shrimps. Sci. Aliment. 15:261–272.
31. Bremer, P. J., and C. M. Osborne. 1995. Efficacy of marinades against *Listeria monocytogenes* cells in suspension or associated with green shell mussels (*Perna canaliculus*). Appl. Environ. Microbiol. 61:1514–1519.
32. Bris, H. le, H. Pouliquen, J. M. Debernardi, V. Buchet, and L. Pinault. 1995. Preliminary study on the kinetics of oxytetracycline in shellfish exposed to an effluent of a land-based fish farm: Experimental approach. Mar. Environ. Res. 40:171–180.

33. Bruhn, C. M. 1995. Strategies for communicating the facts on food irradiation to consumers. J. Food Prot. 58:213–216.
34. Brunn, H., S. Georgii, K. Failing, J. Nilz, and D. Manz. 1990. Polychlorierte Biphenyle (PCB) in Fischen aus Teichwirtschaften und Nachklärteichen. Z. Lebensm.-Unters.-Forsch. 190:104–107.
35. Bryan, F. L. 1980. Epidemiology of foodborne diseases transmitted by fish, shellfish and marine crustaceans in the USA, 1970–1978. J. Food Prot. 43:859–876.
36. Bryan, F. L. 1988. Risks associated with vehicles of foodborne pathogens and toxins. J. Food Prot. 51:498–508.
37. Bryan, F. L. 1988. Risks of practices, procedures that lead to outbreaks of foodborne diseases. J. Food Prot. 51:663–73.
38. Bryan, F. L. 1992. Hazard Analysis Critical Control Point Evaluations: A Guide to Identifying Hazards and Assessing Risks Associated with Food Preparation and Storage. WHO, Geneva.
39. Bryan, F. L. 1996. Hazard analysis: The link between epidemiology and microbiology. J. Food Prot. 59:102–107.
40. Bryan, F. L., C. A. Bartleson, O. D. Cook, P. Fisher, J. J. Guzewich, B. J. Humm, R. C. Swanson, and E. C. D. Todd. 1991. Procedures to implement the hazard analysis critical control point system. International Association of Milk, Food, Environmental Sanitarians, Ames, IA.
41. Buck, W. B. 1978. Toxicity of inorganic and aliphatic organic arsenicals, in F. W. Oehme (ed.), Toxicity of Heavy Metals in the Environment. Marcel Dekker, New York, pp. 357–374.
42. Calvert, G. M. 1991. The recognition and management of ciguatera fish poisoning, in D. M. Miller (ed.). Ciguatera Fish Poisoning. CRC Press, Boca Raton, FL, pp. 1–11.
43. Carmody, E. P., K. J. James, and S. S. Kelly. 1995. Diarrhetic shellfish poisoning: Evaluation of enzyme-linked immunosorbent assay methods for determination of dinophysistoxin-2. J. Assoc. Off. Anal. Chem. Int. 78:1403–1408.
44. Castillo, R., T. Carrilo, C. Blanco, J. Quiralte, and M. Cuevas. 1994. Shellfish hypersensitivity: Clinical and immunological characteristics. Allergol. Immunopathol. 22:83–87.
45. CDC. 1994. Viral gastroenteritis associated with consumption of raw oysters—Florida, 1993. JAMA 272:510–511.
46. CDC. 1996. *Vibrio vulnificus* infections associated with eating raw oysters—Los Angeles, 1996. JAMA 276:937–938.
47. Cendes, F., F. Andermann, S. Carpenter, R. J. Zatorre, and N. R. Cashman. 1995. Temporal lobe epilepsy caused by domoic acid intoxication: Evidence for glutamate receptor–mediated excitotoxicity in humans. Ann. Neurol. 37:123–126.
48. Chai, T. C., and J. Pace. 1994. *Vibrio parahaemolyticus,* in Y. H. Hui, J. R. Gorham, K. D. Murrell, and D. O. Cliver (eds.), Foodborne Disease Handbook, Vol. 1. Marcel Dekker, New York, pp. 395–425.
49. Chalmers, J. W., and J. H. McMillan. 1995. An outbreak of viral gastroenteritis associated with adequately prepared oysters. Epidemiol. Infect. 115:163–167.
50. Chan, T. Y. 1995. Shellfish-borne illnesses. A Hong Kong perspective. Trop. Geogr. Med. 47:305–307.

51. Cliver, D. O. 1994. Epidemiology of foodborne viruses, in Y. H. Hui, J. R. Gorham, K. D. Murrell, and D. O. Cliver (eds.), Foodborne Disease Handbook, Vol. 2. Marcel Dekker, New York, pp. 159–175.
52. Collette, R. 1995. Seafood safety and inspection: Industry's needs and concerns. J. Assoc. Food Drug Off. 59:58–64.
53. Crance, J. M., V. Apaire Marchais, F. Leveque, C. Beril, F. le Guyader, A. Jouan, L. Schwartzbrod, and S. Billaudel. 1995. Detection of hepatitis A virus in wild shellfish. Mari. Pollut. Bull. 30:372–375.
54. Croci, L., L. Toti, D. de Medici, and L. Cozzi. 1994. Diarrhetic shellfish poison in mussels: Comparison of methods of detection and determination of the effectiveness of depuration. Int. J. Food Microbiol. 24:337–342.
55. Cromeans, T., O. V. Nainan, H. A. Fields, M. O. Favorov, and H. S. Margolis. 1994. Hepatitis A and E viruses, in Y. H. Hui, J. R. Gorham, K. D. Murrell, and D. O. Cliver (eds.), Foodborne Disease Handbook, Vol. 2. Marcel Dekker, New York, pp. 1–56.
56. Cross, J. H. 1994. Fish- and invertebrate-borne helminths, in Y. H. Hui, J. R. Gorham, K. D. Murrell, and D. O. Cliver (eds.), Foodborne Disease Handbook, Vol. 2. Marcel Dekker, New York, pp. 279–329.
57. Daul, C. B., J. E. Morgan, and S. B. Lehrer. 1993. Hypersensitivity reactions to Crustacea and mollusks. Clin. Rev. Allergy 11:201–222.
58. Davey, J. T. 1972. The public health aspects of a larval roundworm from the herring. Community Health 4:104–106.
59. Davies, A. R., and A. Slade. 1995. Fate of *Aeromonas* and *Yersinia* on modified-atmosphere-packaged (MAP) cod and trout. Lett. Appl. Microbiol. 21:354–358.
60. DePaola, A., J. T. Peeler, and G. E. Rodrick. 1995. Effect of oxytetracycline-medicated feed on antibiotic resistance of gram-negative bacteria in catfish ponds. Appl. Environ. Microbiol. 61:2335–2340.
61. Desenclos, J. C. 1996. Epidemiology of toxic and infectious risk related to shellfish consumption. Rev. Epidemiol. Sanit. Publique 44:437–454.
62. Desenclos, J. C., K. C. Klonz, L. E. Wolfe, and S. Hoechert. 1991. The risk of *Vibrio* illness in the Florida raw oyster eating population, 1981–1988. Am. J. Epidemiol. 134:290–297.
63. Dienstag, J. L., I. D. Gust, C. R. Lucas, D. C. Wong, and R. H. Purcell. 1976. Mussel-associated viral hepatitis type A: Serological confirmation. Lancet 1 (7959): 561–564.
64. Dodds, K. L. 1994. *Clostridium botulinum,* in Y. H. Hui, J. R. Gorham, K. D. Murrell, and D. O. Cliver (eds.), Foodborne Disease Handbook, Vol. 1. Marcel Dekker, New York, pp. 97–131.
65. Dolin, R., N. R. Blacklow, H. duPont, R. F. Buscho, R. G. Wyatt, J. A. Kasel, R. Hornick, and R. M. Chanock. 1972. Biological properties of Norwalk agent of acute infectious nonbacterial gastroenteritis. Proc. Soc. Exp. Biol. Med. 140:578.
66. Dowell, S. F., C. Groves, K. B. Kirkland, H. G. Cicirello, T. Ando, Q. Jin, J. R. Gentsch, S. S. Monroe, C. D. Humphrey, C. Slemp, D. M. Dwyer, R. A. Meriwether, and R. I. Glass. 1995. A multistate outbreak of oyster-associated gastroenteritis: Implications for interstate tracing of contaminated shellfish. J. Infect. Dis. 171:1497–1503.

67. Doyle, M. P. 1989. Preface, in M. P. Doyle (ed.), Foodborne Bacterial Pathogens. Marcel Dekker, New York, pp. iii–iv.
68. DuPont, H. 1986. Consumption of raw shellfish: Is the risk now acceptable? N. Engl. J. Med. 314:707–708.
69. Embarek, P. K. B., and H. H. Huss. 1992. Growth of *Listeria monocytogenes* in lightly preserved fish products, in Quality Assurance in the Fish Industry. Ministry of Fisheries, Denmark, pp. 293–303.
70. Enriquez, R., G. G. Frosner, V. Hochstein Mintzel, S. Riedemann, and G. Reinhardt. 1992. Accumulation and persistence of hepatitis A virus in mussels. J. Med. Virol. 37:174–179.
71. FAO. 1996. The State of Food and Agriculture. FAO, Rome.
72. Farber, J. M., and P. I. Peterkin. 1991. *Listeria monocytogenes,* a food-borne pathogen. Microbiol. Rev. 55:476–511.
73. FDA. 1995. If you eat raw oysters, you need to know. . . . U.S. Department of Health and Human Services Publication 95-2293. Center for Food Safety and Applied Nutrition, Office of Seafood, Washington, DC.
74. FDA. 1998. Steps in developing your HACCP plan, in Fish and Fisheries Products Hazards and Controls Guide, 2nd ed. U.S. Department of Health and Human Services, Public Health Service, Food and Drug Administration, Center for Food Safety and Applied Nutrition, Office of Seafood, Washington, DC, Ch. 2.
75. FDA. 1998. Pathogens from the harvest area, in Fish and Fisheries Products Hazards and Controls Guide, 2nd ed. U.S. Department of Health and Human Services, Public Health Service, Food and Drug Administration, Center for Food Safety and Applied Nutrition, Office of Seafood, Washington, DC, Ch. 4.
76. FDA. 1998. Natural toxins, in Fish and Fishery Products Hazards and Controls Guide, 2nd ed. U.S. Department of Health and Human Services, Public Health Service, Food and Drug Administration, Center for Food Safety and Applied Nutrition, Office of Seafood, Washington, DC, Ch. 6.
77. FDA. 1998. *Clostridium botulinum* toxin formation, in Fish and Fishery Products. Hazards and Controls Guide, 2nd ed. U.S. Department of Health and Human Services, Public Health Service, Food and Drug Administration, Center for Food Safety and Applied Nutrition, Office of Seafood, Washington, DC, Ch. 13.
78. FDA. 1998. Parasites, in Fish and Fishery Products Hazards and Controls Guide, 2nd ed. U.S. Department of Health and Human Services, Public Health Service, Food and Drug Administration, Center for Food Safety and Applied Nutrition, Office of Seafood, Washington, DC, Ch. 5.
79. Fein, G. G., J. L. Jacobsen, S. W. Jacobsen, P. M. Schwartz, and J. K. Dowler. 1984. Prenatal exposure to polychlorinated biphenyls: Effects on birth size and gestional age. J. Pediatr. 105:315–320.
80. Finelli, L., D. Swerdlow, K. Mertz, H. Ragazzoni, and K. Spitalny. 1992. Outbreak of cholera associated with crab brought from an area with epidemic disease. J. Infect. Dis. 166:1433–1435.
81. Fletcher, G. C., G. Summers, R. V. Winchester, and R. J. Wong. 1995. Histamine and histidine in New Zealand marine fish and shellfish species, particularly kahawai (*Arripis trutta*). J. Aquat. Food Prod. Technol. 4:53–74.
82. Frank, W. 1976. Parasitologie: Lehrbuch für Studierende. Human- u. Veterinärmedizin, d. Biologie u. d. Agrarbiologie. Ulmer, Stuttgart.

83. Gago Martinez, A., J. A. Rodriguez Vazquez, P. Thibault, and M. A. Quilliam. 1996. Simultaneous occurrence of diarrhetic and paralytic shellfish poisoning toxins in Spanish mussels in 1993. Nat. Toxins 4:72–79.
84. Garcia de Fernando, G. D., S. B. Mano, D. Lopez, and J. A. Ordonez. 1995. Effectiveness of modified atmospheres against psychrotrophic pathogenic microorganisms in proteinaceous food. Microbiologia 11:7–22.
85. Garren, D. M., M. A. Harrison, and Y. W. Huang 1995. Growth and production of *Clostridium botulinum* type E in rainbow trout under various storage conditions. J. Food Prot. 58:863–866.
86. Gerba, C. P. 1988. Viral disease transmission by seafoods. Food Technol. 42:99–101.
87. Gessner, B. D., and J. P. Middaugh. 1995. Paralytic shellfish poisoning in Alaska: A 20-year retrospective analysis. Am. J. Epidemiol. 141:766–770.
88. Gibson, D. I., and J. A. Colin. 1982. The *Terranova* enigma. Parasitology 85:36–37.
89. Giddings, G. G. 1984. Radiation processing of fishery products. Food Technol. 38:61–97.
90. Gorham, J. R. 1994. Hard foreign objects in food as a cause of injury, in Y. H. Hui, J. R. Gorham, K. D. Murrell, and D. O. Cliver (eds.), Foodborne Disease Handbook, Vol. 3. Marcel Dekker, New York, pp. 615–626.
91. Gouvea, V., N. Santos, M. do C. Timenetsky, and M. K. Estes. 1994. Identification of Norwalk virus in artificially seeded shellfish and selected foods. J. Virol. Methods 48:177–187.
92. Gray, S. F., and M. R. Evans. 1993. Dose–response in an outbreak of non-bacterial food poisoning traced to a mixed seafood cocktail. Epidemiol. Infect. 110:583–590.
93. Gundavarapu, S., Y. C. Hung, R. B. Brackett, and P. Mallikarjunan. 1995. Evaluation of microbiological safety of shrimp cooked in a microwave oven. J. Food Prot. 58:742–747.
94. Gustafson, P. V. 1953. The effect of freezing on encysted *Anisakis* larvae. J. Parasitol. 39:585–588.
95. Guyer, S. 1989. Viruskontamination der Aare im Zusammenhang mit der Wirksamkeit von Kläranlagen. Diss. Pharm. Bern, Switzerland.
96. Guyer, S., and T. Jemmi. 1991. Behavior of *Listeria monocytogenes* during fabrication and storage of experimentally contaminated smoked salmon. Appl. Environ. Microbiol. 57:1523–1527.
97. Hall, S. 1991. Natural toxins, in D. R. Ward and C. R. Hackney (eds.), Microbiology of Marine Food Products. Van Nostrand Reinhold, New York, pp. 301–330.
98. Hally, R. J., R. A. Rubin, H. S. Fraimow, and M. L. Hoffman Terry. 1995. Fatal *Vibrio parahaemolyticus* septicemia in a patient with cirrhosis. A case report and review of the literature. Dig. Dis. Sci. 40:1257–1260.
99. Halliday, M. L., L. Y. Kang, T. K. Zhou, M. D. Hu, Q. C. Pan, T. Y. Fu, Y. S. Huang, and S. L. Hu. 1991. An epidemic of hepatitis A attributable to the ingestion of raw clams in Shanghai, China. J. Infect. Dis. 164:852–859.
100. Halstead, B. 1994. Fish toxins, in Y. H. Hui, J. R. Gorham, K. D. Murrell, and D. O. Cliver (eds.), Foodborne Disease Handbook, Vol. 3. Marcel Dekker, New York, pp. 463–496.
101. Hansen, J. H. 1994. Nisin as a model food preservative. Crit. Rev. Food Sci. Nutr. 34:69–93.

102. Harada, M. 1978. Methyl mercury poisoning due to environmental contamination (Minamata disease), in F. W. Oehme (Ed.), Toxicity of Heavy Metals in the Environment. Marcel Dekker, New York, pp. 261–302.
103. Harr, J. R. 1978. Biological effects of selenium, in F. W. Oehme (ed.), Toxicity of Heavy Metals in the Environment. Marcel Dekker, New York, pp. 393–426.
104. Harris, L. J., M. A. Daeschel, M. E. Stiles, and T. R. Klaenhammer. 1989. Antimicrobial activity of lactic acid bacteria against *Listeria monocytogenes.* J. Food Prot. 52:384.
105. Harrison, M. A., D. M. Garren, Y. W. Huang, and K. W. Gates. 1996. Risk of *Clostridium botulinum* type E toxin production in blue crab meat packaged in four commercial-type containers. J. Food Prot. 59:257–260.
106. Harvey, D. 1996. New procedures for seafood inspection. Agricultural-Outlook 229:29–31.
107. Hau, L. B., M. H. Liew, and L. T. Yeh. 1992. Preservation of grass prawns by ionizing radiation. J. Food Prot. 55:198–202.
108. Healy, G. R. 1970. Trematodes transmitted to man by fish, frogs, and crustacea. J. Wildl. Dis. 6:255–261.
109. Helbling, A., M. L. McCants, J. J. Musmand, H. J. Schwartz, and S. B. Lehrer. 1996. Immunopathogenesis of fish allergy: Identification of fish-allergic adults by skin test and radioallergosorbent test. Ann. Allergy Asthma Immunol. 77:48–54.
110. Hlady, W. G., R. C. Mullen, and R. S. Hopkin. 1993. *Vibrio vulnificus* from raw oysters. Leading cause of reported deaths from foodborne illness in Florida. J. Fla. Med. Assoc. 80:536–538.
111. Hokama, Y. 1990. Simplified solid-phase immunobead assay for detection of ciguatoxin. J. Clin. Lab. Anal. 4:213–217.
112. Hungerford, J. M., and M. M Wekell. 1992. Analytical methods for marine toxins, in A. T. Tu (ed.), Food Poisoning. Marcel Dekker, New York, pp. 415–473.
113. Huss, H. H., P. K. B. Embarek, and V. Jeppesen. 1995. Control of biological hazards in cold smoked salmon production. Food Control 6:335–340.
114. ICMSF. 1988. HACCP in Microbiological Safety and Control. Blackwell Scientific Publications, Oxford.
115. Ingham, S. C., and N. N. Potter. 1988. Survival and growth of *Aeromonas hydrophila, Vibrio parahaemolyticus,* and *Staphylococcus aureus* on cooked mince and surimis made from Atlantic pollock. J. Food Prot. 51:634–638.
116. Iversen, A. M., M. Gill, C. L. R. Bartlett, W. D. Cubitt, and D. A. McSwiggan. 1987. Two outbreaks of foodborne gastroenteritis caused by a small round structured virus: Evidence of prolonged infectivity in a food handler. Lancet 2 (8558):556–558.
117. Jemmi, T. 1993. *Listeria monocytogenes* in smoked fish: An overview. Arch. Lebensmittelhyg. 44:10–13.
118. Jemmi, T., and A. Keusch. 1994. Occurrence of *Listeria monocytogenes* in freshwater fish farms and fish-smoking plants. Food Microbiol. 11:309–316.
119. Josephson, E. S. 1991. Health aspects of food irradiation. Food Nutr. Bull. 13:40–42.
120. Joy, C. P. H. 1993. An economic and efficient method for treatment of effluent water from seafood processing plants. Seafood Export J. 25:50–52.

121. Kim, C. R., J. O. Hearnsberger, and J. B. Eun. 1995. Gram-negative bacteria in refrigerated catfish fillets treated with lactic culture and lactic acid. J. Food Prot. 58:639–643.
122. Kobayashi, J. 1978. Pollution by cadmium and the itai-itai-disease in Japan, in F. W. Oehme (ed.), Toxicity of Heavy Metals in the Environment. Marcel Dekker, New York, pp. 199–260.
123. Kohn, M. A., T. A. Farley, T. Ando, M. Curtis, S. A. Wilson, Q. Jin, S. S. Monroe, R. C. Baron, L. M. McFarland, and R. I. Glass. 1995. An outbreak of Norwalk virus gastroenteritis associated with eating raw oysters. Implications for maintaining safe oyster beds. JAMA 273:466–471.
124. Kruse, R., K. Boek, and M. Wolf. 1983. Der Gehalt an Organochlor-Pestiziden und polychlorierten Biphenylen in Elbaalen. Arch. Lebensmittelhyg. 34:81–86.
125. Kruse, R., and K. E. Krüger. 1984. Untersuchungen von Nordseefischen auf Gehalte an toxischen Schwermetallen und chlorierten Kohlenwasserstoffen im Hinblick auf lebensmittelrechtliche Bestimmungen. Arch. Lebensmittelhyg. 35:128–131.
126. Kveim Lie, K. 1992. Botulism associated with rakfish in Norway, in WHO Surveillance Programme for Control of Foodborne Infections and Intoxications in Europe. Institute of Veterinary Medicine, Berlin.
127. Le Guyader, F., V. Apaire Marchais, J. Brillet, and S. Billaudel. 1993. Use of genomic probes to detect hepatitis A virus and enterovirus RNAs in wild shellfish and relationship of viral contamination to bacterial contamination. Appl. Environ. Microbiol. 59:3963–3968.
128. Lehrer, S. B. 1993. Seafood allergy. Introduction. Clin. Rev. Allergy 11:155–157.
129. Lewis, R. J., and M. J. Holmes. 1993. Origin and transfer of toxins involved in ciguatera. Comp. Biochem. Physiol. C 106:615–628.
130. Liston, J. 1989. Current issues in food safety—Especially seafoods. J. Am. Diet. Assoc. 89:911–913.
131. Long, A. R., and J. E. Roybal. 1994. Drug residues in foods of animal origin, in Y. H. Hui, J. R. Gorham, K. D. Murrell, and D. O. Cliver (eds.), Foodborne Disease Handbook, Vol. 3. Marcel Dekker, New York, pp. 529–554.
132. Lopez Sabater, E. I., J. J. Rodriguez Jerez, M. Hernandez Herrero, and M. T. Mora Ventura. 1996. Incidence of histamine-forming bacteria and histamine content in scombroid fish species from retail markets in the Barcelona area. Int. J. Food Microbiol. 28:411–418.
133. MacLean, J. D., J. R. Arthur, B. J. Ward, T. W. Gyorkos, M. A. Curtis, and E. Kokoskin. 1996. Common-source outbreak of acute infection due to the North American liver fluke *Metorchis conjunctus.* Lancet 347:154–158.
134. Madden, J. M., B. A McCardell, and J. G. J. Morris. 1989. *Vibrio cholerae,* in M. P. Doyle (ed.), Foodborne Bacterial Pathogens. Marcel Dekker, New York, pp. 525–542.
135. Marcaillou Le Baut, C., Z. Amzil, J. P. Vernoux, Y. F. Pouchus, M. Bohec, and J. F. Simon. 1994. Studies on the detection of okadaic acid in mussels: Preliminary comparison of bioassays. Nat. Toxins 2:312–317.
136. Margolis, L. 1977. Public health aspects of "codworm" infection: A review. J. Fish. Res. Bd. Can. 34:887–898.
137. Marshall, B., A. Pignault, F. Le Querrec, S. Cluzan, and A. Lepoutre. 1992. Les toxi-

infections alimentaires collectives en 1991. Bull. Epidemiol. Hebdom. 32/92:153–155.
138. Martin, J. L., K. Haya, and D. J. Wildish. 1993. Distribution and domoic acid content of *Nitzschia pseudodelicatissima* in the bay of Fundy, in T. J. Smayda, and Y. Shimizu (eds.), Toxic Phytoplankton Blooms in the Sea. Elsevier, Amsterdam, pp. 613–618.
139. Martin, S. E., and E. R. Myers. 1994. *Staphylococcus aureus,* in Y. H. Hui, J. R. Gorham, K. D. Murrell, and D. O. Cliver (eds.), Foodborne Disease Handbook, Vol. 1. Marcel Dekker, New York, pp. 345–394.
140. Maurelli, A. T., and K. A. Lampel. 1994. *Shigella,* in Y. H. Hui, J. R. Gorham, K. D. Murrell, and D. O. Cliver (eds.), Foodborne Disease Handbook, Vol. 1. Marcel Dekker, New York, pp. 319–343.
141. McCaustland, K. A., W. W. Bond, D. W. Bradley, J. W. Ebert, and J. E. Maynard. 1982. Survival of hepatitis A virus in feces after drying and storage for 1 month. J. Clin. Microbiol. 16:957.
142. McDonnell, S., K. B. Kirkland, W. G. Hlady, C. Aristeguieta, R. S. Hopkins, S. S. Monroe, and R. I. Glass. 1997. Failure of cooking to prevent shellfish-associated viral gastroenteritis. Arch. Intern. Med. 157:111–116.
143. McEvily, A. J., R. Iyengar, and S. Otwell. 1991. Sulfite alternative prevents shrimp melanosis. Food Technol. 45:80–86.
144. Meyer, M. C. 1970. Cestode zoonoses of aquatic animals. J. Wildl. Dis. 6:249–254.
145. Millard, J., H. Appleton, and J. V. Parry. 1987. Studies on heat inactivation of hepatitis A virus with special reference to shellfish. Epidemiol. Infect. 98:397–414.
146. Möller, H. 1978. The effect of salinity and temperature on the development and survival of fish parasites. J. Fish. Biol. 12:311–323.
147. Möller, H., and K. Anders. 1986. Diseases and Parasites of Marine Fish. Möller, Kiel.
148. Monk, J. D., L. R. Beuchat, and M. P. Doyle. 1995. Irradiation inactivation of foodborne microorganisms. J. Food Prot. 58:197–208.
149. Morse, D. L., G. S. Birkhead, and J. J. Guzewich. 1994. Investigating foodborne disease, in Y. H. Hui, J. R. Gorham, K. D. Murrell, and D. O. Cliver (eds.), Foodborne Disease Handbook, Vol. 1. Marcel Dekker, New York, pp. 547–603.
150. Morse, D. L., J. J. Guzewich, J. P. Hanrahan, R. Stricof, M. Shayegani, R. Deibel, J. C. Grabaw, N. A. Nowak, J. E. Herrimann, G. Cukor, and N. R. Blacklow. 1986. Widespread outbreaks of clam- and oyster-associated gastroenteritis—Role of Norwalk virus. N. Eng. J. Med. 314:678–681.
151. Muriana, P. M. 1996. Bacteriocins for control of *Listeria* spp. in food. J. Food Prot., suppl. 1996:54–63.
152. Nightingale, S. L. 1990. Seafood safety. Am. Fam. Physician 42:1657–1658.
153. Notermans, S., and A. van de Giessen. 1993. Foodborne diseases in the 1980s and 1990s. Food Control 4:122–124.
154. Oliver, J. D. 1989. *Vibrio vulnificus,* in M. P. Doyle (ed.), Foodborne Bacterial Pathogens. Marcel Dekker, New York, pp. 569–600.
155. Omil, F., R. Mendez, and J. M. Lema. 1996. Anaerobic treatment of seafood processing waste waters in an industrial anaerobic pilot plant. Water Sanit. 22:173–181.

156. Park, E. D., D. V. Lightner, and D. L. Park. 1994. Antimicrobials in shrimp aquaculture in the United States: Regulatory status and safety concerns. Rev. Environ. Contam. Toxicol. 138:1–20.
157. Parry, J. V., and P. P. Mortimer. 1984. The heat sensitivity of hepatitis A virus determined by a simple tissue culture method. J. Med. Virol. 14:277.
158. Perl, T. M., L. Bedard, T. Kosatsky, J. C. Hockin, E. C. Todd, and R. S. Remis. 1990. An outbreak of toxic encephalopathy caused by eating mussels contaminated with domoic acid. N. Engl. J. Med. 322:1775–1780.
159. Peterson, D. A., L. G. Wolfe, E. P. Larkin, and F. W. Deingardt. 1978. Thermal treatment and infectivity of hepatitis A virus in human feces. J. Med. Virol. 2:201–206.
160. PFSE. 1997. Fight Bac: Four Simple Steps to Food Safety. Partnership for Food Safety Education.
www.fightbac.org.
161. Resurreccion, A. V. A., F. C. F. Galvez, S. M. Fletcher, and S. K. Misra. 1995. Consumer attitudes toward irradiated food: Results of a new study. J. Food Prot. 58:193–196.
162. Rippen, T. E. 1980. Understanding contaminants in fish. Michigan Sea Grant Extension Bulletin, E-1434.
163. Schindler, E. 1985. Blei-, Cadmium-, Kupfer- und Quecksilbergehalte in Thunfisch- und Muschelkonserven. Dtsch. Lebensm.-Rundsch. 81:218–220.
164. Schmidt-Lorenz, W. 1990. Is the storage of foods in refrigerators still sufficiently safe? Mitt. Geb. Lebensmittelhyg. 81:233–286.
165. Scholz, E., U. Heinricy, and B. Flehmig. 1989. Acid stability of hepatitis A virus. J. Gen. Virol. 70:2481.
166. Schreiber, W. 1981. Die Belastung von Fischereierzeugnissen mit Schwermetallen—Ein aktueller Überblick. Arch. Lebensmittelhyg. 32:145–149.
167. Scott, E. 1992. Food hygiene information and education—Is the consumer receiving the correct message? Third World Congress on Foodborne Infections and Intoxications 1992, Berlin. Institute of Veterinary Medicine, Berlin, pp. 949–950.
168. Silliker, J. H., R. P. Elliott, A. C. Baird-Parker, R. L. Bryan, J. H. B. Christian, D. S. Clark, J. C. Olson, and T. A. Roberts. 1980. Microbial Ecology of Foods, Vol. II, Food Commodities. Academic Press, New York.
169. Smith, J. W. 1984. The abundance of *Anisakis simplex* L3 in the body cavity and flesh of marine teleosts. Int. J. Parasitol. 14:491–495.
170. Snyder, O. P. 1996. Use of time and temperature specifications for holding and storing food in retail food operations. Dairy, Food Environ. Sanit. 16:374–388.
171. Souci, S. W., W. Fachmann, and H. Kraut. 1986. Food Composition and Nutrition Tables 1986/87, 3rd ed. Wissenschaftliche Verlagsgesellschaft mbH, Stuttgart.
172. Squires, F. H. 1986. Pareto analysis, in L. Walsh, R. Wurster, and R. T. Kimber (eds.). Quality Management Handbook. Marcel Dekker, New York.
173. Stevens, K. A., N. A. Klapes, B. W. Sheldon, and T. R. Klaenhammer. 1991. Nisin treatment for inactivation of *Salmonella* species and other gram-negative bacteria. Appl. Environ. Microbiol. 57:3613.
174. Sugieda, M., K. Nakajima, and S. Nakajima. 1996. Outbreaks of Norwalk-like

virus-associated gastroenteritis traced to shellfish: Coexistence of two genotypes in one specimen. Epidemiol. Infect. 116:339–346.

175. Sun, Y., and J. D. Oliver. 1995. Hot sauce: No elimination of *Vibrio vulnificus* in oysters. J. Food Prot. 58:441–442.

176. Tamplin, M., and C. C. Parodi. 1991. Environmental spread of *Vibrio cholerae* in Peru. Lancet 338:1216–1217.

177. Tanner, P., G. Przzewkas, R. Clark, M. Ginsberg, and S. Waterman. 1996. Tetrodotoxin poisoning associated with eating puffer fish transported from Japan—California 1996. MMWR 45:389–391.

178. Taylor, S. L. 1983. Monograph on histamine poisoning. CX/FH 83/11. Codex Alimentarius Commission, Rome.

179. Taylor, F. J. R. 1984. Toxic dinoflagellates: Taxonomic and biogeographical aspect with emphasis on *Protogonyaulax,* in E. P. Ragelis (ed.), Seafood Toxins. American Chemical Society, Washington, DC, pp. 77–97.

180. Taylor, S. L. 1986. Histamine food poisoning: Toxicology and clinical aspects. CRC Crit. Rev. Toxicol. 17:92.

181. Todd, E. C. D. 1989. Amnesic shellfish poisoning: A new seafood toxin syndrom, in E. Graneli, B. Sundstrom, L. Edler, and D. M. Anderson (eds.), Toxic Marine Phytoplankton. Elsevier, Amsterdam, pp. 504–508.

182. Todd, E. C. D. 1994. Emerging diseases associated with seafood toxins and other water-borne agents. Ann. NY Acad. Sci 740:77–94.

183. Todd, E. C. D. 1994. Surveillance of foodborne disease, in Y. H. Hui, J. R. Gorham, K. D. Murrell, and D. O. Cliver (eds.), Foodborne Disease Handbook, Vol. 1. Marcel Dekker, New York, pp. 461–536.

184. Todd, E. C. D., D. J. Brown, M. Rutherford, A. Grolla, H. Saarkoppel, A. Chan, J. Reffle, S. Ryan, and N. Jerrett. 1992. Illness associated with seafood. Can. Commun. Dis. Rep. 18:19–23.

185. Todd, E. C. D., T. Kuiper-Goodman, W. Watson-Wright, M. W. Gilgan, S. Stephen, J. Marr, S. Pleasance, M. A. Quilliam, H. Klix, H. A. Luu, and C. F. B. Holmes. 1993. Recent illnesses from seafood toxins in Canada: Paralytic, amnesic and diarrhetic shellfish poisoning, in T. J. Smayda and Y. Shimizu (eds.), Toxic Phytoplankton Blooms in the Sea. Elsevier, Amsterdam.

186. Tülsner, M. 1994. Fish Processing, Vol. II, Fish Products and Their Manufacture. Behr's Verlag, Hamburg.

187. Veciana-Nogues, M. T., S. Albala-Hurtado, A. Marine-Font, and M. C. Vidal-Carou. 1996. Changes in biogenic amines during the manufacture and storage of semipreserved anchovies. J. Food Prot. 59:1218–1222.

188. Welsh, P. C., and R. K. Sizemore. 1985. Incidence of bacteremia in stressed and unstressed populations of the blue crab, *Callinectes sapidus.* Appl. Environ. Microbiol. 50:420–425.

189. Werzberger, A., B. Mensch, B. Kuter, L. Brown, J. Lewis, R. Sitrin, W. Miller, D. Shouval, B. Wiens, G. Calandra, J. Ryan, P. Provost, and D. Nalin. 1992. A controlled trial of formaline inactivated hepatitis A vaccine in healthy children. N. Engl. J. Med. 327:453–457.

190. Wessels, S. and H. H. Huss. 1996. Suitability of *Lactococcus lactis* subsp. *lactis*

ATCC 11454 as a protective culture for lightly preserved fish products. Food Microbiol. 13:323–332.

191. Williams, H., and A. Jones. 1994. Parasitic Worms of Fish. Taylor and Francis, London.
192. Wong, H. C., L. L. Chen, and C. M. Yu. 1995. Occurrence of vibrios in frozen seafood and survival of psychrotrophic *Vibrio cholerae* in broth and shrimp homogenate at low temperatures. J. Food Prot. 58:263–267.
193. Work, T. M., A. M. Beale, L. Fritz, M. A. Quilliam, M. Silver, K. Buck, and J. L. C. Wright. 1993. Domoic acid intoxication of brown pelicans and cormorants in Santa Cruz, California, in T. J. Smayda and Y. Shimizu (eds.), Toxic Phytoplankton Blooms in the Sea. Elsevier, Amsterdam, pp. 643–649.
194. Worms, J., N. Bouchard, R. Cormier, K. E. Pauley, and J. C. Smith. 1993. New occurrences of paralytic shellfish poisoning toxins in the Southern Gulf of St. Lawrence, Canada, in T. J. Smayda and Y. Shimizu (eds.), Toxic Phytoplankton Blooms in the Sea. Elsevier, Amsterdam, pp. 613–618.
195. Yamaguti, S. 1975. A Synoptical Review of Digenetic Trematodes of Vertebrates with Special Reference to the Morphology of Their Larval Forms. Keigaku Publishing, Tokyo.
196. Yang, C. C., K. C. Han, T. J. Lin, W. J. Tsai, and J. F. Deng. 1995. An outbreak of tetrodotoxin poisoning following gastropod mollusc consumption. Hum. Exp. Toxicol. 14:446–450.
197. Yang, C. C., S. C. Liao, and J. F. Deng. 1996. Tetrodotoxin poisoning in Taiwan: An analysis of poison center data. Vet. Hum. Toxicol. 38:282–286.
198. Yasumoto, T., et al. 1980. Identification of *Dinophysis fortii* as a causative organism of diarrhetic shellfish poisoning. Bull. Jpn. Soc. Sci. Fish. 46:1405–1411.
199. Yasumoto T., M. Satake, M. Fukui, H. Nagai, M. Murata, and A. M. Legrand. 1993. A turning point in ciguatera study, in T. J. Smayda and Y. Shimizu (eds.), Toxic Phytoplankton Blooms in the Sea. Elsevier, Amsterdam, pp. 455–461.
200. Zahner, P. 1990. Hygienic risks of sous-vide products. Mitt. Geb. Lebensmittelhyg. 81:602–615.
201. Ziprin, R. L. 1994. *Salmonella,* in Y. H. Hui, J. R. Gorham, K. D. Murrell, and D. O. Cliver (eds.), Foodborne Disease Handbook, Vol. 1. Marcel Dekker, New York, pp. 253–318.

5

Safe Handling of Foods for High Risk Individuals

James L. Smith and Pina M. Fratamico
Agricultural Research Service, U.S. Department of Agriculture, Wyndmoor, Pennsylvania

I. INTRODUCTION

Outbreaks and sporadic cases of foodborne disease resulting from ingestion of foods contaminated with bacterial pathogens have had major public health and economic impacts in both industrialized and developing countries. In addition, within the past 20 years, a number of new pathogens have entered the food chain and new foodborne diseases have arisen. Various segments of the population, including those immunocompromised owing to illness or medications, the elderly, pregnant women, and young children, are at a higher risk of contracting foodborne infections. In addition to an increased susceptibility to foodborne infection in these individuals, gastrointestinal disease is usually more severe, and the probability of developing serious complications is increased. Foodborne illnesses caused by bacterial pathogens are usually preventable. Provision of training, education, and information on food handling to food producers, retailers, and foodservice personnel, especially those involved in food preparation for high risk individuals, should have a major impact on reducing the incidence of illness caused by foodborne pathogens.

II. HIGH RISK GROUPS IN THE POPULATION AND SUSCEPTIBILITY TO FOODBORNE DISEASE

The elderly (≥ 65 years), neonates and children (< 5 years), pregnant women, patients with malignancies, residents in nursing hospitals and related care facilities,

Table 1 Populations in the United States That May Be at High Risk for Foodborne Disease

Category	Number of individuals	Percent of total population	Ref.
Elderly, ≥ 65 years (1992)	32,284,000	12.66	130 [Table 13]
Children, < 5 years (1992)	19,512,000	7.65	130 [Table 13]
Pregnant women (1988)	6,341,000	2.49	130 [Table 108]
Cancer cases under care (1992)	4,081,312	1.60	3
Neonates (live births, 1992)	4,065,014	1.59	5
Residents in nursing and related care facilities (1991)	1,729,000	0.68	130 [Table 192]
Total HIV (AIDS) infections (January 1993)	630,000–897,000	0.25–0.35	104
Organ transplant patients (1992)	415,458	0.16	130 [Table 190]
Total population (1992)	255,082,000		130 [Table 13]

HIV/AIDS patients, and patients receiving organ transplants are at increased risk for foodborne disease in the United States (Table 1). These high risk individuals comprise approximately one-fourth of the total population. It is probable that similar percentages of the populations of other developed countries are at risk for foodborne disease.

The number of elderly individuals, constituting approximately 13% of the total population in the United States in 1992 (Table 1), is expected to increase to 20% by the year 2050 (130). Therefore, the number of individuals in elder care and nursing home facilities will rise, and concurrent increase in the incidence of disease is anticipated. The second largest high risk group, children younger than 5 years of age (Table 1), is expected to decrease from 7.65% in 1992 to 6.5% of the population by the year 2050 (130).

Other groups at risk for foodborne disease include individuals who possess the HLA-B27 tissue antigen, individuals with iron overload problems, the malnourished (particularly children and the aged), hypochlorhydric or acholorhydric (decrease in or loss of gastric acidity) individuals, and individuals on opioids (medication with opium derivatives). HLA-B27 positivity is associated with reactive arthritis and Reiter syndrome. These arthritides are triggered by *Campylobacter jejuni, Salmonella typhimurium, S. enteritidis,* and other *Salmonella* species, *Shigella dysenteriae, S. flexneri, S. sonnei,* or *Yersinia enterocolitica* gastroenteritis (114). Approximately 2% of a population that contracts gastroenteritis on exposure to a triggering infection will acquire reactive arthritis or Reiter syndrome, but 10–20% of HLA-B27-positive individuals in that population will contract arthritis (22,114). The presence of the HLA-B27 antigen is not necessary for pathogenesis, since HLA-B27-negative individuals also can become arthritic following episodes of gastroenteritis; however, it does appear that the arthritides in HLA-B27-positive individuals are the more severe (73).

In vertebrates, an iron-withholding system (e.g., iron binding to apoferritin, transferrin, or lactoferrin) has evolved which functions to contain, transport, and detoxify iron. In addition, the iron-withholding system serves to deprive invading microorganisms of the iron necessary for their multiplication in the host (76, 89,132). Hypoferremia limits infection, whereas an excess of iron (iron loading) enhances susceptibility to microbial infection (83). Iron loading can occur in humans as a result of multiple blood transfusions, erythroid hyperplasia (anemias marked with a high degree of ineffective erythropoiesis, e.g., thalassaemia or other diseases in which there is excessive hemolysis), hepatic cirrhosis, or HLA-linked hemochromatosis (137). Studies with animals and humans indicate that iron-loaded individuals are more susceptible than normal persons to infections by foodborne pathogens such as *C. jejuni* (51,72), *Listeria monocytogenes* (122), *Vibrio vulnificus* (19,75,136), *Y. enterocolitica* (39,87,103,120), and *Escherichia coli* (8,10).

Numerous studies indicate that malnutrition can lead to a state of secondary immunodeficiency (33), thus increasing susceptibility to infections. Malnutrition can also predispose an individual to diarrheal diseases (13,54). Malnutrition may exert its effect by compromising the immune and nonimmune defense systems and the regenerative capabilities of the host. Children younger than 2 years of age with moderate malnutrition (low weight for age of 60–75% of normal) have about a 60% increased risk of diarrhea (112). Not only do children with moderate malnutrition experience more diarrhea than their well-nourished counterparts, but they have increased multiple diarrheic episodes (46) and higher risk of the diarrhea becoming persistent (12).

Studies have shown that a large percentage of elderly nursing home residents suffer from protein calorie malnutrition (62). Feeding difficulties, malabsorption, underlying illnesses, and impaired mental function are some of the contributory factors to an increased frequency of malnutrition in these individuals. Chan et al. (33) reported that mice fed a low protein diet succumbed to fulminant and rapidly fatal *Mycobacterium tuberculosis* infection, whereas mice fed a full protein diet survived the infection. Several components of the cell-mediated immune response were compromised in malnourished animals. The synergism between malnutrition and infections suggest that malnourished individuals (children, homeless, poor, aged, and HIV-infected persons, and people suffering from chronic diseases) in both industrialized and developing countries are at increased risk for foodborne diseases.

Individuals with hypochlorhydria or achlorhydria are at increased risk for gastrointestinal diseases. Malnutrition, gastritis, surgery, and certain drugs such as antacids, histamine H_2 receptor antagonists, or marijuana can cause loss of gastric acid (20). As people age, gastric secretion as well as acid production decreases and may be eliminated entirely in the very old (1,35). The newborn produce acid and pepsin normally in response to foods and hormonal stimuli (113); however, premature neonates, infants in intensive care units, and undernourished children have reduced acid and pepsin synthesis.

Hydrochloric acid, pepsin, lysozyme, and mucin make up the gastric juices, which are the chemical barriers that provide a defense against ingested microorganisms (54). In vitro, most bacteria do not survive in normal or artificial gastric juice, but bacteria exposed to achlorhydric gastric juice survive for long periods (40,57). Gastric juice is bactericidal within 30 minutes for *Salmonella* species, *Vibrio cholerae,* and *Shigella* species, and it is likely that other pathogens are also inactivated by gastric juice. After gastric surgery, higher incidences of *Salmonella, Giardia lamblia, V. cholerae,* and *Shigella* infections are seen (54). Ingestion of 10^8 *V. cholerae* cells is necessary to cause disease in normal individuals, but the presence of only 10^4 organisms leads to clinical symptoms when gastric juice is neutralized in vivo (24,67). In cholera outbreaks that occurred in Israel and Italy, individuals who had undergone gastrectomy generally suffered a more severe disease (58,107). Thus, the results of various studies indicate that achlorhydria predisposes to cholera and other gastrointestinal infections.

Opioids are immunosuppressive and have been shown to affect development, differentiation, and function of various immune cells including T and B cells, macrophages, and natural killer cells (105). Eisenstein et al. (45) have shown that morphine increases the susceptibility of mice to *L. monocytogenes* infection. In addition, morphine-treated mice were more likely to develop septicemia from endogenous normal enteric bacteria. The results obtained by Eisenstein et al. (45) suggest that patients who take opioids therapeutically or persons who abuse such substances will be more vulnerable to infections by foodborne pathogens as well as to infections by other organisms.

III. ADVICE FOR SPECIFIC GROUPS OF HIGH RISK INDIVIDUALS

A. Elderly Individuals (> 65 years) in Nursing Homes

Levine et al. (80) discussed 115 recorded foodborne disease outbreaks that occurred in nursing homes for the years 1975–1987. Fifty-two of the outbreaks involved a specific pathogen, such as *Bacillus cereus, C. jejuni, Clostridium perfringens, E. coli* O157:H7, *Salmonella* species, *Staphylococcus aureus,* and *Giardia lamblia. Salmonella* accounted for most of the foodborne illness seen in nursing home patients: 1004 patients out of a total of 4944 (20.3%) suffered from salmonellosis. Death occurred in 51 (1%) of the 4944 cases, and *Salmonella* spp. were responsible for 75.4% of these deaths. *S. enteritidis* was responsible for 50% of the *Salmonella* deaths (80). In outbreaks in which both the causal agent and the food were identified, meat, poultry, eggs, or foods that contained these items were implicated in approximately two-thirds of the cases. Improper storage, improper holding temperature, poor personal hygiene, contaminated equipment, and/or inadequate cooking were found to be the types of food handling errors involved in most of these nursing home outbreaks (80). Mishu et al. (90), analyzing *S. enteritidis* outbreaks for the period 1985–1991, found that 15.5% of foodborne *S. enteritidis* outbreaks occurred in hospitals or nursing homes. Eggs were the food vehicle in 25.4% of the 59 outbreaks in which the causative food was identified. During the 1985–1991 period, *S. enteritidis* was responsible for 50 deaths in all outbreaks; 45 of those deaths (90%) occurred in nursing homes or hospitals (90). The studies by Levine et al. (80) and Mishu et al. (90) indicate that nursing home residents are at high risk for foodborne diseases, particularly those caused by *Salmonella* spp.

The characteristics of nursing home residents render them more susceptible to respiratory and gastrointestinal infections, including foodborne illnesses. Age-related decreased immunity, underlying illnesses, diminished physiological functions, decreased gastric acidity (due to age or use of antacids), and use of antimicrobial agents make the elderly particularly vulnerable to foodborne infections (80,81). Furthermore, in heavily populated wards where residents share the same facilities and sometimes exchange or share food, person-to-person transmission of disease plays an increased role.

Since many ill elderly individuals lack the competence to adequately evaluate a nursing home in terms of cleanliness, food handling, care, etc., it will be necessary for their agents to make such decisions for them. The agent who contemplates placing an elderly person in a nursing home must inspect the facility and the facility kitchen for cleanliness. In addition, the agent should determine whether the kitchen and foodservice staff wear clean apparel, wear gloves when handling food, and present a clean appearance. The agent should read written standard operating procedures (SOPs) concerning food procurement, preparation, and service; procedures for handling a food-borne infection, maintenance of nursing home cleanliness, and personnel cleanliness also should be ascertained. If written SOPs are not available, the nursing home should not be considered. Regular training sessions on food safety, cleanliness, and infection control should be in place. The agent also should do a “toilet check” of the public toilets in the institution to determine their cleanliness.

Other factors to be considered include size of the nursing home, the policy on exchange of staff between different units, and paid leave for sick employees. Li et al. (81) have pointed out that the larger the institution, the more likely it is that infections will occur. This may be attributed to the increased difficulty in managing a larger number of residents and staff. A large facility will have a higher number of residents visiting one another, which in turn may lead to increased infections by person-to-person contact. A large facility will also have increased numbers of outside visitors (friends and relatives of nursing home residents), as well as a larger staff; the increased number of nonresidents may increase the probability of introducing pathogens into the facility (81). Another factor that can increase pathogen introduction into nursing homes is the practice of allowing the staff to shift between different units of the nursing home. Such shifts in working patterns can introduce pathogens from an infected unit to an uninfected one such as the kitchen (81). The granting of paid sick leave to kitchen and foodservice staff suffering from diarrhea could reduce the likelihood of the introduction of foodborne pathogens into foods (81).

The agent should eat several meals at the nursing home before making a decision. Foods that are normally served hot should be hot, and cold foods should be served cold. Meats and meat dishes should not be undercooked, and eggs should be cooked thoroughly. The agent who must select a nursing home does not have an enviable task and simply must do the best that can be done. If there is any doubt about the standards of the nursing home, the agent should reject it from consideration.

B. Elderly Individuals in Elder Care Centers

The competent, healthy elderly person (≥ 65 years) contemplating entering an elder care center where housing, meals, and housekeeping are provided for a fee should inspect the facility in much the same way as the agent acting for an individual entering a nursing home. The facility should be inspected for general cleanliness, and there should be written SOPs for food procurement, preparation, and serving. There should be regular training programs concerning proper food handling for employees of the facility. The individual who lives in an elder care facility should examine all foods to determine whether they have been properly cooked (meats not undercooked and eggs cooked thoroughly) and served (hot foods served hot and cold foods served cold). In other words, a commonsense approach to foods by the elderly resident should help in preventing foodborne illness.

C. Infants and Small Children in Day Care Centers

In North America there has been an increase in single-parent families and an increase in the number of women working outside of the home. Approximately one-half the mothers in the United States who work away from home have children younger than 1 year of age (56,65,79). Therefore, there has been a demand for out-of-home care for infants and young children. Children who attend day care centers have an elevated risk of contracting respiratory and gastrointestinal tract infections. In addition, the child care providers, parents, siblings, and friends of the children in day care are at increased risk for infections (65). The unhygienic behavior of the young child coupled with a naive immune system leads to the ready transmission of pathogens in the crowded day care environment, with possible further spread into the community after an infected child leaves the center. Figure 1 demonstrates how an enteric infection can spread from an infected child throughout the child care center and into the community at large.

Since young children frequently explore their world through oral experience, enteric diseases are quite common in day care centers (65,95). Pathogens that are potentially foodborne and have caused outbreaks in day care centers in-

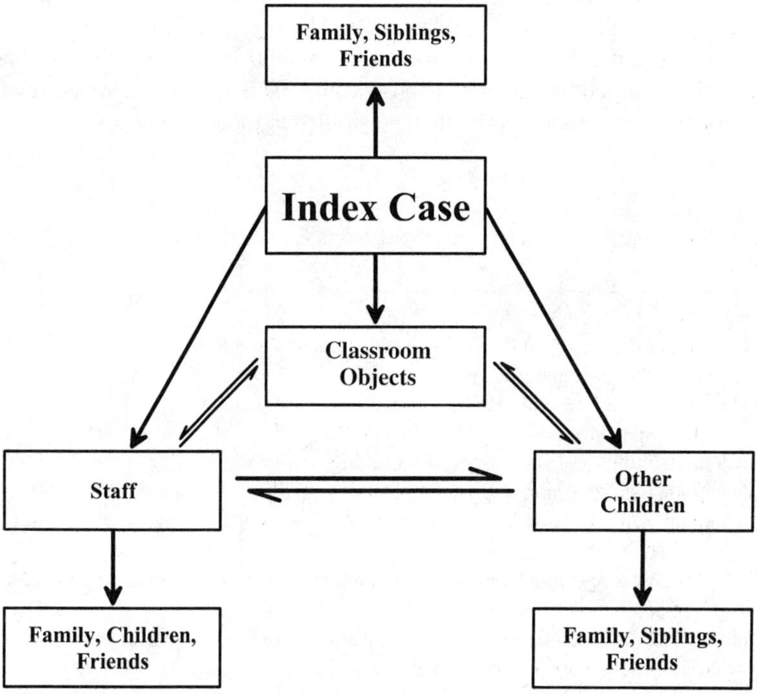

Figure 1 Routes of transmission of an enteric infection in a children's day care center and into the community. (Modified from Ref. 47.)

clude *Aeromonas, Campylobacter, Giardia, Salmonella,* enteropathogenic *E. coli* (65,95), *B. cereus* (27), *Cryptosporidium* (38), enterohemorrhagic *E. coli* (11,99, 121), and *Shigella sonnei* (64,91,124).

In spite of the number of reported outbreaks caused by organisms that are potentially foodborne, transmission of pathogens in day care centers is generally person to person via the fecal–oral route and is rarely due to consumption of contaminated foods (65). However, fried rice served at a day care center was implicated as the source of a *B. cereus* outbreak (27).

The day care center environment, clientele, and staff present a number of attributes that facilitate the spread of enteric infections (Table 2). The presence of children who are not toilet-trained or who are in diapers appears to be the major contributor to the spread of diarrhea in day care centers (95). Day care centers that emphasize proper hand-washing habits and general cleanliness habits in both staff and children have fewer cases of diarrhea (14,38,64,65,91,95).

Other factors that are important in prevention of diarrhea spread include the following: (a) leftover infant formula should never be reused, and fresh formula should be prepared immediately prior to feeding; (b) all surfaces and walls in the day care center should be of material that can be easily washed and disinfected; (c) infants, toddlers, and older children should be segregated, with limited intermixing; (d) diaper-changing areas for infants should be restricted to the infant area, and there should be adequate hand-washing facilities; (e) the diaper-changing and toilet areas of toddlers should be restricted to toddlers and should have ad-

Table 2 Predisposing Characteristics That Lead to Enteric Diseases in Day Care Centers

1. Presence of young, non-toilet-trained children
2. Changing of diapers by staff who also prepare and serve food on a regular basis
3. Day care center operated for profit
4. Sole use of guidelines mandated by state regulatory agencies
5. Inadequate space for number of children served
6. Inadequate number of staff to care for children
7. Inadequate hand-washing and diaper changing facilities
8. Inadequate hand washing by staff, particularly after diaper changing; failure to use disposable gloves
9. Failure of day care center management to emphasize continual education of staff about basic hygiene
10. Failure to disinfect environment of day care center (diaper-changing area, surfaces that children touch, toys, classroom objects)
11. Inadequate toileting facilities for toddlers (both in terms of number of toilets and hand-washing sinks)

Source: Adapted from Refs. 38, 64, 65, 91, and 95.

equate hand-washing facilities; and (f) toilets for older children should be in a separate area and must have adequate hand-washing facilities (94).

Parents should select a day care center with great care, since their children will be spending a large part of each working day in the care of others. Many of the foregoing suggestions for the selection of a nursing home are applicable to day care centers. The day care center should have a clean appearance, and the caretakers should wear clean apparel and have good hygienic habits (regular hand washing, wearing of disposable gloves during diaper changing). The kitchen and serving personnel should present a clean appearance. The individuals who prepare and serve food should not be caretakers of children.

The day care center should have written SOPs for the preparation and serving of food, diaper changing and toileting of children, maintenance of center cleanliness, cleanliness of children's playthings, and for caretaker (and child) hand washing. In addition, there should be written SOPs on infection control and on the manner in which the center handles childhood illnesses. A policy of paid sick leave for employees will aid in prevention of the transfer of infections to children. Regular training sessions on child care and infection control should be in place for the caretaker staff. There should be separate areas for infants, toddlers, and older children; these areas should be spacious and not crowded. Diaper-changing facilities and toilet facilities should be sufficient for the number of children served, and there should be adequate hand-washing facilities for both adults and children.

D. HIV/AIDS Patients

The case definition of AIDS for adolescent and adult patients emphasizes the $CD4^+$ T cell count ($< 200/\mu L$) and a number of clinical conditions (26). Three of the opportunistic clinical infections that define AIDS are caused by pathogens that are potentially foodborne: *Cryptosporidium parvum, Salmonella* species, and *Toxoplasma gondii.* Other opportunistic foodborne infections that may be more common or more severe in AIDS patients are caused by *C. jejuni, L. monocytogenes, Shigella* species, and *V. vulnificus* (2,32,34,119). In comparison to the general population, HIV-infected persons have 20–100 times greater risk of contracting *Salmonella* infections, 39 times greater risk of *C. jejuni* infections, and 150–280 times greater risk of contracting *L. monocytogenes* infections (2).

To prevent salmonellosis as well as other foodborne infections, the HIV-infected person should eat only well-done meat and poultry (the meat center should not be pink and the expressed juices should be clear or pale yellow), should eat thoroughly cooked eggs (yolk is not "runny"), should not eat products prepared with raw eggs, and should eat well-washed vegetables and fruits; fruit juices should be pasteurized. Raw milk or products made with raw milk should not be consumed. Seafood and fish should be thoroughly cooked; eating of raw seafoods such as oysters or sushi is hazardous. Properly chlorinated and treated water should be used for drinking and washing (2,32). Cross-contamination of foods during storage and preparation can be prevented by keeping uncooked meats separate from produce, cooked foods, and ready-to-eat foods. Hands, cutting boards, kitchen counters, knives, and other food preparatory utensils should be sanitized and washed thoroughly after coming in contact with uncooked foods. Hands should be washed before handling any foods and between handling different foods (2).

All HIV-infected individuals and AIDS patients should avoid raw or undercooked meats. In particular, HIV-infected individuals who are seronegative for *T. gondii* should avoid undercooked or raw pork, lamb, and game. Fruits and vegetables should be well washed. Hands should be washed after handling raw meats and after gardening (cats like to defecate in soft garden soils). HIV-infected persons who own cats as pets should not clean the cat litter box. Cats should not be allowed to hunt (it is preferable to keep them indoors), and they should be fed commercial canned or dried food. Under no circumstances should cats be fed raw or undercooked meats (117,118).

In addition to avoiding raw animal products (milk and meat) to prevent listeriosis, the HIV-infected individual should avoid soft cheeses (Mexican-style cheese, feta, Brie, Camembert, and blue-veined cheese); however, hard cheeses, cream cheese, cottage cheese, and yogurt may be consumed. Leftover foods, ready-to-eat foods (hot dogs), and foods obtained from delicatessens should be thoroughly reheated before consumption (31a). Since handling and/or eating of raw or undercooked poultry, drinking of raw milk, and exposure to pets have been correlated with *C. jejuni* infections (2,115), HIV-infected persons, with their increased susceptibility to campylobacteriosis, must exercise caution (2).

Cryptosporidium infections are rarely foodborne, but drinking of raw milk and unpasterized apple cider or eating of salads handled by an infected food handler with poor personal hygiene have been implicated in foodborne disease (28,116). To prevent cryptosporidiosis, HIV-infected persons should avoid drinking raw milk or improperly treated water; also to be avoided are contact with pets and farm animals and contact with both animal and human fecal matter (32). Immunocompromised and HIV-infected individuals appear to have increased susceptibility to *Cyclospora cayetanensis,* a recently emerged food and waterborne protozoan pathogen (48b). Recent outbreaks of *Cyclospora* infections in the United States and Canada due to imported raspberries (31b) suggest that high risk individuals should be cautious about consuming uncooked produce and fruit imported from developing countries.

E. Medically Immunosuppressed Individuals, Iron-Loaded Persons, and Cancer Patients

Transplant recipients and other medically immunosuppressed individuals, cancer patients, and iron-loaded persons are more susceptible to bacterial and parasitic infections than are immunocompetent individuals (44,68,85,111,134). In general, they should follow the recommendations for HIV/AIDS patients as given in Sec. III.D.

F. Pregnant Women, the Fetus, Infants, and Children

Toxoplasmosis and listeriosis are foodborne diseases that have serious consequences for the fetus when a pregnant woman becomes infected. Toxoplasmosis is a common infection of animals and humans caused by the protozoan parasite *T. gondii;* however, disease is rare and clinical signs are generally not evident in immunocompetent hosts (43). Pregnancy in a woman who is seropositive for *T. gondii* poses no disease threat to the woman unless she is seriously immunocompromised. However, if a *T. gondii*–seronegative woman is infected during pregnancy, the fetus is at risk for congenital toxoplasmosis (100). Generally, the mother is asymptomatic or shows only mild illness. Approximately 60% of the fetuses escape infection. Of the infected babies born alive, approximately 26% show subclinical infection, about 10% show clinical signs at birth (6% have mild toxoplasmosis, whereas the remainder have severe disease), and approximately 3% die during the neonatal period (43). If the woman is infected by *T. gondii* late in pregnancy (third trimester), the incidence of transmission of the parasite to the fetus is high (approximately 60%); however, the disease in the neonate will be relatively mild or subclinical (37,135). There is less chance of fetal infection (10–15%) if the woman is infected during the first trimester; however, the risk of miscarriage, stillbirth, or severe clinical disease in the newborn is high. Congenital toxoplasmosis may involve the ocular, auditory, and/or central nervous system. Infants with subclinical toxoplasmosis who appear normal at birth may show neurological and/or ocular abnormalities months or years later (100,135). A minority of congenitally infected babies show severe neurological deficits at birth. Congenital toxoplasmosis poses a grave outcome for newborns; therefore, seronegative women who are pregnant must take special precautions to avoid *T. gondii* infection. Eating raw or undercooked meats and eating unwashed raw or uncooked vegetables or fruits are risk factors for *T. gondii* infections in pregnant women (17,71). In addition, pregnant women should wash their hands after handling raw meats and after gardening. Recommendations regarding cats and cat care are the same as those as for HIV/AIDS patients.

Since *L. monocytogenes* has a predilection for the fetoplacental unit, pregnancy predisposes to listeriosis (15,86). The route of infection for the fetus can be transplacental or ascending. Listeriosis in pregnancy may present as follows: (a) an asymptomatic or mildly ill woman who delivers an infected infant, (b) an actuely ill woman who enters labor prematurely and delivers a stillborn or severely affected infant, or (c) a woman who dies of severe listeriosis after giving birth to an uninfected infant (131). Neonatal listeriosis presents as early-onset (disease onset < 5 days after birth) or late-onset (disease onset > 5 days after birth) infection. In early onset cases, infection has occured in utero, and such infants usually are premature, with low birth weight. Septicemia is a major symptom observed, and the infant has respiratory distress, cyanosis, apnea, pneumonia, and widespread microabscesses. Mortality rates range from 15% to 50% (15,48a, 85,108). Infants with late-onset listeriosis appear normal at birth, and are infected during birth or from environmental sources after birth. Disease presents as meningitis in 94% of cases; mortality is lower than in early-onset listeriosis unless diagnosis is delayed. Long-term neurological sequelae are rare (15,48a,85,108). A woman who is an asymptomatic fecal carrier of *L. monocytogenes* may also have vaginal colonization with the possibility of ascending transmission of bacteria to the fetus or transmission of the organism to the infant during birth (15).

To protect herself and the fetus from *L. monocytogenes* infections during pregnancy, a woman should avoid undercooked or raw meats, unwashed raw vegetables, unpasteurized milk and products made from unpasteurized milk, and soft cheeses. Leftover foods that have been refrigerated should be thoroughly reheated. It is advisable to avoid food obtained from delicatessens, since such food items are commonly handled excessively and may have been subjected to temperature abuse. The pregnant woman should practice strict personal and food hygiene (31).

Neonates (< 1 month of age) and infants aged 1–11 months are at particular risk for salmonellosis. The highest incidence of *Salmonella* infections occur in infants aged 1–4 months (60). In 1994, 12.2% of the total cases of salmonellosis were found in infants younger than 1 year of age, and 14.8% of the cases occurred in the 1–4 year group. The number of cases declined for children aged 5–9 years (Table 3). The peak incidence of *Shigella* infection was found to occur in children aged 1–4 years (96). In 1994, 27.2% of shigellosis cases were reported in children aged 1–4 years; 19.7% of cases occurred in children aged 5–9 years, whereas only 2.2% of the cases occurred in infants (< 1 year of age) (Table 3). Probably as a re-

Table 3 Reported cases of *Escherichia coli* O157:H7, *Shigella,* and *Salmonella* Infections in Children, by Age, in the United States, 1994

		Age[a]		
	Total cases	<1 year	1–4 years	5–9 years
E. coli O157:H7 infections	1,420	28 (2.0)	304 (21.4)	176 (12.4)
Salmonella infections	43,323	5265 (12.2)	6425 (14.8)	2804 (6.5)
Shigella infections	29,769	645 (2.2)	8094 (27.2)	5856 (19.7)

[a]Numbers in parentheses represent percent of total number of cases.
Source: Modified from Ref. 29.

sult of lack of exposure, infants do not appear to be very susceptible to infections caused by *E. coli* O157:H7, whereas 21.4% of the cases reported in 1994 occurred in children aged 1–4 years (Table 3). In children, *E. coli* O157:H7 infections are serious, since in 5–10% of patients younger than 10 years of age, they can lead to hemolytic uremic syndrome. In the United States and Argentina, this syndrome is the leading cause of acute renal failure in children (7,84). It is probable that any serotype of *E. coli* that produces Shiga-like toxins will target young children; outbreaks and cases of bloody diarrhea and hemolytic uremic syndrome have been associated with non-O157 *E. coli* serotypes (70). *Y. enterocolitica* infections appear to be more common in children than in adults (39), with the highest incidence in children 5 years of age or younger (66,125).

Protection of infants and children from infections caused by *E. coli* strains producing Shiga-like toxins, *Salmonella, Shigella,* and *Y. enterocolitica*, demands strict attention to personal hygiene among caretakers and children, since person-to-person contact is important in transmission of these organisms. Advice given for protecting high risk adults from foodborne disease should be followed for young children. Emphasis should be placed on thorough cooking of meat, meat products, eggs, and egg products. In addition, children should not be allowed to consume unpasteurized milk or juices.

IV. GENERIC ADVICE FOR ALL HIGH RISK INDIVIDUALS

A. Advice for Noninstitutionalized High Risk Individuals

In the United States, for the period 1983–1992, 4810 foodborne outbreaks were reported. The home was the locale of 20.7% of the outbreaks, whereas 42.2% occurred in delicatessens, cafeterias, or restaurants (9,30). Todd (128) reported that 8676 foodborne outbreaks occurred in Canada during the period 1975–1984; food mishandling in foodservice establishments accounted for 32.9% of foodborne incidents, whereas 16.4% occurred in homes. In contrast to Canada and the United States, in countries such as England and Wales, Scotland, the Netherlands, Germany, and Spain, foodborne illness most commonly occurs in the home (110). This difference may be due in part to the fact that almost one-half (46%) of the food consumed in the United States in 1993 was not prepared in the home (69). Food consumed in restaurants and hotels accounted for 72% of food not prepared at home (69). Consumption of foods not prepared at home and consumption of food in nonhome locations reduce the consumers' control over how food is prepared, with concomitant increase in the risk of acquiring foodborne illness.

The noninstitutionalized high risk individual is at risk for foodborne disease when eating at home, cafeterias, restaurants, picnics and cookouts, parties, social dinners, banquets, potluck dinners, and other social and business occasions at which food is served. Meals eaten while traveling, especially in underdeveloped countries, may present a special hazard to high risk people (see Chapters 6 and 15).

When a family member is a high risk individual, the home food preparer must be particularly vigilant in the selection of food ingredients and in the handling, preparing, and serving of foods. The data presented in Table 4 indicate factors that have been shown to lead to foodborne outbreaks in homes. To prevent foodborne infections in high risk members of the household, the home food preparer must be aware of the contributory factors listed in Table 4.

The following procedures are important in preventing foodborne illness in homes. Raw vegetables and fruits should be washed thoroughly. Raw meats and poultry should be handled hygienically to avoid cross-contamination of other foods. Frequent hand washing during food preparation also will help prevent cross-contamination. Raw meats should be kept separate from raw vegetables, cooked foods, and ready-to-eat foods. Kitchen countertops and utensils used in food preparation, particularly after poultry preparation, should be disinfected with chlorine bleach after use with a contact time of 1–2 minutes.

All food ingredients and foods should be obtained from reliable suppliers; foods suspected of being temperature abused should not be brought into the home. Perishable foods should be purchased just before going home: that is, shopping should not continue while perishable foods are in the automobile trunk.

Foods should be thoroughly cooked or reheated; a thermometer should be used to verify final cooking or reheating temperatures. Foods that are meant to be kept cold or frozen should be stored at their proper temperatures. Frozen foods should not be thawed at room temperature but rather in the refrigerator or by

Table 4 Factors That Contributed to Foodborne Disease Outbreaks in Homes in the United States, 1973–1982

Contributory factor	Percent[a]
Contaminated raw food or ingredient	42.0
Inadequate cooking, canning, or heat processing	31.3
Food obtained from an unsafe source	28.7
Improper cooling	22.3
Lapse of 12 or more hours between preparation and eating	12.8
Food handled by person with poor personal hygiene	9.9
Inadequate reheating	3.5
Improper hot holding	3.2
Cross-contamination	3.2
Use of left overs 12 or more hours after preparation	2.6

[a]Total exceeds 100% because multiple factors implicated in some outbreaks.
Source: Modified from Ref. 16.

means of microwaving. Cooked food that is not eaten should be quickly cooled and stored at 5°C or below. Hot foods should be kept at 60°C or hotter until served. A good rule for home food preparers to follow is the "5–60" rule: cooked foods should be held at 5°C or less or at 60°C or more until served (Ref. 49, Ch. 3-501.16). Bacterial foodborne pathogens will not grow at or above 60°C, and few pathogens will grow at temperatures at or below 5°C (63). However, there are some foodborne pathogens that can grow, albeit slowly, at or below 5°C; these include *Aeromonas hydrophila, Clostridium botulinum* type E, enterotoxigenic *E. coli, L. monocytogenes,* and *Y. enterocolitica* (93).

Since leftovers are easily subjected to temperature abuse or to overly long storage (with the possibility of growth of pathogens that can multiply at refrigeration temperatures), they probably should not be consumed by high risk individuals. Another good rule to follow is "use food within 4 hours or throw it out." That is, food should be served or discarded within 4 hours after it has been removed from temperature control (Ref. 49, Ch. 3-501.19B). For individuals who are at high risk, however, it is preferable to freeze or refrigerate food within 2 hours after serving.

High risk individuals should not eat raw or undercooked meats. Heat denaturation of heme pigment defines the extent of "doneness" of meat. Well-done meat has a brown center, and the expressed juices are colorless or pale yellow (53). Raw or undercooked eggs should not be eaten; the yolk should not be "runny" in thoroughly cooked eggs. Only pasteurized milk should be consumed. The consumption of raw animal products (meat, eggs, milk) or foods containing raw animal products poses particular hazards for young, elderly, and immunocompromised people.

In approximately 10% of home-related foodborne outbreaks, poor personal hygiene of food handlers was a contributory factor (Table 4). The home food preparer must have a strict sense of personal hygiene; that is, hands should be thoroughly washed with soap and water after defecation; if there is an infant in the home, hands should be washed after changing diapers. It would be preferable for individuals suffering from diarrhea not to prepare food; this may not always be possible, however, and gloves should be worn when the rule must be ignored.

In noncommercial environments where food is prepared and eaten (social gatherings, receptions, parties, picnics, cookouts, etc.), the high risk individual must be very cautious about which foods are consumed. For example, sporadic infections of *E. coli* O157:H7 correlated with eating ground beef in private homes, at noncommercial picnics, or at barbecues, but did not correlate with eating ground beef in commercial establishments such as fast food outlets or restaurants (78). Undercooked ground beef was more commonly served in noncommercial environments owing to food handlers' lack of training and lack of equipment appropriate for cooking food for large groups. In addition, guests may be unwilling to criticize the host's cooking or food handling standards and may accept undercooked ground beef that they normally would not eat (78). The high risk individual may find it necessary to decline dining invitations if the food handling standards of the host are not acceptable. At times, it may even be necessary to refuse to eat foods the high risk individual feels may be unsafe. Being impolite may prevent a serious foodborne illness when one is at high risk.

How can the high risk individual select a restaurant (or other commercial eating establishment) that should be a reasonably safe place to dine? If the restaurant does not appear to be clean (dirty windows and floors, no clean table covers, etc.) or if the serving personnel are not wearing clean apparel, the high risk person should not patronize the establishment. The "toilet check" is another good way to evaluate the cleanliness of a restaurant. A clean public toilet generally indicates that the restaurant management is committed to cleanliness. Good food handling is indicated when hot foods are served hot and cold foods are served cold, when meats are not undercooked and when eggs are thoroughly cooked. When consuming food, the noninstitutionalized high risk individual must practice enlightened self-interest combined with common sense.

B. Advice for Institutionalized High Risk Individuals or Their Agents

Competent individuals (e.g., ≥ 65 aged person seeking an elder care facility), the agents for noncompetent ill individuals (e.g., seeking nursing homes for sick elderly patients, AIDS patients, etc.), or the parents of minor children (seeking a day care center) must evaluate the care facility in terms of cleanliness, food handling, care, safety, etc. It is not unreasonable for the evaluator to check the facility's kitchen for cleanliness and to check both the kitchen and serving personnel for a clean appearance. Written SOPs for food handling, cleanliness, food handling training sessions, shifting of employees from one area of the facility to another, and granting of paid sick leave to employees who are food handlers should be available. A policy on stool testing of staff who have recently suffered from gastrointestinal infections to determine whether they are shedding or are carriers of pathogens is a useful practice; however, it may not be feasible. By following these guidelines, the evaluator can be assured of selecting an institution that is safe and clean. The Handbook of Child and Elder Care Resources (92) presents a list of questions that should prove valuable for individuals selecting nursing homes or child day care centers.

V. ECONOMIC ASPECTS OF FOODBORNE DISEASE IN HIGH RISK GROUPS

The number of cases of foodborne illness due to bacteria, viruses, parasites, seafood toxins, plant poisons, and chemical poisons in the United States has been estimated to be 12.6 million annually, with a cost of $8,426 million (4). The number of cases for Canada has been estimated at 2.1 million annually with a cost of $1,330 million (129). These estimates are probably conservative, and the actual number of cases and costs are likely to be higher. In these times of fiscal constraints, foodborne illness represents a significant drain on the economy in both Canada and the United States as well as in other countries of the world. It is probable that costs of foodborne illness in immunocompromised high risk groups are higher than the costs seen in immunocompetent individuals. However, there is a paucity of data concerning costs of foodborne disease to high risk groups. Several studies have been performed on determining the costs of toxoplasmosis and listeriosis to immunocompromised individuals.

A. Toxoplasmosis and AIDS

An AIDS patient who has acquired toxoplasmosis either as a new infection or from reactivation of an earlier infection must receive lifelong therapy to prevent relapse (32). Toxoplasmic encephalitis is the second most common opportunistic infection of the central nervous system in AIDS patients. Reactivation occurs in 10–50% of AIDS patients who are seropositive for *T. gondii* and have $CD4^+$ lymphocyte counts below $100/mm^3$ (101). Pulmonary and ocular toxoplasmosis in AIDS is less common than toxoplasmic encephalitis, but both the former condition require lifelong treatment (36,97). Gable et al. (55) have estimated that over $17,000 (1995 dollars) is needed to treat toxoplasmosis in each affected AIDS patient annually.

B. Congenital Toxoplasmosis

The number of congenital toxoplasmosis cases that occur each year in the United States is not accurately known; however, it has been estimated to range from 1 to 10 cases per 10,000 live births (135). In Canada, the estimate is 4–40 per 10,000 live births (23), and worldwide, the estimated incidence of congenital toxoplasmosis ranges from under 10 to 30 cases per 10,000 live births (77).

Roberts et al. (102) stated that the number of congenital toxoplasmosis cases in the United States ranges from 420 to 10,920 yearly, depending on the methodology used for the estimation. Thus, there is a large range in estimated costs depending on the number of cases that actually occur. Roberts et al. (102) estimated that the combined costs of medical treatment, loss in productivity, and special education and residential care due to congenital toxoplasmosis ranges from $0.4 to $8.8 billion (1992 dollars).

C. Listeriosis

Foods contaminated with *L. monocytogenes* are responsible for 85–95% of listeriosis seen in the United States (21). While the incidence of listeriosis has de-

Table 5 Economic Losses Due to Listeriosis in the United States, 1993

Category	Estimated number of cases	Estimated number of deaths	Estimated cost ($\$ \times 10^6$)	Cost/case ($\$ \times 10^3$)
Maternal	252	0	3.4	13.5
Fetal/new born	338–403	14–79	45.6–97.3	134.9–241.4
Adult (nonpregnant)	1248	431	163.7	660.1
Totals	1838–1903	445–510	212.7–264.4	

Source: Data from Ref. 21.

creased by about 40% in the United States since 1988 (123), *Listeria* infections still cause severe economic losses in the range of approximately $0.2 billion annually (Table 5).

Five studies (18,52,59,82,109) involving 478 cases of listeriosis in nonpregnant adults indicated that 91.8% (range 87.6–100%) of these adults were more susceptible to listeriosis because of an underlying disease or because they were over 65 years of age. Thus, most of the adult listeriosis cases in Table 5 occurred in immunocompromised, aged, or listeriosis-predisposed individuals.

VI. FURTHER RECOMMENDATIONS

Procedures could be introduced that would aid in preventing foodborne infections in high risk groups. These would include radiation pasteurization and/or sterilization of foods, inclusion of probiotics in the diets of high risk individuals, use of disposable foodservice items, education of both high risk individuals and their caretakers about proper food handling and hygiene, and finally, increased emphasis on proper hand washing.

A. Food Irradiation

Ionizing radiation of foods would appear to be an underutilized technique for the elimination of pathogenic microorganisms in foods. Ideally, radiation processing could be used to prepare microbiologically safe foods for high risk individuals, and Thayer et al. (127) have suggested that irradiation of food could be used as a means of protecting children and other high risk consumers from foodborne pathogens. Low dose irradiation (≤ 3 kGy) will eliminate most non-spore-forming pathogenic bacteria and parasites found in meat (98,126,127). In addition, low dose radiation can sensitize bacteria to subsequent heat treatment (98). Much higher radiation doses are necessary to achieve complete sterilization of foods (126,127). Foods irradiated under good manufacturing practices (GMP) are nutritionally adequate; toxic products are not induced by radiation; and radiation does not appear to induce in food microflora mutations that could result in microbiological risk (42).

B. Probiotics

Probiotics may be defined as viable bacteria, included as part of the diet, that influence the health of the host in a beneficial manner (106). Yasui and Ohwaki (138) demonstrated enhancement of both B-cell proliferation and antibody production in mouse Peyers patch cells cultured with *Bifidobacterium breve.* Lactic acid bacteria can stimulate nonspecific immunity by inducing the production of cytokines. Live lactic acid bacteria (*Bifidobacterium, Lactobacillus,* or *Lactococcus* spp.) were found to be more potent than gram-negative bacterial lipopolysaccharide as inducers of tumor necrosis factor α and interleukin 6 by human peripheral blood monocytes (88). There was a significant decrease in the incidence of acute diarrhea in young children who were administered probiotics (6). Administration of probiotics to individuals on antibiotic therapy was found to reduce the incidence of diarrhea by one-half (6). Antimicrobial treatment can alter the microflora of the intestinal tract and can cause diarrhea.

The feeding of a lyophilized lactic acid bacteria supplement (*Lactobacillus bulgaricus* and *Streptococcus thermophilus*) to normal human subjects resulted in potentiation of interferon gamma (IFN-γ) and increased the numbers of B cells and natural killer cells (41). In addition to producing a number of cytokines, natural killer cells have antiviral, antibacterial, and antitumor activity. Oral administration of Labre (*L. brevis* subsp. *coagulans*) to healthy humans resulted in enhanced production of IFN-α in a dose-dependent manner (74). Kishi et al. (74) suggested that administration of Labre in severely immunocompromised people would stimulate host immunity; however, they did not actually test its effects in these individuals.

Although probiotics partially survive transit through the gastrointestinal tract, they do not implant, and there is also little evidence that they carry out metabolic activity (133). Controlled studies are needed to determine whether probiotics, by stimulating the cellular and/or humoral immune systems, can protect the high risk population from foodborne pathogens.

C. Foodservice Disposables

Felix (50) has suggested that the use of disposable cups, plates, cutlery, etc., can make a positive contribution to public health. These one-time, one-person use items, since they are sanitary products, would eliminate bacterial contamination problems due to improper dishwashing and improper sanitizing of dishware and cutlery. The use of foodservice disposables by high risk individuals may be a useful adjunct to regimens for the prevention of foodborne disease.

D. Education

At any given time, approximately one-fourth of the general population belongs to a high risk group (Table 1). Therefore, the initiation of an educational program concerned with proper handling and hygiene of foods would enable all individuals, including those at high risk, to cope with problems of microbial food safety. If such educational programs were begun with children in kindergarten and continued through the high school years, future generations of consumers would become knowledgeable about food handling and food hygiene problems. In addition, it is imperative that public health departments provide continuing education about proper food handling for consumers who are no longer in the educational system. Such programs could include providing food safety information at supermarkets, in recipes, and through broadcast media.

There is a need for education of employees in foodservice areas. Employees in foodservice present a special problem: wages are low, benefits such as health insurance, paid sick leave, and paid vacations are lacking, and advancement opportunities are few. In 1988, 38% of employees in foodservice businesses had not completed their high school education, and in 1990, 42% of employees in foodservice businesses had worked less than a 1 year for their current employer (61). Thus, the lack of educational skills combined with high turnover rates makes it difficult to provide on-the-job training in food handling and hygiene.

Interestingly, Genshimer (56) pointed out that child care personnel, in many ways, have similar problems to those of employees in foodservice businesses. Child care employees are generally young women, most with a high school education but not trained in child development or the health aspects of child care. Few have training in hygienic food handling (even though more meals may be served in child care centers than in restaurants). Child caretaking is generally underpaid and is unattractive work to people who can find better paying jobs. Gensheimer (56) suggested that it is time to end placing small children, one of our most valuable assets, in the hands of undertrained and undersupervised caretakers. There is need to create a body of professional or paraprofessional individuals with the necessary training to properly care for young children—that is, to move the concept of the child caretaker from that of a babysitter to that of a skilled professional. A similar approach for people who work in foodservice areas would create a cadre of foodservice personnel who are professionals in food handling and food hygiene. The same approach to proper training would also be desirable for nursing home personnel.

E. Hand Washing

While the importance of proper hand washing has been emphasized in this chapter, it seems necessary to repeat it again as a final recommendation in the protection of high risk individuals from foodborne pathogens. In September 1996, the American Society for Microbiology initiated "Operation Clean Hands," an educational program designed to provide individuals with information demonstrating the importance of hand washing as a means of infection control (25). The program was conceived to protect pregnant women, children, the elderly, and the immunocompromised—the segments of the population who are most vulnerable to infections resulting from hand-to-hand, environment-to-hand, or hand-to-food contamination. Proper hand washing must always be viewed as the first line of defense against foodborne infections.

VII. CONCLUSIONS

To prevent foodborne-related illnesses, segments of the population at higher risk and those who care for these individuals require specific information concerning proper food handling and preparation. This chapter has addressed these issues. Adequate cooking and proper storage of foods and avoidance of raw animal products, unwashed raw vegetables, unpasteurized milk, and products made from unpasteurized milk should help to prevent foodborne diseases. Of paramount importance for high risk individuals is cleanliness in the handling, preparation, and serving of food. Perhaps the simplest and most important practice that high risk individuals and their caretakers should exercise is regular hand washing while handling foods.

REFERENCES

1. Abrams, W. B., M. H. Beers, and R. Berkow (eds.) 1995. The Merck Manual of Geriatrics, 2nd ed. Merck Research Laboratories, Whitehouse Station, NJ.
2. Angulo, F. J., and D. L. Swerdlow. 1995. Bacterial enteric infections in persons infected with human immunodeficiency virus. Clin. Infect. Dis. 21 (suppl. 1):S84–S93.
3. Anonymous. 1992. Cancer Facts & Figures—1992. American Cancer Society, Atlanta, table p. 15.
4. Anonymous. 1994. Foodborne pathogens: Risks and consequences: Table 6.6. Task Force Report 122, September 1994. Council for Agricultural Science and Technology, Ames, IA.
5. Anonymous. 1996. World Almanac and Book of Facts 1996. World Almanac Books, Funk & Wagnalls, Mahwah, NJ, p. 961.
6. Anonymous. 1996. Probiotics: Putting the good microbes to good uses. ASM News 62:456–457.
7. Ashkenazi, S. 1993. Role of bacterial cytotoxins in hemolytic uremic syndrome and thrombotic thrombocytopenic purpura. Annu. Rev. Med. 44:11–18.
8. Barry, D. M. J., and A. W. Reeve. 1977. Increased incidence of gram-negative neonatal sepsis with intramuscular iron administration. Pediatrics 60:908–912.
9. Bean, N. H., P. M. Griffin, J. S. Goulding, and C. B. Ivey. 1990. Foodborne disease outbreaks, 5-year summary, 1983–1987. J. Food Prot. 53:711–728.

10. Becroft, D. M. O., M. R. Dix, and K. Farmer. 1977. Intramuscular iron–dextran and susceptibility of neonates to bacterial infections. Arch. Dis. Child. 52:778–781.
11. Belongia, E. A., M. T. Osterholm, J. T. Soler, D. A. Ammend, J. E. Braun, and K. L. MacDonald. 1993. Transmission of *Escherichia coli* O157:H7 infection in Minnesota child day-care facilities. JAMA 269:883–888.
12. Bhandari, N., M. K. Bhan, S. Sazawal, J. D. Clemens, S. Bhatnagar, and V. Khoshoo. 1989. Association of antecedent malnutrition with persistent diarrhoea: A case-control study. Br. Med. J. 298:1284–1287.
13. Black, R. E., and C. F. Lanata. 1995. Epidemiology of diarrheal diseases in developing countries, in M. J. Blaser, P. D. Smith, J. I. Ravdin, H. B. Greenberg, and R. L. Guerrant (eds.), Infections of the Gastrointestinal Tract. Raven Press, New York, pp. 13–36.
14. Black, R. E., A. C. Dykes, K. E. Anderson, J. G. Wells, S. P. Sinclair, G. W. Gary, M. H. Hatch, and E. J. Gangarosa. 1981. Handwashing to prevent diarrhea in day-care centers. Am. J. Epidemiol. 113:445–451.
15. Bortolussi, R., and W. F. Schlech. 1995. Listeriosis, in J. S. Remington and J. O. Klein (eds.), Infectious Diseases of the Fetus and Newborn Infant, 4th ed. Saunders, Philadelphia, pp. 1055–1073.
16. Bryan, F. L. 1988. Risks of practices, procedures and processes that lead to outbreaks of foodborne diseases. J. Food Prot. 51:663–673.
17. Buffolano, W., R. E. Gilbert, F. J. Holland, D. Fratta, F. Palumbo, and A. E. Ades. 1996. Risk factors for recent toxoplasma infection in pregnant women in Naples. Epidemiol. Infect. 116:347–351.
18. Bula, C. J., J. Bille, and M. P. Glauser. 1995. An epidemic of food-borne listeriosis in western Switzerland: Description of 57 cases involving adults. Clin. Infect. Dis. 20:66–72.
19. Bullen, J. J., P. B. Spalding, C. G. Ward, and J. M. C. Gutteridge. 1991. Hemochromatosis, iron, and septicemia caused by *Vibrio vulnificus*. Arch. Intern. Med. 151: 1606–1609.
20. Butterton, J. R., and S. B. Calderwood. 1995. *Vibrio cholerae* O1, in M. J. Blaser, P. D. Smith, J. I. Ravdin, H. B. Greenberg and R. L. Guerrant (eds.), Infections of the Gastrointestinal Tract. Raven Press, New York, pp. 649–670.
21. Buzby, J. C., T. Roberts, C.-T. J. Lin, and J. M. MacDonald. 1996. Bacterial foodborne disease. Medical costs & productivity losses. Agricultural Economic Report No. 741. Economic Research Service, U.S. Department of Agriculture, Washington, DC, pp. 1–81.
22. Calin, A. 1988. Ankylosing spondylitis and the spondylarthropathies, in H. R. Schumacher (ed.), Primer on the Rheumatic Diseases, 9th ed. Arthritis Foundation, Atlanta, pp. 142–147.
23. Carter, A. O., and J. W. Frank. 1986. Congenital toxoplasmosis: Epidemiologic features and control. Can. Med. Assoc. J. 136:618–623.
24. Cash, R. A., S. I. Music, J. P. Libonati, M. J. Snyder, R. P. Wenzel, and R. B. Hornick. 1974. Response of man to infection with *Vibrio cholerae*. I. Clinical, serologic, and bacteriologic responses to a known inoculum. J. Infect. Dis. 129:45–52.
25. Cassell, G., and M. Osterholm. 1996. A simple approach to a complex problem. ASM News 62:516–517.

26. Centers for Disease Control and Prevention. 1992. 1993 revised classification system for HIV infection and expanded surveillance case definition for AIDS among adolescents and adults. MMWR 41 (No. RR-17):1–19.
27. Centers for Disease Control and Prevention. 1994. *Bacillus cereus* food poisoning associated with fried rice at two child day care centers—Virginia, 1993. MMWR 43:177–178.
28. Centers for Disease Control and Prevention. 1996. Foodborne outbreak of diarrheal illness associated with *Cryptosporidium parvum*—Minnesota, 1995. MMWR 45: 783–784.
29. Centers for Disease Control and Prevention. 1994. Summary of notifiable diseases, United States, 1994. MMWR 43 (53):10.
30. Centers for Disease Control and Prevention. 1996. Surveillance for foodborne-disease outbreaks—United States, 1988–1992. MMWR 45 (No. SS-5):1–66.
31. (a) Centers for Disease Control and Prevention. 1992. Update: foodborne listeriosis—United States, 1988–1990. MMWR 41:251, 257–258. (b) Centers for Disease Control and Prevention. 1996. Update: Outbreaks of *Cyclospora cayetanensis* infection—United States and Canada, 1996. MMWR 45:611–612.
32. Centers for Disease Control and Prevention. 1995. USPHS/IDSA guidelines for the prevention of opportunistic infections in persons infected with human immunodeficiency virus: A summary. MMWR 44 (No. RR-8):1–34.
33. Chan, J. Y. Tian, K. E. Tanaka, M. S. Tsang, K. Yu, P. Salgame, D. Carroll, Y. Kress, R. Teitelbaum, and B. R. Bloom. 1996. Effects of protein calorie malnutrition on tuberculosis in mice. Proc. Natl. Acad. Sci. USA 93:14857–14861.
34. Chang, T.-I., S. I. Peldton, and H. W. Winter. 1995. Enteric infections in HIV-infected children, in M. J. Blaser, P. D. Smith, J. I. Ravdin, H. B. Greenberg, and R. L. Guerrant (eds.), Infections of the Gastrointestinal Tract. Raven Press, New York, pp. 499–510.
35. Christiansen, J. L., and J. M. Gryzbowski. 1993. Biology of Aging. Mosby–Year-Book, St. Louis, MO.
36. Cochereau-Massin, I., P. LeHoang, M. Lautier-Frau, E. Zerdoun, L. Zazoun, M. Robinet, P. Marcel, B. Girard, C. Katlama, C. Leport, W. Rozenbaum, J. P. Coulaud, and M. Gentilini. 1992. Ocular toxoplasmosis in human immunodeficiency virus-infected patients. Am. J. Ophthalmol. 114:130–135.
37. Cook, G. C. 1990. *Toxoplasma gondii* infection: A potential danger to the unborn fetus and AIDS sufferer. Q. J. Med., New Ser. 74, No. 273:3–19.
38. Cordell, R. L, and D. G. Addiss. 1994. Cryptosporidiosis in child care settings: A review of the literature and recommendations for prevention and control. Pediatr. Infect. Dis. J. 13:310–317.
39. Cover, T. L. 1995 *Yersinia enterocolitica* and *Yersinia pseudotuberculosis,* in M. J. Blaser, P. D. Smith, J. I. Ravdin, H. B. Greenberg, and R. L. Guerrant (eds.), Infections of the Gastrointestinal Tract. Raven Press, New York, pp. 811–823.
40. Dare, R., J. T. Magee, and G. E. Matheson. 1972. In-vitro studies on the bacterial properties of natural and synthetic gastric juices. J. Med. Microbiol. 5:395–406.
41. De Simone, C., R. Vesely, B. Bianchi Salvadori, and E. Jirillo. 1993. The role of probiotics in modulation of the immune system in man and animals. In. J. Immunother. 9:23–28.

42. Diehl, J. F., and E. S. Josephson. 1994. Assessment of wholesomeness of irradiated foods (a review). Acta Alim. 23:195–214.
43. Dubey, J. P., and C. P. Beattie. 1988. Toxoplasmosis of Animals and Man. CRC Press, Boca Raton, FL.
44. Dummer, S., and B. M. Allos. 1995. Gastrointestinal infections in transplant recipients, in M. J. Blaser, P. D. Smith, J. I. Ravdin, H. B. Greenberg, and R. L. Guerrant (eds.), Infections of the Gastrointestinal Tract. Raven Press, New York, pp. 511–525.
45. Eisenstein, T. K., M. E. Hilburger, A. L. Truant, T. J. Rogers, J. J. Meissler, B. Satishchandran, and M. W. Adler. 1996. Opioids sensitize to endogenous and exogenous bacterial infection. J. Neuroimmunol. 69:41–42.
46. El Samani, E. F. Z., W. C. Willett, and J. H. Ware. 1988. Association of malnutrition and diarrhea in children aged under five years. A prospective follow-up study in a rural Sudanese community. Am. J. Epidemiol. 128:93–105.
47. Erkanem, E., H. L. Dupont, L. K. Pickering, B. J. Selwyn, and C. M. Hawkins. 1983. Transmission dynamics of enteric bacteria in day-care centers. Am. J. Epidemiol. 118:562–572.
48. (a) Farber, J. M., and P. I. Peterkin. 1991. *Listeria monocytogenes,* a food-borne pathogen. Microbiol. Rev. 55:476–511. (b) Farthing, M. J. G., M. P. Kelly, and A. M. Veitch. 1996. Recently recognized microbial enteropathies and HIV infection. J. Antimicro. Chemother. 37 (suppl. B):61–70.
49. FDA. 1993. Food Code 1993. U.S. Public Health Service, Food and Drug Administration, U.S. Department of Health and Human Services, Washington, DC.
50. Felix, C. W. 1990. Foodservice disposables and public health. Dairy Food Environ. Sanit. 10:656–660.
51. Field, L. H., V. L. Headley, S. M. Payne, and L. J. Berry. 1986. Influence of iron on growth morphology, outer membrane protein composition, and synthesis of siderophores in *Campylobacter jejuni.* Infect. Immun. 54:126–132.
52. Fleming, D. W., S. L. Cochi, K. L. MacDonald, J. Brondum, P. S. Hayes, B. D. Plikaytis, M. B. Holmes, A. Audurier, C. V. Broome, and A. L. Reingold. 1985. Pasteurized milk as a vehicle of infection in an outbreak of listeriosis. N. Engl. J. Med. 312:404–407.
53. Fox, J. B. 1987. The pigments of meat, in J. F. Price and B. S. Schweigert (eds.), The Science of Meat and Meat Products, 3rd ed. Food and Nutrition Press, Westport, CT, pp. 193–216.
54. Furuta, G. T., and W. A. Walker. 1995. Nonimmune defense mechanisms of the gastrointestinal tract, in M. J. Blaser, P. D. Smith, J. I. Ravdin, H. B. Greenberg, and R. L. Guerrant (eds.), Infections of the Gastrointestinal Tract. Raven Press, New York, pp. 89–97.
55. Gable, C. B., J. C. Tierce, D. Simison, D. Ward, and K. Motte. 1996. Costs of HIV^+/AIDS at $CD4^+$ counts disease stages based on treatment protocols. J. AIDS Hum. Retrovirol. 12:413–420.
56. Gensheimer, K. F. 1994. A public health perspective on child care. Pediatrics. 94 (6 Suppl. S):1108–1109.
57. Giannella, R. A., S. A. Broitman, and N. Zamcheck. 1973. Influence of gastric acidity on bacterial and parasitic enteric infections. Ann. Intern. Med. 78:271–276.

58. Gitelson, S. 1971. Gastrectomy, achlorhydria and cholera. Isr. J. Med. 7:663–667.
59. Goulet, V., and P. Marchetti. 1996. Listeriosis in 225 non-pregnant patients in 1992: Clinical aspects and outcome in relation to predisposing conditions. Scand. J. Infect. Dis. 28:367–374.
60. Hargrett-Bean, A. T. Pavia, and R. V. Tauxe. 1988. *Salmonella* isolates from humans in the United States, 1984–1986. CDC Surveillance Summaries, June 1988. MMWR 37 (No. SS-2):25–31.
61. Hedberg, C. W., K. L. MacDonald, and M. T. Osterholm. 1994. Changing epidemiology of food-borne disease: A Minnesota perspective. Clin. Infect. Dis. 18:671–682.
62. Henderson, C. 1988. Nutrition and malnutrition in the elderly nursing home patient. Clin. Geriatr. Med. 4:527–547.
63. Hobbs, B. C. 1974. Food Poisoning and Food Hygiene, 3rd ed. Edward Arnold, London.
64. Hoffman, R. E., and P. J. Shillam. 1990. The use of hygiene, cohorting, and antimicrobial therapy to control an outbreak of shigellosis. Am. J. Dis. Child. 144:219–221.
65. Holmes, S. J., A. L. Morrow, and L. K. Pickering. 1996. Child-care practices: Effects of social change on the epidemiology of infectious diseases and antibiotic resistance. Epidemiol. Rev. 18:10–28.
66. Hoogkamp-Korstanje, J. A. A., and V. M. M. Stolk-Engelaar. 1995. *Yersinia enterocolitica* infection in children. Pediatr. Infect. Dis. J. 14:771–775.
67. Hornick, R. B., S. J. Music, R. Wenzel, R. Cash, J. P. Libonati, M. J. Synder, and T. E. Woodward. 1971. The Broad Street pump revisited: Response of volunteers to ingested cholera vibrios. Bull. NY Acad. Med. 47:1181–1191.
68. Israelski, D. M., and J. S. Remington. 1993. Toxoplasmosis in the non-AIDS immunocompromised host. Curr. Clin. Top. Infect. Dis. 13:322–356.
69. Jensen, H., and L. Unnevehr. 1995. The economics of regulation and information related to foodborne microbial pathogens, in T. Roberts, H. Jensen, and L. Unnevehr (eds.), Tracking Foodborne Pathogens from Farm to Table: Data Needs to Evaluate Control Options. U.S. Department of Agriculture, Economic Research Service, Food and Consumer Economics Division, Miscellaneous Publication No. 1532. USDA, Washington, DC, pp. 124–139.
70. Johnson, R. P., R. C. Clarke, J. B. Wilson, S. C. Read, K. Rahn, S. A. Renwick, K. A. Sandhu, D. Alves, M. A. Karmali, H. Lior, S. A. McEwen, J. S. Spika, and C. L. Gyles. 1996. Growing concerns and recent outbreaks involving non-O157:H7 Serotypes of verotoxigenic *Escherichia coli*. J. Food Prot. 59:1112–1122.
71. Kapperud, G., P. A. Jenum, B. Stray-Pedersen, K. K. Melbey, A. Eskild, and J. Eng. 1996. Risk factors for *Toxoplasma gondii* infection in pregnancy. Results of a prospect case-control study in Norway. Am. J. Epidemiol. 144:405–412.
72. Kazmi, S. U., B. S. Roberson, and N. J. Stern. 1984. Animal-passed, virulence-enhanced *Campylobacter jejuni* causes enteritis in neonatal mice. Curr. Microbiol. 11:159–164.
73. Kingsley, G., and J. Sieper. 1996. Third International Workshop on Reactive Arthritis. An overview. Ann. Rheum. Dis. 55:564–570.

74. Kishi, A., K. Uno, Y. Matsubara, C. Okuda, and T. Kishida. 1996. Effect of the oral administration of *Lactobacillus brevis* subsp. *coagulans* on interferon-α producing capacity in humans. J. Am. College Nutr. 15:408–412.
75. Koenig, K. L., J. Mueller, and T. Rose. 1991. *Vibrio vulnificus.* West. J. Med. 155: 400–403.
76. Kontoghiorghes, G. J., and E. D. Weinberg. 1995. Iron: Mammalian defense systems, mechanisms of disease, and chelation therapy approaches. Blood Rev. 9:33–45.
77. Lappalainen, M., M. Koskiniemi, V. Hiilesmaa, P. Ämmälä, K. Teramo, P. Koskela, M. Lebech, K. O. Raivio, K. Hedman, and the Study Group. 1995. Outcome of children after maternal primary *Toxoplasma* infection during pregnancy with emphasis on avidity of specific IgG. Pediatr. Infect. Dis. J. 14:354–361.
78. Le Saux, N., J. S. Spika, B. Friesen, I. Johnson, D. Melnychuck, C. Anderson, R. Dion, M. Rahman, and W. Tostowaryk. 1993. Ground beef consumption in noncommercial settings is a risk factor for sporadic *Escherichia coli* O157:H7 infection in Canada. J. Infect. Dis. 167:500–502.
79. Levine, M. M., and O. S. Levine. 1994. Changes in human ecology and behavior in relation to the emergence of diarrheal diseases, including cholera. Proc. Natl. Acad. Sci. USA. 91:2390–2394.
80. Levine, W. C., J. F. Smart, D. L. Archer, N. H. Bean, and R. V. Tauxe. 1991. Foodborne disease outbreaks in nursing homes, 1975–1987. JAMA 266:2105–2109.
81. Li, J., G. S. Birkhead, D. S. Strogatz, and F. B. Coles. 1996. Impact of institution size, staffing patterns, and infection control practices on communicable disease outbreaks in New York State nursing homes. Am. J. Epidemiol. 143:1042–1049.
82. Linnan, M. J., L. Mascola, X. D. Lou, V. Goulet, S. May, C. Salminen, D. W. Hird, L. Yonekura, P. Hayes, R. Weaver, A. Audurier, B. D. Plikaytis, S. L. Fannin, A. Kleks, and C. V. Broome. 1988. Epidemic listeriosis associated with Mexican-style cheese. N. Engl. J. Med. 219:823–828.
83. Litwin, C. M., and S. B. Calderwood. 1993. Role of iron in regulation of virulence genes. Clin. Microbiol. Rev. 6:137–149.
84. López, E. L., M. M. Contrini, S. Devoto, M. F. deRosa, M. G. Graña, L. Aversa, H. F. Gomez, M. H. Genero, and T. G. Cleary. 1995. Incomplete hemolytic uremic syndrome in Argentinean children with bloody diarrhea. J. Pediatr. 127:364–367.
85. Lorber, B. 1990. Clinical listeriosis—Implications for pathogenesis, in A. J. Miller, J. L. Smith, and G. A. Somkuti (eds.), Foodborne Listeriosis. Elsevier, New York, pp. 41–49.
86. Luft, B. J., and J. S. Remington. 1982. Effect of pregnancy on resistance to *Listeria monocytogenes* and *Toxoplasma gondii* infections in mice. Infect. Immun. 38: 1164–1171.
87. Melby, K., S. Slørdahl, T. J. Guttenberg, and S. A. Nordbø. 1982. Septicaemia due to *Yersinia enterocolitica* after oral overdoses of iron. Br. Med. J. 285:467–468.
88. Miettinen, M., J. Vuopio-Varkila, and K. Varkila. 1996. Production of human tumor necrosis factor alpha, interleukin-6 and interleukin-10 is induced by lactic acid bacteria. Infect. Immun. 64:5403–5405.
89. Mietzner, T. A., and S. T. Morse. 1994. The role of iron-binding proteins in the survival of pathogenic bacteria. Annu. Rev. Nutr. 14:471–493.

90. Mishu, B., J. Koehler, L. A. Lee, D. Rodigue, F. H. Brenner, P. Blake, and R. V. Tauxe. 1994. Outbreaks of *Salmonella enteritidis* infections in the United States, 1985–1991. J. Infect. Dis. 169:547–552.
91. Mohle-Boetani, J. C., M. Stapleton, R. Finger, N. H. Bean, J. Poundstone, P. A. Blake, and P. M. Griffin. 1995. Communitywide shigellosis: Control of an outbreak and risk factors in child day-care centers. Am. J. Public Health 85:812–816.
92. Office of Personnel Management. 1993. Handbook of Child and Elder Care Resources. Superintendent of Documents, Washington, DC.
93. Palumbo, S. A. 1987. Can refrigeration keep our foods safe? Dairy Food Sanit. 7:56–60.
94. Petersen, N. J. and G. K. Bressler. 1986. Design and modification of the day care environment. Rev. Infect. Dis. 8:618–621.
95. Pickering, L. K., A. V. Bartlett, and W. E. Woodward. 1986. Acute infectious diarrhea among children in day care: Epidemiology and control. Rev. Infect. Dis. 8:539–547.
96. Pickering, L. K., R. L. Guerrant, and T. G. Cleary. 1995. Microorganisms responsible for neonatal diarrhea, in J. S. Remington and J. O. Klein (eds.), Infectious Diseases of the Fetus and Newborn Infant, 4th ed. Saunder, Philadelphia, pp. 1142–1222.
97. Pomeroy, C., and G. A. Filice. 1992. Pulmonary toxoplasmosis: A review. Clin. Infect. Dis. 14:863–870.
98. Radomyski, T., E. A. Murano, D. G. Olsen, and P. S. Murano. 1994. Elimination of pathogens of significance in food by low-dose irradiation: A review. J. Food Prot. 57:73–86.
99. Reida, P., M. Wolff, H.-W. Pöhls, W. Kuhlmann, A. Lehmacher, S. Aleksic, H. Karch, and J. Bockemühl. 1994. An outbreak due to enterohaemorrhagic *Escherichia coli* O157:H7 in a children day care centre characterized by person-to-person transmission and environmental contamination. Zbl. Bakt. 281:534–543.
100. Remington, J. S., R. McLeod, and G. Desmonts. 1995. Toxoplasmosis, in J. S. Remington and J. O. Klein (eds.), Infectious Diseases of the Fetus and Newborn Infant. Saunders, Philadelphia, pp. 140–267.
101. Richards, F. O., J. A. Kovacs, and B. J. Luft. 1995. Preventing toxoplasmic encephalitis in persons infected with human immunodeficiency virus. Clin. Infect. Dis. 21 (suppl. 1):S49–S56.
102. Roberts, T., K. D. Murrell, and S. Marks. 1994. Economic losses caused by foodborne parasitic diseases. Parasitol. Today. 10:419–423.
103. Robins-Browne, R. M., and J. K. Prpic. 1985. Effect of iron and desferrioxamine on infections with *Yersinia enterocolitica.* Infect. Immun. 47:774–779.
104. Rosenberg, P. S. 1995. Scope of the AIDS epidemic in the United States. Science 270:1372–1375.
105. Roy, S., and H. H. Loh. 1996. Effect of opioids on the immune system. Neurochem. Res. 21:1375–1386.
106. Salminen, S., E. Isolauri, and E. Salminen. 1996. Clinical uses of probiotics in stabilizing the gut mucosal barrier: Successful strains and future challenges. Antonie van Leeuwenhoek 70:347–358.

107. Schiraldi, O., V. Benvestito, C. di Bari, R. Moschetta, and G. Pastore. 1974. Gastric abnormalities in cholera: Epidemiological and clinical considerations. Bull WHO 51:349–352.
108. Schuchat, A., B. Swaminathan, and C. V. Broome. 1991. Epidemiology of human listeriosis. Clin. Microbiol. Rev. 4:169–183.
109. Schwartz, B., C. A. Ciesielski, C. V. Broome, S. Gaventa, G. R. Brown, B. G. Gellin, A. W. Hightower, L. Mascola, and the Listeriosis Study Group. 1988. Association of sporadic listeriosis with consumption of uncooked hot dogs and undercooked chicken. Lancet. ii:779–782.
110. Scott, E. 1996. Foodborne disease and other hygiene issues in the home. J. Appl. Bacteriol. 80:5–9.
111. Sepkowitz, K. A., and D. Armstrong. 1995. Gastrointestinal infections in neutropenic patients, in M. J. Blaser, P. D. Smith, J. I. Ravdin, H. B. Greenberg, and R. L. Guerrant (eds.), Infections of the Gastrointestinal Tract. Raven Press, New York, pp. 527–534.
112. Sepúlveda, J., W. Willett, and A. Muñoz. 1988. Malnutrition and diarrhea. A longitudinal study among urban Mexican children. Am. J. Epidemiol. 127:365–376.
113. Sherman, P. M., and S. N. Lichtman. 1995. Pediatric considerations relevant to enteric infections, in M. J. Blaser, P. D. Smith, J. I. Ravdin, H. B. Greenberg, and R. L. Guerrant (eds.), Infections of the Gastrointestinal Tract. Raven Press, New York, pp. 143–152.
114. Smith, J. L. 1994. Arthritis and foodborne bacteria. J. Food Prot. 57:935–941.
115. Smith, J. L. 1995. Arthritis, Guillain–Barré syndrome, and other sequelae of *Campylobacter jejuni* enteritis. J. Food Prot. 58:1153–1170.
116. Smith, J. L. 1993. *Cryptosporidium* and *Giardia* as agents of foodborne disease. J. Food Prot. 56:451–461.
117. Smith, J. L. 1993. Documented outbreaks of toxoplasmosis: Transmission of *Toxoplasma gondii* to humans. J. Food Prot. 56:630–639.
118. Smith, J. L. 1991. Foodborne toxoplasmosis. J. Food Saf. 12:17–57.
119. Smith, P. D. 1995. Intestinal infections in HIV-1 disease, in M. J. Blaser, P. D. Smith, J. I. Ravdin, H. B. Greenberg, and R. L. Guerrant (eds.), Infections of the Gastrointestinal Tract. Raven Press, New York, pp. 483–498.
120. Smith, R. E., A. M. Carey, J. M. Damare, F. M. Hetrick, R. W. Johnston, and W. H. Lee. 1981. Evaluation of iron dextran and mucin for enhancement of the virulence of *Yersinia enterocolitica* serotype O:3 in mice. Infect. Immun. 34:550–560.
121. Spika, J. S., J. E. Parsons, D. Nordenberg, J. G. Wells, R. A. Gunn, and P. A. Blake. 1986. Hemolytic uremic syndrome and diarrhea associated with *Escherichia coli* O157:H7 in a day care center. J. Pediatr. 109:287–291.
122. Sword, C. P. 1966. Mechanisms of pathogenesis in *Listeria monocytogenes* infection. I. Influence of iron. J. Bacteriol. 92:536–542.
123. Tappero, J. W., A. Schuchat, K. A. Deaver, L. Mascola, and J. D. Wenger. 1995. Reduction in the incidence of human listeriosis in the United States. JAMA 273:1118–1122.
124. Tauxe, R. V., K. E. Johnson, J. C. Boase, S. D. Helgerson, and P. A. Blake. 1986. Control of day care shigellosis: A trial of convalescent day care in isolation. Am. J. Public Health. 76:627–630.

125. Tauxe, R. V., G. Wauters, V. Goossens, R. van Noyen, J. Vandepitte, S. M. Martin, P. de Mol, and G. Thiers. 1987. *Yersinia enterocolitica* infections and pork: The missing link. Lancet i:1129–1132.
126. Thayer, D. W. 1990. Food irradiation: Benefits and concerns. J. Food Qual. 13:147–169.
127. Thayer, D. W., E. S. Josephson, A. Brynjolfsson, and G. G. Giddings. 1996. Radiation pasteurization of food. Council for Agricultural Science and Technology Issue Paper No. 7, April 1996. Ames, IA.
128. Todd, E. C. D. 1992. Foodborne disease in Canada—A ten-year summary from 1995–1984. J. Food Prot. 55:123–132.
129. Todd, E. C. D. 1989. Preliminary estimates of costs of foodborne disease in Canada and costs to reduce salmonellosis. J. Food Prot. 52:586–594.
130. U. S. Bureau of Census. 1994. Statistical Abstracts of the United States: 1994 (114th ed.). Washington, DC.
131. Weinberg, E. D. 1984. Pregnancy-associated depression of cell-mediated immunity. Rev. Infect. Dis. 6:814–831.
132. Weinberg, E. D., and G. A. Weinberg. 1995. The role of iron in infection. Curr. Opin. Infect. Dis. 8:164–169.
133. Wilson, K. H. 1995. Ecological concepts in the control of pathogens, in J. A. Roth, C. A. Bolin, K. A. Brogden, F. C. Minion, and M. J. Wannemuehler (eds.), Virulence Mechanisms of Bacterial Pathogens, ASM Press, Washington, DC, pp. 245–256.
134. Witte, D. L., W. H. Crosby, C. Q. Edwards, V. F. Fairbanks, and F. A. Mitros. 1996. Hereditary hemochromatosis. Clin. Chim. Acta 245:139–200.
135. Wong, S.-Y., and J. S. Remington. 1994. Toxoplasmosis in pregnancy. Clin. Infect. Dis. 18:853–862.
136. Wright, A. C., L. M. Simpson, and J. D. Oliver. 1981. Role of iron in the pathogenesis of *Vibrio vulnificus* infections. Infect. Immun. 34:503–507.
137. Wurapa, R. K., V. R. Gordeuk, G. M. Brittenham, A. Khiyami, G. P. Schechter, and C. Q. Edwards. 1996. Primary iron overload in African Americans. Am. J. Med. 101:9–18.
138. Yasui, H., and M. Ohwaki. 1991. Enhancement of immune response to Peyer's patch cells cultured with *Bifidobacterium breve.* J. Dairy Sci. 74:1187–1195.

6

Safe Food Handling in Airline Catering

Young-jae Kang
Asiana Airlines, Inc., Seoul, Korea

I. HISTORICAL AND CURRENT INCIDENTS OF FOODBORNE ILLNESS

Foodborne illnesses on air flights are rare but can have a devasting impact, such as the 1984 salmonellosis outbreak that affected 177 international airline passengers, leading to 37 hospitalizations and one fatality (32). Twenty-three foodborne outbreaks aboard commercial aircraft have been documented from 1947 to 1984 (Table 1). Fourteen (61%) of these occurred in the 1970s, with five in 1976 alone. Overall, seven outbreaks were caused by *Salmonella,* and five each by *Staphylococcus aureus* and *Vibrio* spp. Since 1984, several more incidents have occurred.

In October 1988, 30 of 725 passengers on 13 flights who had eaten cold sandwiches prepared by a caterer in the Twin Cities area contracted shigellosis, involving a single strain of *Shigella sonnei.* This outbreak was identified only because of its impact on a professional football team (18). In August 1991 there was an outbreak of viral gastroenteritis as a result of consumption of orange juice contaminated with a Norwalk-like agent supplied to caterers in Melbourne, Australia. More than 3053 individuals, including some nontravelers, were affected (23). This outbreak is probably the largest in airline catering history. It was quite an unusual outbreak because the vector was viral and the source of the problem was the orange juice manufacturer, not the caterer. In February 1992 an outbreak of cholera caused 75 cases of illness for international airline passengers; 10 were hospitalized, and one died. The agent suspected was cold seafood salad, served by a caterer in Lima, Peru, where there was a cholera epidemic (14). In November 1996 there was an outbreak of salmonellosis onboard an international flight from

Table 1 Foodborne Outbreaks Aboard Commercial Aircraft, 1947–1984

Outbreaks				Attack rate[a]		
Date	Agent	Implicated food	Number of flights	Passengers	Crew	Food origin
7/47	*Salmonella typhi*	Sandwiches	1	4/17	0/4	Anchorage, AK
1/61	*Staphylococcus aureus*	Chicken	1	13/128	0/?	Vancouver, BC
1/67	?	Oyster	1	0/?	23/?	London
5/69	Multiple	?	1	21/42	?	Hong Kong
7/69	Multiple	?	1	24/59	?	Hong Kong
11/70	*Clostridium perfringens*	Turkey	8	3/18	22/62	Atlanta, GA
9/71	*Shigella*	Seafood cocktail	1	19/43	?	Bermuda
2/72	*Vibrio parahaemolyticus*	Seafood hors d'oeuvres	1	12/?	3/?	Bangkok
11/72	*Vibrio cholerae* O1	Hors d'oeuvres	2	47/357	0/19	Bahrain
6/73	*Vibrio cholerae* non-O1	Egg salad	1	64/?	2/?	Bahrain
10/73	*Staphylococcus aureus*	Custard	3	247/440	0/?	Lisbon
10/73	*Salmonella thompson*	Breakfast	1	17+/117	?	Denver, CO
2/75	*Staphylococcus aureus*	Ham	1	196/343	1/20	Anchorage, AK
2/76	*Salmonella typhimurium*	Cold salads	11	550/2500	?	Las Palmas, Spain
4/76	*Salmonella brandenburg*	Multiple	45	232/?	58/?	Paris
6/76	*Staphylococcus aureus*	Eclair	1	28/185	?	Rio de Janeiro
10/76	*Salmonella typhi*	Tourist-class menu	1	13/225	?	New Delhi ?
12/76	*Vibrio parahaemolyticus*	?	1	28/?	?	Bombay
?/78	*Vibrio cholerae* non-O1	Sandwiches	1	23/?	?	Dubai, United Arab Emirates
8/82	*Staphylococcus aureus*	Custard	2	6/502	10/?	Lisbon
5/83	*Salmonella enteritidis*	Swiss steak ?	2	12/?	?	New York
10/83	*Shigella*	?	1	42/48	?	Acapulco
3/84	*Salmonella enteritidis*	Hors d'oeuvres	32	177/?	9/?	London

[a]Ill/at risk
Source: Ref. 32.

Cairns, Australia, to several Asian cities. Sweet desserts from the bakery goods supplier were suspected as the source. Investigation documentation is still restricted because of litigation (19).

II. COMMON TRENDS AND FACTORS LEADING TO FOODBORNE ILLNESS

A. Common Trends in Inflight Foodborne Illness

International travel has increased dramatically during the twentieth century. Aircraft size has also increased with the emergence of large jets holding several hundred passengers. The production volume of airline caterers has also increased dramatically. The majority of inflight meals, which are for the economy class, are produced by a small number of skilled chefs and food handlers in a relatively short period. Because of this, the characteristics of inflight foodborne outbreaks usually appear simultaneously and in masse. Deaths are unusual but were recorded from illnesses arising meals contaminated with *Salmonella, S. aureus,* and *Vibrio* spp. from 1947 to 1984 (32). In 18 of the 23 outbreaks reviewed by Tauxe et al. (32), contaminated food was reported, with the ratio of cold food to hot food of 14:4. Cold hors d'oeuvres and custard desserts/éclairs were the causative agents in six and four outbreaks, respectively. In seven outbreaks, the implicated food had been held in the kitchen too long at an inappropriate temperature during and after preparation. Inadequate cooking and contamination by the food handler was also identified as the cause in some outbreaks. No waterborne outbreaks have been reported.

B. Investigation of Inflight Foodborne Outbreaks

Foodborne outbreaks related to inflight meals are difficult to detect and to investigate, particularly when the incubation period is longer than the flight time. Exposed persons may disperse widely, and most ill passengers may not know that others have been affected. Illness may also be attributed to other travel exposures, and travelers may not have ready access to physicians. When an outbreak is suspected, the passengers are difficult to identify and locate, the aircraft may have departed, and the suspected food may have been discarded. The attack rates may be so high and the exposure of passengers so similar that food-specific attack rates may not implicate a particular food item. Because of the short incubation period, outbreaks of *S. aureus* intoxications are usually recognized before landing in flights more than a few hours, and the investigations can begin within hours. An outbreak of *S. aureus* caused by ham served by an Anchorage kitchen in February 1975, on a flight from Tokyo to Paris with a stopover in Alaska, is such an example; passengers who had become ill were taken off the plane in Copenhagen (15,16).

C. Factors Leading to Foodborne Illness

Factors that contributed to outbreaks of foodborne illness stemming from inflight meals may be the same as those involving other foodservice establishments, since there are no major differences in food production except for a lapse of 12 or more hours between preparation and eating in airline catering. For such catering, because of its operational characteristics, 48 hours is allowed for cold meals from preparation to delivery to the aircraft, and 72 hours for hot meals. This delay makes hygiene control in airline catering extremely important. There are many steps in the production of a meal with potential contributing factors for contamination and growth. Thus to prevent outbreaks from occurring, these steps must be perfectly controlled (Table 2). Usually more than one contributing factors is implicated in an outbreak.

III. PRACTICAL ADVICE TO FOOD PREPARERS ON FOODBORNE ILLNESS PREVENTION

Inflight foodservice is particularly susceptible to food safety problems because of extensive handling of food products and because of the extended time from pro-

Table 2 Factors That Contributed to Outbreaks of Foodborne Illness

Improper cooling
Lapse of 12 or more hours between preparing and eating
Poor personal hygiene/infected persons
Incorporation contaminated raw food/ingredients into food that received no further cooking
Inadequate cooking/canning/heat processing
Improper hot holding
Inadequate reheating
Obtaining food from unsafe source
Cross-contamination
Improper cleaning of equipment/utensils
Use of leftovers
Containers/pipelines adding toxic chemicals
Intentional additives
Improper thawing
Flying insects
Others

Source: Refs. 5, 6, and 12.

duction to service. In addition, most foods produced by the caterer are potentially hazardous foods (PHFs). Therefore, food safety is considered a prime quality parameter of airline catering meals. Many efforts for improving hygiene and assuring food safety have been tackled at various levels in the food chain. Among them, the hazard analysis and critical control point (HACCP) system is widely implemented by most of the leading airline caterers and successfully prevents onboard outbreaks.

A. Emergence of Inflight Catering HACCP Systems

A generic model for inflight catering HACCP is advocated by many groups of professionals. International organizations such as the International Inflight Foodservice Association (IFSA) (10), the International Flight Catering Association (IFCA) (3), and the Association of European Airlines (AEA) (8) have issued HACCP-based food safety manuals. An ad hoc committee of the U.K. Airline Caterers have also produced an HACCP-based Code of Good Hygiene Practice for Airline Catering (2). It is unlikely that airline caterers will be able to use a formal generic HACCP system because of the wide range of specialty foods produced by technical personnel. Most caterers and airlines have adapted generic HACCP systems and have developed a customized HACCP system that suits their own production and operation conditions. Caterers should have their own hygiene manual and microbial standards based on these guidelines and local regulations.

Even though the caterer may implement a perfectly sound HACCP system, it is not a guarantee for perfect food safety, since a critical control point (CCP) such as personal hygiene, especially hand washing, cannot be monitored, measured, and corrected throughout the whole operation.

1. Hazard Analysis of Inflight Catering

Since hazard analysis of inflight catering is based on epidemiological data from actual foodborne outbreaks recorded in the food processing industry, it is unlike its counterparts in other food industries. The contributing factors leading to outbreaks of foodborne illness have identified the source of hazards in food processing, and these make it possible to set up inflight catering HACCP. These factors can be controlled by implementing the HACCP system.

2. Hazards and CCPs in Airline Catering

Hazards in airline catering are the contributing factors that lead to outbreaks of foodborne illness in which undesirable microorganisms can contaminate foods and survive, and multiply there. The following is a general summary of the hazards and CCPs that may be found in airline catering.

All raw food is assumed to be contaminated by pathogens.

Properly cooked or sanitized foods that have pathogens at safe level should be regarded as "clean."

All steps in the process designed to kill or reduce bacteria in the food (e.g., cooking or chemical sanitization) must be regarded as CCPs.

Any work surfaces or equipment coming into direct contact with food, particularly "clean" food, must be regarded as a possible source of contamination.

Any procedure that involves direct handling of "clean" food by catering staff also presents a high risk of contamination. Personal hygiene is of prime importance.

Any time and temperature control step (cooking or chilling) is a CCP.

Any measure to prevent cross-contamination is a CCP.

3. Criteria of Each CCP

Each CCP should have criteria that assure the safety of food. These criteria should be based on scientific studies and should be measurable. The criteria in airline catering are upper or lower limits, not upper and lower limits (e.g., only lower limits for cooking temperature and sanitizer concentration, and only upper limits for cold storage temperature, blast chilling time/temperature, and time control). The definitions and descriptions in this chapter are to be considered a framework with regard to set criteria in the sense that criteria may be tightened if required by a regulatory agency, by a customer, or by internal policy. All criteria should be compared internally and with the customer's national criteria, since criteria for various items can differ among countries.

4. Checklists and Corrective Action

A checklist of each CCP must be designed to cover all criteria, but it should also be easy to fill out. Monitoring must be done by the worker who is doing the job; therefore, this system is called a "self-inspection system." Monitoring results are recorded on the spot, on the checklist of each CCP. Immediate corrective action must be taken if deviations from the established standards are identified. They should be highlighted on the record and the section manager should be notified immediately.

Proper corrective action should be clearly described on the checklist; otherwise no action or wrong action may be taken. The person responsible for monitoring and evaluation should be clearly listed on each definition sheet. A sample CCP checklist is given in Table 3, which is associated with the Definition of CCP 3: Cold Storage (see Sec. III.C.4.a).

Table 3 Example of a CCP Checklist

CHECKLIST CCP 3. FOOD COLD STORAGE SECTION: ______

WEEK COMMENCING: Refrigerator/Freezer Number: ______

DAY	TIME	GAUGE TEMP.	FOOD TEMP.	SEGREGATION[a]	DATE MARK[b]	COVER[c]	COMMENTS	SIGNATURE
MON	am							
	pm							
TUE	am							
	pm							
WED	am							
	pm							
THU	am							
	pm							
FRI	am							
	pm							
SAT	am							
	pm							
SUN	am							
	pm							

*Corrective Action: When wrong item, uncovered item, no date marked or outdated item is found, correct it.
When the refrigerator/freezer temp. is over 5/−18°C, call section head and maintenance team.

CHECKED BY: (USER) FILING TO: HYGIENE CONTROL OFFICE
DATE:

[a]Check the presence of wrong item in that specific cold holding room.
[b]Check the presence of datemark missing items or outdated items.
[c]Check the presence of uncovered or inadequately covered items.

ASIANA CATERING

B. Prerequisite HACCP Program

1. Building Layout

The unit layout is very important in the production of wholesome inflight meals. To prevent cross-contamination, the layout should permit effective segregation of clean and unclean materials/processes and should flow in one direction. To prevent airborne contamination, the preparation of pastry and bread dough, which generate high amounts of dust containing bacteria, yeast, and molds, should be carried out in areas separate from the rest of the meal production areas.

2. Cleaning and Sanitizing

Cleaning in the airline catering kitchen must be accompanied with sanitizing. Food-handling areas of the building should be designed to be easily cleaned and sanitized. For example, walls, floors, and ceilings should be constructed of a material impervious to water, fats, and chemicals. Working surfaces should be of smooth, durable, and readily cleansible materials. A cleaning program must be setup and used which itemizes the area/equipment to be cleaned, the person responsible, and the frequency with which cleaning must occur. The chemicals and the method of cleaning and sanitizing must also be specified. Chemicals employed must be suitable for use in catering premises and should be stored separately from foods. Staff must be fully trained in cleaning methods and chemical usage.

3. Hygiene Facilities

To control flying insects, electric fly killers should be installed and maintained properly at strategic locations. The locations of electric fly killers must be carefully decided for maximizing efficiency and preventing food contamination. Some models of electric fly killers should not be located less than 6 feet from the exposed food because electrocution creates aerosols of insect parts. UV lamps with the sticky board type are especially recommended for food preparation areas.

The air curtain at every outside entrance should have a blower strong enough to prevent flying insects from entering. High pressure hoses, sometimes used for cleaning trolleys, meal carts, or floors, produce aerosol mists that may spread contaminants around the production areas. There can be special problems when such hoses are used in areas of heavy contamination sources, such as drains. Unrestricted use of high pressure hoses should be avoided. Drains should be opened for regular cleaning and then flushed with a bactericidal agent. Cleaning equipment (mops, squeegees, etc.) should be kept in sets, dedicated to specific areas. Implements should be cleaned after use and kept in separate storage areas, where they can be allowed to dry.

4. Temperature Measurement

Liquid-filled glass thermometers are not allowed for taking food temperatures because of the danger of breakage. Bimetallic dial thermometers are not suitable because of poor accuracy and slow response (28). They are also not appropriate for thin foods. These thermometers can be used for refrigerators and freezers with a wall mount, however. A wide variety of electronic instruments are available, such as simple pen-type digital thermometers (thermistors) and probe-exchangeable, handheld digital thermometers (thermocouples). The thermocouple and the thermistor can quickly measure temperature even in thin foods. Waterproof digital thermometers are more suitable for chefs, since thermometers can easily get wet by water or oil. In addition, the noncontact type of infrared (IR) surface thermometer is recommended for some purposes. Measuring equipment should be accurate to, at most, 0.5°C or 1°F. Technical information on kitchen thermometers is issued by the U.S. Food Safety and Inspection Service (FSIS) (9). All thermometers should be properly calibrated at certain intervals, as recommended by the manufacturer, and records of this should be kept. Thermometer probes may act as a source of cross-contamination if they remain dirty or are used for both raw and cooked foods. After each use, probes should be wiped and disinfected with a 70% alcohol pad or equivalent.

C. The Airline Catering HACCP System

1. Flow Diagram of Production Processes

Airline catering consists of many different systems, as can be seen from the simplified flow diagram (Fig. 1). During meal production, some foods may undergo many different processes before assembly into airline tray sets, while others are subject to no further manipulation and are dispatched to the aircraft in the same state in which they were delivered to the catering unit. The shaded boxes in Fig. 1 are the CCPs, which have possible problem sources, although corresponding preventive measures exist, as well.

2. Menu Review

Some meal components available at restaurants or homes should not be used as in-flight foods. Menus or recipes must be thoroughly scrutinized for such restricted meal components, since they may contain pathogens or parasites. The caterer should have the list of restricted meal components. Precautions are especially necessary when ethnic foods are requested by airlines, since handling and cooking of some ethnic foods do not follow safe food-handling practices. By doing a menu review, the caterer can eliminate the contributing factor "Incorporating contaminated raw food/ingredient into food that received no further cooking."

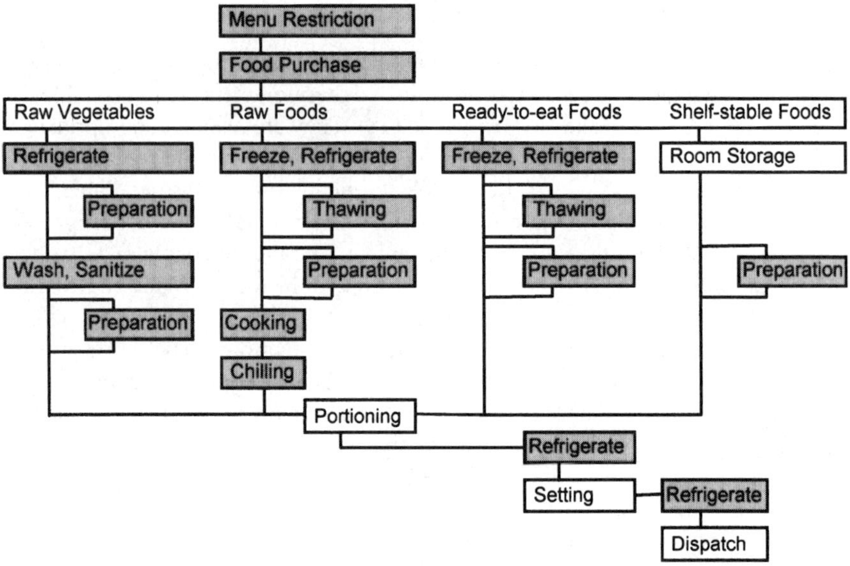

Figure 1 Simplified flowchart of airline catering and CCPs (shaded boxes).

Definition of CCP 1. Meal Component Restrictions

Hazard: Survival of undesirable microorganisms.

Criterion: Absence of restricted meal components in the menu.

Restricted meal components: Raw or unpasteurized foods such as undercooked poultry, eggs, ground meat/raw fish or shellfish, raw vegetable sprouts, raw eggs, soft cheese made from unpasteurized milk, mayonnaise made from raw eggs, and marinated fish from fresh, unfrozen fish.

Monitoring: Observation of relevant menu, recipes, and production procedures.

Frequency of monitoring: New menu preparation, menu or recipe changes.

Corrective action: Make new menu or change recipes.

3. Food Purchase

Raw materials delivered to an airline caterer can range from unclean, possibly pathogen-carrying foods to clean, ready-to-eat (RTE) foods, from PHFs to commercially sterile foods, and from frozen foods to shelf-stable foods. All foods, especially RTE foods, must be purchased from reputable and reliable suppliers with acceptable hygiene standards. Suppliers' premises and processes should be controlled by the HACCP system and audited by competent persons reporting to the catering company routinely, to ensure that the level of hygiene is the same as that of the caterer. Deliveries should be inspected to ensure that they meet specifications. Any goods not meeting specification should be returned to the supplier. The reasons for failure should be investigated promptly, and corrective action must be taken.

All goods must have suitable wrapping and packaging to prevent contamination. Food must be received at the proper temperature to prevent or minimize multiplication of pathogens during delivery. In measuring temperature, care must be taken not to damage the package or contaminate the thermometer probe or hand. The IR surface thermometer is strongly recommended for measuring the temperature of receiving goods. When the purchasing manager inspects the quantity and temperature, the quality and freshness should be also inspected by a chef who knows the product well.

Fresh shell eggs, which may be contaminated with *Salmonella* and other pathogens, should not be brought into the airline catering kitchen. Pasteurized liquid egg is recommended for use. When fresh shell eggs must be used for specific products, sanitized shell eggs should be purchased. If sanitized shell eggs are not available at the local market, eggshell sanitizing should be performed at the receiving area. Because eggs may contain *Salmonella enteritidis* (22), fresh shell eggs are considered to be a PHF, and their temperature must be monitored at receiving and storage. Shell eggs should be used as ingredients only for food that is thoroughly cooked.

Random samples for microbiological analysis should be taken periodically by a hygiene controller.

Definition of CCP 2. Potentially Hazardous Food Purchase

Hazards: Survival, contamination, and multiplication of undesirable microorganisms.

Criteria: 1. Refrigerated foods should be delivered at a temperature below 5°C at delivery.

2. Frozen foods should be frozen solid and without signs of previous thawing.

3. Product should bear valid label, with agreement on residual shelf life at delivery.

4. Microbial quality of purchased products at delivery point should meet caterer's microbial standards.

Monitoring: 1. Measure temperature of refrigerated foods.

2. Visual assessment and measure temperature of frozen foods.

3. Visual assessment of label.

4. Routine sampling from hygiene laboratory.

Frequency of monitoring: all refrigerated/frozen products in receiving.

Monitoring tool: IR surface thermometer.

Corrective action: Reject receiving or issue a warning letter to supplier to ensure preventing recurrence after consulting with hygiene team.

4. Storage

a. Cold Storage. Temperature-controlled stores should be fitted with accurate thermometers of either the direct reading or recording type, as well as alarms to signal system failure. Checks should be made at least twice daily on the efficient operation of all storage units and written records kept for at least 3 months. The temperature for the refrigerator and the freezer should be colder than 5°C (41°F) and −18°C (0°F), respectively. Installation of an air curtain or plastic strip curtain at the refrigerator door is recommended to minimize temperature fluctuations.

Food is stored in the refrigerator to control the growth of pathogens and spoilage bacteria. However, it is now known that a number of pathogens can multiply in foods at temperature below 7.2°C (45°F). Bacterial pathogens known to grow at below 5°C (41°F) include *Yersinia enterocolitica, Listeria monocytogenes, Aeromonas hydrophila, Clostridium botulinum* type E, and *Bacillus cereus.* Pathogens that can grow between 5°C (41°F) and 7.2°C (45°F) include *Salmonella heidelberg, Salmonella typhimurium,* and *Staphylococcus aureus.* However, most serotypes of salmonellae fail to grow at temperatures below 7°C, and *S. aureus* can produce toxin only at or above 10°C. Psychrotrophic spoilage bacteria and molds can also multiply at refrigeration temperatures (29,22). Some of the new caterers have adopted superchilling storage, which is −1 to 1°C (30–34°F), hence extending the refrigerated shelf life of products.

Time control is another important factor in preventing pathogen multiplication even at cold storage. Since the time and temperature for bacterial multiplication are inversely correlated, the higher the storage temperature, the shorter the shelf life. In the 1997 FDA Food Code (7), the standard for cold holding of an RTE, PHF is 5°C (41°F) for 7 days or 7.2°C (45°F) for 4 days. Airline catering should have stricter time control at each step of the process. Ready-to-use raw materials and/or prepared meals can be stored within 48 hours at 5°C (41°F). Foods should be taken from chilled storage for assembly in strict rotation, first-in, first-out (FI FO), not exceeding the time limit referred to in Sec. III.C.16. The use of a color-coded date mark is very helpful for distinguishing outdated items at a glance.

To prevent cross-contamination, all food must be placed in physically segregated storage areas, clearly designated according to product type. As soon as practicable after delivery, PHFs should be placed in a segregated walk-in cooler for meat, seafoods, poultry, vegetables, fruits, and RTE foods. At least two coolers are needed, one for raw and one for RTE foods. Separate freezers for raw and cooked frozen foods are required. Storage of different foods in one chilled unit should be organized to avoid any risk of cross-contamination.

For intermediate cold storage during food production, food should be properly protected from condensate drip and airborne contamination, since the cooling unit and evaporating pan can accumulate dust and harbor microorganisms. Such stores should be dedicated to storage of foods ready for assembly.

All foods must be stored clear of the ground. For clear viewing and easy cleaning, the gap between the floor and the lowest shelf should be greater than 15 cm (6 in.). Temperature-controlled stores should be adequately designed to provide access for mobile trolleys and racking or shelving as required, sufficient space should be available for proper stock rotation and cleaning. Positioning of food in chilled stores should not impede good air circulation. In particular, food must not be stacked too high or immediately in front of the fans of the evaporator fan unit.

Definition of CCP 3. Cold Storage

Hazards: Contamination and multiplication of undesirable microorganisms.

Criteria: 1. Segregation of raw and cooked food.
2. Covering.
3. Date mark.
4. Refrigeration and freezer temperatures must be below 5°C (41°F) and −18°C (0°F), respectively.

Monitoring: 1, 2, 3. Visual assessment.
4. Monitoring of storage temperature.

Frequency of monitoring: Once per shift.

Monitoring tool: IR surface thermometer.

Corrective action: When wrong item, uncovered item, non-date-marked, or outdated item is found, correct it. When refrigerator/freezer temperature is abnormal, call section head and maintenance team. Section head will contact hygiene controller to check the duration exposed in the temperature danger zone and decide usage.

b. Dry Storage. Shelf-stable food items such as canned food or dried food items must be stored in a cool and dry storeroom. The optimum storage temperature is 10–21°C (50–70°F), and relative humidity is 50–60%. Humidity control is very important because molds can grow in a humid area. The storeroom must be adequately proofed to prevent the entry of vermin, and all storage racks must be at least 15 cm (6 in.) from the floor and wall. Stores should be managed systematically for FIFO and expiration control.

Most canned goods are stored in a dry storeroom, however, cans with a "Keep Refrigerated" or "Keep Frozen" label must be stored in cold storage.

Canned goods that will not be heat-treated must be transferred to the refrigerator for prechilling before use. Defective cans should be thrown out or put on the "Do Not Use" shelf for credit and pickup by supplier.

Cleaning and sanitizing chemicals should be stored in a separate room. Each container should have an easily distinguishable label on it. Material Safety Data Sheets should be available on-site.

5. Preprocessing

a. Decanting. By minimizing outer packaging materials such as cardboard, cans, bottles, and pails in the food-handling area, airborne contamination and vermin problems can be reduced. Where possible, bulk food items should be decanted to a clean sanitized pan before being taken into food production areas. A decanting room must be equipped with a can opener, sink, hand-washing station, and drain. Trays, containers, and other utensils used for the handling of raw food must be differentiated from those used for cooked food (e.g., by color-coding).

b. Thawing of Frozen Foods. Precooked, frozen foods that require no further heat treatment may be decanted and put on plates directly from the frozen state. Frozen vegetables may also be blanched directly from a frozen state. However, raw meat, poultry, and fish supplied frozen must be properly thawed under controlled temperature conditions before cooking. If incompletely thawed foods are cooked, heat treatment may not reach pasteurization temperatures, a situation that may allow for the survival and multiplication of pathogens.

Safe thawing conditions are decided by temperature and time combination: low temperature/long time (in the refrigerator), high temperature/short time (at ambient temperature), or moderate temperature/moderate time (in the thawing room). The refrigerator is good for small-sized frozen foods but impractical for large-sized frozen foods, which take a very long time to completely thaw. Thawing at ambient temperature is fast but dangerous. By the time the core of the food has thawed, the surface has been exposed to ambient temperatures for too long. Therefore, special controlled thawing rooms or cabinets, which maintain a temperature of 10°C (50°F), are recommended for thawing large items (13). At this temperature, complete thawing will be done moderately quickly and safely. However, the FDA Food Code (1997) allows thawing in the refrigerator at 7.2°C (45°F) or below (7).

Alternative methods for fast thawing,—using running cold water (21°C/70°F or below; in accordance with the FDA Food Code 1997) or microwaves,—can be used only for emergencies or small portions. Frozen fish may be thawed in potable, cold, running water; raw meat and poultry may be thawed similarly, provided they are suitably packed to prevent cross-contamination by the environment and other foods (from splashing). When a microwave device is used for thawing, confirm complete thawing by probing with a thermometer in many places. Heat generated by microwaves does not penetrate evenly in the solid food. Completely thawed food should be transferred to the refrigerator immediately.

Definition of CCP 4A. Thawing

Hazards: Survival, contamination, and multiplication of undesirable microorganisms.

Criteria: 1. Food temperature must not exceed 10°C (50°F) during thawing.
2. For fast thawing, foods packed in watertight material are thawed in cold running water.
3. Raw foods should be segregated from cooked foods.
4. Thawing of raw foods should be completed prior to cooking.

Monitoring: 1. Product temperature upon completion of thawing and thawing room temperature are monitored.
2. 3. Visual assessment.
4. Probing for possible ice detection in food by thermometer probe.

Frequency of monitoring: When the foods are brought into and taken out of thawing room.

Monitoring tool: Probe thermometer, alcohol pad.

Corrective action: If thawing room temperature is over 10°C (50°F), call the maintenance team. If ice is detected inside the food, leave in thawing room until thawing is complete or use a fast thawing method.

c. Vegetable Preparation. All vegetables and fruits used in airline catering kitchens must be washed and sanitized because their surfaces may carry pathogens from many sources such as soil, feces, insects, and people (11). Bacteria attached on the surface of raw whole vegetables and fruits cannot multiply greatly, even in the temperature danger zone, which is between 5°C (41°F) and 60°C (140°F). However, when pathogens come in contact with protein foods such as ham or salad, they can multiply quickly and cause problems. For this reason, all salad items, vegetable crudités, and fruits should be treated with a suitable sanitizing solution (e.g., 50–75 ppm total chlorine or equivalent). However, even with sanitizing, the physical structure of the produce surface does not allow for complete removal all the attached microflora. A standard procedure for washing lettuce leaves in tap water was reported to remove 92.4% of the microflora (1), and 100 ppm free chlorine in wash water reduced the total bacterial count by 97.8% (11).

Sanitation cannot be applied where the treatment will affect the quality (flavor or texture) of the food. For example, chlorine will damage soft fruits such as strawberries and raspberries, as well as some fresh herbs (dill, mint). Liquid sanitization cannot be applied to dried herbs. Prescreening of good microbial quality products during purchasing is extremely important for such items. This may not apply to some countries whose national legislation does not allow the use of chemical sanitizers.

Definition of CCP 4B. Disinfection of Raw Vegetables and Fruits
Hazards: Survival of undesirable microorganisms and foreign bodies.
Criteria: 1. Leaves of lettuce should be separated and washed individually to ensure effective cleaning.
2. Concentration of hypochlorites should be 75 ppm with a reaction time of 5 minutes.
3. Disinfected vegetables/fruits should be visually inspected for foreign bodies.
Monitoring: 1,3. Visual assessment.
2. Monitoring of concentration of free chlorine with test strip.
Frequency of monitoring: Every batch of washing.
Monitoring tool: Chlorine test strip.
Corrective action: Chlorine must be added with a measuring cup. Rinse again if foreign bodies are found.

6. Preparation of Raw and Ready-to-Eat Foods

Preparation of raw foods should be carried out in areas physically separate from those designated for handling cooked and sanitized foods. Poultry processing areas should be separated from other areas, since raw poultry is considered to be highly contaminated. In the United States, about 35% of the raw chicken on the market is contaminated with *Salmonella* (10) and 63% with *Campylobacter* (25). In the United Kingdom, the *Campylobacter* contamination rate for raw poultry can be between 40 and 90% (25). Ideally, there should be physically separated areas for meat, seafoods, poultry, fruits and vegetables, hot kitchen, cold kitchen, bakery, and slicing room.

Use of a color-coded cutting board is recommended for preventing cross-contamination (e.g., red for meat, blue for seafoods, green for fruits and vegetables, yellow for poultry, and white for cooked or RTE foods). If such a selection is not available, a white cutting board can be used for cooked or RTE foods and a colored cutting board for raw foods. Equipment and location for preparing raw and RTE foods should be determined and posted.

IFCA's criterion on the food temperature and time out of the refrigerator during food preparation is "must not exceed 15°C (59°F) and must be less than 45 min" (3). PHFs that have been closely temperature-controlled can withstand slightly higher temperatures for short periods without promoting bacterial growth. This lag in bacterial growth can be quite long if temperatures remain below 15.5°C (60°F). As temperatures increase to the 21.1–48.8°C (70–120°F) range, the bacterial lag phase shortens and multiplication becomes much more rapid (4).

Food temperature will rise during preparation at room temperature. To minimize multiplication of microorganisms in food, exposure time and temperature increase should be controlled. When the temperature difference between food and ambient is smaller, the food temperature rise is delayed. So the lowest room temperature setting that can be tolerated by workers is recommended for food preparation, portioning, and tray setting areas. In practice, temperatures from 15 to 18°C (59–64.4°F) are widely used.

Definition of CCP 5. Temperature Control, Segregation of Raw and RTE Foods in Preparation

Hazards: Contamination and multiplication of undesirable microorganisms.

Criteria: 1. Equipment and location for preparing raw and ready-to-eat foods, respectively, should be determined and posted.

2. Chilled food temperature must not exceed 15°C, 45 minutes during preparation.

3. Room temperature should be maintained at 18 ± 1°C.

Monitoring: 1. Presence of instruction. Visual assessment of compliance with instruction.

2. Food surface temperature is monitored at completion of preparation.

3. Room temperature is monitored.

Frequency of monitoring: Every 2 hours.

Monitoring tool: IR surface thermometer.

Corrective action: If not properly segregated, foods and equipment must be returned to a designated location. If food temperature is over 15°C (59°F), return to the refrigerator and handle smaller amounts at a time. If room temperature is over 19°C (66°F), call the maintenance team.

7. Personal Hygiene

Food handlers are one of the major contamination sources of foods. Pathogens on hands originating from raw foods, feces, saliva, any part of body, unclean equipment and utensils, or other sources can be transferred into clean foods. Discharged body fluids, such as through sneezing and coughing, can also mediate the transfer of pathogens onto clean foods.

a. Preemployment Health Screening. Preemployment health screening is considered to have limited value; however, many airline caterers still require health cards for screening carriers of *Salmonella typhi* and other contagious microorganisms because of company policy or national regulation. The Code of Good Catering Practice (2) also requires preemployment health screening.

All applicants who work with food or in the food-handling area should be screened before employment. Other persons who work outside the food-handling and washing area may be employed at the employer's discretion depending on the precise nature of their job, but they should not have any contagious diseases. A medical examination including tests for tuberculosis and hepatitis can be a useful tool for screening. Applicants who have allergic rhinitis are not suitable to work as food handlers because of frequent sneezing and nasal discharge.

All applicants should complete a questionnaire detailing their relevant medical history, such as the one used by Asiana Airlines Catering Services (21) (Table 4). The completed questionnaire should be reviewed by a medical expert to determine the candidate's suitability.

b. In-Service Health Monitoring. Any food handlers suffering from gastroenteritis must be excluded from work immediately and referred to a physician; they must not be reengaged until medically cleared. Staff with other conditions referred to in Table 5 must be excluded from work until they are symptomless. Instructions should be given in writing to all management and staff, reminding them of their obligation to report any illness. Staff who have traveled abroad, where public hygiene is poor, should be required to complete the health questionnaire and, if necessary, submit fecal samples before returning to their food handling duties.

For preventing *S. aureus* food poisoning, any staff member suffering from ear discharge, acne, infected cuts, or similar septic lesions should not be employed in food handling until the infections have cleared up.

c. Staff Hygiene Rules. Personal hygiene must play a major part in the day-to-day running of any safe food production operation, and this policy must be duly emphasized in all training functions throughout the work life of each employee. Employees are requested to practice basic personal hygiene such as daily bathing or showering, regular brushing of teeth, keeping hair and beards clean and groomed, and wearing clean personal clothing. While at work, personnel should wear the complete protective uniform provided and change it regularly, especially when visibly soiled. Uniforms should be laundered by professional cleaners whose service includes a sanitizing step. Personnel should also abide by the following.

Hair should be completely covered by a coverall type of cap to prevent hair from dropping into food.

Prevent coughing or sneezing over food. Otherwise, microorganisms containing saliva droplets will spread all over the food and food contact surfaces.

Minimize direct contact foods with hands: wear disposable gloves on clean hands, use scoops, tongs, and similar implements wherever possible.

Jewelry and wristwatches that prevent proper hand and arm washing should not be worn during food handling. It is permissible to wear plain wedding bands and plain stud earrings (sleepers), but not nose studs.

Table 4 Preemployment Health Questionnaire

ASIANA CATERING

PREEMPLOYMENT HEALTH INTERVIEW AND AGREEMENT FOR FOOD HANDLERS

Name	Sex
Date of birth	Department
Address	

INTERVIEW	YES	NO
Have you ever had typhoid or paratyphoid fever?		
Have you ever had tuberculosis?		
Are you suffering from:		
1. Diarrhea now or within the past 2–3 weeks?		
2. Discharge from eyes, ears, or nose?		
3. Skin infection (boils, suppurating wounds, or cuts, etc.)?		
4. Skin rash/eczema?		
5. Any type of allergy?		

AGREEMENT WITH EMPLOYER

I agree to report immediately to the employer when suffering from an illness involving any of the following:

— Diarrhea	— Fever
— Vomiting	— Skin rash/Eczema
— Sore throat	— Tuberculosis
— Discharge from eyes, ears, or nose	— Allergy
— Skin infection (boils, suppurating wounds and cuts, etc.)	

I agree to report to my head of department before commencing work if having suffered from any of the above conditions while on holiday.

I accept that I may be required to inform my head of department of my health condition after returning from abroad.

I understand that failure to comply with this agreement could lead to disciplinary action.

Date: Signature:

Source: Ref. 21.

Smoking and eating are allowed only at designated places separate from the food handling area. At the beginning of each shift, at the entrance to the kitchen, each worker's appearance should be visually inspected.

d. Hand Hygiene. Most foods produced by airline caterers are PHFs and handled extensively by hands, so hand hygiene is one of the most important

Table 5 Health Questionnaire for Food Handlers

ASIANA CATERING

PERIODIC HEALTH INTERVIEW AND AGREEMENT
FOR FOOD HANDLERS

Name		Sex
Employee No.		Department
Purpose	Annual ☐	Follow-up ☐

INTERVIEW	YES	NO
Are you suffering from:		
1. Diarrhea now or within the past 2–3 weeks?		
2. Discharge from eyes, ears, or nose?		
3. Skin infection (boils, suppurating wounds, or cuts, etc.)?		
4. Skin rash/eczema?		
5. Any type of allergy?		

AGREEMENT WITH EMPLOYER

I agree to report immediately to the employer when suffering from an illness involving any of the following:

— Diarrhea
— Vomiting
— Sore throat
— Discharge from eyes, ears, or nose
— Skin infection (boils, suppurating wounds and cuts, etc.)
— Fever
— Skin rash/Eczema
— Tuberculosis
— Allergy

I agree to report to my head of department before commencing work if having suffered from any of the above conditions while on holiday.

I accept that I may be required to inform my head of department of my health condition after returning from abroad.

I understand that failure to comply with this agreement could lead to disciplinary action.

Date: Signature:

Source: Ref. 21.

factors to be controlled (31). In airline catering, gastrointestinal infection is considered to be a very serious disease and employees who have loose bowels should not be allowed to work in food preparation areas. However, by nature, it is not easy to identify employees who still have mild infections. Nor is it possible without laboratory testing to distinguish diarrhea caused by pathogens from that due to a non microbial source. Therefore, preventing fecal–oral transmission of pathogens by proper hand washing is a more practical means of control.

There are two types of microflora on human skin, resident and transient. Fecal bacteria, which are transferred from feces or toilet environments to skin, comprise an example of transient microflora. Resident microflora, such as *Lactobacillus* spp. and *S. aureus,* which reside under the epidermis, cannot be completely removed by hand washing and sanitizing. *Lactobacillus* spp. are the dominant microflora that protect our skin from the transient microflora and should not be completely removed from skin. This is why some dermatologists do not recommend the use of antibacterial soap or hand disinfectant chemicals. Introduction of *Lactobacillus* spp. to food does no harm to the foods, but *S. aureus* may cause acne, inflammation in cuts and burns, and foodborne intoxication if it produces toxin in the food. Therefore, action must be taken to control *S. aureus* so that it is not introduced to the food during processing. Direct food contact by hand should be minimized. Even the use of plastic gloves does not completely prevent contamination because of defects in the gloves such as pinholes (31); frequent changing of disposable gloves on washed hands is recommended. Nondisposable gloves are not recommended, since gloved hands usually contaminate foods by touching anything that has not been properly washed or sanitized. Minor cuts and abrasions should be covered with suitable brightly colored, waterproof dressings and plastic gloves.

Normal hand washing (single washing method) cannot remove soils under and around the nails. The double washing method with a fingernail brush should be used whenever the hands are suspected of being heavily soiled. To facilitate proper hand washing, fingernails should be kept short, and rings and wristwatches should not be worn.

Hands must be washed by a double washing method that includes the use of a nail brush, especially:

- Whenever entering a food production area
- After every visit to the toilet
- After handling raw or unwashed foodstuffs
- After coughing or sneezing into the hands
- After blowing one's nose, or scratching the head or any other part of the body

Hands may be washed by a single washing method.

- After any cleaning jobs
- Between different job functions
- After handling equipment that has not been disinfected
- After changing disposable gloves

Every food handling area must have a properly equipped hand-washing basin available. The location must be close to the work area and easy to access. Suitable hand-washing basins, which are not used for any other purpose, should be provided with a constant supply of hot water and ideally with "no touch" mixer taps supplying water at around 43°C (110°F). At this temperature, employees will feel comfortable when washing their hands, and any grease present will melt easily. Liquid soap and paper towels should be supplied from dispensers. A hot air blow-dryer is not recommended because it dries too slowly and tempts employees to dry hands on their uniforms. Complete drying of the hands is important to prevent bacterial transfer (26). An iodine-based postwash hand dip may also be provided. Hand-washing signs should be posted at all relevant locations such as the toilet, entrance ways, and lunch room, to remind employees to wash their hands before returning to work. Hand-washing training and microbial analysis of food handlers' hands should be done routinely to verify the hand-washing behavior and cleanliness of these personnel.

Definition of CCP 6. Personal Hygiene and Infection Control

Hazards: Contamination of undesirable microorganisms.

Criteria: 1. Hygienic hands, include proper hand washing and use of disposable gloves.
2. Proper wearing of clean protective clothes and cap without jewelry, watch, and hairpins.
3. Valid health card holder.
4. Food safety training performed.

Monitoring: 1. Visual observation and microbial analysis.
2. 3. Visual assessment.
4. Check of training protocol.

Frequency of monitoring: Before each shift for worker's appearance. Once every month for a microbial test.

Monitoring tool: Microbiological analysis.

Corrective action: Change to clean protective clothing. Remove any personal ornaments. Train staff in food safety as soon as possible. Staff with poor hand hygiene or no health card must be excluded from work.

8. Cleaning and Disinfection of Food Contact Surfaces

All food contact surfaces must be free from microorganisms to prevent contamination of clean food. All equipment and utensils should be readily cleanable and maintained in a clean and sound condition. Even though normal washing with detergent can reduce microorganisms by a factor of 10^5 (29), airline catering kitchens are required to apply a disinfectant on food contact surfaces during or after washing equipment and utensils (3,4,7,8). Because bacteria may reside in grooves or holes with food debris, wooden tables or wooden equipment shall not be used in any section of the kitchen or tray-setting area except for the bakery.

a. Manual Washing. Airline catering kitchen utensils are usually washed manually. A three-compartment sink can be used for detergent, rinse, and disinfection. Chlorine (100–200 ppm), quaternary ammonium compounds, or iodine (25 ppm) is usually used as the disinfectant. Sanitizer concentration test kits must be available around the sink. Sufficient contact time at proper concentrations should be allowed for effective disinfection. After washing, equipment and utensils should be dried or stored in a way that ensures quick drying and prevents accidental contamination.

Cutting boards should be soaked in chlorine and iodine-based disinfectant after washing, since these surfaces are hard to clean and disinfect. For knives, an iodine-based disinfectant soaking tank with a knife holder on top of the working table can be set up; this is helpful in preventing cross-contamination. Knives can be placed in a soaking tank every morning and between uses until the end of the shift. Chlorine-based disinfectants are not recommended because they can corrode knives.

The use of disposable dishcloths such as paper towels is strongly recommended because reusable dishcloths can transfer pathogens to foods or food contact surfaces (17). Staff must be fully trained in cleaning methods and chemical usage.

b. Machine Washing. Inflight service equipment such as china dishes, melamine resin bowls, trays, and cutlery are usually washed by a flight-type dishwasher. Disinfection of this equipment is done by hot water at the final rinse step. Therefore, temperature control of the dishwashing operation is very important to achieve proper disinfection. Since the temperature gauge on the machine panel indicates the water temperature in the pipe, not the temperature of the water sprayed onto the equipment, the final rinse water temperature must be monitored by means of a thermolabel, attached to the surface of metal equipment, which changes color from silver to black when exposed to temperatures over 71°C (161°F). To facilitate fast drying, a surface-active chemical is added to the final rinse water.

Definition of CCP 7. Cleaning and Disinfection of Food Contact Surfaces

Hazards: Survival and contamination of undesirable microorganisms.

Criteria: For dishwasher:

1. Temperature of each step shall be within the recommended range.
2. Dishwashing machines shall provide pasteurization by final rinse.
3. Cleaned equipment shall be stored as to ensure quick drying.
4. Cleaned equipment and utensils shall be stored properly to prevent accidental contamination.

For a three-compartment sink, cutting board, and knife-soaking tank:

1. Concentration of disinfectant must be within effective range.

Monitoring: For dishwasher;
1. Monitoring temperature of each step.
2. Positive reaction of 71°C (161°F) thermolabel on metal surface (once per week).
3. 4. Visual assessment.
For a three-compartment sink, cutting board, and knife-soaking tank:
1. Monitoring concentration of disinfectant solution.

Frequency of monitoring: Once per shift.
Monitoring tools: thermolabel, disinfectant concentration test paper.
Corrective action: For dishwasher, if displayed temperature on the machine is out of the recommended range or if thermolabel did not change to black, call the maintenance team. If washing and drying is poor, clean nozzle, change water, and check surfactant. For a three-compartment sink, cutting board, and knife-soaking tank: if the disinfectant concentration is out of effective range, chemicals should be added to increase the concentration.

9. Cooking and Reheating

From a hygiene point of view, the purpose of cooking is to inactivate or the pathogen load reduce it to a safe level. Cooking should be done under the assumption that all raw food is contaminated by pathogens. Pathogens that can be eliminated by proper cooking include bacteria such as *Salmonella* spp., *Shigella, Campylobacter, Listeria monocytogenes, Streptococcus, E. coli, Vibrio* spp., *Yersinia enterocolitica, S. aureus, Brucella,* vegetative forms of spore-forming bacteria, parasites such as *Trichinella* and *Anisakis,* and viruses such as hepatitis A virus and Norwalk virus. Opportunistic pathogens such as *Enterobacter* spp. can also be eliminated. Therefore, cooking is one of the most important ways to achieve food safety objectives. But normal cooking conditions cannot eliminate all microorganisms in foods. Spores of *C. perfringens, C. botulinum,* and *B. cereus* cannot be eliminated, even by boiling. In addition, heat-stable toxins from *S. aureus* and *B. cereus* are not destroyed at normal cooking conditions, and thermoduric bacteria can survive pasteurization temperatures.

The minimum cooking time/temperature guides for different foods animal origin have been published by the FDA (7). These are 68°C (154.4°F) for 15 seconds for ground beef, pork, seafood, shellfish, egg and egg products; 74°C (165.2°F) for 15 seconds for poultry and stuffed food; and 63°C (145.4°F) for 3 minutes or 54°C (129.2°F) for 121 minutes for roast beef. IFSA has published similar cooking temperature guidelines for HACCP programs as follows (10):

Raw poultry: 73.8°C (165°F) internal temperature
Raw pork: 65.5°C (150°F) internal temperature, seared on surface

Raw ground pork, raw hamburger: 68.3°C (155°F) internal temperature
Raw shellfish: 60°C (140°F) internal temperature
Raw red steaks: seared on surface
Multiple raw foods: cooking temperature reflecting most hazardous ingredient
Previously cooked foods: 73.8°C (165°F) unless part of HACCP system
Other potentially hazardous commodities: conduct hazard analysis, cook accordingly

If there is no cross-contamination between raw foods and if the chef can control the cooking temperature, this guide can be used in airline catering. But for safety reasons, applying simpler and stricter minimum cooking temperature criteria for most foods of animal origin, such as "cook to a core temperature of 75°C (167°F)," is recommended. The more thorough the cooking of raw foods, the higher the food safety of the cooked product, notwithstanding quality.

Meats that will be subjected to preparation (slicing, etc.) should be pasteurized. Such meats should be chilled prior to preparation. The USDA/FDA pasteurization standards for whole beef roasts and corned beef roasts are as follows (7):

Temperature (°C/°F)	Time (min)
54/130	121
56/132	77
57/134	47
58/136	32
59/138	19
60/140	12
61/142	8
62/144	5
63/145	3

Usually, no reheating is carried out in the airline catering kitchen. All hot meals are reheated onboard. If reheating is required, the time and temperature conditions must be the same as cooking to destroy pathogens that may have recontaminated the food. Chefs tend to undercook food that will be reheated onboard. Reheating stored, undercooked foods may cause problems, however, since the toxins produced by *S. aureus* and *B. cereus* are resistant to heat. This is why all hot meals that are to be reheated onboard must be cooked to a specific temperature and rapidly chilled with minimum contamination.

Definition of CCP 8. Cooking
Hazards: Survival of undesirable microorganisms.

Criteria: 1. Raw poultry and eggs, and products containing raw poultry, eggs, and minced meat, should be cooked to a minimum core temperature of 75°C.
2. Whole pieces of meat should be evenly cooked to a minimum surface temperature of 75°C.
Monitoring: 1. Core temperature is monitored by a probe thermometer on completion of cooking.
2. Surface temperature is monitored by a surface thermometer on completion of cooking.
Frequency of monitoring: All cooked items except those completely boiled.
Corrective action: Extend cooking time until temperature rises to 75°C.

10. Rapid Chilling

If cooked foods are cooled at ambient temperatures and then put in the refrigerator for further cooling, they will remain in the temperature danger zone for a very long time. During this uncontrolled cooling, spores of *B. cereus* (often in many cereals, grains, beans, pasta) and of *C. perfringens* (often in beef and poultry dishes) can germinate and multiply rapidly. These coupled with other surviving bacteria and postcooking contaminants may create conditions leading to an outbreak.

For cooling hot foods after cooking, the FDA Food Code requires that foods be cooled from 60°C (140°F) to 21.1°C (70°F) within 2 hours and from 21.1°C (70°F) to 5°C (41°F) within 4 hours by using a refrigerator (7). For safety as well as quality reasons, all airline catered food must be transferred to suitable containers and rapidly chilled immediately after cooking to preserve the organoleptic quality as well as for safety reasons. Chilling should start within 30 minutes unless the temperature is maintained at or above 65°C (149°F). The chilling capacity and production scheduling must be carefully correlated, especially at rush periods. Chilling should be finished when the core temperature of the food reaches 5°C (41°F) or colder within 4 hours. To achieve this rapid chilling, blast chillers should be used (30). The speed of chilling foods will be affected by a combination of factors including the size or weight of the food, the physical state of a food, food thickness, the thermal conductivity of the food container, and the pattern of use of the blast chiller. Most modern blast chillers can cool food to 5°C (41°F) within 90 minutes if the thickness is less than 5 cm (2 in.). However, this mode of operation must be consistent with the manufacturer's operating criteria for rapid chill equipment.

The blast chiller should have not only a good cold air current, but also a clean air supply to minimize airborne contamination. Some models are equipped with UV lamps to sterilize surface, air, and eventual residual condensation. Measurements of the temperature in different parts of the food should be taken to ensure that chilling has been effective. If a thermometer probe is fitted in the chiller unit, care must be taken that the probe and holder are kept clean and sanitized before each use. Records of time in and out, and temperature in and out, should be reviewed routinely for confirming chilling efficiency. After blast chilling, chilled foods must be transferred immediately to a refrigerator with proper protection from airborne contamination. If a blast chiller is not available, a refrigerator or ice bath can be used. In this case, the cooling rate of the food can be monitored based on the physical state, type, size, and depth of food in the metal container.

Definition of CCP 9. Blast Chilling

Hazards: Multiplication of undesirable microorganisms.

Criteria: Cooked food temperature should pass the temperature interval from 60°C to 5°C (140°F to 41°F) within 90 minutes. For this, the food layer thickness should be a maximum of 5 cm.

Monitoring: Time and temperature before and after chilling process. Visual assessment of food layer thickness.

Frequency of monitoring:
All cooked foods. If one trolley is filled with the same item, measure the temperature in the top, middle, and lowest portions.

Monitoring tool: Probe thermometer, alcohol swab.

Corrective action: If more than 90 minutes is needed for chilling, call the maintenance team. When a malfunction of the blast chiller is detected, put hot foods temporarily into the refrigerator.

11. Bakery

All controls in the bakery are aimed at the reduction of cross-contamination by bacteria and molds that may be present in bakery ingredients. The baking section handles some powdered ingredients, and oven operations generate dust and heat, both creating potential sources of aerosol contaminants. Since the pastry section handles PHFs such as dairy cream, whipped butter, and custard cream, it needs to be set apart from the baking section to prevent contamination and also to control the room temperature. Disposable piping bags should be used in the pastry section.

Fondant and chocolate kept molten in tempering machines should be discarded at the end of each production period. The water in the tempering machine should also be drained and fresh water used on each run. Special attention should be given to all aspects of pest control in bakery areas.

12. Meal Assembly

During meal assembly, there is a chance for pathogen contamination and temperature abuse. Portioning should be done by clean and sanitized tools or with clean plastic disposable gloves. Food temperatures must be below 15°C (59°F) until the end of portioning. For food assembly and preparation, the room temperature should be kept low, and only small quantities of food should be handled at any one time. If necessary, equipment such as refrigerated tables should be used to avoid any unnecessary rise in food temperature.

13. Tray Setting and Packing

The tray-setting and packing operation handles foods that are refrigerated and packaged. This operation usually entails no direct hand contact with food items. For this reason, the rising temperature during tray settings and packing meals is a primary hazard that must be controlled. The production system must be designed and operated so that food will be processed as rapidly as possible.

14. Final Cold Holding and Distribution to Aircraft

Aircraft trolleys (carts) should be stored at 5°C (41°F) or colder as soon as possible after filling. Both sides of the trolley door should be left open and positioned to allow free circulation of chilled air. The doors should be closed just before the trolley is taken from the final holding refrigerator for distribution, unless security regulations dictate otherwise. Even though perishable chilled food should leave the catering unit at a temperature of 5°C (41°F) or colder, small rises in temperature during delivery and service are expected. Dry ice may be a useful aid in maintaining a low temperature of food during distribution. The amount of dry ice should be adjusted after the loading temperature of meals has been monitored. Oven racks filled with chilled entrées usually do not have a physical barrier to a temperature increase during delivery to the aircraft. One block of dry ice in an oven rack covered with a plastic bag is helpful for keeping meals cold for short-distance delivery. The objective should be to deliver cold food to the aircraft at or below 10°C (50°F), or at or below the temperature stipulated in the relevant legislation.

According to the IFSA guidelines (4), the PHF should ideally arrive at the aircraft at a temperature of 7.2°C (45°F). The temperatures should be no higher than 10°C (50°F) at the conclusion of catering, and cold foods should be no higher than 13°C (55.4°F) prior to onboard service. All PHFs must be within the total time-out-of-temperature equivalents for onboard service, under the assumption that foods are slowly increasing in temperature during transportation, boarding, and onboard storage. Useful cooled postkitchen shelf life for food is 9 hours if boarded at 10°C (50°F) and/or prior to service reaches 15.5°C (60°F), and at 12 hours if boarded at 7.2°C (45°F) and/or prior to service reaches 13°C (55.4°F), respectively. PHFs stored beyond these times and temperatures should be discarded.

This guideline is useful to decide what to do with onboard meals in case of heavy delays by aircraft maintenance, bad weather, aircraft change, etc. Hot foods, such as hot soup in flasks, should be delivered to the aircraft at a minimum temperature of 60°C (140°F) as stipulated by IFCA (3) and IFSA (4), and 65°C (149°F) as outlined by AEA Hygiene Guidelines (8).

Definition of CCP 10. Final Cold Holding, Dispatch, and Loading
Hazards: Multiplication of undesirable microorganisms.
Criteria: 1. Food temperature must be cooled to below 5°C before dispatch and must not exceed 10°C during dispatch and loading to aircraft.
2. Check insertion of dry ice in cart drawer.
3. Check plastic cover on oven rack with dry ice when ambient temperature is over 10°C.
Monitoring: 1. Monitoring food surface temperatures at dispatch and loading to aircraft.
2, 3. Visual assessment.
Frequency of monitoring: Six flights per shift.
Monitoring tool: IR surface thermometer.
Responsible for monitoring/confirmation: checker/operation manager.
Corrective action: additional dry ice to be inserted or reduce the handling time.

15. Hygiene Guidelines for Delayed Flights

IFSA guidelines for postkitchen shelf life, referred to in Sec. III.C.14, can be used for delayed flight meal control (4). However, AEA guidelines include procedures to follow if flights are delayed after meals have been loaded (8). If there are no active cooling facilities on the aircraft, meals that have been loaded at a temperature not exceeding 12°C (53.6°F) should have their temperatures monitored every 2 hours. If the temperature of the PHF rises above 20°C (68°F), it is recommended that the food be replaced. If the delay exceeds 6 hours and the temperature of PHF rises above 15°C (59°F) in that time, it is recommended that the food be replaced. Where active cooling facilities are available on the aircraft, and are in operation, the temperature of the food should be monitored every 4 hours to ensure that it is maintained below 12°C (53.6°F). Should a temperature increase occur, and if the delay exceeds 6 hours with the food temperature of the PHF rising above 15°C (59°F) during that time, it is recommended that the food be discarded.

16. Time Control

IFCA describes time control as CCP 11 in its HACCP guidelines (3). Time control means the "time span between food preparation and delivery to the aircraft shall consistently be kept short in order to minimize the hazard of multiplication." But the criteria on each in-house cooked food, as well as the preparation of RTE foods and sanitized fruit and vegetables, should be set up based on regulatory agencies, customer requirements, or internal policy. For example, the Gate Gourmet International Hygiene Manual states the following (13): "When the rest of the CCPs are properly controlled, the maximum time span for purchased RTE foods is 24 h from preparation to portioning and 24 h from portioning to delivery. For in-house cooked foods, the maximum time should be 48 h from preparation to portioning and 24 h from portioning to delivery."

Definition of CCP 11. Time Control
Hazards: Multiplication of undesirable microorganisms.
Criteria: 1. The maximum time span for purchased ready-to-eat foods is 24 hours from preparation to portioning and 24 hours from portioning to delivery
2. For in-house cooked foods, times are 48 hours from preparation to portioning and 24 hours from portioning to delivery.
Monitoring: 1,2. Monitoring time span.
Frequency of monitoring: All products at the end of portioning and at dispatch.
Monitoring tool: Time punch or clock.
Responsible for monitoring/confirmation: Chef, checker/section head, operation manager.
Corrective action: Discard all foods that are out of time limits.

For most caterers, the food containers and trolleys are changed at each step of processing. Therefore, it is extremely difficult to trace back to when the preparation or cooking of a specific meal was started, even though a date mark is used. To do time control, the time span from preparation or cooking to the scheduled time of departure (STD) should be measured by reviewing the production plan. It should take less than 48 hours, since the preliminary meal information (PMI) is released 48 hours before STD. Then, a decision on how long leftover food can be used may be based on set criteria, such as a half-day or one day. This can be easily controlled by color-coded date marking with a proper checklist.

Definition of CCP 11. Time Control (Modified)
Hazards: Multiplication of undesirable microorganisms.
Criteria: 1. Time span from preparation of cold meal and hot meal cooking to STD on production plan sheet is less than 24 and 36 hours, respectively.
2. At tray setting, before noon, today and yesterday's date-marked foods can be used; only the same day date-marked foods can be used during the afternoon setting.
Monitoring: 1. At production, review the production plan sheet.
2. Visual assessment.
Frequency of monitoring: Daily review of production plan sheet. At tray setting, check date mark before each shift start.
Monitoring tool: N/A.

Responsible for monitoring/confirmation: Production planner, section head/production manager, operation manager.
Corrective action: Discard all foods that are out of time limits.

17. Reuse of Returned Meals

Meals that for one reason or another are returned to the catering unit, with the intention of reuse at a later stage, may constitute a safety hazard if the food temperature has reached the danger zone or if time of reuse exceeds time control limits. Some caterers consider this to be a CCP, but others consider this to be an extension of time and temperature control. The guideline for this is as follows: meals must be returned to the catering unit for reuse only if the food temperature is below 10°C (50°F) at the time of return; reuse should take place within the time frames as outlined in CCP 11, Time Control.

18. Onboard Food Storage, Reheating, and Service

Onboard food storage, reheating, and service to passengers are the airline company's responsibility, not the caterer's. The IFCA designates this step as CCP 12 (3). There are no IFCA temperature criteria for cold foods stored onboard the aircraft (3), but the AEA hygiene guidelines suggest that temperatures below 12°C (53.6°F) (4). For reheating, both associations suggest a minimum core temperature of 72°C (161.6°F), except for whole steaks and beef and lamb roasts, which may be served rare.

Definition of CCP 12. On board Food Storage, Reheating, and Service
Hazards: Multiplication and survival of undesirable microorganisms.
Criteria: 1. Food to be served within 2 hours from the time of departure; need no cooling during storage onboard.
2. Food to be served later than 2 hours from time of departure; need cooling (e.g., galley chiller, dry ice) during storage onboard and should be maintained below 12°C.
3. Chilled and frozen hot meals should be heated to a temperature not less than 72°C.
Monitoring: 1. Visual assessment
2, 3. Monitor food temperature
Frequency of monitoring: Every flight.
Monitoring tool: Thermometer.
Responsible for monitoring/confirmation: Crew in charge/purser.
Corrective action: Heating until temperature reaches 72°C.

AEA hygiene guidelines concerning food safety also identify as the airline's responsibility menu planning, receiving, handling/service, training of crew, galley cleaning, and infection control of crew (4).

19. Garbage Disposal

Trash bins throughout the catering kitchen must be well maintained, emptied, and cleaned by cleaners when necessary. Since airline catering kitchens should not allow any flying and/or crawling insects, lids are not required. If lids are fitted, they must be foot-operated or equipped with self-closing doors to prevent food handlers' hand contamination. There should be regular collection of waste material from production rooms whenever processing takes place. Wastes that have been removed from the production area and are awaiting collection and removal from the site must also be kept in suitable containers to keep thing tidy and to discourage vermin. Waste storage areas, compactors, etc., must be kept clean by regular disinfection and hosing down. The method for disposal of food wastes removed from the aircraft arriving from overseas poses specific problems related to quarantine control, and local regulations must be observed. Steam sterilization of waste before disposal into trash compactors or by incineration is commonly requested.

D. Support Programs

1. Staff Training

Training must be carried out at a level appropriate to the work of each employee. All new employees should be given a brief orientation before starting work. This will inform them of the basic responsibilities attached to food handling work. It will also introduce the basic rules relating to standards of dress, personal hygiene (especially the need for frequent hand washing), the prohibition of smoking, spitting, and eating in food rooms, and other points as outlined in the staff hygiene rules (see Sec. III.C.7.c). Employees should also be reminded of their responsibility to report certain illnesses as outlined in Table 5.

Each operation should also have a program of further training and refresher training. Further training should ensure that staff appointed to specific tasks or given extra responsibilities receive the necessary hygiene training. Periodic refresher training should be given to staff at all levels to ensure continued awareness of hygienic food practices.

All training material must be comprehensible to the particular workforce, taking into account the individuals' educational skills and language difficulties. Personal hygiene should obviously feature in all orientation training programs with refresher courses at suitable intervals. As the career of an employee advances, further training will be necessary to introduce more advanced concepts. On-site training is very effective whenever food safety violations are detected by the management group. Some caterers call this program "Five-Minute Training" (FMT), which can be performed by hygiene controllers or team leaders. This training covers specific wrong activities only and is restricted to the specific group that would be assigned to correct the problem (13).

2. Microbiological Analysis

The data from microbiological analysis provides useful information on the effectiveness of the HACCP system and other control measures (27). If the number of bacteria detected from a meal is higher than expected, there may be unknown contamination sources or CCPs not being properly controlled. If this is the case, samples should be taken at each step of processing and handling, and the problem source verified and corrected. For this, an in-house laboratory or commercial laboratories can be used.

AEA has issued Routine Microbiological Standards for Aircraft-Ready Food (Table 6). The Code of Good Catering Practice (2) also covers microbiological standards, but they are stricter than AEA's. Some of the major carriers have their own microbiological guidelines. These are intended to apply to meals ready for dispatch to the aircraft, not standards for acceptance or rejection of raw materials. Microbiological classification has been divided into three groups: "Safe," "Investigation Advised," and "Unsafe." Food will be considered "safe" for consumption if microbiological analysis results are within the standards shown in Table 6. Investigation of food production methods is advised when the microbiological results are higher than the standards with respect to total viable counts (including lactic acid bacteria) and the presence of *Enterobacteriacae* and coliforms. Food are considered "unsafe" when microbiological examination reveals the presence of pathogenic bacteria, including *Salmonella, S. aureus, B. cereus, Campylobacter, C. perfringens,* and *E. coli,* at any level above the standard.

Airline meals often involve composites of a variety of meal components, both cooked and raw. Thus the AEA microbiological guidelines identify eight different food categories. It is important to carefully identify the meal components and method of preparation. Minor ingredients may have a significant effect on the overall microbiology of the meal (e.g., sliced green onion on a selection of cooked meats; cold cuts or salads on top of lettuce). It is recommended that individual components of the plated meal be tested separately, although interpretation of the results must take into account the possibility of transfer of contaminants from one component to another.

For selecting a raw food material supplier, microbial analysis data gives brief descriptions of harvest, handling, storage, and transportation conditions. For inspection on receiving processed PHFs, the microbial standards must be stricter than aircraft-ready foods, since the PHFs need extra preparation and time. If these foods are exclusively made for the airline caterer, the microbial standards need to be agreed to with the supplier.

Table 6 Routine Microbiological Standards for Aircraft-Ready Food

Food category	Total count	Coliforms	*E. coli*	*Salmonella*	*S. aureus*[a]	*B. cereus*	*Campylobacter*[b]	*Clostridium perfringens*	Yeast and molds[c]
Bulk items									
Items that have not been manipulated after heat treatment (e.g., hot meats, gravies); mayonnaise	100,000/g	1000/g	10/g	0/25 g	100/g	1000/g	0/25 g	1000/g	10,000/g
Items that have been portioned after heat treatment	500,000/g	1000/g	10/g	0/25 g	100/g	1000/g	0/25 g	1000/g	10,000/g
Items that have been manipulated after heat treatment (e.g., sandwiches, starters, snacks, plates, desserts—all cold)	1,000,000/g	10,000/g	10/g	0/25 g	100/g	1000/g	0/25 g	1000/g	10,000/g
Undercooked items (e.g., vegetables, deep-frozen blanched vegetables, steaks that will receive no more heat treatment before leaving the flight kitchen)	—	—	10/g	0/25 g	100/g	—	—	—	—
Cold-smoked or cold-cured fish, meat, or poultry	—	—	10/g	0/25 g	100/g	—	0/25 g	—	—
Water and wet ice (i.e., cold water as supplied to the flight kitchen taps; wet ice made in unit or brought in)	—	1/100 mL	0/100 mL	—	—	—	—	—	—
Raw vegetables or raw fruits (or items containing them) sampled when ready for use in aircraft meal	—	—	10/g	0/25 g	—	—	—	—	100,000/g
Acid foods (e.g., yogurt, fruit juices, and fruit segments)	—	—	—	—	—	—	—	—	100,000/g
Cheeses	—	—	10/g	0/25 g	100/g	—	—	—	—

[a]Need be carried out routinely only on cereal, pasta, and egg products.
[b]Need be carried out routinely only on poultry items.
[c]Need be carried out only if excessive aging is suspected.
Source: Ref. 8.

Food handlers' hands, food contact surfaces, washed equipment, and air quality can be monitored by microbial analysis. There are no standards in these areas, but some caterers set their own standards based on data they have collected.

3. Visitors to the Kitchen

All visitors to a food production unit, whatever their status, must comply with the personal hygiene rules. Visitors entering the production area must wear suitable protective clothing and headwear. Supply and delivery personnel must be dressed appropriately for the products they are handling and delivering.

IV. CONCLUSION

Because of extensive handling of potentially hazardous foods and extended hours from production to service, airline catering food is very susceptible to conditions that may lead to massive foodborne outbreaks. By virtue of HACCP system, the safety of inflight foods has greatly improved recently. Hygienic design and construction are extremely important in the successful implementation of an HACCP system in airline catering. Kitchen lay out must support physical segregation during storage and preparation, as well as one-way flow of the food processing. Facilities should be installed hygienically.

Even when an HACCP system is implemented, it cannot guarantee absolute safety because of the human factor. Personal hygiene is included as a CCP in generic airline catering HACCP systems. However, this factor cannot be monitored and controlled throughout food handling, especially in respect to hand hygiene. If training has been effective, employees should be washing their hands voluntarily whenever they touch unclean things. Since pathogens are not visible, the results of microbiological analysis should be used as a tool for confirming or revising the HACCP system and part of the incoming goods inspection. Ready-to-eat foods, raw foods, ice, food contact surfaces, and workers' hands must be analyzed regularly.

REFERENCES

1. Adams, M. R., A. D. Hartley, and L. J. Cox. 1989. Factors affecting the efficiency of washing procedures used in production of prepared salads. Food Microbiol. 6:69–77.
2. Anonymous. 1990. Airline Catering—Code of Good Catering Practice. Silverdale Press, Silverdale Road, Hayes, Middx. UB3 3BH U.K.
3. Anonymous. 1994. An Introduction to Food Safety in Airline Catering Based on HACCP. International Flight Catering Association, Surrey, England. 40 pp.
4. Anonymous. 1994. Food Safety Quality Assurance Manual. International Inflight Foodservice Association, Louisville, KY.

5. Anonymous. 1995. Foodborne outbreaks in California, 1993–1994. Dairy, Food Environ. Sanit. 15:611–615.
6. Anonymous. 1995. Prevention of food-borne illness. Dairy, Food Environ. Sanit. 15:357–367.
7. Anonymous. 1997. Food code. U.S. Public Health Service. Department of Health and Human Services, Food and Drug Administration, Washington, DC.
8. Anonymous. 1997. Hygiene Guidelines. Association of European Airlines, Brussels, Belgium.
9. Anonymous. 1997. Kitchen thermometers, Technical Information from FSIS. U.S. Department of Agriculture, Food Safety and Inspection Service, Washington, DC.
10. Anonymous. 1998. USDA unveils new way to fight *Salmonella.* CNN. March 19.
11. Beuchat L. R., and J.-H. Ryu. 1997. Produce handling and processing practices. Emerg. Infect. Dis. 3:459–465.
12. Bryan, F. L. 1988. Risks of practices, procedures and processes that lead to outbreaks of foodborne diseases. J. Food Prot. 51:663–673.
13. Christensen, S. 1998. Global Quality Manual, Vol. 1, Hygiene. Version 2.0. Gate Gourmet International, Zurich.
14. Eberhart-Phillips, J., R. E. Besser, M. P. Tormey, D. Feikin, M. F. Araneta, J. Wells, L. Kilman, G. W. Rutherford, P. M. Griffin, R. Baron, and L. Mascola. 1996. An outbreak of cholera from food served on an international aircraft. Epidemiol. Infect. 116:9–13.
15. Effersoe, P., and K. Kjerulf. 1975. Clinical aspects of outbreak of staphylococcal food poisoning during air travel. Lancet. ii599–600.
16. Eisenberg, M., K. Gaarslev, W. Brown, M. Horwitz, and D. Hill. 1975. Staphylococcal food poisoning aboard a commercial aircraft. Lancet. ii595–599.
17. Enriquez, C. E., R. Enriquez-Gordillo, D. I. Kennedy, and C. P. Gerba. 1997. Bacteriological survey of used cellulose sponges and cotton dishcloths from domestic kitchens. Dairy, Food Environ. Sanit. 17:20–24.
18. Hedberg, C. W., W. C. Levine, K. E. White, R. H. Carlson, D. K. Winsor, D. N. Cameron, K. L. MacDonald, and M. T. Osterholm. 1992. An international foodborne outbreak of shigellosis associated with a commercial airline. JAMA 268:3208–3212.
19. Heggie S. 1998. Personal communication. Tropical Public Health Unit, Cairns, Australia.
20. Jay, J. M. 1991. Modern Food Microbiology, 4th ed. Chapman & Hall, London.
21. Kang, Young-jae. 1998. Hygiene Manual. Asiana Airlines Catering Services. Seoul, Korea.
22. Kornacki, J. L., and D. A. Gabis. 1990. Microorganisms and refrigeration temperature. Dairy, Food Environ. Sanit. 10:192–195.
23. Lester, R., T. Stewart, J. Carnie, S. Ng, and R. Taylor. 1991. Air travel–associated gastroenteritis outbreak, August 1991. Commun. Dis. Intell. (Aust.) 1517:292–293.
24. Madden, J. M. 1990. *Salmonella enteritidis* contamination of whole chicken eggs. Dairy, Food Environ. Sanit. 10:268–270.
25. Nuki, P. 1998. Killer bug spreads among chickens. Sunday Times. March 15, London, U.K.
26. Patrick, D. R., G. Findon, and T. E. Miller. 1997. Residual moisture determines the

level of tough-contact-associated bacterial transfer following hand washing. Epidemiol. Infect. 119:319–325.
27. Silliker, J. H. 1995. Microbiological testing and HACCP programs. Dairy, Food Environ. Sanit. 15:606–610.
28. Snyder, O. P. 1996. Limitations of bimetallic-coil thermometers in monitoring food safety in retail food operations. Dairy, Food Environ. Sanit. 16:300–304.
29. Snyder, O. P. 1997. The microbiology of cleaning and sanitizing cutting board. *http://www.hitm.com/Documents/Cutboard.html*
30. Snyder, O. P. 1997. Two-inch and four-inch food cooling in a commercial walk-in refrigerator. Dairy, Food Environ. Sanit. 17:398–404.
31. Snyder, O. P. 1998. Hand washing for retail food operations—A review. Dairy, Food Environ. Sanit. 18:149–162.
32. Tauxe, R. V., M. P. Tormey, L. Mascola, N. T. Hargrett-Bean, and P. A. Blake. 1987. Salmonellosis outbreak on transatlantic flights; foodborne illness on aircraft: 1947–1984. Am. J. Epidemiol. 125:150–157.

7
Food Safety in Catering Establishments

Chris Griffith
University of Wales Institute, Cardiff, South Wales, United Kingdom

I. INTRODUCTION

Everyone must eat to survive, and safe food is a basic human right. It is paradoxical that when science and technology have eliminated some major infectious diseases, gastrointestinal infections constitute a growing problem worldwide (7). While some enteric disease may be transmitted directly from person to person, those involving food are also a major concern (2,7). People may accept foodborne illness if they themselves are responsible as a result of poor home food preparation, but if they become ill as a result of eating at catering establishments of various types, their level of "outrage" increases. This outrage may then be directed, sometimes with the help of the mass media, toward the catering industry, the food manufacturing industry, or even governments. The degree of outrage felt is important in claims for compensation and in how consumers perceive risk. With an identified upward trend in notifications, in many countries, the avoidance of foodborne illness is a topical subject of debate (2,24). What is clear is that all participants in the food chain have a role to play (Fig. 1).

II. THE RANGE OF CATERING ESTABLISHMENTS AND HOW THEY DIFFER FROM FOOD MANUFACTURING ENTERPRISES

The catering industry has always been diverse, but with changes in society the size and scope has increased dramatically over the past 30 years. The overall term

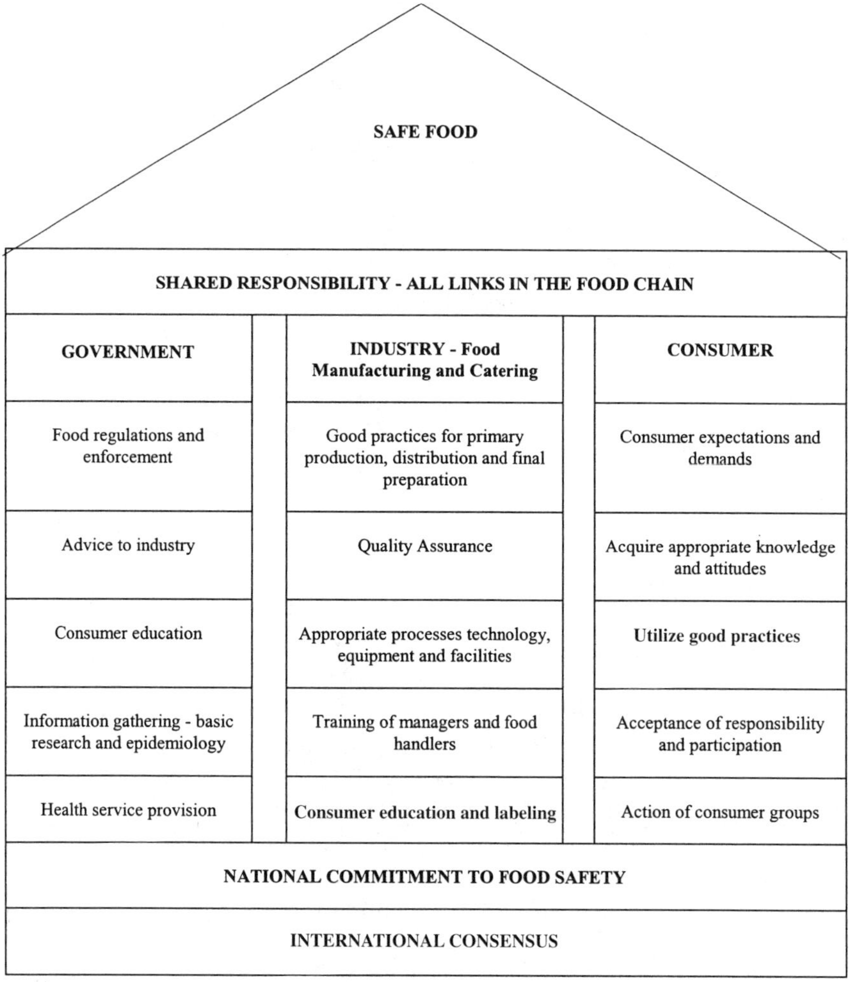

Figure 1 Shared responsibility in the food chain.

"catering" now encompasses many different types of business, although all have one thing in common: they provide a "service" that relates to food. Thus while terminology may vary between countries, at the simplest level food manufacturers produce but do not serve food, whereas catering establishments may prepare, assemble, or just serve individual foods or whole meals. Because of the service element, some countries prefer to use the term "foodservice" rather than catering.

The differences between a caterer and a food manufacturer are summarized

Table 1 Difference Between Food Manufacturing and Catering Operations

Catering	Food manufacture
Produce and serve	Produce only
Preparation in peaks and troughs; low volume	Even production runs; higher volume
Usually whole meals	Often single food item only
Little scientific backup	Good scientific backup
Ill-defined product	Well-defined product
Little or informal quality assurance	Formal, comprehensive quality assurance
Wider product range; smaller scale	Limited product range, larger scale
Little research and development	Considerable research and development
Small-scale equipment	Large-scale equipment

in Table 1. The differences outlined are not always easily distinguished, however, since catering evolves with some cook–chill operations more like food manufacturing. In turn, food manufacturers, traditionally producing single food items, may now produce complete meals.

Various attempts have been made to structure or classify businesses within catering such as institutional, contract, fast food, and so on, as they apply to different sectors of the industry. This chapter attempts to provide a general overview to food safety in catering establishments; for specific advice on different sectors, the reader may consult other chapters. However caterers are classified, one aspect of importance is the predominance of small independent businesses (2,7). Typically 75–85% of catering businesses are likely to be owner-managed (7) and this coupled with the industry's diversity can make communication difficult (27). Owners often enter the industry with little or no previous food knowledge or hygiene training. This and the lack of structured channels of food safety communication, coupled with the need to prepare often quite large quantities of food in advance of service, contribute to the relatively large proportion of foodborne illness outbreaks attributed to catering establishments.

III. CATERING ESTABLISHMENTS AND FOODBORNE ILLNESS

A. Statistics and Risk

Hazards can be defined as "things" that can cause harm, with *risk* the probability of a hazard occurring (7). One common calculation that can be used to express risk is the number of cases of general or specific foodborne illness per 100,000 of population. This allows trend analysis within countries and also comparisons between different countries. For example in 1992/1993, New Zealand had a record rate of 240 notifications of campylobacterosis per 100,000 of population. For the comparable period in the United Kingdom, the rate was 76 per 100,000. However, the latest provisional U.K. data for 1997 indicate that this is figure now nearer 100 per 100,000 population.

Statistics from a number of countries, including the United States (33), the Netherlands (30), and the United Kingdom (11), indicate that general outbreaks of foodborne illnesses are commonly associated with catering or foodservice locations, which may be the origin of up to 70% of general outbreaks (2). Different terminology and methodology are used to collect data, and variations occur between countries and the time periods studied; however, the generalized data in Table 2 can be used as a guide. A more quantitative expression of risk (7) indicated that 1 in every 1527 catering establishments in England and Wales was likely to be implicated in an outbreak of foodborne illness in a given year.

If the trend toward eating out or away from the home continues to increase, then unless additional precautions are taken, catering locations are even more likely to be identified as the source of outbreaks in the future. The costs to a catering establishment that is identified as the source of an outbreak are likely to be many and increasingly punitive. Loss of business, claims for compensation, and/or fines are increasing, and in some countries the estimated bankruptcy rate for such premises is in the order of 30%.

B. Food Safety Implementation Problems Within the Catering Industry

Some or all of the following conditions and trends may be added to the problems faced by caterers as described in Sec. III.A:

Table 2 Types of Catering Locations Implicated (as %) in Outbreaks of Foodborne Illness

Institutional catering	36
Restaurants	25
Hotels	14
Private residence	8
Others/unknown	17

Source: Generalized data simplified and adapted from Phillips, B., N. Rutherford, T. Gorsuch, M. Mabey, N. Looker, and R. Boggiano. 1995. How indicators can perform for hazard and risk management in risk management of food premises. Food Sci. Technol. Today 9(1):19–30, and Refs. 11, 28, and 30.

High turnover of staff
Low staff pay
Low status of staff
Large number of part-time workers
Staff language problems and/or low educational levels
Often little attention to quality assurance
Large number of complex meals
Majority of food often served/prepared to meet short periods of high demand at mealtimes
Current fashion for visually "artistic" dishes requiring much handling
Provision of food to large numbers of vulnerable consumers (e.g., the elderly)
Poor access to food safety information
Facilities and equipment often cramped inadequate

Cumulatively, these problems can lead to difficulties in providing accurate and appropriate hygiene information to all food handlers, who in turn may lack interest in food safety or have negative attitudes toward it. Food handlers' knowledge and attitudes combine a complex framework of social and personality factors to result in food-handling behavior. The importance of this matrix should not be underestimated. It has been suggested that improper food handling behavior may be implicated in up to 97% of foodborne illness outbreaks associated with catering (18) and can be due to either lack of knowledge and/or failure to implement known food hygiene practices (36). Training may or may not remedy knowledge deficiencies, and even when it does, the behavioral changes required to reduce foodborne illness will not necessarily follow.

C. Outbreaks of Foodborne Illness Associated with Catering Establishments

Outbreaks of foodborne illness in catering establishments can involve different pathogens in a wide range of circumstances and occasions (see Table 3). The two examples described illustrate faults typical of many outbreaks due to faulty catering practices.

At a function attended by 350 people over 100 became ill within 2 days, all suffering from gastroenteritis due to *Salmonella enteritidis*. Four were admitted to hospital, and an examination of possible foods centered on egg products supplied by an outside caterer, Some of these foods had been refrigerated overnight and others stored at ambient temperature prior to rewarming. This example illustrates a twofold fault found in many catering operations, namely, preparation of food in advance coupled with inadequate temperature control.

Table 3 Selected U.K. Examples of General Outbreaks of Foodborne Illness Involving Catering Establishments in July and August 1998

Organism	Type of catering establishment	Number ill	Food involved[a]
Salmonella enteritidis PT4	Hotel	16	Chocolate mousse
Salmonella enteritidis PT4	Restaurant	3	Chicken
Salmonella enteritidis PT4	Hospital	13	Eggs
Salmonella enteritidis PT4	Restaurant	2	Not indicated
Salmonella enteritidis PT4	Funeral	33	Eggs and chicken rolls
Salmonella enteritidis PT4	Public house	10	Lemon meringue pie made with shell eggs
Salmonella enteritidis PT4	Hotel	15	Not indicated
Salmonella enteritidis PT4	Party	4	Not indicated
Salmonella enteritidis PT4	Restaurant	12	Cannelloni
Salmonella enteritidis PT4	Hospital	2	Not indicated
Salmonella enteritidis PT5a	Canteen	40	Not indicated
Salmonella enteritidis PT34a	Residential institution	2	Not indicated
Salmonella heidelberg	Christening	>1	Tandoori chicken
Salmonella typhimurium DT49	Community hall	3	Eggs
Bacillus cereus	Restaurant	2	Chinese meal
Bacillus cereus	Restaurant	3	Egg fried rice
Clostridium perfringens	Hotel	10	Beef and vegetable casserole
Clostridium perfringens	Residential institution	>10	Not indicated
Scambrotoxin	Restaurant	2	Canned tuna
Unknown	Restaurant	2	Frozen shrimp

[a]Identification of the food involved based upon statistical, microbiological, or other evidence.
Source: Data from Commun. Dis. Rep. 8(37), 1998.

Campylobacter is a very common cause of gastroenteritis, with the majority of cases recorded as sporadic. One general outbreak involving over 20 diners occurred in the United States. The victims ate pathogen-laden lettuce that had been cross-contaminated from raw poultry. This case indicates the common mistake of using the same area and equipment for the preparation of both raw and ready-to-eat foods.

Studies of such outbreaks are important because they allow investigators to identify patterns of handling malpractice and the frequency of involvement of specific foods.

D. Risk Factors Associated with Foodborne Illness in Catering Establishments

Foodborne illness should be managed on a risk-based approach, and a range of reviews have attempted to quantify the relative importance of different food handling practices (4,5,6,28,33). These reviews assess and record events in different ways and direct comparisons are difficult, although an adapted summary of some of these can be found in Table 4. While different practices are more likely to be risky when associated with specific foods or types of premises, a number of points are worth considering in greater detail.

Data collection practices are important, and the normal method is investigation by food surveillance/enforcement personnel after an outbreak has occurred. One of the most variable risk factors in the different studies was cross-contamination. Some studies implicated cross-contamination in only 5% of cases, others in as many as 28%. This variability is perhaps not surprising given the different manners of data collection. When questioned about their practices, food handlers may not be able to remember minor behavioral details likely to have caused cross-contamination. Therefore in this type of investigative approach cross-contamination is likely to be underreported. Observational studies of food preparation behavior (35) indicate that the potential for cross-contamination is much greater than epidemiological data might suggest.

Different types of catering establishment are likely to have their own specific problems with respect to cross-contamination, for example;

Contaminated cloths used in busy hotels/restaurants
Inability to separate properly raw and cooked foods in very small establishments
Lack of washing facilities in street vending

Table 4 Major Risk Factors Associated with General Outbreaks of Foodborne Illness

Risk factor	Implication in outbreaks (%)[a]
Inadequate heating	33
Incorrect storage	28
Cross-contamination	15
Infected food handler	12

[a]Figures do not add up to 100% because more than one factor may be implicated in any one outbreak.

Source: Data adapted from Refs. 5, 28, and 33.

Pest contamination of foods (e.g., flies in hot countries or summer months)
Poor or inappropriate cleaning regimes in a wide variety of premises

Foods will often undergo a series of preparative stages and may become part of more complex dishes. A catering or domestic kitchen is also subject to contamination from a range of sources, including the many types of raw foods imported into them. Studies in a number of countries using a range of methods have demonstrated rapid dissemination of pathogens around kitchens (19). Therefore it is not surprising that potential food pathogens have been isolated from a range of microbiological studies in which kitchen surfaces were examined (see Table 5). Along with the many handling activities and contact surfaces needed to prepare a whole menu of often quite complex dishes in restaurants and hotels, caterers may experience difficulties in preventing cross-contamination. In addition, some pathogens (e.g. *E. coli* O157) have a very low minimum infective dose. Avoiding cross-contamination and instituting procedures for the correct handling of ready-to-eat or previously cooked foods are particularly important in preventing outbreaks caused by such organisms.

Reported separately in some reviews of risk factors (5) and in some instances found in over 50% of general outbreaks is "preparation of food in advance." Many types of catering establishment do not "serve to order," hence must prepare food for large numbers, often well in advance of consumption. This is particularly true in so-called function catering, such as wedding receptions, large "sit-down" banquets, or even buffets. In these cases, the kitchen staff may be called on to produce foods for much larger numbers than usual. This places a strain on preparation space (and an increased likelihood of cross-contamination); refrigeration space also may prove to be inadequate for storing the foods after production and prior to consumption. These problems are compounded if several functions are held on the same day!

Preparation, especially if done many hours in advance of consumption coupled with inappropriate storage, allows any pathogens that may have survived cooking, an opportunity to grow or produce toxin. An additional problem is post-cooking contamination: the greater the time for preparation in advance, the more likely this is to occur. There should also be an awareness that psychrotrophic pathogens can cause problems if the food is held for too long at cool or even refrigerated temperatures.

Cooking is an important component of most catering operations. It makes food more attractive (depending on the skill of the chef), more digestible, and above all safe. However, chefs select cooking times more by guesswork or sensory considerations than by concern for food safety. The heat resistance of pathogens can vary depending on the food environmental conditions, and cooking times should be more than adequate to destroy the most heat-resistant pathogen likely to be present. Heat penetration times in food will depend on the size of the food, its

Table 5 Reported Isolations of Different Potential Pathogens from Specific Environmental Sites Within Food Preparation Areas

Environmental site	*Campylobacter* spp.	*Salmonella* spp.	*Y. enterocolitica*	*S. aureus*	*E. coli*	*Bacillus* spp.	*B. cereus*	*L. monocytogenes*
Dish cloth				•	•	•		•
Cleaning cloth		•		•			•	
Wash-up sponge		•		•				
Wash-up brush					•			•
Wash cloth		•						•
Floor mop					•	•		
Tea/hand towel				•	•	•		
Sink		•	•		•		•	•
Taps				•	•		•	
Refrigerator/Door				•	•		•	•
Waste/Pedal bin				•	•	•		
Chopping boards					•			
Works surfaces	•				•	•		
Floor	•				•			

Source: Data adapted from Finch, J. E., J. Prince, and M. Hawksworth. 1978. A bacteriological study of the domestic environment. J. Appl. Bacteriol, 45:357–364. Griffith, C. J., C. Davidson, and A. Peters. The Microbiology of Domestic and Catering Kitchens and Implications for Risk Assessment. Unpublished data. Mendes, M. F., and D. J. Lynch. 1978. A bacteriological survey of kitchens. Environ. Health October, pp. 227–231. Scott, E., and S. F. Bloomfield. 1990. The survival of microbial contamination on work surfaces and cloths and its transfer via cloths, hands and utensils. Environ. Health February, pp. 34–37.

initial temperature, the type of container, and the performance characteristics of the cooking equipment. Heat penetration is especially important in re-formed or reconstituted foods (e.g., hamburgers), as well as in foods (e.g., rolled roasts) where microorganisms may gain access to the center or coldest point within the food.

Perhaps one of the most important current needs in catering is, wherever possible, to validate cooking times and temperatures (i.e., to demonstrate that they are effective). These considerations should be based on the latest information on the heat resistance of likely pathogens. If customers wish to have the food less well cooked, even after contrary advice, any adverse consequences become their own responsibility.

Once cooking has been appropriately validated, it requires monitoring. Probe thermometers are useful for this purpose, and provided they are sanitized appropriately, they are important hygiene aids. However, it may be necessary to check the adequacy of cooking some time after the cooking has finished, and a temperature measurement at this point could reflect a storage temperature. A variety of enzyme tests can now be used to determine the degree of heat processing that a food has received. An example using alkaline phosphatase ("chef test") is illustrated in Fig. 2.

IV. STRATEGIES FOR PREVENTING FOODBORNE ILLNESS IN CATERING ESTABLISHMENTS

A. Introduction

The aim of all successful catering operations should be to produce high quality food, and one component of this goal is safety. Safety requires management, and it is unlikely to happen by accident. Managers must lead by example and demonstrate appropriate food safety attitudes and practices. Managers have a responsibility to both comply with legislation and train their workforce appropriately: fulfilling these two responsibilities is a good start toward producing safe food.

B. Legislation

Food safety legislation around the world is being updated in an attempt to halt the increase in foodborne illness and to meet international trade requirements. Much of the legislation reflects the need for organizations doing business in the European community to conform to hygiene directive 93/43/EEC. A key feature of the newer types of legislation is the incorporation of a risk-based approach, including the identification of steps in the production of food that are critical to safety. Catering firms also need to be able to demonstrate "due diligence," and this means both having appropriate food safety management systems in place and implementing them.

Failure to comply with legislation can result in fines, closure, and even imprisonment in some countries. Results of inspections may be published in the local press. In many countries, however, the appropriate resources cannot be brought to bear to implement the legislation. Provision for implementing food safety legislation has not kept pace with the expansion in catering establishments (4).

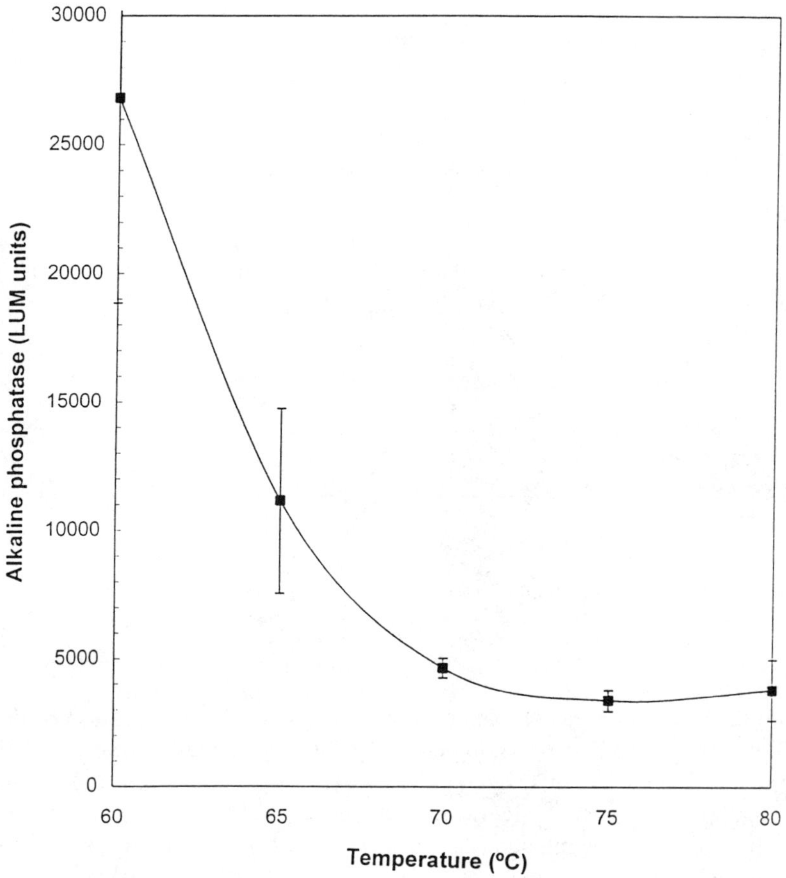

Figure 2 In situ analysis of the effect of heat on alkaline phosphatase, measured using a Charm "chef test," in pork sausages: $n = 10$; mean + SEM. LUM = luminiometer units. (From Redmond, E. C., Griffith, C. J., and Peters, A. C., Validating alkaline phosphatase reduction as an indicator for meat cooking efficiently. Presented at Microbial Control in the Food Industry Conference, Cardiff, July 1998.)

C. Quality Management Systems

All companies are likely to have a formal or informal (more likely in small companies) quality management system. This can be defined as the organizational structure, responsibilities, procedures, and resources for implementing quality management. Quality management systems differ between companies and are related to such factors as size, organization, finances, and personnel structure, but

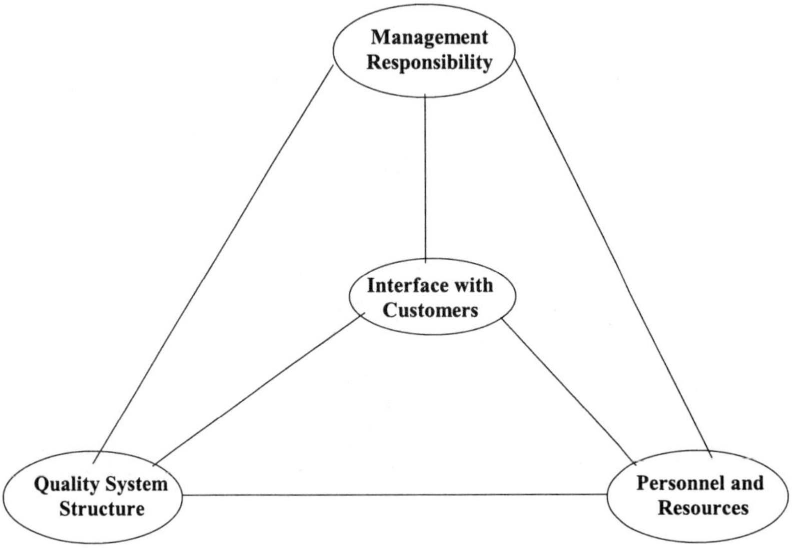

Figure 3 Key factors of a quality system.

they are likely to represent the main cost of achieving quality. It is, however, increasingly recognized that delivering quality is the responsibility of everyone in the company or organization (Fig. 3).

Quality management systems can be organized in different ways using different philosophies, although traditional approaches are based very much on quality control. This involves checking and inspection, with particular emphasis on the end product; any product failing such checks and inspections is then rejected.

Modern approaches are based on quality assurance, which is much broader than quality control. Quality assurance attempts to achieve zero defects by "getting it right first time," every time. All factors about the product are considered and measures taken to ensure that appropriate quality will be obtained. Whereas quality control checks to verify that certain standards have been met, quality assurance attempts to stop things from going wrong in the first place or to design quality into the product.

The quality assurance philosophy can be specifically adapted to food safety, identifying what can go wrong and then ensuring that it does not. This is the basis of the hazard analysis–critical control point (HACCP) approach (see Sec. IV, E).

Different (although not mutually exclusive) approaches can be taken to implement quality assurance. One is the ISO 9000 approach, which is very much based on quality manuals and specifications and involves a considerable amount of paperwork. The other concentrates more on collective responsibility and employee attitudes, behavior, and motivation. This approach, called total quality management (TQM), entails a management philosophy that calls for a commitment at all levels, as well as trust and communication, and seeks to continually improve the effectiveness and competitiveness of the business as a whole. Both types of approach to quality assurance can be found in the catering industry. The collective combination of catering and quality procedures aimed at ensuring that foods are consistently prepared, cooked, and served to an appropriate standard can be referred to as good catering practice (GCP) and is analogous to good manufacturing practice (GMP) in food production (20).

D. Food Safety and Hygiene

Understanding the meaning and structure of food hygiene (see Fig. 4) is important in implementing the many guidelines for the safe handling of food (see, e.g., Ref. 14).

Hygiene has to do with health, and food hygiene deals with ensuring that food is safe to eat. Food hygiene covers all aspects of processing, preparing, transport, handling, or serving of food to ensure that it is safe to eat.

Food hygiene is vital throughout the food chain from primary production via processing and distribution through to serving—that is, from farm to fork. Like any other chain, the food hygiene chain is only as strong as its weakest link.

Many food hygiene practices and procedures (14) contribute to preventing microbial food poisoning, although they all ultimately rely on one of two fundamental principles:

Preventing contamination of food
Preventing the growth or survival of any contaminants

Caterers unsure of whether a particular action is hygienic should ask themselves the following:

Will it increase the risk of contamination?
Will it increase the chances of growth/survival?
If the answer to either question is yes, the chances are the action contemplated is unhygienic.

The traditional strategic approach to food safety management in catering has been to implement hygienic practices, often in a relatively uncoordinated way, based on food storage, cleaning, pest control, personal hygiene, and so on. Often management responsibility for the practices was been shared among various staff members, and food safety was based on a traditional "floors, walls, and ceilings" approach.

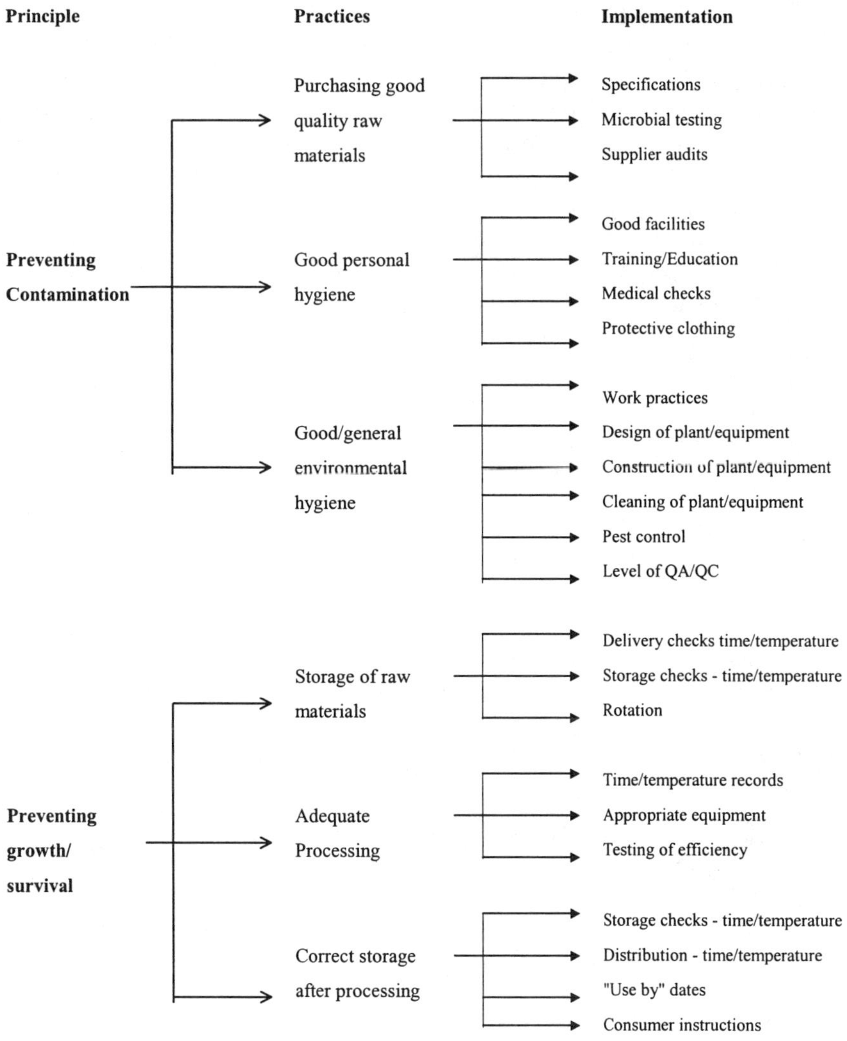

Figure 4 Principles, practices, and implementation of food hygiene.

E. HACCP, ASC, and SAFE

It is not the function of this chapter to describe HACCP in detail, and the reader should consult one of the many books on the subject (e.g., Ref. 9) for further information.

Since its origin approximately 30 years, ago the HACCP system has become the universally accepted method of assuring food safety (23). However, there may be real as well as perceived barriers to implementing of HACCP, and many catering and other food operations have not embraced HACCP with the anticipated enthusiasm (12).

The barriers, real and perceived, have been reviewed (12). For example, there has been concern about the applicability of HACCP, originally developed for food manufacturing, to the catering industry. The approach needed to introduce an HACCP system into small catering operations must be different from that taken in a large manufacturing unit. Nevertheless the basic HACCP principles can be applied, and this has led to the development of the assured safe catering (ASC) and systematic assessment of food environment (SAFE) approaches (3,8). Both these systems allow the HACCP principles to be adopted for even small-scale catering operations.

Recently the U.S. Food and Drug Administration (FDA) issued a draft guidance note on the application of HACCP in retail food establishments (i.e., catering firms). These guidance notes make clear that HACCP is not the total solution and emphasize the role of prerequisite programs, including pest control and good personal hygiene.

Smaller operations are likely to find beneficial the use of a generic HACCP approach to similar groupings or categories of products (21). For example, one might start with a number of recipes (e.g., chicken curry, chicken supreme, chicken and parsley casserole, chicken cacciatore), which all contain poultry and are prepared similarly, and then develop one generic HACCP plan with variations. It is likely that generic flow charts, rather than the specific diagrams used in food manufacture, will need to be used (see Fig. 5). While in some ways this approach is contrary to the original HACCP concept of product and line specificity, it will probably be essential if HACCP is to be introduced into catering.

Nevertheless, given the large number of catering operations and the even larger number of employees in most countries, informing the owners/managers about HACCP is a long-term project. After implementation, one of the greatest hazards then may be complacency (24), and caterers will always have to be alert to maintaining high standards of food safety.

F. Food Safety and Training—A Paradox?

Gilbert in 1983 (16) recommended education and training as one of the main ways of reducing the level of foodborne illness, and these measures have been supported, at regular intervals, by a range of influential reports (25,27,34). Furthermore, industry itself perceives training to be highly important and feels that food handlers should be better trained (24). As stated by Todd (32), "Control measures cost money and unless [control] is perceived to be cost effective no serious effort will be made to implement them." There is no doubt that training is perceived to be important, and a training requirement is part of food hygiene legislation in

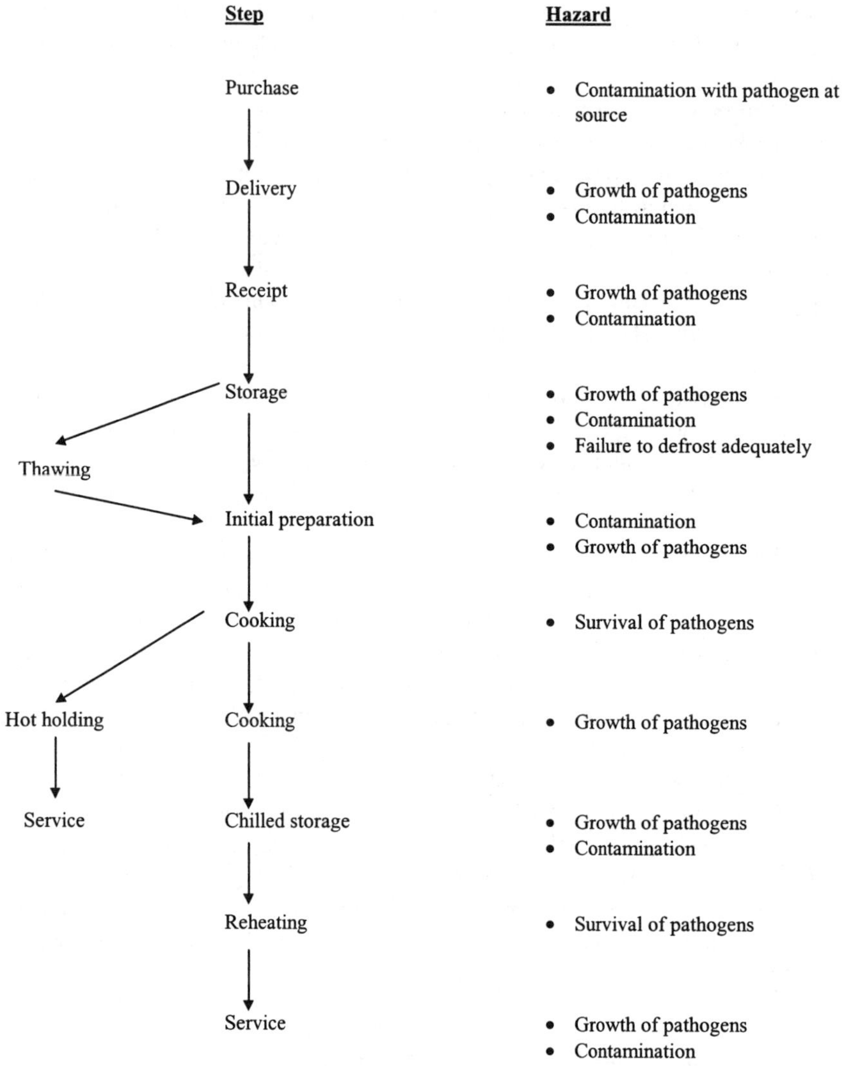

Figure 5 Generic flow diagram for catering operations.

some countries, although data on its cost-effectiveness is largely lacking. Guidelines concerning compliance with U.K. legislation have been published (15), along with a matrix of training levels in relation to the food handler's catering function and responsibility.

The problems associated with the training of the large number of food handlers working in catering include high job turnover and large numbers of part-time workers. Although a large number of catering premises in the United Kingdom proudly display certificates of success in hygiene training courses, it has been estimated that only 46% of food handlers have received hygiene training (1). Even after training, caution against complacency has been urged, with recommendations to include aspects of the behavioral sciences in implementing food safety education and training (17). Little has been done to assess the effectiveness of training, and the research that has been carried out supports the cautious approach.

One study (13) reported no significant improvement in knowledge as a result of training; another (18) showed that while aspects of knowledge improved, food handling behavior did not improve and in some instances even deteriorated posttraining. These studies and others (17,26) have all advocated the application of health education/psychological models to training. The simplest perhaps of these, called KAP (knowledge, attitudes, practice), has had only very limited success (26). The KAP approach relies on the provision of information coupled with assumption that people will act on the information and behave rationally (13). Numerous examples, both from health education in general and food safety specifically, indicate that this sequence often does not play out (13). Training based on the use of social cognition models, "whilst not a panacea, deserves a higher priority" (17). This type of approach argues that behavioral change (i.e., the implementation of required hygiene practices) is not easily achieved and that consideration must be given to motivation, constraints, barriers, and facilities, as well as to cultural aspects of the appropriate practices. Undesirable behavior is often long established, "deeply rooted in the culture of the kitchen," and not easily overturned (29). The following list summarizes factors likely to make a training course successful.

Make training targeted and specific. This is often best done in the workplace, using specific work-related examples. Locate the information within an HACCP framework dealing with specific menu items. Training should concern theory and practice, and the trainer should be able to relate directly to the trainees and be perceived as trustworthy and knowledgeable.

Provide information that is accurate, clear, and concise and at the appropriate educational level of the employees.

Examine and identify likely barriers to food handlers implementing good hygiene behavior. The training should include ways in which the barriers can be overcome.

Provide *all* the necessary facilities to ensure that the food handlers can implement the knowledge they have acquired.

Provide a motivational framework to encourage implementation of learned behavior: What incentives can be provided?

Ensure that prevailing "cultural norms," as evidenced by the behavior of the owner/manager and other important food handlers, demonstrate appropriate practices. Indicate that inappropriate food safety behavior will not be tolerated. Insist that the owner/manager also undergo training successfully and ensure that he or she understands the employees' perspective.

Reinforce messages concerning hazards and risks. Emphasize the longer term social, medical, and financial costs of foodborne illness (e.g., the effect of pathogenic *E. coli* on children, claims for compensation).

Training is not a "one-off" event; ensure regular updating, and keep a training log. Make the individual handlers proud of the training records.

Evaluate the training in terms of knowledge and behavior.

G. Self-Assessment of Risk

The EC hygiene directive requires businesses to identify any step in the activities of the food business that is critical to ensuring food safety, and to ensure that adequate safety procedures are identified implemented, maintained, and reviewed. This requirement, which has been adopted as regulation 4 (3) within the 1995 Food Safety (General Food Hygiene) Regulations, requires caterers to undertake a hazard analysis regime in the production of food they serve. This must have taken place prior to a visit by an enforcement officer, who will assess compliance with the legislation. This hazard analysis requirement is new but is deemed to be important in the management of food safety, and attempts are being made to ensure its consistent enforcement. Thus proprietors are required to undertake a review of their operations and to assess risks posed by the methods and types of food used. Unfortunately, there is evidence (6,10) that caterers frequently underestimate the level of risk posed, lack current information on hazards, lack information on regulation 4 (3), or even lack the time to undertake the required review of their operations.

As an initial stage in this process, self-assessment models have been devised to help caterers assess the level of risk posed by their premises (6). This type of approach is one of a range of quantitative approaches to internal and external auditing that can be used in catering operations (10). The self-assessment approach requires caterers to think in detail about how and why they produce foods and their relative safety. This thought process is a useful educational experience.

V. PRACTICAL TIPS FOR CONSUMERS EATING OUT

One thing is certain. There is no way of looking at, smelling, or tasting that will tell the purchaser whether every food is safe to eat under all conditions. Consumers trust that the caterer will provide them with safe and wholesome food every time they eat out. They can however minimize risk by asking themselves the following questions, based on the two principles of hygiene, namely, preventing contamination and preventing growth and survival of harmful microorganisms.

Are there good facilities for staff to practice hygiene?*
Are hygiene certificates on display?
Are staff wearing adequate protective clothing?
Are kitchen/preparation areas visible?
Are surroundings clean, tidy, and in good repair?
Is there evidence of cleaning programs?
Is there evidence of separation of raw food from cooked foods?
Are high risk food items served by means of clean equipment, rather than hands?
Is food on display protected from contamination?
Does the food look cooked?
Is food hot or cold when served? Warm food allows the growth of pathogens and should be avoided.
Does the food on display look hot or cold? (Are the storage temperatures on display?)
Are display refrigerators overloaded?
Is hot food on display in shallow trays and regularly stirred?
Is there evidence of temperature monitoring?
How long has the food been on display?

VI. SPECIAL CONCERNS RELATING TO CATERING IN DOMESTIC PREMISES

Outbreaks of foodborne illness are generally associated with the commercial preparation of food served in hotels, restaurants, or institutions served by catering firms. However, a significant proportion of general outbreaks—up to 16% in England and Wales (28)—are attributed to domestic kitchens. Usually, while commercial kitchens are subject to legislation and inspection, domestic kitchens are exempt. If, however, domestic kitchens are used as the basis of a commercial business, they are likely to be subject to legislation. Domestic kitchens may be used to prepare food for quite large numbers for birthday celebrations, engagement parties, wedding receptions, barbecues, and other functions. However, they are unlikely to be adequate in size or facilities or ventilation, a shortcoming that poses a particular problem if the ambient temperature is high. The problems of using domestic kitchens for catering purposes can be summarized as follows:

*According to one school of thought, it is necessary only to check the staff toilets and changing areas to determine standards of hygiene.

Inadequate size and space leading to cross-contamination
Inadequate cleaning regimes leading to cross-contamination
Lower specification cooking equipment
Inadequate monitoring facilities (e.g., temperature readings)
Inadequate refrigeration space and ability to store prepared foods separate from raw foods
Inadequate personal hygiene facilities
Possible presence of domestic pets and other sources of pathogens (e.g., diapers)

The reality of these problems was illustrated by a U.K. study of risk factors associated with general outbreaks of foodborne illness linked to domestic catering (28). Inappropriate storage was identified in 24% of cases, inadequate heating in 23%, cross-contamination in 22%, and infected food handlers in 14%; the most commonly implicated pathogen comprised *Salmonella* spp. These figures may underrepresent the scale of the problem (28), and people preparing food for large numbers in a domestic kitchen need to consider the suitability of the premises in relation to the type and quantity of food prepared as well as the frequency of preparation.

VII. SUMMARY AND CONCLUSION

Catering establishments are recognized as a major location for general outbreaks of foodborne illness. Therefore they must treat food as a major factor in the enterprises' survival and success. It is recognized that introducing and maintaining new HACCP-based approaches into catering operations is not without problems. Nevertheless, data (22) suggest that such improvements can be achieved within the financial means of small catering premises, with nonrecurring start-up costs as low as 0.33% and running costs of 5.3% of weekly turnover. However, many of these latter costs are not unique to HACCP itself and include general management costs. Caterers therefore should heed the financial message—in the long run, it is cheaper to manage food safety correctly than to ignore it. Other than financial factors, caterers also have a moral responsibility to ensure that their customers do not become ill as a result of eating catered foods.

REFERENCES

1. Anonymous. 1998 Survey reveals training shortfall. Initial Food News 11:1.
2. Border, P., and M. Norton. 1997, Safer Eating: Microbiological Food Poisoning and Its Prevention. Parliamentary Office of Science and Technology, London.

3. British Hospitality Association. 1995. S.A.F.E. Systematic Assessment of the Food Environment. British Hospitality Association, London.
4. Bryan, F. L. 1982. Foodborne disease risk assessment of food service establishments in a community. J. Food. Prot. 45:93–100.
5. Bryan, F. L. 1995. Hazard analysis of street foods and considerations for food safety. Dairy, Food Environ. Sanit. 15:64–69.
6. Coleman, P., and C. J. Griffith. 1998. Risk assessment. A diagnostic self-assessment tool for caterers. Int. J. Hosp. Manage. 17:289–291.
7. Coleman, P., and C. Griffith. 1997. Food safety legislation, risk and the caterer. Hyg. Nut. Food Service Catering. 1:231–244.
8. Department of Health 1993. Assured Safe Catering. Her Majesty's Stationery Office, London.
9. Dillon, M., and C. J. Griffith. 1996. How to HACCP: An Illustrated Guide. M. D. Associates, Grimsby, U.K.
10. Dillon, M., and C. J. Griffith. 1997. How to Audit: Verifying Food Control Systems. M. D. Associates, Grimsby, U.K.
11. Djiretic, T., P. G. Wall, M. J. Ryan, H. S. Evans, G. K. Adak, and J. Cowden. 1996. General outbreaks of infectious intestinal disease in England and Wales, 1992–1994. Commum. Dis. Rep. 6:R57–R63.
12. Ehiri, J. E., G. P. Morris, and J. McEwen. 1995. Implementation of HACCP in food businesses: The way ahead. Food Control 6:341–345.
13. Ehiri, J. E., G. P. Morris, and J. McEwen. 1997. Evaluation of a good hygiene training course in Scotland. Food Control 8:137–147.
14. Farber, J. M., and A. Hughes. 1995. General guidelines for the safe handling of foods. Dairy, Food Enviro. Sanit. 15:70–78.
15. Food Safety and Hygiene Working Group. 1997. Industry Guide to Good Hygiene Practice: Catering Guide. Chadwick House Group, London.
16. Gilbert, R. J. 1983. Food-Borne Infections and Intoxications—Recent Trends and Prospects for the Future. Food Microbiology Advances and Prospects for the Future. Academic Press, London.
17. Griffith, C. J., B. Mullan, and P. E. Price. 1995. Food safety: Implications for food, medical and behavioural scientists. Br. Food J. 97:23–28
18. Howes, M., S. McEwen, M. Griffiths, and L. Harris. 1996. Food handler certification by home study: Measuring changes in knowledge and behaviour. Dairy Food Sanit. 16:737–744.
19. Humphrey, T. J., K. W. Martin, and A. Whitehead. 1994. Contamination of hands and work surfaces with *Salmonella enteritidis* PT4 during the preparation of egg dishes. Epidemiol, Infect 113:403–409.
20. Institute of Food Science and Technology. 1992. Guidelines to Good Catering Practice. IFST, London.
21. Microbiology and Food Safety Committee of the National Food Processors Association 1993. HACCP Implementation: A generic model for chilled foods. J. Food Prot. 56:1077–1084.
22. Mortlock, M. P, A. C. Peters, and C. J. Griffith. 1998. Applying HACCP to small retailers and caterers: A cost benefit approach. Conference Proceedings: The Economics of HACCP: New Studies of Costs and Benefits. Washington, DC.

23. Motarjemi, V., F. Kaferstein, G. Moy, S. Miyagawa, and K. Miyagishima. 1996. Importance of HACCP for public health and development. The role of the World Health Organisation. Food Control 7:77–85.
24. Penner, M. P., C. W. Shamlin, and A. Thomson. 1997. Food safety in food service: Exploring public policy options. Dairy, Food Environ. Sanit. 17:781–787.
25. Pennington, T. H. 1997. The Pennington Group Report. Her Majesty's Stationery Office, Edinburgh, Scotland.
26. Rennie, D. M. 1995. Health education models and food hygiene education. J. R. Soc. Health. 115:75–79.
27. Richmond, M. (ed.). 1991. The Microbiological Safety of Food. Part Two. Her Majesty's Stationery Office, London.
28. Ryan, M. J., P. G. Wall, R. J. Gilbert, M. Griffin, and B. Rowe. 1996. Risk factors for outbreaks of infections intestinal disease linked to domestic catering. Commun. Dis. Rep. Rev. 6:R179–R182.
29. Sheppard, J., M. Kipps, and J. Thomson. 1990. Hygiene and hazard analysis food service, in C. P. Cooper (ed.), Progress in Tourism, Recreation and Hospitality Management. Behaviour Press, London.
30. Simone, E., M. Goosen, S. H. W. Notermans, and M. W. Borgdorff. 1997. Investigations of foodborne diseases by food inspection services in the Netherlands, 1991–1994. J. Food Prot. 60:442–446.
31. Smith, P. 1996. The hazard of complacency. Food Process. 65:46.
32. Todd, E. C. 1989. Costs of acute bacterial food-borne disease in Canada and the United States. Int. J. Food Microbiol. 9:313–326.
33. Weingold, S. E., J. J. Guzewich, and J. K. Fudala. 1994. Use of foodborne disease data for HACCP risk assessment. J. Food Prot. 57:820–830.
34. WHO. 1988. Health Education in Food Safety. World Health Organisation, Geneva.
35. Worsfold, D., and C. J. Griffith. 1996. Cross-contamination in domestic food preparation. J. Hyg. Nut. Food Serv. Catering 1:151–162.
36. Worsfold, D., and C. Griffith. 1997. Assessment of the standard of consumer food safety behaviour. J Food Prot. 60:399–406.

8

Safe Preparation of Foods at the Foodservice and Retail Level: Restaurants, Take-Out Food, Churches, Clubs, Vending Machines, Universities, Colleges, Food Stores, and Delicatessens

John J. Guzewich
U.S. Food and Drug Administration, Washington, D.C.

I. INTRODUCTION

This chapter, which covers factors and concepts for assuring the safety of foods prepared and served at foodservice (restaurants) and retail food (food stores) establishments, also provides practical advice for safe food preparation at home. The chapter begins with a discussion of the size and nature of the industry and problems the industry faces that impact on food safety. Next, the chapter describes the epidemiology of foodborne disease in the industry, food microbiology, laboratories, new ideas to aid in implementing hazard analysis–critical control point (HACCP) systems, and the use of foodborne disease surveillance findings for HACCP systems. The chapter concludes with practical advice to food prepares in all settings and practical advice to consumers on foodborne illness prevention.

II. BACKGROUND

A significant and growing percentage of ready-to-eat foods consumed in both the industrialized nations and the developing world is prepared outside the home. In the United States, the number of persons who consume at least one food or beverage away from home each day has increased 33% since the late 1970s. At the same time the consumption of foods prepared outside the home is growing and the population of high risk individuals is also increasing—up to 25% of the total U.S. population by some estimates (5). High risk individuals for foodborne disease include pregnant women, neonates, children under 5 years of age, elderly persons (>65 years), residents of nursing homes, cancer patients, organ transplant recipients, and AIDS patients (see Chapter 5).

The nature of the foodservice and retail food industries is constantly evolving. Establishments in both these industries are today more likely to engage in activities previously seen in the food processing industry, such as smoking meats, poultry, fish, and other foods; bottling of sauces and salad dressings; and packaging ready-to-eat foods in reduced oxygen containers to lengthen shelf life and preserve quality.

Foodservice/retail food establishments have many inherent characteristics that make implementation of food safety programs difficult in this labor-intensive industry. The number and variety of foods prepared and the frequent change in menus means that managers must constantly adapt food safety principles to the new items and train staff in implementing them. This task is made more difficult by the need to rely on poorly trained managers and on a workforce of young, inexperienced, poorly trained, and undersupervised employees who are in low paying, low esteem jobs, with a high rate of turnover. This is the first steady job of many such employees. The situation is no better in temporary establishments that operate at festivals organized by community groups, such as religious or fraternal organizations, where the workforce is made up of volunteers. In these instances, the workers are not familiar with the hazards, let alone means of their control, associated with preparing and serving large quantities of food.

The food supply today is a global one. Foods prepared and sold can originate from sources far removed from the site of preparation. Raw foods of plant or animal origin may originate in locations where public health protection guidelines or standards either do not exist or are not enforced. Today's rapid transportation system enables perishable foods to be transported half-way around the world and still be served fresh. These foods may arrive with microbiological, chemical, or physical contaminants. When such foods are served raw or lightly cooked in an industrialized country, consumers may be exposing themselves to risks they would seek to avoid if dining in a developing country. Similarly, processed foods may originate in settings where standards we less stringent than or different from those that exist in the place of consummation; for example, raw milk cheeses are sold in countries whose consumers would assume that cheese is made from pasteurized milk.

Historically, food safety regulatory programs have focused their attention on the physical environment. But whereas walls, floors, ceilings, and general cleanliness merit aesthetic consideration, they are of limited public health significance. While government and industry have touted the need to redirect efforts toward recognized contributing factors to foodborne disease, implementation of the (HACCP) approach has been slow.

III. EPIDEMIOLOGY OF FOODBORNE DISEASE

Epidemiology is the basic science of public health, and therefore it is the basic science of food protection. Through the systematic surveillance, investigation, and reporting on the incidence of foodborne illness and its causes, patterns that point to the factors contributing to those illnesses can be identified. By studying these contributing factors and the frequency and circumstances of their occurrence, one can determine how best to prevent them in the future and to set priorities for prevention efforts. One reason control is difficult in developing countries is that surveillance is inadequate, and, therefore, the burden of foodborne disease is not fully understood by policy makers (11). Foodborne disease surveillance in industrialized countries varies greatly in how it is carried out, in how the findings are reported, and in the level of commitment of the designated regulators. As a result, it is very difficult to compare data from within or among various countries during different time periods. Summaries and analyses of foodborne disease surveillance data are periodically published in the United States and Europe (1–3,6,10).

IV. FOODBORNE DISEASE INVESTIGATION FINDINGS

Summaries of findings from foodborne disease outbreak investigations can identify agents, vehicles (significant ingredients), and contributing factors. Foodborne disease surveillance is generally poor, and the summaries of investigations are often not compiled in ways that are useful for developing prevention strategies. Data from the New York State Department of Health (NYSDOH) are reported here because the information has been carefully collected and compiled in ways that are useful for our purposes.

A. Agents

Table 1 lists the frequency with which agents were reported by the NYSDOH foodborne disease surveillance program for the period 1980–1993. Viral agents including nonspecific viral gastroenteritis, Norwalk gastroenteritis, hepatitis A, and rotavirus comprise 35% of all confirmed or suspected outbreak etiologies dur-

Table 1 Frequency (%) of Foodborne Disease Outbreaks in New York State for Which Agents Were Reported: 1980–1993

Agent	Frequency (%)
Nonspecific viral gastroenteritis	25.7
Salmonellae	23.6
Clostridium perfringens	8.3
Scombrotoxin	7.2
Gastroenteritis from Norwalk Virus	6.6
Staphylococcus aureus	5.9
Other chemical sources	5.3
Bacillus cereus	4.6
Campylobacter spp.	2.4
Hepatitis A	1.6
Mushrooms	1.4
Rotavirus	1.1
Heavy metal	0.9
Trichinella spiralis	0.86
Other	4.54

ing the reporting period. These agents were associated with the consumption of raw or lightly cooked molluscan shellfish or with the consumption of foods that had been contaminated because an infected food worker had prepared the food vehicle. Salmonellosis was the second most frequently reported foodborne disease agent, having been reported in 23.6% of the outbreaks. Salmonellosis is most often associated with foods of animal origin, including eggs and poultry, or with infected food workers. Inadequate cooking and/or inadequate refrigeration of foods and contaminated ingredients are often contributing factors. *Clostridium perfringens,* reported in 8.3% of the outbreaks, is most often associated with large masses of food that are cooled slowly after a cooking step. Scombrotoxin was reported in 7.2% of outbreaks. This agent is most often associated with fresh ocean fish that were not properly refrigerated at some point after the catch. *Staphylococcus aureus* was reported in 5.9% of the outbreaks. This agent usually is associated with contamination of food during preparation by a food worker who either is infected with the agent or carries it. Subsequent time and temperature abuse allows the bacteria to grow and produce toxin in the food.

B. Significant Ingredients

A "significant ingredient" refers to a food item eaten alone or combined in a mixed food that is the origin of the agent in a foodborne disease outbreak. Table 2 lists the

Table 2 Frequency (%) of Foodborne Disease Outbreaks in New York State for Which Significant Ingredients Were Reported: 1980–1993

Significant ingredient	Frequency (%)
No specific ingredient	33.3
Shellfish	17.3
Finfish	7.8
Beef	7.1
Poultry	6.3
Eggs	6.0
Infected worker	4.8
Starchy food	4.2
Fruits	2.5
Beverage	2.3
Pork	2.1
Dairy	1.9
Other seafood	1.5
Mushrooms	1.3
Other	1.6

significant ingredients reported in foodborne disease outbreaks in New York State between 1980 and 1993. No specific ingredient could be identified in 33.3% of the outbreaks. Molluscan shellfish were reported in 17.3% of the outbreaks. Molluscan shellfish become fouled in the waters where they grow as the result of human sewage contamination transported by rivers and streams that discharge into the growing area, or by human sewage discharged by boats into the growing waters. Finfish were reported in 7.8% of the outbreaks; the outbreaks usually involved fish in the scombroid family or other ocean fish that are improperly refrigerated after being caught, a practice that allows certain bacteria to produce histamine in their flesh. Beef was a significant ingredient 7.1% of the time. *C. perfringens* and salmonellae were the most frequently reported agents, and inadequate cooking, improper hot holding, and improper cooling were the most frequently reported contributing factors in beef outbreaks. Poultry was reported 6.3% of the time. *Salmonella* and *C. perfringens* were the predominant agents, and inadequate cooking, inadequate refrigeration, and improper cooling were the most frequently reported contributing factors in those outbreaks. Eggs were the significant ingredients in 6.0% of the outbreaks. *Salmonella* was the agent reported in almost all of the egg outbreaks. Inadequate refrigeration, contaminated ingredients, and inadequate cooking were the contributing factors most often reported.

Table 3 Frequency (%) of Foodborne Disease Outbreaks in New York State for Which Contributing Factors Were Reported: 1980–1993

Contributing factor	Frequency (%)[a]
Inadequate refrigeration	26.7
Contaminated ingredient	23.2
Infected person	17.6
Unapproved source	16.8
Consumption of raw or lightly cooked food of animal origin	16.5
Inadequate cooking	16.3
Inadequate hot holding	15.9
Improper cooling	11.9
Natural toxicant	10.9
Cross-contamination	9.1
Preparation several hours before service	8.8
Inadequate reheating	8.4
Bare hand contact with ready-to-eat food	6.3
Unclean equipment	6.2
Added chemical	4.0
Other	2.3
Toxic container	0.5
Anaerobic package	0.2

[a]Numbers total more than 100% because some outbreaks had more than one reported source.

C. Contributing Factors

Table 3 lists the contributing factors reported in the NYSDOH study. "Inadequate refrigeration," which refers to either malfunctioning refrigeration equipment or food held out of refrigeration, was reported in 26.7% of the outbreaks. Inadequate refrigeration permits bacterial pathogens to grow and multiply to numbers sufficient to cause an outbreak. "Contaminated ingredient," indicating an ingredient that arrived at the point of preparation already contaminated with the agent, was reported in 23.2% of the outbreaks; eggs and molluscan shellfish were most often involved. An infected person was cited in 17.6% of the outbreaks; that is, a food worker who was carrying the agent either on or in his or her body had contaminated food with the agent. Most often the agent is a fecal–oral one and the worker is symptomatic when the food is being prepared. "Unapproved source" refers to food that is contaminated with the agent at its source (e.g., growing waters or farm). This factor was reported 16.8% of the time. The agents usually involved were viruses or *Salmonella enteritidis,* and the significant ingredients were shellfish or eggs.

Consumption of raw or lightly cooked food of animal origin was reported in 16.5% of the outbreaks. This factor most often involved consumption of raw or lightly cooked shellfish, which is a consumer preference, or the consumption of lightly cooked eggs, although in this case consumers may or may not realize that the problem exists. Consumers tend to be unaware that many dishes, including sauces and deserts, are made with uncooked eggs. Inadequate cooking was reported in 16.3% of the outbreaks. Thorough cooking will destroy vegetative bacteria and viruses, but when the intended cooking process is incomplete, pathogens may survive. Inadequate hot holding, reported in 15.9% of the outbreaks, involves food that is cooked and then held at too cool a temperature until served. Inadequate cooling was reported 11.9% of the time: that is, cooked foods were cooled too slowly, allowing spores or bacteria that survived the cooking step, or were introduced after cooking, to grow and multiply.

By becoming familiar with the agents, significant ingredients, and contributing factors, one can begin to anticipate problems with different foods and develop procedures to prevent those problems.

V. MICROBIOLOGY

Epidemiologic information regarding foodborne disease indicates that most foodborne illness is associated with microbial agents, therefore a knowledge of food microbiology is needed. Many factors influence the growth, survival, and destruction of microorganisms in foods, as well as the growth of bacteria in any environment. Bacterial pathogens are responsible for most of the best-known foodborne diseases, and insight into the growth and control of bacterial growth in foods can be obtained from an understanding of the factors enumerated next.

1. *Initial numbers.* When bacteria are introduced into a food, they must adjust to their new surroundings. The more rapidly they can adjust, the more rapidly they will grow to significant numbers. Good sanitation practices are aimed at reducing initial numbers.

2. *Competition.* Many foods, especially raw foods of animal origin, contain many different bacterial, mold, and yeast species. All these organisms are competing with each other for available nutrients. The organisms that are most effective in competing at a given moisture level and temperature will be able to grow to the greatest numbers. If a pathogen is a successful competitor, it can multiply to an infectious dose. Pathogens that do not compete well (e.g., *E. coli* and salmonellae) will not multiply to sufficient numbers to cause an illness unless there are few other microorganisms present.

3. *Moisture.* Organisms require moisture for growth, to bring nutrients to the cell wall, to remove wastes, and to maintain an osmotic balance (the relationship of the concentration of water and solutes inside the cell to the concentration of water and solutes outside the cell). Microbiologists use the term water activity

(a_w) to represent the amount of free available water in a food. For pure water a_w = 1.0. The water activity of foods is expressed as a decimal fraction of the pure water value. For example, fresh red meat has a_w of 0.95–0.99. Processed cheese and high salt bacon have a_w values in the range of 0.90–0.95. Sweetened condensed milk, Hungarian salami, jams, and margarine have a_w values in the 0.80–0.90 range. At the other end of the a_w spectrum, dried milk, dried vegetables, and dried walnuts have values around 0.20. A common method of reducing water activity is through the addition of salt or sugar. Common pathogens require relatively high water activity levels to grow. The U.S. Food and Drug Administration (FDA) definition of potentially hazardous food excludes foods with an a_w of 0.85 or less, because water activities below 0.90 are inhibitory to the foodborne bacterial pathogens or their toxin production.

4. *Nutrients.* Some bacteria require the presence of certain amino acids, carbohydrates, or certain other vitamins or minerals to multiply. Other organisms can survive and grow in a much wider range of environments. *Salmonella,* for instance, is usually associated with outbreaks in foods of animal origin, but it has caused outbreaks in everything from chocolate to tomatoes to watermelon.

5. *Oxidation–reduction (OR) potential and oxygen tension.* Organisms that grow exclusively in the presence of oxygen are referred to as aerobic; those that can grow only in the absence of oxygen are said to be anaerobic. The most infamous anaerobic foodborne bacterium is *Clostridium botulinum,* which causes the disease botulism. Most other foodborne bacterial pathogens are facultative anerobes, meaning that they can grow in the presence or absence of oxygen.

6. *pH.* The acidity or alkalinity of food is measured in terms of pH values. A neutral pH value is stated as a 7.0. Acidic compounds have pH values between 1 and 7, while alkaline compounds have pH values between 7 and 14. A food with a pH of 6.0 is 10 times more acidic than one with a pH of 7.0 A pH of 1.0 is 1,000,000 times more acidic than a pH of 7.0 Similarly, a pH of 9.0 is 100 times more alkaline than one of 7.0.

The pH value is a very important factor in food microbiology; bacteria can grow only within a certain pH range. The FDA definition of potentially hazardous food excludes foods with a pH of 4.6 or less, since bacterial pathogens are not able to grow well, if at all, at those levels. Naturally acidic foods, such as most fruits, or fermented foods such as vinegar, sauerkraut and acidophilus milk, have pH values below 4.5. Foods of animal origin and many vegetables generally have pH values between 4.5 and 7.0, and as such lack natural pH protection from pathogens. Foods can be protected from bacterial pathogen growth by lowering the pH (e.g, by adding vinegar, pH 2.9).

7. *Inhibitory substances.* Certain naturally occurring and man-made substances in food can retard or prevent the growth of bacteria. Some microorganisms produce compounds that can inhibit the growth of competing microorganisms. The drug penicillin, produced by a mold, is such a compound.

8. *Temperature.* Most foodborne pathogens grow best at about human body temperature, 98.6°F (37°C). The bacterium *C. perfringens* is unusual in that it grows best at relatively high temperatures of around 115°F (46°C). Refrigeration is an effective disease-preventing step because temperatures of 41°F (5°C) or less inhibit the growth of most of the foodborne disease pathogens. Potentially hazardous foods that are held hot are required to be held at temperatures of 140°F (60°C) or greater, as these temperatures destroy vegetative cells of bacterial pathogens.

9. *Time.* Bacteria require time to adjust to a new environment and additional time to grow to sufficient numbers to cause an illness. A lapse of a day or more between preparation and service is a commonly reported contributing factor to foodborne disease outbreaks. Food managers should prepare food as close to the time of service as possible, and they should reduce or eliminate leftovers that would provide bacteria with sufficient time to grow.

Whenever a bacterium is placed in a new environment it must adjust to that environment-too hostile an environment will destroy the cell, but a less hostile one will allow for survival. A still more favorable environment will allow for very slow growth after an extended period of adjustment (lag phase of growth). Optimal conditions consist of a short lag phase and a maximum period of accelerated growth (log phase of growth). A bacterium must deal with each of the growth factors discussed above in every new environment. The relationship between a bacterium and its environment is dynamic, and so as the organism adjusts to environmental changes, the lag phase and growth phase also change. A bacterium expands energy as it develops the proper metabolic processes and physical relationships necessary for growth. Each of these adjustments takes time and reduces energy reserves. While no single environmental factor may be at a level to prevent growth, it is very possible for an organism to encounter two or more environmental factors such as a_w and pH that, in combination, will inhibit growth.

Bacteria exist in two life forms. Vegetative cells are the growth form of bacteria, they produce toxins and/or invade the host, depending on the bacterium involved. Since vegetative cells grow and multiply, this is the form most influenced by environmental factors.

Spores are the other bacterial life form. They are dormant in a shell that is highly resistant to adverse environmental factors, reverting to the vegetative stage when a favorable environment exists. Only some bacterial pathogens form spores, which function to protect the microorganism. Important among the spore-forming bacteria are *C. perfringens, Bacillus cereus,* and *C. botulinum.*

Foodborne viruses are a less well appreciated group of pathogens. Hepatitis A is the best known of the foodborne viruses. Several different viruses are known to cause gastroenteritis, and Norwalk agent is the best known in this group. Foodborne viruses are obligate intracellular parasites, with typical sizes in the 27–70 nm range. They consist of a nucleic acid (RNA or DNA) and a protein coat. Foodborne viruses are host specific, and they harm but usually do not destroy host cells. Viruses are generally not destroyed by temperatures that would destroy vegetative bacteria, and they are often more resistant to chemical sanitizers than are bacteria.

A gram of feces from an infected person may contain 10^8–10^{10} viral particles, while as few as 1–10^2 particles can be an infective dose. Since viruses only grow in host cells, laboratory identification and study of them has been very difficult.

Parasites are less commonly reported agents of foodborne illness, however, outbreaks have been documented with *Giardia lamblia, Cryptosporidium parvum,* and *Cyclospora cayetanensis.* These agents are most commonly reported in contaminated water supply outbreaks, child day care centers, and immunocompromised individuals (e.g., AIDS patients). When foodborne illness has been documented, suggested sources of the pathogen have included cross-contamination from water, an infected worker, and manure contamination of field crops. Prevention of contamination and thorough cooking are the most important control mechanisms involved.

VI. LABORATORIES

Laboratory support is essential to many parts of a food protection program, providing identification and confirmation of agents in foodborne disease outbreak investigations. New agents are identified, methods for their identification are improved, and the biology of agents is determined. Baseline microbial populations are determined as part of the process of developing microbial guidelines or standards and for performance criteria. Laboratories also are used in developing HACCP systems and in verifying that those systems are being implemented.

VII. HAZARD ANALYSIS–CRITICAL CONTROL POINT

The use of the HACCP system in food services and retail settings has been described in many publications. A few examples are provided (4,5,8–10). More general guidelines for safe preparation of foods can be found in references such as the one by Farber and Hughes (7).

While so-called HACCP principles have been incorporated into some regulatory agency programs and chain retail food operations, much of the industry still has not implemented HACCP systems. When people are first exposed to HACCP, they tend to view it as far too complex, especially if that exposure deals with the traditional seven-step food processing based approach. HACCP does not need to be overwhelming. A HACCP system can be seen as analogous to the chain of infection spoken of in public health programs for the prevention of disease. The chain of infection refers to the many steps or links that must come together in sequence for an infectious agent to cause disease. Those links involve the agent, the environment, and the host. As with HACCP, public health programs are designed to prevent infectious diseases by breaking multiple links of infection, or to put it another way, by creating multiple barriers to infection by implementing steps (cleaning, cooking, refrigeration, etc.) that act on the agent, the host, or the environment. The monitoring of critical limits in HACCP is analogous to implementing multiple barriers in the chain of infection. The concept that measuring a temperature can be one barrier to a foodborne infection may be easier to understand than terminology like "monitoring a critical limit at a critical control point."

For a HACCP system to be effective, it needs to be built into the routine of the food preparation process, and it must be simple rather than being seen as imposing a seemingly meaningless extra task. If cooking is the critical control point and the critical limit is to reach a measured internal temperature of 165°F (74°C), then the kitchen staff needs to know that the measuring of an end-point cooking temperature is a routine part of cooking that is needed to assure the quality of the product served. Temperature measuring should be written into the recipe. This approach of making monitoring a part of "normal operations" can be accomplished at any critical control point (e.g., cooling, reheating, preventing bare hand contact with ready-to-eat foods). The staff also needs to know what to do if the critical limit has not been met. Staff responsibilities are assigned not in terms of elaborate record keeping, but in terms of simple, straightforward actions such as "continue cooking until required temperature is reached," or "report situation to supervisor immediately."

One of the most controversial aspects of HACCP implementation in the foodservice and retail food industries has to do with record keeping. Keeping records of the findings of HACCP monitoring is often viewed as needless paperwork with no benefit. If establishments are keeping records only to satisfy a regulatory requirement, then that assertion is correct. The employees writing any records, and the supervisor who checks that work, need to know what those records are showing. This knowledge is best conveyed if product quality and customer satisfaction can be readily linked to a given monitoring procedure. If cooking temperature is involved, then choose a temperature that satisfies product quality (e.g., level of doneness) as well as the regulatory parameter. This may actually mean exceeding the regulatory requirement in some cases—for example, cooking poultry to 180°F (82°C), since that level of doneness is preferred by most customers even though regulations require only 165°F (74°C).

If cooling is involved, choose a procedure that improves quality while satisfying requirements. Then the staff will be checking for a uniform quality product that will assure customer satisfaction. For these reasons, kitchen staff will be much more motivated to do the monitoring and to remedy any problem that they find.

HACCP systems at foodservice and retail food establishments need to address the safety of the water supply, as well as that of the food itself. Water is used as an ingredient; it is used to wash food, equipment, and utensils; and it is usually served as a beverage. If the establishment receives water from a municipal water supply, the operator needs to be sure that that supply complies with all applicable drinking water standards. If the water supply is on-site, then the operator needs to assure that the source is properly protected, that the water is properly disinfected (if disinfection is required), and that monitoring of source protection and disinfection is being conducted in accordance with all applicable water quality standards.

VIII. USE OF FOODBORNE DISEASE SURVEILLANCE FINDINGS FOR HACCP

Foodborne disease surveillance summary information can be used to identify new agents, vehicles, and contributing factors, as well as trends and predominance for known agents, vehicles, and contributing factors. For example, the New York State Department of Health has a foodborne disease surveillance program that compiles data into a series of tables, which then can be used in the state's HACCP program (12). This system includes grouping each foodborne disease vehicle into one of 16 methods of preparation categories, as well as one of 18 significant ingredient categories. Methods of preparation categories group vehicles by similarities in the way the foods are prepared. For example, the "Foods Eaten Raw or Lightly Cooked" category is defined as follows:

> These are served uncooked or after a heating that would not destroy vegetative pathogens. Preparation steps involve cold storage, cleaning, opening, steaming, or other light cooking and service. This category does not include commercially canned foods, e.g., hard-shell clams, oysters, mussels (consumed whole, raw, or steamed), steak tartar, Caesar salad with raw egg, lightly cooked eggs and hollandaise sauce (12).

Significant ingredient categories identify the ingredient that was determined to have introduced the agent and/or characterizes the vehicle. An example is the shellfish category, which consists of "raw clams, steamed clams, fried clams, clam strips, raw oysters, steamed mussels, chowder, scallops, and winkles" (12). In addition to the two-category system of grouping vehicles, the New York system a lists 19 contributing factors, which provide a high degree of specificity on the errors that led to an outbreak.

The vehicle, contributing factor, and agent categories are grouped into a series of tables that provide insight into the relationship between pairs of these epidemiologic features. An overall summary table groups method of preparation, significant ingredient, agent, and contributing factors in a way that allows the user to determine the frequency with which contributing factors and agents occur when various significant ingredients are prepared in the methods of preparation categories. For example, when eggs are served as cook/serve, (e.g., fried, scrambled, poached) one would want to know what agents have been identified (e.g., *Salmonella*) and what contributing factors were revealed (e.g., inadequate refrigeration, inadequate cooking, contaminated ingredients) by the department's investigations.

Findings from New York State's surveillance system show *Salmonella* to be the predominant bacterial agent, with *C. perfringens, Staphylococcus aureus,* and *B. cereus* also occurring, but much less frequently. Scombrotoxin is the major chemical agent, with mushroom poisoning being reported in a second, but far less frequent position. Viral agents are reported very frequently, but because of the unavailability of confirmatory laboratory tests, the most frequent citation is "nonspecific viral gastroenteritis."

Shellfish have been the most frequently reported specific ingredients, reflecting mainly the large number of outbreaks that occurred in the early to mid-1980s. Other frequently identified vehicles include finfish, beef, poultry, eggs, and starchy foods. The predominant preparation conditions involve foods eaten raw or lightly cooked, solid masses of potentially hazardous foods, multiple foods, cook/serve foods and natural toxicants (e.g., scombrotoxin or mushroom toxins). The following are the most frequently reported sites of food contamination: foodservice establishment, stream or bay, home, farm, and processing plants. The major contributing factors have been contaminated ingredients, consumption of a raw or lightly heated food of animal origin, food from an unapproved source, inadequate refrigeration, infected person, inadequate cooking, inadequate hot holding, and improper cooling.

These data can be referenced when one is developing a hazard analysis–critical control point system for a food. Foodborne disease surveillance data can be used in setting priorities for foods that should be set up on an HACCP system. The methods of preparation and specific ingredients that have been involved in the largest number of outbreaks and are included on the menu of the establishment involved should be selected first. At the hazard analysis step, the agent(s) of concern, and the frequency at which they were reported, for the given method of preparation and specific ingredient, can be determined from the table that groups the methods of preparation, specific ingredients, agents, and contributing factors. The critical control points can be determined by reviewing the list of contributing factors involved in outbreaks with the specific ingredient and method of preparation. Steps in the preparation of the food that relate to the contributing factors are likely critical control points for that food.

The same foodborne disease surveillance data can be used by government agencies in their routine regulatory programs. Establishments that serve menu items that are most frequently identified as vehicles in foodborne disease outbreaks should be scheduled for greater regulatory oversight and should receive the most attention during regulatory visits. Similarly, industry quality assurance programs can assess menus for priority in HACCP development.

In all these applications, steps in food preparation that relate contributing factors should be carefully evaluated to determine whether they can be eliminated, thereby simplifying the process and making it safer. Monitoring incidence trends in these agents, vehicles, and factors over time allows newly emerging problems to be identified and the effect of interventions to be measured.

IX. PRACTICAL ADVICE TO FOOD PREPARERS IN ALL SETTINGS

A. Major Concepts

The first thing food preparers should have in mind is the potential impact of their own state of health on customers. As cited above, agents carried by employees who prepared the food caused approximately 18% of the outbreaks reported in New York State during the period from 1980 to 1993. In a few cases those agents were carried on the skin or in the throat of the employee's (e.g., *S. aureus,* group A *Streptococcus*). In most cases those agents (e.g., Norwalk virus, rotavirus, hepatitis A, salmonellae, *Shigella*) were in the gastrointestinal tract and stool of the employees. The workers involved usually were symptomatic, suffering from diarrhea, abdominal cramps, vomiting, nausea, fever, etc. when they prepared the food. They probably did not wash their hands well, or they likely would not have spread the infection. Most food safety regulations prohibit food workers who are carrying such an infectious agent or have the symptoms from preparing food, and they also require thorough hand washing after using the toilet. Food workers must honor these requirements by not working when they are ill and by always thoroughly washing their hands after using the toilet. In addition, some regulations prohibit bare-hand contact with ready-to-eat foods. Such requirements are a further barrier to the spread of fecal–oral agents and should be followed regardless of whether such a regulation exists.

The other major concept in food safety has to do with the importance of food temperatures and limiting the amount of time foods are held at temperatures that permit bacterial growth. The temperature of refrigeration is important to prevent growth. The temperature of hot holding both inhibits growth and destroys microorganisms. The cooking and hot holding temperature requirements that exist are designed to destroy the pathogens of concern in particular foods. Cooling of foods is a function of both time and temperature. Large masses of food tend to cool slowly, therefore foods should be cooled in small quantities, or volumes, by a method that will cause the food temperature to be between 120 and 41°F (49 and 5°C) for as little time as possible.

B. Steps in Implementing Food Safety Concepts

Food safety professionals relate food safety measures to the flow of food or the steps food passes through from purchasing to service. When purchasing food, always bear in mind that not all suppliers and manufacturers are equal when it comes to food quality and safety, and the lowest price is not the only factor to consider when making a selection. When possible, purchase foods from manufacturers that have implemented an HACCP system. Ask to review inspection reports by food regulatory agencies. Visit the manufacturer's operation, or contract with a third-party company to do such inspections. These steps may sound new or impractical for a foodservice operator, but such attention to the quality for incoming ingredients/products has become standard operation procedure for large foodservice chains, for food manufacturers, and in international trade. Experienced company agents stop using suppliers who fail to measure up, and their current suppliers are more careful with the quality of the products they supply. As a result, operators who pay close attention to the quality and safety of ingredients they purchase are able to serve a higher quality and more consistent quality product.

Receiving is the next step in the food flow. When food is received, it should be examined to assure that it came from the supplier specified and that it is in sound condition and of good quality. A refrigerated product, should be at 41°F (5°C), or less. Frozen products should be frozen solid, and there should be no signs of thawing and refreezing. Canned food containers should not be bulging, and there should be no rust or pits in the metal. Milk and dairy products should be pasteurized. Meat and poultry should be from inspected plants. Molluscan shellfish should be from legally open waters and should bear an identification tag where required. Dry goods (grain, cereals, breads, etc.) should be in sound packages without tears, holes, or evidence of water exposure, all of which could have led to contamination. Such products should be free of evidence of insect or rodent damage. Any products that have "sell by" or "best by" dating on the package should be within the recommended period for use. Fruits and vegetables should be free from excessive rotting, spoilage, and insect damage, or excessive bruising. Foods that do not meet the firm's standards should be rejected.

Once foods have been received, they need to be stored in a manner that will maintain their quality until used. Store dry foods in a cool location, protected from flooding and overhead drippings. The storage area should be kept clean, and entry of insects and rodents should be prevented. Perishable foods should be stored at 41°F (5°C), except for smoked fish products, which should be stored at 38°F (3.3°C) or less. Raw meats, poultry, and fish should be stored in leak-proof containers and away from or below raw produce and ready-to-eat foods, to prevent accidental spillage or dripping onto these foods. Ready-to-eat foods and fruits and vegetables should be wrapped or covered to prevent contamination. (except for previously cooked foods, while they are cooling to refrigeration temperatures.) Freezers should operate at 0°F (−17.8°C) or less, and frozen foods should be tightly wrapped to prevent freezer burn. All products should be marked with the date they are placed in storage to facilitate product use on a first-in, first-out basis.

The preparation step is critical for foods that will not be cooked prior to service (e.g., fresh fruits and vegetables, molluscan shellfish, previously cooked or processed foods). Food workers, work surfaces, and equipment can introduce contamination at this step that will not be destroyed because no cooking step will follow. Food workers must not have gastrointestinal symptoms (e.g., vomiting, diarrhea), and they must wash their hands thoroughly before they prepare the food and after they contaminate their hands (by going to the toilet, touching raw animal foods, etc.). Food workers should prevent bare-hand contact with food by using utensils, deli paper, or disposable gloves. Fresh fruits and vegetables should be washed with cool water and leaf lettuce leaves should be individually cleaned, since dirt is trapped in the multiple layers of leaves involved. Work surfaces, equipment, and utensils should be washed, rinsed, and sanitized before and after use.

Foods that will be cooked prior to service receive a significant protective effect from cooking. Thorough cooking destroys most pathogens of concern for foodborne illness. Therefore, cooking is one of the most important critical control points. Where regulations specify cooking temperatures for various foods, make sure that those temperatures are achieved throughout the food for any required time interval. Use a metal stem bayonet thermometer to measure the final cooking temperature for most foods. A needle-type thermocouple thermometer will be required to measure the center temperature of thin foods, such as hamburgers.

Food that has been prepared, or cooked may be held for a period of time before service. Hot foods being held for service should be held at or above 140°F (60°C) and cold foods being held for service should be held at 41°F (5°C) or less. Use a thermometer to periodically check the temperature of foods being held for service to assure that the recommended temperatures are being maintained.

Cooked foods that are being cooled to refrigeration temperatures should be cooled rapidly to prevent the growth of spores that might have survived the cooking step. Such foods should be cooled from 140°F (60°C) to 41°F (5°C) within 6 hours to prevent the growth of these spores. This can be accomplished by dividing large masses of food into smaller ones, by cutting roasts of meat or poultry into pieces weighing 6 lb or less, and by portioning semisolid and liquid food into smaller shallow containers. Other effective means of rapid cooling include cooling foods by immersion in an ice water bath and placing them in special rapid cooling refrigeration units.

Previously cooked and cooled foods that are being reheated should be heated to a temperature of 165°F (74°C) within 2 hours to assure the destruction of microbial pathogens that have survived earlier steps or have been newly introduced. The reheat temperature should be measured with a thermometer to assure that 165°F (74°C) has been achieved in all portions.

X. PRACTICAL ADVICE TO CONSUMERS ON PREVENTING FOODBORNE ILLNESS

Consumers often ask what practices should they follow to prevent foodborne illness in the home. Information prepared for foodservice establishments is perceived as being not relevant, or too complicated, for consumers to understand and implement. The following key concepts provide practical actions for consumers. By faithfully following these recommendations, consumers can significantly reduce the chance of having a foodborne illness in their homes.

A. High Risk Consumers and High Risk Foods

Certain foods pose a greater risk for foodborne illness for the general public and more particularly for high risk individuals. High risk individuals include young children, the elderly, immuno compromised persons (e.g., patients undergoing cancer therapy, persons with HIV/AIDS, persons with liver disease). High risk consumers should assure that all animal foods they eat have been thoroughly cooked. High risk foods include raw or lightly cooked animal foods (e.g., eggs, molluscan shellfish, milk, and dairy products, fish, meat, and poultry). These foods might be either eaten raw or lightly cooked by themselves, or as a "hidden" raw ingredient in a recipe. Consumers should assure that shell eggs and recipes that call for raw eggs are thoroughly cooked. Shellfish should be thoroughly cooked. Milk and dairy products should be pasteurized, and meats and poultry thoroughly cooked.

B. Four Important Steps

1. Clean

Persons preparing food in the home should avoid preparing food if they are suffering from vomiting or diarrhea. All food preparers should thoroughly wash their hands before they begin food preparation and after they possibly contaminate their hands (e.g., by using the toilet, touching raw animal food, caring for an ill person in the home, changing a diaper). All work surfaces, equipment, and utensils should be thoroughly washed prior to beginning food preparation. Surfaces, equipment, and utensils that have been in contact with raw animal foods should be thoroughly washed before they are used in the preparation of ready-to-eat foods. Raw meat, poultry, fish, and molluscan shellfish are often contaminated with pathogens. Fresh fruits and vegetables should be washed with clean, cool water, and produce such as leaf lettuce should have the leaves separated and individually flushed with clean water. These cleaning practices can significantly reduce the chances that food will become contaminated during preparation, as well as reducing levels of contamination found on raw food products.

2. Separate

Raw animal foods should be stored in leak-proof containers on shelves below ready-to-eat foods to prevent cross-contamination of pathogens on the raw animal foods from accidentally dripping, splashing, or otherwise spreading on to the ready-to-eat food. Separate work surfaces should be used to prepare raw animal food and ready-to-eat food. Where that is not possible, thoroughly wash the surfaces after use for raw animal foods and before use for ready-to-eat foods.

3. Chill

Refrigeration can slow the growth of any pathogens that might be in foods. Purchase foods that require refrigeration last at the market. Arrive home after shopping as soon as possible and return these foods to refrigeration right away. Minimize the amount of time refrigerated foods are held at room temperature during preparation and during and after meal periods. Large masses of cooked foods, such as roasts of meat, casseroles, stews, soups, and gravies that are to be refrigerated should be cooled as rapidly as practical. This may require cutting solid foods into smaller pieces or dividing semisolid and liquid foods into smaller containers before placing them in the refrigerator. Do not hold these foods at room temperature any longer than is absolutely necessary. Placing a container of hot food into an ice water bath and stirring the cooling food frequently can increase the cooling rate. Once the food temperature has dropped close to refrigeration temperatures, cover the container and place it in the refrigerator.

4. Cook

Thorough cooking of foods is one of the most important and practical ways to assure food safety. Raw animal foods alone, and foods that include raw animal food as ingredients, should be thoroughly cooked. Many recipes call for the use of partially cooked ingredients, (e.g., hollandaise sauce and Caesar salad containing raw egg). Modify these recipes to use an egg substitute or pasteurized egg. Use a metal stem bayonet thermometer, or a smaller metal probe called a thermocouple thermometer, to measure the center temperature of cooked foods prior to ending the cooking process. All cooked foods should be cooked to at least 140°F (60°C). Egg-containing foods should be cooked to 145°F (62.7°C), beef to 145°F (62.7°C) [except for ground beef (hamburger), which should be cooked to 160°F (71°C)], pork to 150°F (65.5°C), poultry to 165°F (74°C), as well as any foods previously cooked at home to 165°F (74°C). Commercially cooked foods that are being reheated may be reheated to a lower temperature if the package directions so indicate.

XI. CONCLUSIONS

Food safety programs for restaurants and food stores need to be based on scientific principles. Setting of priorities, as well as the steps for achieving those priorities, should begin with a review of epidemiologic findings regarding agents, vehicles, and contributing factors to determine what are the problems and what is leading to those problems. This information, along with knowledge of food microbiology, can then be used to develop and implement HACCP systems, working from the highest risk situations to those of minimal public health risk.

REFERENCES

1. Bean, N. H., J. S. Goulding, M. T. Daniels, and F. S. Angulo. 1997. Surveillance for foodborne-disease outbreaks—United States, 1988–1992. J. Food Prot. 60:1265–1286.
2. Bean, N. H., J. S. Goulding, C. Lao, and F. J. Angulo. 1996. Surveillance for foodborne-disease outbreaks—United States, 1988–1992. CDC Surveillance Summaries, October 25, 1996. MMWR 1996:45 (No. SS-5).
3. Bean, N. H., and P. M. Griffin. 1990. Foodborne disease outbreaks in the United States, 1973–1987: Pathogens, vehicles, and trends. J. Food Prot. 53:804–817.
4. Bryan, F. L. 1989. HACCP Hazard Analysis Critical Control Point Manual. Food Marketing Institute, Washington, DC.
5. Committee on Communicable Diseases Affecting Man. 1991. Procedures to Implement the Hazard Analysis Critical Control Point System. International Association of Milk, Food, and Environmental Sanitarians, Ames, IA.
6. Council for Agricultural Science and Technology. 1994. Foodborne pathogens: Risks and consequences, Report 122. CAST, Ames, IA.
7. Farber, J. M., and A. Hughes. 1995. General guidelines for the safe handling of foods. Dairy Food Environ. Sanitat. 15:70–78.
8. National Advisory Committee on Microbiological Criteria for Foods. 1992. Hazard analysis and critical control point system. Int. J. Food Microbiol. 16:1–23.
9. National Restaurant Association. 1993. HACCP Reference Book. National Restaurant Association, the Educational Foundation, Chicago.
10. Schmidt, K. (ed.). 1995. WHO Surveillance Programme for Control of Foodborne Infections and Intoxications in Europe, Sixth Report, 1990–1992. Federal Institute for Health Protection of Consumers and Veterinary Medicine, Berlin.
11. Todd, E. C. 1996. Worldwide surveillance of foodborne disease. The need to improve. J. Food Prot. 59:82–92.

12. Weingold, S. E., J. J. Guzewich, and J. K. Fudala. 1994. Use of foodborne disease data for HACCP risk assessment. J. Food Prot. 57:820–830.
13. U. S. Department of Health and Human Services. 1997. HACCP Regulatory Applications in Retail Food Establishments, 2nd ed. U.S. Department of Health and Human Services, Public Health Service, Food and Drug Administration, Rockville, MD.

9

Food Safety in Institutions: Health Care Institutions, Schools, and Correctional Facilities

Marilyn B. Lee
Ryerson Polytechnic University, Toronto, Ontario, Canada

I. HISTORICAL AND CURRENT INCIDENTS OF FOODBORNE ILLNESS

A. Introduction

When foodborne illness occurs in health care institutions, hospitals, or long-term care facilities (LTCFs), large numbers of people are usually affected because food is prepared en masse. An equally important contributing factor is the increased susceptibility of the residents. In LTCFs, nursing homes, and homes for the aged, the elderly are often debilitated with chronic diseases, immunologically weak, and/or, occasionally, undernourished. In developed countries more than half of the documented deaths from gastroenteritis (30), some of them due to foodborne illness, occur in the elderly. In hospitals, patients may be in a weakened state after having undergone surgery or chemotherapy. However, foodborne illness is largely preventable if proper control strategies are followed by staff in the kitchen and by the health care providers. These strategies are increasingly important in reducing morbidity and mortality in our growing population of elderly (> 65 years), which is projected to increase from 32 million in 1990 to 76 million by 2020 in the United States (72).

Another major factor, particular to health care institutions but not always well documented, is person-to-person spread arising from initial cases of foodborne illness. Statistics on person-to-person spread are not directly included here because the discussion is limited to "foodborne" illness. Admittedly because of variation in reporting practices around the world, cases reported as foodborne may in fact have resulted from person-to-person spread.

B. Incidence

Many outbreaks in institutions are unrecognized (96), and of those that are detected only some are investigated and documented (76). Many incidents cannot be traced to their cause for reasons such as time delay in reporting to health authorities, poor recall of foods eaten, or failure to find a pathogen in food or fecal samples. Hence numbers reported represent only a small proportion of those actually becoming ill.

In most countries institutional foodborne outbreaks represent only a small fraction of the total reported outbreaks: in Ontario, outbreaks in health care facilities (hospitals and LTCFs), 5.9% (Fig. 1); in Canada, institutions (single and multiple cases included as incidents) 6.0% (Fig. 2); in the United States, health care facilities (hospitals and residential homes) 3.4% (80); in England and Wales, health care facilities, 12.1% (Fig. 3); in Australia, institutions, 10.5% (Fig. 4)), and in Argentina, outbreaks of salmonellosis in hospitals 3.4% (8). In Canada, LTCFs account for more outbreaks and cases than other institutions (Fig. 5).

Although the number of institutional outbreaks is small, the number of cases from each outbreak is high in comparison to other eating establishments where people acquire foodborne illness. In Ontario (1993–1996), for example, although there were only 12 outbreaks of documented foodborne illness in health care facilities, 352 people became ill for an average of 29 cases per outbreak (Fig. 1). For restaurants, fast food outlets, and private homes, the average number of cases per outbreak was only 5. However, when food is prepared in bulk, such as at catered events, the case-per-outbreak average, at 34, is even higher than that at health care facilities.

Similar averages have been observed elsewhere. In Canada (1990–1993), the average number of cases per incident for institutions was twenty seven, compared with only 8.4 for restaurants/hotels in the same time period (Fig. 2). In the United States (1975–1992), the average number of cases per outbreak for confirmed foodborne illness was 46 in health care facilities and 33 in all other locations (80). In Italy (1991–1994) the average number of cases per outbreak was 58 for institutions (nursing homes, schools, and hospitals), 4 for homes, and 15 for restaurants (79).

Foodborne illness appears to be increasing in health care institutions. For example, in England and Wales although notifications from hospitals rose slightly from 55 per 100 000 population in 1991–1992 to 63 per 100 000 in 1993–1994, outbreaks in nursing homes and geriatric hospitals almost doubled from 79 in 1992 to 152 in 1994 (21).

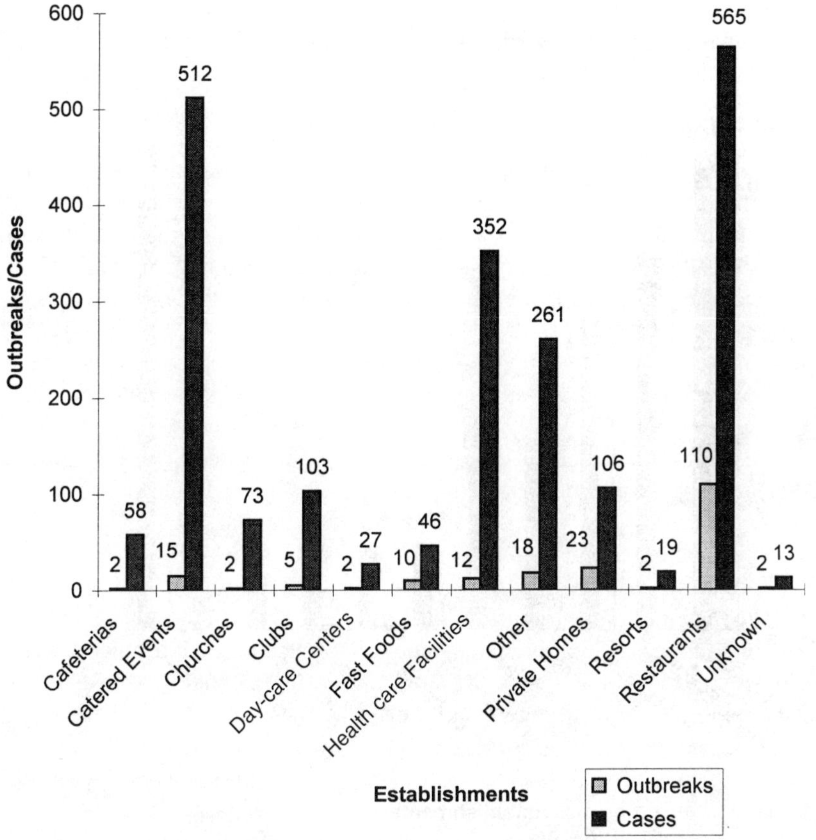

Figure 1 Confirmed foodborne outbreaks/cases from microbiological sources according to establishments. Ontario, 1993–1996. (Data from Ref. 62.)

C. Foodborne Pathogens

The bacterial foodborne pathogen most often discovered as the cause of illness has historically been *Salmonella,* and health care institutions are generally no exception. From a total of 27 known-cause foodborne outbreaks reported in LTCFs in the United States (1975–1987), 52% were caused by *Salmonella* spp. (48). *S. typhimurium* was the most common serotype identified. The second leading cause of bacterial foodborne illness was *Staphylococcus aureus* (13%), followed by *Clostridium perfringens* (11%). In U.S. health care facilities, which include hospitals and nursing homes (1975–1992), the leading cause of outbreaks was *Sal-*

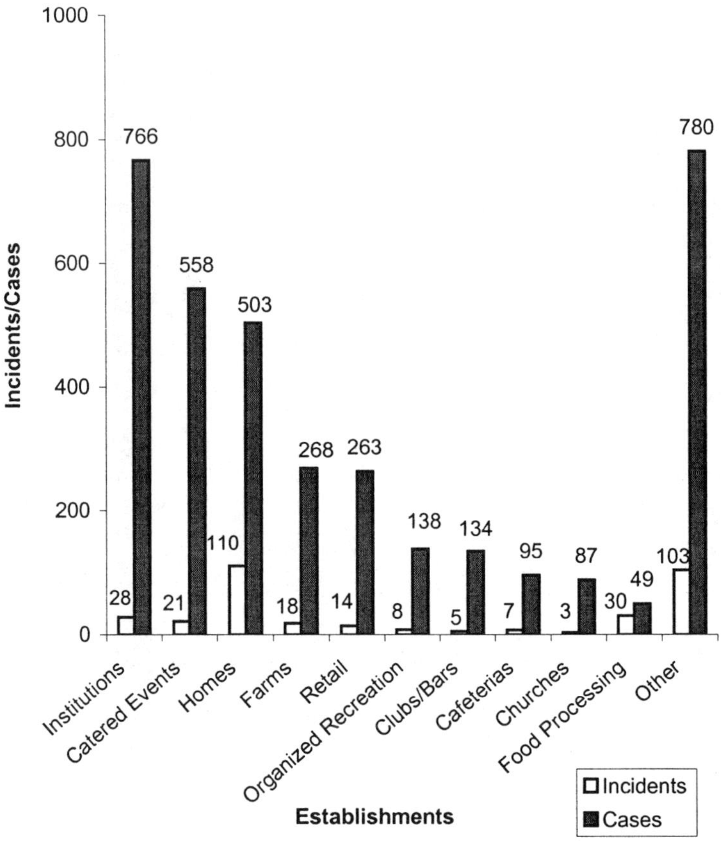

Figure 2 Confirmed foodborne incidents/cases from microbiological sources according to establishments, Canada, 1990–1993. (Data from Refs. 92 and 93.)

monella (60%), followed by *C. perfringens* (11%), and *S. aureus* (8%) (48). In 58 foodborne outbreaks in residential institutions, mostly nursing homes in England and Wales (1992–1994), the most common pathogens isolated were *Salmonella* (43%), *C. perfringens* (40%), and small round structured viruses (SRSV, 10%) (76). However, of 48 outbreaks in hospitals in Scotland (1978–1987), 52% arose from *C. perfringens* and 35% from *Salmonella* contamination (17). In 28 institutional foodborne outbreaks in Canada (1990–1993), the most common pathogens were *Salmonella* (43%), and *Staphylococcus aureus* (18%) (92, 93).

D. Case Fatality Rates

The case fatality rate (CFR) for residents in nursing homes who suffer from foodborne bacterial gastroenteritis is 10 times greater than for the general population

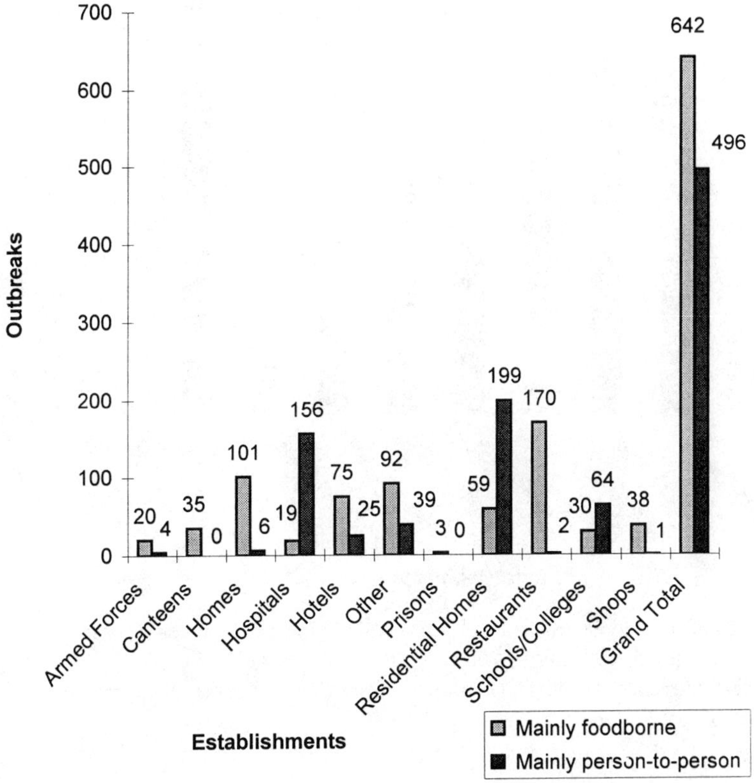

Figure 3 Settings of outbreaks by mode of transmission, England and Wales, 1992–1994. (Patrick Wall, Consultant Epidemiologist, Public Health Laboratory Service, U.K., 1997, personal communication.)

(30). Specifically, in U.S. nursing homes (1975–1992) the CFR was 2.2% when the agent was identified, compared with 0.17% in hospitals and 0.15% elsewhere (80). Although CFRs for bacterial foodborne gastroenteritis are generally only slighter higher in hospitals than in the general population, a large outbreak of *Salmonella enteritidis* PT8 in a New York City hospital had a case fatality rate of 2.2% (88). As a result, the New York State Department of Health issued recommendations to all health care facilities to eliminate raw or undercooked eggs from the diets of persons who are institutionalized, elderly, or immunocompromised. *S. enteritidis* may have higher potential for invasiveness than other *Salmonella* serotypes (87).

Salmonella has been and continues to be a common cause of fatalities. In LTCFs in the United States (1975–1987), 81% of the fatalities were due to

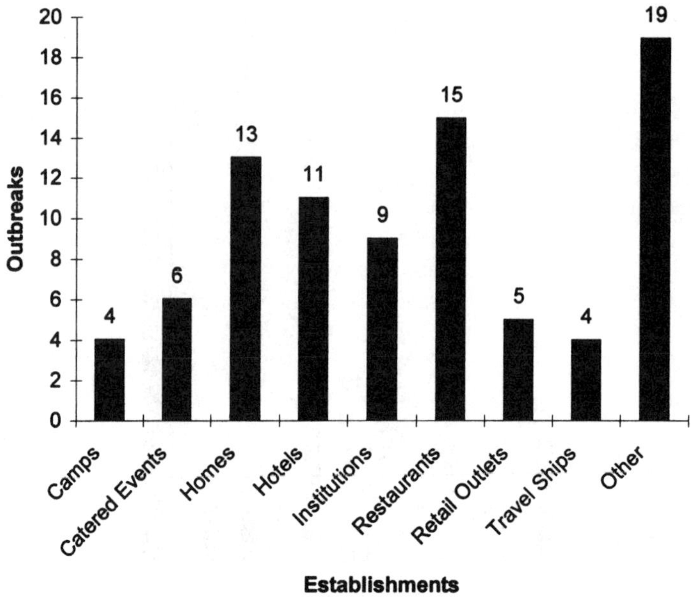

Figure 4 Confirmed foodborne outbreaks from microbiological sources according to establishments, Australia, 1980–1995. (From Scott Crerar, Epidemiology Registrar, Commonwealth Department of Health and Family Services, Canberra, Australia, 1997, personal communication.)

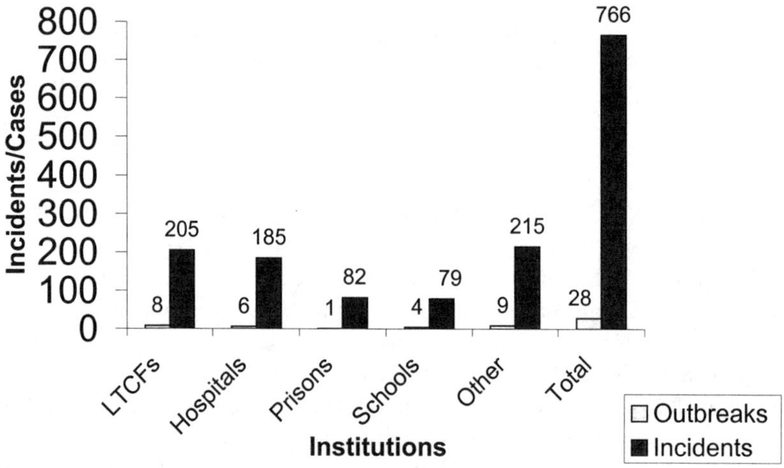

Figure 5 Confirmed institutional foodborne incidents/cases from microbiological sources, Canada, 1990–1993 (Data from Refs. 92 and 93.)

salmonellosis (48). The CFR for salmonellosis in residential facilities in England and Wales (1992–1994) from both food and person-to-person spread was 3% (76).

Since 1982 when *E. coli* O157:H7 was first identified, many LTCF outbreaks that resulted in high CFRs have been brought on by the complication of hemolytic uremic syndrome (HUS), characterized in part by acute renal failure (11). For example, in a London (Ontario, Canada) nursing home outbreak (1985), the case fatality rate was an astonishing 35% (Table 1), with 12 out of 19 deaths attributable to HUS. Undercooked hamburger seems to be the major culprit in many outbreaks (Table 1). Lower CFRs are reported for *C. perfringens* and *S. aureus*. Out of 899 victims of *C. perfringens* illness in hospital outbreaks in Scotland (1978–1987), the case fatality rate was 0.3% (71). For *S. aureus* outbreaks in U.S. nursing homes, the CFR is similarly low at 0.4% (48).

E. Food Vehicles

The most common food vehicles in U.S. nursing home outbreaks (1975–1987) were poultry (8.7%), eggs (6.1%), salads that contained no beef, eggs, or poultry (5.2%), and beef (3.5%) (48). However, the food vehicles in most of these outbreaks remain unknown (59%). More recently (1985–1991), *S. enteritidis* has been a major cause of illness in the United States. Of the 380 outbreaks due to this pathogen, 16% occurred in nursing homes and hospitals (80). Shell eggs were implicated in most, as illustrated by the following well-documented case. In a large nosocomial outbreak of *S. enteritidis* in a New York City hospital in 1987, 404 patients out of 965 (42%) were affected (88). The kitchen staff had prepared mayonnaise from shelled grade A eggs to be used in tuna–macaroni salad the following day. Five hours before service the mayonnaise was mixed with the tuna–macaroni ingredients and left at room temperature, allowing the salmonellae present to multiply. *S. enteritidis* PT8 was identified from leftover salad, victims' stool samples, the raw shell eggs, and an ovary from a hen from the farm at which the eggs had been collected.

II. COMMON TRENDS AND FACTORS LEADING TO FOODBORNE ILLNESS

A. Food Mishandling Problems

Factors that lead to outbreaks of foodborne illness are varied, but at the top of the list are improper temperature controls. Undercooking hamburgers in nursing homes appears to be a common cause of illness (34, 84). In a nursing home in Ottawa in 1982, in one of the first documented outbreaks of gastroenteritis due to *E.*

Table 1 Case Fatality Rates[a] (CFR) for Some *Escherichia coli* O157:H7 Outbreaks in Health Care Institutions

Place	Type of institution	Year	Food vehicle	Cases: Patients	Cases: Staff	Deaths	Patient CFR (%)	Ref.
Ottawa (Canada)	Home for the aged	1982	Hamburger	30	2	1	3	84
Nebraska (U.S.)	Nursing home	1984	Hamburger	34		4	12	77
London (Canada)	Nursing home	1985	Sandwiches	55	18	19	35	11,12
Alberta (Canada)	Nursing home complex	1987	Hamburger	15		2	13	34
Utah (U.S.)	Home for mentally handicapped	1988?	Ground beef	20	30	4	20	68
Kingston (Canada)	Nursing home	1989	Untreated lake water	6	2	1	17	67
Lanarkshire (Scotland)	Psychiatric hospital	1990	*E. coli* in food brought into hospital, spread person to person	9	2	4	44	40
Levis, Quebec (Canada)	Home for the aged	1991	Ground beef	13		3	23	92

[a]Outbreaks with unusually high CFRs.

coli O157:H7, 31 residents became ill. Although in many cases the infection was spread from person to person, insufficiently cooked hamburger was likely the originating source (84).

In November 1983, *Salmonella reading* was isolated from 20 foodservice staff and 3 hospital patients in Connecticut. A food-specific attack rate table identified turkey as the likely food vehicle. Frozen 18–20 lb turkey breasts were cooked in a slow roaster for 5 hours at 250°F (121°C), then for 10 hours at 160°F (71°C). No thermometer probe was used, and the turkey was probably eaten undercooked. The hospital changed its cooking routine thereafter by cooking in a standard oven to an internal temperature of 165°F (74°C) (78). No further outbreaks have been reported.

In an investigation of a hospital outbreak in Ontario, 1986, six turkeys were placed in an oven at 450°F (232°C) at 6:00 A.M. (69). Five were removed at 11:00 A.M. and served at noon. The sixth turkey was left in the oven 90 minutes longer because when probed at 11:00 it had been found to be undercooked. This sixth carcass was then stripped, whereupon the turkey meat was first refrigerated, then moved to a warming oven until service from 5:00–6:30 P.M.. Ten patients and six staff came down with gastroenteritis caused by *S. hadar,* phage type 2. The sixth turkey had likely been undercooked and not held at a high enough temperature during rewarming to destroy the salmonellae.

In an outbreak of *S. enteritidis* PT4 at a 300-bed hospital for mentally handicapped persons in Wales, 101 residents and 8 staff were affected when rissoles, made from minced beef, reconstituted dried vegetables, and shell eggs, were deep-fried to an internal temperature of only 119–140°F (48–60°C). Inadequate cooking was subsequently identified as the problem, along with preparation of food 3 days in advance of serving (26). The rissoles were apparently judged to be cooked when they turned golden brown and floated in the heated oil. Doneness was not confirmed by a thermometer.

Other outbreak investigations have revealed other causative factors: improper refrigeration (41, 71), improper reheating (14, 71), improper hot storage (5, 15), cross-contamination (83), and preparing food far in advance of service (17, 41, 71). From 1975 to 1992 in the United States, food handling errors were reported in 212 foodborne outbreaks occurring in hospitals or nursing homes. The most frequently observed problem was holding food at improper temperatures (hot or cold), followed by inadequate cooking, poor personal hygiene, use of food from unsafe sources, and use of contaminated equipment (80). Similarly, faults thought to contribute to foodborne outbreaks in residential institutions in England and Wales (1992–1994) were inadequate heating (50%), improper storage (hot and cold), (19%), cross-contamination from raw food (9%), and the infected food worker (9%) (76).

Often outbreaks occur because of a combination of food mishandling problems. In a Sydney, Australia, institution in 1995, 17 staff became ill from eating contaminated sandwiches. One of the food handlers in the cafeteria had continued to work with illness, characterized by abdominal cramps, vomiting, and diarrhea. Microbiological examination of the sandwich fillings showed evidence of widespread fecal contamination, and *Salmonella ohio* was detected in the corned beef. Storage temperatures of some of the fillings were as high as 68°F (20°C) (100)!

B. Equipment Maintenance

With technological advances, more sophisticated equipment is being used in institutional kitchens. It is not surprising that outbreaks occasionally occur because of inability or failure to dismantle and clean this equipment properly on a regular basis. Such was the case in a small, but new hospital in Ontario when a food mixer was found to harbor *S. enteritidis* (42, 54). Investigators thought raw egg was the source of the organism, which had entered a hollow shaft through a loose gasket and subsequently seeded tuna, turkey, and other sandwich fillings. The hollow shaft could be cleaned only with a specialized brush, and this cleaning was not being done by kitchen staff. Investigators further postulated that some other unreported outbreaks might have occurred because of this practice. Managers of foodservice operations must have in place a policy and procedures manual describing when and how equipment is to be dismantled, cleaned, and sanitized. New workers should be correctly taught, then monitored to ensure that they are carrying out the procedures correctly. For convenience, and as a reminder, some institutions post the procedures on the wall behind the equipment.

C. Recurrence of an Outbreak

As indicated in Sec. II. B, outbreaks may go undetected only to be discovered as a follow-up to another outbreak at the same facility. Such was the case in a *C. perfringens* outbreak in a correctional institution in Florida in 1984 (86). On March 19, the county health department was notified of an outbreak of gastroenteritis. Even though epidemiologic analysis implicated roast beef, and bacteriological examination of stool samples from ill inmates detected *C. perfringens,* no corrective action was taken in the kitchen. Eight days later, on March 27, a similar outbreak occurred. The second investigation revealed that the walk-in cooler could not maintain large amounts of food at temperatures below 45°F (7°C). At the time of the inspection, the internal temperature of the supposedly cooled meat was 85°F (29°C). In addition, meat was being reheated in overcrowded steam tables, with pans of food stacked three layers high. If the investigation of the first outbreak had been thorough and problems corrected, the second outbreak would not have occurred.

In another instance, 40 confirmed cases of foodborne illness and two deaths due to *Salmonella hadar* occurred between September 28 and November 2, 1990, at a psychiatric hospital in Kingston, Ontario (57). The hospital was closed to admissions while confirmed and symptomatic cases were moved to isolation wards to control this propagated outbreak, which was officially declared over on November 12. But 10 days later, on November 22, hospital staff again noted a number of patients with diarrhea. By the next morning 22 patients were exhibiting diarrhea, and eventually 88 cases of *S. berta* were confirmed. An in-depth investigation of the outbreak revealed that a cold tuna salad was the food vehicle. The hospital's food preparation procedures called for defrosting of frozen chicken breasts in a square roasting vessel in which the cold tuna salad was later prepared. There were no procedures in place for cleaning and sanitizing the roasting vessel between uses. Since both outbreaks took place at the same point in the menu cycle, the origin of the first outbreak, even though the infection was spread from person to person, was likely the same.

D. Temporary Kitchens

Another factor contributing to outbreaks arises when a kitchen is renovated and food must be temporarily handled in a different manner. A major outbreak occurred in London, Ontario, in 1985 at a nursing home where 71 people met the case definition for *E. coli* O157:H7 foodborne illness (11, 12). Nineteen residents died (Table 1) (see also Sec. II. E). For 2 weeks prior to the outbreak the kitchen had been undergoing renovations, and a temporary kitchen had been set up in the dining room. All meat for this 2-week period had been precooked and frozen except for some veal patties, which were cooked raw. Meals were prepared by microwaving the cooked, frozen meat. On September 5, the day the presumably contaminated food (sandwiches) was consumed, kitchen renovations were completed and there was interference in the regular food preparation procedures. The source of the *E. coli* O157:H7 was thought to be a box of veal patties packed in ice. The food had been placed on a wooden preparation table, and the melting ice in the box may have deposited the pathogens onto the table where sandwiches were later made (Jo-Ann Powell, Environmental Health Officer, Middlesex-London Health Unit, 1997, personal communication).

E. Puréed and Minced Diets

More often than regular diets, a minced or puréed diet is the culprit in outbreaks of foodborne disease. Fifteen cases of *E. coli* O157:H7 were identified in an Ottawa, Ontario, nursing home outbreak (5). Although a pathogen was not identified in the suspected food, residents on a minced diet were at greatest risk ($p = 0.0006$) of becoming ill. Inspection of the kitchen revealed potential for cross-contamination and a faulty warming cart that may have allowed pathogens to grow to infectious levels. In an outbreak of *E. coli* O157:H7 in a nursing home in Prince Edward Island (Canada), all seven residents affected had been served soft or blended hamburger 2 days before the first person became ill (85).

In an outbreak of diarrhea in a psychogeriatric hospital in the Greater London (England) district, 58 elderly residents developed *C. perfringens* foodborne illness and two died after eating minced beef that had been inadequately chilled and reheated in a steam unit (71). The beef had been cooked 2 days before. In another *C. perfringens* outbreak in England in 1989, 50 elderly residents in a long-stay psychiatric hospital became ill and 2 died. Analysis of food questionnaires implicated a minced beef meal (70). In an outbreak of *C. perfringens* affecting 379 patients resident in a large British hospital, minced ham was found to be the food vehicle. Large cuts of meat had been kept too long at room temperature before refrigeration and were cross-contaminated when cooked meat was put through mincers also used for raw meat (89).

Because blended or minced foods are often made from food left over from the previous day's service, two problems are posed. Cross-contamination can occur from the surfaces of the blender or mixer (53), while the time lapse provides an opportunity for microbial growth if the cooling is slow. Therefore, blended or minced food should be cooled quickly, and close attention should be paid to the proper cleaning and sanitizing of blenders to prevent cross-contamination.

F. Paid Leave for Ill Employees

Food handlers who are ill with gastroenteritis should refrain from working in a kitchen to prevent transmission of a disease agent from fecal matter to food via inadequately washed hands. Numerous outbreaks of viral and bacterial agents have occurred where temptation to "tough it out" won and workers prepared food while ill or recovering but still shedding pathogens (100). In an outbreak of severe dysentery at a U.S. Naval Hospital in 1983, 107 became ill from *Shigella dysenteriae* (50). Investigation implicated the salad bar. In the 3 weeks preceding this outbreak, gastroenteritis had been a common cause of absenteeism among the food handlers, who were in the convalescent stage and were therefore presumed to be source of the organism, which spread from fecally contaminated fingers to salad ingredients.

The presence of a policy for financial compensation that allows those who are ill to stay home is the most significant determining factor on whether a worker works when he or she is ill. Most hospitals in Ontario, as a matter of policy, pay full-time kitchen staff to stay home if they have gastroenteritis. However, for part-time or temporary workers there is no such compensation and thus no incentive to stay home. Rather, these employees would have to sustain a financial loss if they stayed away from work. In investigating risk factors in New York State nursing homes Li et al. (49) found that nursing homes with paid employee sick leave were less likely to have outbreaks of gastrointestinal and respiratory disease. They concluded that it should be possible to lower the rate of outbreaks if important risk factors are recognized and acted upon. All kitchen workers should be excused from duties and financially compensated if they have gastroenteritis.

G. Routine Stool Screening

It was once common for employees at an institution to submit stool samples once a year for screening of enteric pathogens. However, routine stool screening for food handlers is now not recommended (19, 59, 60, 98). Random sampling may be misleading (3) because excretion of pathogens is intermittent, producing a "snapshot" picture on one day that might not be not indicative of what the same employee is carrying the next day. The futility of routine stool screening cannot better be demonstrated than in the occurrence of an outbreak at a Jordan university hospital in September 1989, when 183 people became ill from *S. enteritidis.* A food-specific attack rate table implicated the mashed potatoes ($p < 0.01$). During the investigation it was found that one culture-positive employee had prepared the mashed potatoes; yet this person was asymptomatic and had had a negative stool sample 3 months earlier, as part of a policy for routine stool surveillance of kitchen employees every 3 months (39)! Furthermore, the cost involved in gathering samples and processing by the laboratory is horrendous (96). In one study that attempted to identify typhoid carriers, only one was found among a million food handlers examined at a cost of U.S. $2.5 million (18). In place of such screening, encouraging proper hand washing, soaping up for 10 seconds, and rinsing lifted materials away under tepid water, followed by hand drying, has now been adopted as a more practical strategy to prevent fecal pathogens from entering food.

A preemployment questionnaire administered by a trained health professional, which will identify illness without need for expensive lab tests, is also a more cost-effective strategy (4). Unfortunately, routine fecal screening is part of existing regulations in many jurisdictions.

H. Impact of Cost-Cutting

A trend in the developed world is the reduction of hospital funding, resulting in fewer hospital beds, fewer services, and fewer hospitals. Most patients are discharged earlier, while only the sickest patients who may be in a more vulnerable condition are kept (21, 76). An increase in the percentages of nosocomial infections for these patients, who may suffer severely, is therefore, predictable.

Foodservice directors by necessity are becoming creative at cost-cutting. One response has been to outsource food so that less preparation (less labor) is required in the hospital kitchen. For example, a hospital in Guelph (Ontario,

Canada) receives frozen lasagnas and casseroles that need only thawing and reheating. No large cuts of meat such as roast beef or turkey are prepared in the kitchen (Sue MacNeil, Manager, Dietetics, Guelph General Hospital, 1997, personal communication). It has been found to be cheaper to buy salad ingredients washed and precut from an outside supplier than to do the preparation on site. Foodservice managers have raised concerns about receiving contaminated products. Lack of refrigeration during transport and possible failure to exercise due diligence in selecting the supplier are additional issues.

With these cost-cutting measures, the major food preparation site is shifting from the hospital kitchen to the supplier. If the supplier makes a mistake in food handling, a greater number of people may become ill. It is not unheard of for a supplier to be the source of a problem. This was the case in 1981 when 43 hospital employees and 3 patients in Vermont experienced diarrheal illness from precooked roast beef supplied by a New York processor. The roast beef had been contaminated with three *Salmonella species (S, chester, S. tennessee,* and *S. livingston)* before packaging. One patient died. Inadequate refrigeration at the hospital [i.e., one refrigeration unit read 60°F (16°C)] likely allowed the salmonellae to reach infective doses. Thereafter, all precooked roast beef was required to be recooked at this hospital (83). Hospitals may not have the capacity to make decisions on the wholesomeness of the food from their suppliers. To overcome this difficulty, Geller et al. (29) have suggested that the food sanitation inspection records of suppliers be made available to hospital staff for better assessment and selection of appropriate suppliers.

Understaffing arising from cost-cutting may be instrumental in lowering staff morale. An investigation into an outbreak of *C. perfringens* in England in 1989 revealed problems in food handling by the hospital kitchen staff who often worked double shifts and on weekends because of recruitment problems. Recruitment was poor because of low wages, lack of recognition by management, and a poor image of the hospital in the community (70). This was not the first time understaffing led to additional workloads resulting in food handling errors. Understaffing has been an unfortunate outcome of competitive tendering in the United Kingdom and was thought to have contributed to a hospital outbreak of salmonellosis that affected more than 460 staff and patients and led to 19 deaths in 1984 at Stanley Royd Hospital (56, 70, 96).

Shortage of nurses in LTCFs may also lead to a greater risk of communicable disease transmission due to failure to promptly isolate patients ill with gastroenteritis (49).

A cost-efficient trend gaining momentum in North America but common in Europe is cook–chill processing. Meals can be prepared in large volume far in advance of service (i.e., enough food for 7 days can be prepared in a 4- or 5-day kitchen workweek) (52). In one type of cook–chill processing, blast chilling, food is cooked conventionally then placed in shallow pans 2.5 in. (6 cm) deep. Super-cooled air is blown over and around the pans to lower temperature rapidly. To bring the temperature up rapidly before service, the food is reheated in rethermalization units, which may be located distant from the kitchen at a satellite facility. Major advantages of cook–chill operations are the labor savings due to the efficiency of preparing many meals at once, as well as consistency of product and lower likelihood of microbial growth due to rapid cooling.

Although the U.K. practice of blast chilling or rapid cooling of foods to below 37.4°F (3°C within 90 minutes (4) will minimize microbial growth, if food is cross-contaminated or if temperature controls fail, illness can still result. This happened in a *Salmonella heidelberg* and *Campylobacter jejuni* outbreak that affected 119 residents in a nursing home in New York (45). Here, chicken livers were cooked and blended, then placed in a small bowl where chicken juices remained. The chopped liver salad was blast-chilled, but a coolant failure in the storage refrigerator caused the salad's temperature to reach 50°F (10°C) by the next day.

Blast chilling did not prevent *S. enteritidis* from being distributed to hospital patients in the U.K., although no one became ill. The *Salmonella* was found during routine sampling. During subsequent investigation, it was found that raw eggs had been used in preparing a fried food, which had not been thoroughly cooked (43).

III. PRACTICAL ADVICE TO FOOD PREPARES ON FOODBORNE ILLNESS PREVENTION

A. HACCP Concept

Policies and procedures need to be in place to prevent patients or residents from becoming ill (4). Good preventive measures in the kitchen revolve around the hazard analysis–critical control points (HACCP) concept to monitor the steps in food preparation that have historically lead to foodborne illness: inadequate cooling, inadequate cooking, inadequate hot storage, inadequate reheating, and consumption of raw foods (58). Most of these problems center around improper temperature controls. The HACCP procedure requires (a) diagramming the stages in the processing of the food, (b) identifying and labeling critical control points (CCPs:, points in the process at which contamination could occur or microbes could grow or survive), (c) determining how the CCPs will be monitored (usually with a thermometer), and (d) deciding beforehand what to do if there is a food handling mistake (throwing food out or reheating are common options) (98).

Not all steps in the process can be labeled CCPs, for having too many "critical" points dilutes the significance of the HACCP concept, which emphasizes the greatest risks of mishandling that might lead to illness. Perceived problems such as lack of a hairnet, a missing floor tile, or a dirty wall rarely lead to foodborne illness and are not CCPs.

A complete HACCP program for all foods is time-consuming to plan. In Ontario at best some institutions are using a "modified HACCP," which relies on an environmental health officer (EHO: equivalent to a public health inspector or sanitarian) to assist the foodservice director in following a food item from beginning to end. Ideally, the concepts learned by the kitchen staff in this approach will be used in all other food preparation. Most hospitals try to build HACCP principles into their menus.

It is important that institutional food handlers be careful in completing the following tasks, which are most closely related to causing illness in institutions if not done successfully.

1. The number-one cause of foodborne outbreaks throughout the world is inadequate refrigeration: that is, leaving hazardous food at room temperature for too long or putting food into the refrigeration unit in too bulky a container. This allows any bacterial pathogens present such as *Salmonella* and *C. perfringens,* the two most common causes of institutional outbreaks, to multiply readily. The goal is to discourage bacterial growth by bringing the food through the danger zone [140–40°F (60–5°C)] quickly—that is, in less than 6 hours total (95), as measured from the time the food enters the upper limit of the danger zone until the center of the food reaches 40°F (5°C). In fact bacteria grow best between 140 and 68°F (60 and 20°C), and food should be in this narrower zone for a maximum of only 2 hours. To achieve this, food must not be left at room temperature more than 1.5 hours (33), and preferably even less. Then, the food must be placed into containers such that it is no higher than 2 in. (5 cm), which may require slicing or reapportioning into smaller containers, and put into a refrigeration unit having a temperature of 40°F (5°C) or lower. Food should not be overcrowded in the refrigeration unit, for this would impede cooling. Stockpots should be discouraged for the same reason. Determining whether the food is cooled within the 6-hours total time frame or the narrower 2-hour time frame requires the use of a thermometer. If it takes any longer, ways to reduce the temperature of the food more rapidly should be explored, such as adding ice to a gravy at the end or stirring a large pot of food in an ice bath before refrigeration.

2. Food should be prepared as close to serving as possible. Minced and puréed foods are often leftovers that have been temperature abused. If prompt service after cooking is not possible, cooling foods quickly through the danger zone or storing hot are the two options.

3. When food is stored hot, the temperature in the center of the food needs to be checked periodically to verify that it is at least 140°F (60°C). Adjustment of controls on the steam table or in the cooker may be necessary.

4. Food taken from the refrigerator to be served hot must be reheated to an internal temperature of 165°F (74°C), to kill any bacteria that have multiplied or produced heat-sensitive toxins. Food must never be put on a steam table directly

from the refrigeration unit. This will not bring the temperature up high enough, and certainly not rapidly enough. Reheating is considered the "last line of defense" because many food handling mistakes that have led to bacterial growth can be negated by this final heating. Probably a third of all outbreaks could be prevented by reheating hot food sufficiently.

5. Food should be cooked to the minimum internal temperatures taken from the thickest part of the food. See Table 2.

6. Many foodborne outbreaks in LTCFs arise from egg salad sandwiches made from insufficiently cooked eggs. Eggs are common sandwich fillings because they are nutritious, easy to prepare, and soft to chew (an important consideration in feeding the elderly, who often have dentures). Unfortunately, sandwiches are frequently left unrefrigerated, permitting any *Salmonella* that survive to multiply. *Salmonella* can survive in the egg yolk or mixtures with egg yolk if the temperature does not reach 178°F (80°C) (35). Therefore to ensure safety, eggs should be boiled for a minimum of 9 minutes (35).

The U.K. Advisory Committee on the Microbiological Safety of Foods (58) and the U.S. Centers for Disease Control and Prevention (CDC) and Food and Drug Administration (FDA) (48) recommend wider use of pasteurized eggs. Because hospitalized patients and LTCF residents are at high risk for foodborne illness, pasteurized eggs should be substituted for shell eggs whenever possible.

7. Any surface in contact with raw meats (e.g., cutting board, counter, knives, slicers, mincers) must be washed with hot sudsy water, rinsed, sanitized with a disinfectant (100 ppm free available chlorine) (27) such as 1 T of chlorine bleach/gal water (2 mL/L), and rinsed or sanitized in a dishwasher at 180°F (82°C)

Table 2 Minimum Internal Temperatures for Cooking Hazardous Food

	Temperature		
Hazardous food	°C	°F	Ref.
Poultry	74	165	13, 28, 95
Ground beef	71	160	—[a]
Beef, lamb, pork and veal (medium)	71	160	28
Fish (flesh is opaque, flakes easily)	71	160	28
Alternatives			
158°F (70°C) for at least 2 min			4
150°F (65°C) for 10 minutes			3
176°F (80°C) for a few seconds			3

[a]Although the U.S. FDA recommends cooking hamburgers until (155°F) (68°C) (20), it is probably prudent to err on the side of safety for susceptible populations and cook to a higher temperature. The Food Safety Inspection Service (USDA) recommends 160°F (71°C) for hamburgers (28).

(27). Cutting boards that are color-coded and dedicated to a specific activity (e.g., red for raw meats, green for salads, etc.) are recommended. This is a very inexpensive way of minimizing the risk of cross-contamination.

8. Food handlers should never work with food when they have diarrhea. Someone else may become ill or even worse, die.

9. Hands need to be washed after using the toilet, handling raw meats, smoking, mopping, and taking out garbage. Management should be willing to experiment with different hand soaps to find one that is not drying to the skin. This may help promote hand washing.

B. Design Features

1. Blast Cooling and Freezing

Some hospitals utilize cook–chill (blast-chilling) systems and/or cook–freeze methods to bring foods through the danger zone quickly. All new kitchens for health care facilities should be designed with blast chillers because it is difficult to bring the temperature of a food down through the danger zone [140–40°F (60–5°C)] within 6 hours by conventional means. Rollin and Matthews found that a food mass 3 in. (7.5 cm) high could not be sufficiently cooled in a conventional walk-in cooler in less than 7 hours; dense foods such as beef stew sometimes took as long as 11 hours (75). In blast–chill systems, food is processed within 30 minutes of cooking and allowed to fall to 32–37°F (0–3°C) within 90 minutes (4). In cook–freeze systems the freezing starts within 30 minutes of cooking and must reach (23°F (−5°C)) within 90 minutes and 0°F (−18°C) for storage (4). In addition to minimizing bacterial growth, these systems offer the advantage of a saving on labor because large amounts of food are processed at one time. However, in addition to the disadvantage of the large initial investment in equipment, inadequate reheating may still lead to a large foodborne outbreak.

2. Other

Because hand-washing faucet handles are touched first by unclean hands then by cleaned ones, hand-washing sinks should be foot-or elbow-operated. These sinks must always be free for hand washing and designated for no other activities (e.g., vegetable washing). I have observed equipment storage in sinks and a coat draped over a basin!

Poor product flow can also increase the risk of cross-contamination. Raw meats/foods should be handled in an area separate from where the ready-to-eat product is served. Likewise, soiled utensils should not be placed adjacent to clean ones.

C. Routine Food Sampling

Since in many outbreak investigations the food vehicle no longer exists, having been either eaten up or destroyed, it is not always possible to link the pathogen found in feces to the food source. Keeping samples of hazardous food from all meals served in institutions for a period of 7 days has been recommended (46, 73). This practice would prove useful in an outbreak situation, but too much valuable freezer space may be wasted. Although an outbreak may not be evident for 5 days or more after meal service, keeping samples for 5 days offers the best compromise.

IV. CONTROL

Control of foodborne outbreaks is important to reduce illness and to prevent placing an economic burden on society. Although there are few recent cost estimates on institutional outbreaks, a *Salmonella* outbreak at a Los Angeles nursing home in 1987, affecting 44 residents, was estimated to cost over $98,000 (15). The largest costs were for vacant beds in the nursing home ($20,000), acute hospital stay ($19,500), and skilled nursing care beds ($28,000). An Ontario hospital outbreak occurring in 1982 caused salmonellosis in 196 patients and staff and was estimated to cost $480,310 (Can) (90). In an Ontario LTCF, 101 residents and 14 staff became ill from *S. infantis* in 1984, when raw chicken was allowed to cross contaminate ready-to-eat foods from a shared preparation table. This outbreak was estimated to cost $522,981 (Can) (90). Foodborne illness is preventable and is largely an unnecessary cost to society.

A. Role of Infection Control Practitioner

Good infection control practices in institutions can play a major role in decreasing the incidence of foodborne outbreaks. As residential care expands in developed countries to accommodate the growing elderly population, staff need to be adequately trained in infection control. Ideally, outbreaks should be minimized by proper food handling in the kitchen first, however, proper surveillance to catch an outbreak early and institute control measures rapidly is also important.

Facilitating prompt identification of outbreaks is the job requirement of the infection control practitioner (ICP), as defined by hospital accreditation boards such as the U.S. Joint Commission on Accreditation of Healthcare Organizations and the Canadian Council on Health Services Accreditation. Recently LTCFs have shown increased interest in infection control; however, the job of the ICP in Ontario is usually only part-time. The ICP, usually a registered nurse on staff, works in coordination with an infection control committee (ICC) made up of various professionals from the facility [director of nursing, medical staff, employee health nurse, foodservice director, administrator, laboratory chief etc. (36)], with some representation from public health personnel (9), either the medical officer of health or the EHO.

It is important for the ICP to collect baseline data on such occurrences as diarrhea (82). In LTCFs, differentiating infectious diarrhea from diarrhea induced by laxatives or stool softeners is not always easy. When the rate of symptoms exceeds the expected, notification of the local health department for assistance with an investigation is essential to prevent an outbreak from getting out of hand. Therefore, it is beneficial for the ICP to have a good working relationship with the local EHO (63). In some instances, the ICP may attempt to solve the problem internally without reporting, but sometimes an outbreak turns out to be much larger than it might have been if help had been sought earlier (47). The ICP should not delay reporting any suspected foodborne outbreak to the local health department (6, 61, 63).

B. Outbreak Management

All institutions should have in place an outbreak contingency plan (47) that describes what to do in the event of an outbreak: who to contact, the responsibilities of those designated to assist, a diagram to show where all food is stored and prepared in the facility, a list of any locations served outside the facility, a list of types of sample to be collected, instructions for collecting them and so on. At the Stanley Royd Hospital outbreak that affected over 460 patients and staff in 1984, no outbreak management plan was in place (38). Having a plan to follow avoids much confusion and saves time in the event of an outbreak.

After the local health authority has been notified of a possible outbreak, an outbreak management team (OMT), often the ICC, should be formed. The OMT needs to confirm that there is indeed an outbreak and work on the assumption that the source is foodborne until proven otherwise. The OMT establishes a case definition (symptoms within a time frame), then proceeds to determine how cases should be handled to avoid mortality and a secondary wave of person-to-person spread, especially with *E. coli O*157:H7 and *Salmonella.* Hospitals cases may be moved into "enteric isolation" rooms, where staff and visitors who enter wash hands and don gloves and gown. When they are ready to leave, they place the gown into a "enteric isolation" laundry bag, throw away the gloves, and wash their hands again. This laundry should not be re-sorted before washing. Isolation may significantly limit the secondary wave of spread (47). Cohort nursing should be in effect such that staff serve either noncontagious or the sick, but not both. In LTCFs isolation is a much greater challenge, since movement to another room may be traumatic for the elderly. However, shared staff in nursing homes have been shown to increase the risk of disease transmission as they move from patient to patient (49). Strict hand washing is the key to control (6).

In an outbreak, symptomatic (diarrheatic) staff who may be shedding large numbers of pathogens (68), and asymptomatic staff with positive cultures, should remain off work until two negative stool samples have been collected at least 24 hours apart (61). The focus of the inspection will be on the food handling in the kitchen and what may have gone awry; but nonfood premises areas such as the laundry, water supply, and pharmacy should also be checked. Additional considerations include closing the institution to visitors if cases increase and hiring temporary staff to assist with isolation and cohorting (85). Control measures should be in place one week after the return of formed stools, since some patients positive for *E. coli* may have relapses of diarrhea (84).

In two *E. coli O*157:H7 institutional outbreaks, one in Utah and one in London, (Canada), improvement in infection control techniques such as strict hand washing, exclusion of ill employees, and isolation of ill residents was credited with having decreased the number of propagated cases (12, 68).

The EHOs on the OMT will want to submit fecal specimens to the laboratory, ask questions on food histories, construct food-specific attack rate tables to identify the food vehicle, and ensure that control methods are in place. Once a food vehicle has been implicated, more specific questions on how that food was prepared, or perhaps a reenactment, will be necessary to pinpoint "what went wrong," to prevent a recurrence. It is important not to place blame on an individual but to find the problem and correct it.

C. Risk of Transmission While Working When Ill

Whether a food handler should work in foodservice while apparently healthy (asymptomatic) yet known to carry a pathogen is hotly debated. It is thought that in the absence of an outbreak, the risk of transmission of some disease agents, such as noninvasive *Salmonella,* is minimal provided hand washing is well practiced (9, 18, 59, 60, 100). However,there are documented cases of asymptomatic food handlers who were responsible for transmitting a bacterial disease agent to food and thus to consumers (23, 39, 96). For example, in a 1993 outbreak of 22 patients and 7 staff in two U.K. hospitals, transmission was believed to be due to an *S. enteritidis* carrier who intermittently contaminated food (23). Despite the controversy, it is generally agreed that food handlers should not work if they have any of the following conditions: diarrhea or vomiting (since pathogens may be excreted in large numbers), the outbreak pathogen (since this entails the risk that the pathogen may be further spread), *Shigella,* Norwalk virus, or hepatitis A (because these organism are highly infectious), and *Salmonella typhi* and paratyphi A and B (invasive salmonellae, because of the seriousness of the illness). It remains controversial whether food handlers should work if they are carriers of protozoan parasites such as *Giardia* or *Entamoeba histolytica* or the bacteria *E. coli O*157:H7, *Campylobacter,* and other pathogens for which minimum infective doses are low. The Ontario Hospital Association and Ontario Medical Association (59) and

Cruickshank and Humphrey (19) recommend no exclusion for these carriers if hand washing is thorough. However, authors of some parasitology textbooks (51) recommend exclusion until after successful treatment of protozoan parasites because the asymptomatic carrier may be shedding millions of cysts a day. Because a highly susceptible population resides in hospitals and LTCFs, it may be prudent to exclude food handlers who are recognized as carriers of organisms that can be transmitted in low doses. One would not want to establish a screening program to identify them unless there was an outbreak, however because it would not be cost-effective.

Without dispute, food handlers who have invasive salmonellae should not return to work until the carrier state has been eradicated. For hepatitis A the food handler must remain off work during the first 2 weeks of illness, and if not identified until jaundice develops, the food handler should be off work until 7 days following the onset of jaundice (59, 60). For Norwalk virus the food handler must remain off work until symptom-free for 48 hours (59, 60). For *Shigella* or the outbreak pathogen the food handler must produce two negative stool samples, taken 24 hours or more apart, beginning at least 24 hours after diarrhea ends (59, 60). If the patient was treated with an antibiotic, the first stool sample must be submitted at least 48 hours after the antibiotic has last been administered (59, 60).

D. Treatment

Oral rehydration therapy (ORT) is vital for treatment of patients with diarrhea in LTCFs since "its use can minimize the morbidity and mortality associated with gastroenteritis"(6). Antimotility agents such as diphenoxylate hydrochloride may increase the mortality and morbidity of *E. coli O*157:H7 infections by decreasing intestinal motility, which facilitates multiplication of the organism and its verotoxin (77).

Antibiotic treatment for salmonellosis remains controversial. In a nursing home outbreak involving fourty four people there was an increased duration of *Salmonella* stool excretion for residents who were treated with antibiotics (15 vs. 6 weeks), although the results were not statistically significant (15). In contrast, shigellosis in nursing homes is always treated with antibiotics (6).

V. EDUCATION

A. Role of Management

The core of a food safety program must include the commitment of management to train employees and to serve as a resource and role model. If management is not committed to preventing foodborne illness, their employees also will lack such commitment. Both parties need to understand the basics in food safety. First, there should be an "open-door policy" such that a food handler feels free to ask any questions and report gastroenteritis or open lesions without fear of reprisal. A policy for financial compensation for staying home when ill, as discussed earlier (Sec II.F) is particularly effective. Second, there needs to be an organized food safety training program. In some jurisdictions food handler education courses are mandatory, but this is the exception. Food safety programs for managers should emphasize the following (from WHO recommendations) (98):

1. Foodborne illness (reporting requirements, some common pathogens, how illness is caused, factors that affect growth/survival of microorganisms)
2. Factors that contribute to outbreaks (inadequate temperature controls, cross-contamination, etc.)
3. How to prevent foodborne illness

Many front-line health care institution workers in kitchens are only marginally literate. Staff with poor educational backgrounds may not understand why things are done or even how things should be done. Managers should monitor their employees' behavior in the kitchen to ensure understanding of the material presented and to encourage workers to put into practice what has been learned.

Some procedures, such as final cooking temperatures, can be built right into the recipes. Log books can provide a record of final internal temperatures and hot storage temperatures, helpful to managers and EHOs in assessing proper food handling techniques. There also has to be a plan of action if the criteria are not met. A colleague was inspecting a hospital kitchen and noticed in the log book that the chicken was only cooked to to only 112°F (50°C), with holding at the same temperature. Although cooking to 165°F (74°C) was required (and stated), the staff did not know what to do when the requirement was not met (Joanne Braithwaite, Public Health Inspector, City of Toronto, 1997, personal communication).

Employees also need training at the start of their employment and periodic refresher courses. Implementation of training programs in food safety is far more effective than laboratory screening of stool samples (18, 98). Often local health units present food safety courses at night or are willing to give courses when numbers warrant at the institution. Community colleges may also offer food safety courses. Learning proper food handling is a lifelong undertaking and should not be limited to one course.

B. Schools

1. Historical and Current Incidents of Foodborne Illness in Schools

a. Pathogens of Concern. Foodborne outbreaks in schools are not as common as in health care institutions (Fig. 5) because school-aged children are generally not a predisposed group for gastrointestinal illness. An exception is seen in very young children and for a few infectious agents: *Campylobacter, E. coli* *O*157:H7, hepatitis A, and *Shigella.* The last two agents are primarily spread from person to person rather than by food and are especially likely to occur when insufficient hand washing plays a major role in a fecal–oral route of transmission.

In a U.S. study of *Campylobacter* infections, 75% of outbreaks were associated with drinking raw milk on school field trips, and 70% of the young victims were in kindergarten through grade 3 (97). In Ontario (Canada), 43 kindergarten children developed illness from *E. coli O*157:H7 after visiting a dairy farm (22). Epidemiological evidence strongly implicated unpasteurized milk as the most likely source. A review of outbreaks in schools in England and Wales (1979–1988) found that 3600 children fell victim to *Campylobacter,* while the organism affected only 114 adults (37). In five of the nine outbreaks considered, the vehicle of infection was contaminated or unpasteurized milk. Wood et al.(97) were not able to determine whether younger children are more likely to go on field trips and thus have a greater potential for exposure, or whether their young age predisposes them to *Campylobacter* infections.

Hepatitis A virus, a pathogen commonly encountered in day care facilities, can also be found in the school environment. In England and Wales (1979–1988) there were eight outbreaks of hepatitis A affecting 175 children and 34 adults (37). Three of these outbreaks were from a common source, followed by propagated spread. In only one was the food vehicle identified, frosted buns, which the index case had helped prepare in cooking class. Thirty-nine point source cases resulted, with another 58 occurring by person-to-person spread.

More recently a large hepatitis A outbreak involved school children in Michigan. There were 153 cases, with 151 from students and staff in four different school districts. Preliminary analysis has linked illness to frozen strawberries originating in Mexico. The berries were packed in 30 lb containers for distribution in lunch programs sponsored throughout the country by the U.S. Department of Agriculture (USDA) (32). Although immunoglobulin administered to contacts within 14 days of exposure can reduce the risk of illness, dissemination of virus to family members and the community at large is still a concern (32).

An even larger and more devastating outbreak occurred in primary school settings in several cities in Japan from mid-July to August 8, 1996 (24, 64). Most of the cases occurred in Sakai City, where 5727 people were affected in an outbreak of *E. coli* O157;H7. White radish sprouts were suspected to be the principal food vehicle. The organism was never found on the sprouts, however, although it was found on salad and seafood sauce that were served in school lunches in outbreaks other than Sakai City. In all, 9451 cases were reported; 1808 persons were hospitalized, and 12 died (99).

In Taiwan in 1994, 54% of large outbreaks (affecting more than five people) were associated with commercial lunchboxes supplied to elementary and junior high schools (66).

b. Incidence. In foodborne outbreaks in the United States (1988–1992) only 4% (32 of 796) occurred from mishandlings of food in schools (94). In Canada (1990–1993) the percentage was also 4% (4/103) (Figs. 2 and 5). In England and Wales (1992–1994) the incidence of foodborne outbreaks in schools was similarly 4.7% (Fig. 3).

2. Common Trends and Factors Leading to Foodborne Illness in Schools

School kitchens, like other institutional kitchens, are not immune to mishandling problems. In 1985, at an elementary school in Georgia, 351 children and staff developed febrile gastroenteritis associated with the eating of turkey salad (81). Culture of leftover food yielded *S. enteritidis,* with each child estimated to have received a total of 5.0×10^7 organisms. After cooking, the turkey had been deboned and refrigerated overnight in a pan 8 in. (20 cm) deep. Inadequate cooling was a factor here; the center of the stored turkey meat did not cool quickly, thus allowing bacteria to multiply.

In Oklahoma, in 1986, a salmonellosis outbreak affecting an estimated 202 children was associated with the eating of chicken from a school cafeteria (10). *S. heidelberg* was isolated from 32 patients and *S. stanley* from 5. The chicken had been was cooked in an oven for 2 hours at 350°F (177°C); then the heat was turned off and the chicken left overnight in the warm oven. Another chicken was cooked in a steam cooker and left overnight at the lowest setting. In a reenactment using beans, the temperature only reached (84°F) (29°C) at the center of the beans in the oven, while the steam cooker at the lowest temperature setting produced a holding temperature of only (110°F)(43°C). Internal cooking temperatures were not high enough to destroy *Salmonella,* and the low holding temperature allowed them to multiply. However, when queried, cafeteria workers were unable to identify any food handling problems!

In 1990, 65 children suffered from staphylococcal food poisoning from lunch served at an elementary school in Rhode Island (74). A food worker, who later tested positive for the enterotoxigenic strain of *S. aureus,* had removed casings from two of nine warm ham rolls 48 hours prior to service and inadvertently contaminated them. The ham had been held between 50 and 120°F (10–49°C) for a minimum of 15 hours, allowing *S. aureus* to grow and produce enterotoxin.

Although inadequately cooked foods and temperature-abused foods commonly lead to outbreaks in schools, we are seeing another risk factor emerging: the production of large quantities of food at centralized facilities. The Taiwanese lunchbox outbreaks (1994), the *E. coli* O157:H7 outbreaks in Japan (1996), and the hepatitis A outbreak in the United States (1997) are prime examples of this trend. Preparation of such large quantities requires food to be prepared far in advance of service, and if contamination is present, harmful bacteria have ample time to multiply. Centralization saves administrative and production costs but is disastrous if food is mishandled. Because the trend is toward consolidating food handling at large facilities, there probably will be more and larger outbreaks not only in schools, but in other institutions as well.

3. Practical Advice on Controlling Foodborne Outbreaks in Schools

Like foodborne outbreaks in other establishments, many outbreaks in schools go unreported because school staff fail to recognize the need to report illness. Cooperation between teachers, administrators, parents, and the local health authorities is essential to ensure opportunity for investigation and follow-up.

Primary teachers can support public health by encouraging children to wash their hands, particularly after using the toilet and before eating. This is a lifelong habit that can keep an individual healthy and protect the community, although unfortunately even adults forget or dismiss it. The time to start is in grade school—school staff can support proper behavior by example and by encouragement. Warm water, soap, and paper towels (or an air dryer) should be available in student washrooms.

On field trips, children should not even taste unpasteurized milk. In rural schools on nonmunicipal water, the drinking water should be tested twice a year for potability. In one private school outbreak in the United Kingdom affecting 234 pupils and 23 staff, *Campylobacter* was thought to have originated from an open water storage tank (65). Birds or bats roosting nearby may have been the reservoir. In another waterborne *Campylobacter* outbreak in the United Kingdom, 32 pupils and 4 staff were affected (25). Upon investigation, fecal coliform organisms and the epidemic strain of *Campylobacter coli* were found in the water supply.

Cafeteria staff should be mindful of potential problems in food handling in their kitchens, particularly undercooking food, not cooling foods quickly, and not storing at high enough temperatures before service, as discussed in connection with health care institutions (Sec. III. A).

4. Education

a. Staff Training Food safety training for school cafeteria employees varies widely (10). In most jurisdictions there is no legislation requiring adequate training of foodservice workers employed in schools (10). Often, kitchen staff are unaware that inadequate cooling and reheating of foods are major risks of foodborne illness (74). Across the United States and Canada and in most other countries, there is no uniformity in the educational training or materials available to school cafeteria staff. Local health department staff regularly deliver food safety courses for diverse groups and will even deliver a course on-site if numbers warrant. Perhaps boards of education should be more proactive and invite health department staff to present food safety education to their assembled cafeteria staff. A separate course for a larger number of people might be justifiable. However, it is usually left to management to train their staff and acquire resource materials.

b. Food Safety for Youth. Large numbers of cases of foodborne illness take place in homes (Figs. 1–4). What better way to reduce these numbers than by training our youth to handle foods safely early in life? Bryan (7) and Todd (91) have stated that education of our youth offers the most promise in reducing the overall incidence of foodborne illness, since we all become food handlers for ourselves, our families, and/or professionally.

Local health departments may be required by legislation to be involved in development of food safety curricula in school in accordance with state or provincial requirement. Even so, curricula become fractured, with each local authority "doing its thing." Within the school, food safety education may take any form a teacher sees fit. A more structured program to be created by public health teams and teachers, perhaps family studies teachers/home economics teachers (2) at the junior highs school level (ages 13–15 years) or even earlier, might reach more children and have a bigger impact.

C. Correctional Facilities

1. Historical and Current Incidents of Foodborne Illness in Correctional Facilities

a. Introduction. Correctional facilities often provide care for a large number of inmates who have HIV infection, AIDS, tuberculosis, and other chronic infections or conditions (31), or are alcoholics. Any one of these conditions may predispose the inmate to the risk of communicable disease, including foodborne illness. Another risk factor in this setting is the large number of inmates who help in the kitchen, often with only minimal training and supervision (44). With lack of skills and little motivation to do a good job, food handlers in such settings are prone to generate problems.

b. Problems in Investigations. As with schools, the frequency of foodborne illness reports from jails and prisons is low because such incidents are not always reported. Indeed, many probably go unreported, and findings of investigations often are inconclusive because of impediments to thoroughness inherent in the institutions themselves. For example, inmates were uncooperative in their recall of foods eaten in an outbreak of *S. typhimuirum* and *enteritidis* at a New York City jail affecting 145 persons over a 2-month period, and a case-control study had to be abandoned (1). Here, delaying inspection of the kitchen until almost a month after cases first appeared certainly impaired the ability to identify the specific problems leading to the outbreak. The division of jurisdictional responsibility between health services and the correctional institution hampered the investigation because health services staff did not have the authority to canvass the inmate population for potential cases. All modes of secondary transmission could not be tracked because all persons interviewed denied having oral sex, which was an infraction of institutional rules, despite known occurrences of this practice. Although the pathogens were identified and control measures were established to minimize secondary spread, the contaminated food was not identified and the food mishandling problem(s) remained unknown.

Responding rapidly in an outbreak and completing a thorough investigation can lead to identification of problems, and strategies can be put in place to prevent a reccurrence. In a foodborne outbreak in a Florida state prison in 1990, 215 residents had diarrhea, vomiting, or both. *S. infantis* and enterotoxigenic *S. aureus* were found in samples of leftover turkey and stool specimens (53). Major contributing factors to this outbreak were the holding of the turkey at room temperature and in a food warmer, which allowed multiplication of pathogens and toxin elaboration. Temporary removal of infected and colonized food handlers from kitchen duties and a policy of reheating foods were recommended to prevent further disease transmission here.

c. Incidence. In a U.S. study (1974–1991) of 88 desmoteric (illness arising from within a correctional facility) foodborne outbreaks, the average number of cases per outbreak was 163 (16). *Salmonella* spp. accounted for most outbreaks (37%), followed by *C. perfringens* (34%) and *S. aureus* (22%). No fatalities resulted. Of the food vehicles known, beef and poultry were implicated in 20% of outbreaks, followed by fish or poultry salads (7%) and Mexican food (7%). Ninety percent of outbreaks were due to improper food storage.

2. Common Trends and Factors Leading to Foodborne Illness in Correctional Institutions

a. Food Mishandling Problems. In 1984 two *C. perfringens* outbreaks occurred at a Florida correctional facility: 74 inmates became ill in the first wave, followed 8 days later by another 100 (86). Although food vehicles were different, contributing factors in both outbreaks were inadequate cooling of meats, inadequate reheating, and improper holding temperatures before service. Meat had an internal temperature of 85°F (31°C) even after being in the cooler for 5 hours. Moreover, meat was "reheated" on steam tables with pans of food stacked three layers high. Obviously, neither the proper internal reheating temperature of 165°F (74°C) nor the prescribed hot storage temperature in the stacked pans [140°F (60°C)] had been reached.

Although many food handling errors may contribute to an outbreak, inadequate cooling often plays a key role. In an outbreak at a Georgia prison in 1992, 113 residents met the case definition for illness (16). Rice pudding, the only statistically implicated food, was made from pooled shelled eggs, the likely source of *S. infantis.* The pudding had been left to cool at room temperature for about 5 hours before serving.

At a Delaware prison in 1992, 281 residents sought medical attention for their diarrhea (16). Chicken salad was the only food statistically associated with illness. In this *S. enteritidis* outbreak, chicken had been boiled, deboned, and refrigerated. Three hundred eggs were hardboiled, cooled, shelled, and added to the chicken along with commercial mayonnaise. Unfortunately, the salad was held at room temperature for 4.5 hours before serving.

The use of pooled shell eggs and the decision to leave the food at room temperature appear to be the most likely factors in the cause of prison foodborne outbreaks. "Power drinks" made from shell eggs are popular among inmates but might lead to illness and should be discouraged.

b. Impact of Cost-Cutting. In 1990 the average daily population of jail and prison inmates in the United States had increased to 1.2 million (300/100,000), up from 0.5 million (150/100,000) in 1980 (33). The (U.S.) inmate population rose by 8% in 1996, but food purchases only by 4% (43). Overcrowding in correctional facilities is now commonplace because funding has not kept pace with inmate numbers. Yet, penal institutions are having to provide more services while costs for goods and services rise. This is one of the main challenges facing correctional facilities today, and it has a direct impact on the direction of correctional foodservice.

On September 12, 1996, the Ontario Solicitor General and Minister of Correctional Services announced the forthcoming replacement of 14 existing correctional institutions with 5, larger modern institutions or "superjails." After restructuring, the incarceration costs per inmate per day were to be reduced from $124 to $76, predicted by the minister to become the lowest in Canada. Correctional facility food is to be prepared in a centralized commissary by a type of cook–chill process called tumble-chilling. In tumble-chilling or ice-chilling, cooked viscous foods such as soups, gravies, and salad dressings are pumped into plastic casings that are vacuumed sealed and labeled. The casings are then tumbled in an ice-water bath until the temperature is below 38°F (4°C). Once chilled, the casings are removed from the bath and stored under refrigeration (52). Non viscous foods, such as meats, are prepared for cooking by vacuum sealing in plastic, cooking in hot water slowly in a steam kettle, and then in circulating ice water to cool quickly.

Cook–chill operations are gaining popularity in U.S. correctional facilities primarily because of their cost savings. In the Philadelphia prison system, for example, meal costs have been reduced from $1.31 to $1.18 and labor costs by 30–50% (44). Since there is little contact with the food and cooling is rapid, the risk of foodborne illness is minimized, but not eliminated.

3. Practical Advice on Controlling Foodborne Outbreaks of Illness in Correctional Institutions

Adequate refrigeration facilities are a must when one is attempting to control pathogen growth. Large amounts of food require much refrigeration space, so there is a temptation to leave food out at room temperature. As we have seen, pudding and chicken salad with eggs can harbor *Salmonella,* which readily grow to infective doses in just 4.5–5 hours at room temperature. Therefore, if the food is not to be served within an hour it should be refrigerated. As in the recommendation for health care facilities, use of pooled shell eggs should be discouraged and pasteurized eggs substituted in jails and prisons (55).

4. Education

In most situations, civilian food workers are encouraged to attend food safety workshops, which may be offered by health department staff or professional organizations. However, the correctional facility inmates who work in the kitchen outnumber civilian staff by about 1:4. Obviously, a whole group of persons needing training is missed. It is my recommendation that more attention be paid to this forgotten group by "train the trainer" strategies, whereby a civilian kitchen employee is specifically trained to train the inmates in food safety. This may be done informally but could benefit from a set of standardized procedures.

VI. CONCLUDING REMARKS

Foodborne illness is on the rise in health care institutions. The current trend to cut costs by centralizing food services poses the risk of outbreaks affecting larger numbers of people, since a food handling mistake affects enormous amounts of food. This was seen in 1996 in the *E. coli* O157:H7 outbreaks in Japanese schools, which affected thousands of children who apparently had eaten contaminated radish sprouts, and in the hepatitis A outbreak in U.S. schools, which affected over 150 children who had eaten contaminated strawberries. A positive aspect of centralized facilities, however, is that investment in expensive, specialized equipment, such as blast chillers, will now be cost-effective because of larger-scale operations. Since inadequate cooling of food is the major cause of foodborne disease, being able to chill food rapidly is certainly desirable from a risk management point of view. We should not become complacent, however, because food may become contaminated or pathogens may grow at many other processing steps: before chilling, after chilling during transport, or as a result of temperature abuse in a late processing step. Educating food handlers and our youth about proper food handling techniques will continue to be our best weapon against foodborne illness.

ACKNOWLEDGMENTS

The author thanks Sandy Isaacs, field epidemiologist, Laboratory Centre for Disease Control, Guelph, Ontario, Canada, for her assistance in managing the Reportable Disease Inspection Systems database, and Dr. Eng-Hong Lee for his critical review of the manuscript.

REFERENCES

1. Alcabes, P., B. O'Sullivan, D. Nadal, and M. Mouzon. 1988. An outbreak of *Salmonella* gastroenteritis in an urban jail. Infect. Control Hosp. Epidemiol. 9:542–547.
2. Al-Zubaidy, A. A., H. E. el Bushra, and M. Y. Mawlawi. 1995. An outbreak of typhoid fever among children who attended a potluck dinner at Al-Mudhnab, Saudi Arabia. East Afr. Med. 72:373–375.
3. Ayliffe, G. A. J., B. J. Colins, and L. J. Taylor. 1990. Hospital-Acquired Infection Principles and Prevention, 2nd ed. Wright, London.
4. Barrie, D. 1996. The provision of food and catering services in hospital. J. Hosp. Infect. Control 33:13–33.
5. Basrur, S., B. Reeder, J. A. Carlson, R. Fralick, M. Jabar, P. DiBatista, D. Fralick, T. Heap, and H. Lior. 1987. An outbreak of *E. coli* O157:H7 diarrhea in a nursing home—Ontario. Can. Dis. Wkly. Rep. 13:205.
6. Bennett, R. G. 1993. Diarrhea among residents of long-term care facilities. Infect. Control. Hosp. Epidemiol. 14:397–404.
7. Bryan, F. 1981. Current trends in foodborne salmonellosis in the United States and Canada. J. Food Prot. 44:394–402.
8. Caffer, M. I., and T. Eiguer. 1994. *Salmonella enteritidis* in Argentina. Int. J. Food Microbiol. 21:15–19.
9. Canadian Council in Health Sources Accreditation (formerly Canadian Council on Health Facilities Accreditation). 1993. Accreditation Standards. Management Services. Standard Area XI. Infection Control, MGT-39–43. (Ottowa, Ont.)
10. Carr, R., S. Brown, A. Goodall, D. Head, B. Stacy, R. Bryce, T. Hill, and G. Istre. 1987. Salmonellosis in a school system—Oklahoma. MMWR 36:74–75.
11. Carter, A. O., A. A. Borczyk, J. A. K. Carlson, B. Harvey, J. C. Hockin, M. A. Karmali, C. Krishnan, D. A. Korn, and H. Lior. A severe outbreak of *Escherichia coli* O157:H7–associated hemorrhagic colitis in a nursing home. 1987. N. Engl. J. Med. 317:1496–1500.

12. Carter, A. O., and J. A. K. Carlson. 1985. Extendicare London nursing home outbreak of diarrheal disease—September 1985. Ont. Dis. Surv. Rep. (now PHERO) 6:531–543.
13. Castellani, A. G., R. R. Clarke, M. I. Gibson, and D. F. Meisner. 1953. Roasting time and temperature required to kill food poisoning microorganisms introduced experimentally into stuffing turkeys. Food Res. 18:131–138.
14. Chan, J., and S. Pollock. 1987. Multi-*Salmonella* food poisoning outbreak in a home for the aged—Etobicoke. Ont. Dis. Surv. Rep. 8:273–275.
15. Choi, M, T. T. Yoshikawa, J. Bridge, A. Schlaifer, D. Osterweil, D. Reid, and D. C. Norman. 1990. *Salmonella* outbreak in a nursing home. J. Am. Geriatr. Soc. 38:531–534.
16. Cieslak, P. R., M. B. Curtis, D. M. Coulombier, A. L. Hathcock, N. H. Bean, and R. V. Tauxe. 1996. Desmoteric foodborne outreaks in the United States, 1974–1991. Arch. Intern. Med. 156:1883–1888.
17. Collier, P. W., J. C. M. Sharp, A. F. MacLeod, G. I. Forbes, and F. Mackay. 1988. Food poisoning in hospitals in Scotland, 1978–1987. Epidemiol. Infect. 101:661–667.
18. Cruickshank, J. G. 1990. Food handlers and food poisoning. B. Med. J. 300:207–208.
19. Cruickshank, J. G., and T. J. Humphrey. 1987. The carrier food-handler and nontyphoid salmonellosis. Epidemiol. Infect. 98:223–230.
20. Davis, M., C. Osaki, D. Gordon, and M. W. Hinds. 1993. Update: Multistate outbreak of *Esherichia coli* O157:H7 infections from hamburgers—Western United States, 1992–1993. MMWR 42:258–263.
21. Djuretic, T., M. J. Ryan, D. M. Fleming, and P. G. Wall. 1996. Infectious intestinal disease in elderly people. CDR Rev. 6:R107–R112.
22. Duncan, L., A. Carter, J. A. K. Carlson, and A. Borczyk. 1987. Outbreak of gastrointestinal disease—Ontario. Can. Dis. Wkly. Rep. 13:5–8.
23. Dryden, M. S., N. Keyworth, R. Gabb, and K. Stein. 1994. Nosocomial salmonellosis and food handlers. J. Hosp. Infect. 24:195–208.
24. Enterohaemorrhagic *Escherichia coli* infection. 1996. Wkly. Epidemiol. Rec. 35: 267–268.
25. Evans, H. S., and H. Maguire. 1996. Outbreaks of infectious intestinal disease in schools and nurseries in England and Wales 1992–1994. CDR Rev. 6(7):R103–R108.
26. Evans M. R., P. G. Hutchings, C. D. Ribeiro, and D. Westmoreland. 1996. A hospital outbreak of salmonella food poisoning due to inadequate deep-fat frying. Epidemiol. Infect. 116:155–160.
27. Food Premises Regulation. 1990. Ontario Regulation 562/90, made under the Health Protection and Promotion Act.
28. Food Safety Inspection Service, U.S. Department of Agriculture. 1991. Cooking foods safely. Food News for Consumers 8:11.
29. Gellert, G. A., M. Tormey, G. Rodriguez, G. Brougher, D. Dassey, and C. Pate. 1989. Foodborne disease in hospitals: Prevention in a changing food service environment Am. J. Infect. Control 17:136–140.
30. Gerba, C. P., J. B. Rose, and C. N. Haas. 1996. Sensitive populations: Who is at the greatest risk? Int. J. Food Microbiol. 30:113–123.

31. Glaser, J. B., and R. B. Greifinger. 1993. Correctional health care: a public health opportunity. Ann. Intern. Med. 118:139–145.
32. Hepatitis A associated with consumption of frozen strawberries—Michigan, March. 1997. MMWR 46:288, 296.
33. Hobbs, B. C., and D. Roberts. 1987. Food Poisoning and Food Hygiene, 5th ed. Edward Arnold, London.
34. Hockin, J., H. Lior, L. Mueler, C. Davidson, E. Ashton, and F. Wu. 1987. An outbreak of *E. coli* O157:H7 diarrhea in a nursing home. Can. Dis. Wkly. Rep. 13:206.
35. Humphrey, T. J., M. Greenwood, R. J. Gilbert, B. Rowe, and P. A. Chapman. 1989. The survival of salmonellas in shell eggs cooked under simulated domestic conditions. Epidemiol. Infect. 103:35–45.
36. Joint Commission on Accreditation of Healthcare Organizations. 1990. Joint Commission Standards—1990. Hospital Infection Control. (Oakbrook Terrace, IL) January.
37. Joseph, C., N. Noah, J. White, and T. Hoskins. 1990. A review of outbreaks of infectious disease in schools in England and Wales, 1979–88. Epidemiol. Infect. 105: 419–434.
38. Kapila, M., and R. Buttery. 1986. Lessons from the outbreak of food poisoning at Stanley Royd Hospital: What are health authorities doing now? B. Med. J. 293: 321–322.
39. Khuri-Bulos, N. A., M. A. Khalaf, A. Shehabi, and K. Shami. 1994. Foodhandler-associated *Salmonella* outbreak in a university hospital despite routine surveillance cultures of kitchen employees. Infect. Control Hosp. Epidemiol. 15:311–314.
40. Kohli, H. S., A. K. R. Chaudhuri, W. T. A. Todd, A. A. B. Mitchell, and K. G. Liddell. 1994. A severe outbreak of *E. coli* O157 in two psychogeriatric wards. J. Public Health Med. 16:11–15.
41. Kolbe, F. 1997. Outbreak of *C. perfringens* in a hospital. Public Health Epidemiol. Rep. Ont. 8:29–32.
42. LCDC (Canada), Ontario Ministry of Health, and Bruce-Grey-Owen Sound Health Unit. 1992. Hospital outbreak of *Salmonella enteritidis* infection—Ontario. 1992. Can. Commun. Dis. Rep. 18:57–60.
43. Lacey, S. L., and S. E. Buckingham. 1993. Isolation of *Salmonella enteritidis* from cook–chill food distributed to hospital patients. J. Hosp. Infect. 25:133–136.
44. LaVecchia, G. 1996. New demands of captive customers. Food Manage. 31: 38,40,42.
45. Layton, M. C., S. G. Calliste, T. M. Gomez, C. Pattton, and S. Brooks. 1996. A mixed foodborne outbreak with *Salmonella heidelberg* and *Campylobacter jejuni* in a nursing home. Infect. Controe Hosp. Epidemiol. 18:115–121.
46. Lee, S. 1985. *Salmonella* outbreak in a hospital: a case study. Environ. Health Rev. (Can.) 29:26–27.
47. LeRiche, H. 1986. Address to the jury at the coroner's inquest to the Extendicare food poisoning epidemic, London, Ontario, May–June 1986.
48. Levine, W. C., J. F. Smart, D. L. Archer, N. H. Bean, and R. T. Tauxe. 1991. Foodborne disease outbreaks in nursing homes, 1975 through 1987. JAMA 266:2105–2109.
49. Li, J., G. S. Birkhead, D. S. Strogatz, and F. B. Coles. 1996. Impact of institution

size, staffing patterns, and infection control practices on communicable disease outbreaks in New York state nursing homes. Am. J. Epidemiol. 143:1042–1049.
50. Longfield, R., E. Strohmer, R. Newquist, J. Longfield, J. Coberly, G. Howell, and R. Thomas. 1983. Hospital-associated outbreak of *Shigella dysenteriae* type 2—Maryland. MMWR 32:250–252,257.
51. Markell, E. K, D. T. John and W. A, Krotoski. 1999. Medical Parasitology, 8th ed. Saunders, Philadelphia.
52. Mathews, L. E. 1993. Cook–chill centralized food service in corrections. Am. Jails July/August 1993:59–60.
53. Meehan, P. J., T. Atkeson, D. Kepner, and M. Melton. 1992. A foodborne outbreak of gastroenteritis involving two different pathogens. Am. J. Epidemiol. 36(1):611–616.
54. Middleton, B. Summary of an outbreak of *Salmonella* at an acute care hospital. 1992. Ontario Branch, Canadian Institute of Public Health Inspectors. Ontario Branch News (Clinton, Ont.) 13:31–33.
55. Mishu, B. J. Koehler, L. A. Lee, D. Rodriguez, F. H. Brenner, P. Blake, and R. V. Tauxe. 1994. Outbreaks of *Salmonella enteritidis* infections in the United States, 1985–1991. J. Infect. Dis 169:547–52.
56. Morris, P. 1986. Errors that led to tragedy. Nurs. Times 82:18–19.
57. Mowat, D. L., G. F. Hutchings, D. MacWiliam, J. Zalewski, and P. Carr. 1992. Institutional outbreaks of *Salmonella* gastroenteritis in Kingston in 1990. Public Health Epidemiol. Rep. Ont. 3:74–76.
58. National Advisory Committee on Microbiological Criteria for Foods. 1992. Hazard analysis and critical control point system. Int. J. Food Microbiol. 16:1–23.
59. Ontario Hospital Association and Ontario Medical Association. 1989. Enteric Diseases Surveillance Protocol for Ontario Hospitals. OHA-OMA, Toronto, Ont, Canada.
60. Ontario Ministry of Health. Public Health Branch. 1990. Enteric Disease Screening Recommendations and Case Management Guidelines on Foodhandlers and Patient Care Workers. OMH, Toronto, Ont., Canada.
61. Ontario Ministry of Health. Public Health Branch. 1993. A Guide to the Control of Enteric Disease Outbreaks in Health Care Facilities. OMH, Toronto, Ont., Canada.
62. Ontario Ministry of Health. 1997. Reportable Disease Information Systems (RDIS) database, 1993–1996. Toronto, Ont., Canada.
63. Ontario Ministry of Health, Residential Services Branch. 1992. Outbreak Control Guidelines for Long Term Care Facilities. OMH, Toronto, Ont., Canada.
64. Outbreaks of enterohemorrhagic *Escherichia coli* O157:H7 infection, 1996, Jpn. Infectious Agents Surveillance Report 17(8)
65. Palmer, S. R., P. R. Gully, J. M. White, A. D. Pearson, W. G. Suckling, D. M. Jones, J. C. L. Rawes, and J. L. Penner. 1983. Waterborne outbreak of *Campylobacter* gastroenteritis. Lancet i:287–290.
66. Pan, T. M., C. S. Chiou, S. Y. Hsu, H. C. Huang, T. K. Wang, S. I. Chiu, H. L. Yea, and C. L. Lee. 1996. Foodborne disease outbreaks in Taiwan, 1994. J. Formos. Med. Assoc. 95:417–420.
67. Panaro, L., D. Cooke, and A. Borczyk. 1990. Outbreak of *Escherichia coli* O157:H7 in a nursing home—Ontario. Can. Dis. Wkly. Rep. 16:89–92.
68. Pavia, A. T., C. R. Nichols, D. P. Green, R. V. Tauxe, S. Mottice, K. D. Greene, J. G.

Wells, R. L. Siegler, E. D. Brewer, D. Hannon, and P. A. Blake. 1990. Hemolytic–uremic syndrome during an outbreak of *Escherichia coli* O157:H7 infections in an institution for mentally retarded persons:Clinical and epidemiologic observations. J. Pediatr. 116:544–551.

69. Pollett, G., and B. A. Reeder. 1987. An outbreak of salmonellosis at a district hospital—Halton. Ont. Dis. Surv. Rep. 8:268–272.
70. Pollock, A., and P. M. Whitty. 1990. Crisis in our hospital kitchens: Ancillary staffing levels during an outbreak of food poisoning in a long stay hospital. B. Med. J. 300:383–385.
71. Pollock, A. M., and P. M. Whitty. 1991. Outbreak of *Clostridium perfringens* food poisoning. J. Hosp. Infect. 17:179–186.
72. Preston, S. 1993. Demographic change in the U.S., 1970–2050, in K. G. Manton, B. H. Singer, and R. M. Suzman (eds.), Forecasting the Health of the Elderly. Springer-Verlag, New York, Ch. 3.
73. Reffle, J. 1989. Food sampling guidelines for institutions. Ont. Dis. Wkly. Rep. 10:22.
74. Richards, M. S., M. Rittman, T. T. Gilbert, S. M. Opal, B. A. DeBuono, R. J. Neill, and P. Gemski. 1993. Investigation of a staphylococcal food poisoning outbreak in a centralized school lunch program. Public Health Rep. 108:765–771.
75. Rollin, J. L., and M. E. Matthews. 1977. Cook/chill foodservice systems: Temperature histories of a cooked ground beef product during the chilling process. J. Food Prot. 40:782–784.
76. Ryan, M. J., P. G. Wall, G. K. Adak, H. S. Evans, and J. M. Cowden. 1997. Outbreaks of infectious intestinal disease in residential institutions in England and Wales 1992–1994. J. Hosp. Infect. 34:49–54.
77. Ryan, C. A., R. V. Tauxe, G. W. Hosek, J. G. Wells, P. A. Stoesz, H. W. McFadden, P. W. Smith, G. F. Wright, and P. A. Blake. 1986. *Escherichia coli* O157:H7 diarrhea in a nursing home: Clinical, epidemiological, and pathological findings. J. Infect. Dis. 154:631–638.
78. Sabetta, J. R., S. Hyman, J. Smardin, M. L. Carter, and J. L. Hadler. 1991. Foodborne nosocomial outbreak of *Salmonella reading*—Connecticut. MMWR 40:804–806.
79. Scuderi, G., M. Fantasia, E. Filetici, and M. P. Anastasio. 1996. Foodborne outbreaks caused by *Salmonella* in Italy, 1991–4. Epidemiol. Infect. 116:257–265.
80. Slutsker, L., M. E. Villarino, W. R. Jarvis, and J. Goulding. 1997. Foodborne disease prevention in health care facilities, in J. V. Bennett and P. S. Brachman (eds.), Hospital Infections, 4th ed. Philadelphia, Lippencott-Raven.
81. Smith, M. W. Fancher, R. Blumberg, G. Bohan, D. Smith, T. McKinley, and R. K. Sikes. 1985. Turkey-associated salmonellosis at an elementary school—Georgia. MMWR 34:707–708.
82. Soule, B. M. (ed.). 1983. The APIC curriculum for infection control practice. Vols. I and II. Kendall/Hunt, Dubuque, IA.
83. Spitalny, K. C., E. N. Okowitz, and R. L. Vogt. 1984. Salmonellosis outbreak at a Vermont hospital. South. Med. J. 77:168–172.
84. Stewart, P. J., W. Desormeaux, and J. Chene. 1983. Hemorrhagic colitis in a home for the aged—Ontario. Can. Dis. Wkly. Rep. 9:29–32.
85. Sweet, L., R. Davies, E. MacLeod, L. P. Abbott, and G. Feetham. 1991. Control of

Escherichia coli O157:H7 in nursing homes—Prince Edward Island. Can. Dis. Wkly. Rep 17:63–7.

86. Tavris, D. R., R. P. Murphy, J. W. Jolley, S. M. Harmon, C. Williams, and C. L. Brumback. 1985. Two successive outbreaks of *Clostridium perfringens* at a state correctional institution. Am. J. Epidemiol. 75:287–288.
87. Taylor, J. L., D. M. Swyer, C. Groves, A. Bailowitz, et al. 1993. Simultaneous outbreak of *Salmonella enteritidis* and *Salmonella schwarzengrund* in a nursing home. J. Infect. Dis. 781–782.
88. Telzak, E. E. L. D. Budnick, M. S. Z. Greenberg, S. Blum, M. Shayegani, C. E. Benson, and S. Schultz. 1990. A nosocomial outbreak of *Salmonella enteritidis* infection due to the consumption of raw eggs. N. Engl. J. Med. 323:394–397.
89. Thomas, M., N. D. Noah, G. E. Male, M. F. Stringer, M. Kendall, R. J. Gilbert, P. H. Jones, and K. D. Phillips. 1977. Hospital outbreak of *Clostridium perfringens* food-poisoning. Lancet i:1046–1048.
90. Todd, E. C. D. 1985. Economic loss from foodborne disease outbreaks associated with foodservice establishments. J. Food Prot. 48:169–180.
91. Todd, E. C. D. 1989. Preliminary estimates of costs of foodborne disease in Canada and costs to reduce salmonellosis. J. Food Prot. 52:586–594.
92. Todd, E. C. D. 1997. Foodborne and Waterborne Disease in Canada, 1990–1991. Polyscience Publications, Morin Heights, Quebec.
93. Todd, E. C. D. 1998. Foodborne and Waterborne Disease in Canada. 1992–1993. Polyscience Publications, Morin Heights, Quebec.
94. U.S. Department of Health and Human Services. 1996. Surveillance for foodborne-disease outbreaks—United States, 1988–1992. MMWR Surveillance Summary SS-5.
95. U.S. Department of Health and Human Service. 1999. Food Code. Public Health Service, Food and Drug Administration, Washington, DC.
96. Wall, P. G., M. J. Ryan, L. R. Ward, and B. Rowe. 1996. Outbreaks of salmonellosis in hospitals in England and Wales: 1992–1994. J. Hosp. Infect. 33:181–190.
97. Wood, R. C., K. L. MacDonald, and M. T. Osterholm. 1992. *Campylobacter* enteritis outbreaks associated with drinking raw milk during youth activities. JAMA 268: 3228–3230.
98. World Health Organization. 1989. Health Surveillance and Management Procedures for Food Handling Personnel. WHO, Geneva.
99. World Health Organization. 1997. Prevention and control of enterohaemorrhagic *Escherichia coli* (EHEC infections). Report of a WHO Consultation, Geneva, April 28–May 1, 1997.
100. Yankos, P., and J. McAnulty. 1996. Gastroenteritis outbreak linked to food handler. N. South Wales Public Health Bull. 7:101, 107–8.

10
Food Safety in the Home

Elizabeth Scott
Consultant in Food and Environmental Hygiene, Newton, Massachusetts

I. INTRODUCTION

It has been said that the last line of defense against foodborne illness is in good food hygiene practice in the home. This comes as a surprise to many consumers, who are largely unaware that the domestic kitchen can be at the center of food poisoning incidents. Consumers do play an essential role in the prevention of foodborne illness and require clear and consistent information on all aspects of food handling, from selecting food at the store to dealing with leftovers, constituting what is in effect a food hygiene advisory for the home.

A. The Data

Traditionally, it has generally been assumed that most cases of foodborne illness result from the consumption of contaminated food served outside the home, for example, in restaurants and takeout establishments. However, in contrast to this assumption, many countries have recently noted the high proportion of outbreaks affecting single homes (23). Much of the available data refers only to the incidence of illness caused either by *Salmonella* or *Campylobacter*, the two organisms responsible for most foodborne illness. It is widely accepted that because of underreporting, the real figure for foodborne illness may be many times higher than is indicated by the recorded data. The World Health Organization has estimated that in most European countries only 10% of incidents that occur are reported (31). In the United States, the numbers of sporadic illness are estimated to be 30–100 times higher than those reported in outbreak data, and sporadic illness is mostly related to food prepared and eaten at home (11). Unreported cases usually represent persons with mild illness who treat themselves or, if they consult a doctor, for some reason their condition is not reported. Outbreaks of foodborne illness occurring in private homes are less likely to be reported than those occurring in commercial and public premises.

Data for England and Wales, published in the Communicable Disease Report Review (28), indicates that a total of 2766 *Salmonella* outbreaks were reported between 1989 and 1991, 86% of which were classed as family outbreaks (i.e., only members of a single household were affected). A total of 252 outbreaks of food poisoning due to other bacterial causes were also recorded, and only 31 of these affected members of single households. In the same period, a total of 1097 *Campylobacter* outbreaks were recorded, 97% of which were classed as family outbreaks. The authors noted that increased reports in family outbreaks of *Salmonella* and *Campylobacter* infection reflect an improved reporting system. In addition, it should be noted that not all outbreaks classed as "family outbreaks" are necessarily accurately defined. Despite these improvements, it is widely accepted that home-based outbreaks remain underreported.

Data for Scotland, provided by the Communicable Diseases (Scotland) Unit, indicate that the most common place of food consumption leading to illness is the private household, and this is also the most common known place in which contamination or mishandling of food occurred (31). Detailed investigations carried out in Spain show that private homes are the location of nearly 50% of all outbreaks (31).

Questionnaire data collected during a sentinel study in the Netherlands suggest that 80% of *Salmonella* and *Campylobacter* infections arise within the home (13). Likewise, trends in the data from Germany (15, 26, 31) and from France (12) indicate that single cases and household outbreaks also play a major role in the foodborne illness burden of those two countries. Data from Italy (27a) indicate that private homes are the location of more than 70% of *Salmonella* infections.

Data from Canada for 1990 and 1991 identifies the home as the place in which food was mishandled in 18–22% of known microbiological incidents (30). These figures exclude the data for unknown causes and therefore may be an underestimate.

In the United States, estimates for foodborne *Campylobacter* and *Salmonella* total more than 4 million cases annually (6), and such infections are among the most common causes of foodborne illness in the country. Many experts now estimate that more than 50% of cases of foodborne illness originate from foods prepared at home. In a New Jersey study of the so-called "hamburger disease," caused by *E. coli* O157, it was found that 80% of suspect hamburgers were eaten at home and food preparers in those homes were significantly less likely to report washing their hands or work surfaces (16a). Other factors, such as the popular practice of home canning, have led to a situation in which approximately 92% of the 231 outbreaks of botulism in the United States between 1973 and 1987 were associated with food prepared in the home (3).

In addition to the toll of illness and death, there is a huge economic burden associated with foodborne illness. For example, the annual cost of foodborne illness to the U.S. economy approaches $10 billion (1). These costs are borne not only by national agencies and industry, but also by the individual at home. The proportion of total costs that are directly attributable to the occurrence of foodborne disease in the home is not yet known, however. The actual costs per episode resulting from breakdowns in hygienic practice in the home are likely to be less than similar breakdowns in the food and catering industry. Nevertheless, total costs related to illness associated with the home must still be substantial and would include items such as cancelled activities (weddings and vacations) and unforeseen expenses (medical bills; costs of special foods and cleaning materials), as well as loss of income (27). In addition, it has been shown that just over half of the cost to the national economy in England and Wales is due to lost production resulting from absence from work because of illness or time off work to care for relatives stricken by foodborne illness (27). These costs are incurred regardless of the place of origin of illness, indicating reductions in the occurrence of foodborne disease in the home could bring about significant reductions in these particular costs. For example, a detailed analysis of the costs of foodborne disease in Canada and the United States (29) compared the costs for incidents of specific foodborne diseases arising from mishandling at a range of settings including the home, with incidents arising from mishandling at food processing establishments. Data relating directly to the home setting were not given, but the costings for incidents of illness involving *Campylobacter jejuni* are of interest because this organism is usually associated only with sporadic incidents of the type that occur in the home. Therefore, the costs are probably similar to those that are incurred when illness arises within the home, in that they do not involve the legal costs and the losses to food companies that occur when faults are attributed to food processing establishments. The average cost per case of *Campylobacter* infection in the United States in 1989 was found to be $1331. This cost was generally determined from factors such as the incident investigation costs, loss of income due to illness, medical costs, and loss of business (29).

B. Consumer Concerns and Knowledge

Consumers around the globe are beginning to register their concerns about the microbial quality of their food. In a 1996 survey of consumers in Australia, 73% responded that microbial contamination of foodstuffs was the primary threat to public health, ahead of such issues as chemical contamination (2). Public concern in the United Kingdom, America, Japan, and Australia is underscored by regular articles concerning food hygiene and foodborne illness in the media, with both regional and national coverage. But despite their broader concerns, consumers remain largely unaware of their own role in contributing to foodborne illness, and although some may be better informed today, a 1990 survey of British consumers reported that 20% of respondents knew that foodborne illness occurs in the home (18), compared to 11% in 1988 (17). In a 1991 survey of American consumers, only 16% of particpants thought that food safety problems were most likely to occur in the home (32). The proportion of respondents who identified the home in a 1993 survey remained almost the same at 17% (11). These data indicate that perhaps less than one-fifth of the population is aware of importance of proper food hygiene in the home to prevent foodborne illness.

Attempts to inform and educate consumers on the subject of foodborne illness are unlikely to be successful unless the current level of consumer knowledge, as well as preexisting perceptions and cultural practices, are addressed. For example, without an awareness of the size of the foodborne illness problem associated with the home, consumers will not be sufficiently alerted to receive and act upon the relevant information. In a 1995 survey of American consumers (1), 80.2% of the total sample claimed to have heard of *Salmonella* spp., whereas only 4.7% had heard of *C. jejuni.* Clearly, anyone desiring to convey information and advice concerning the prevention of *Campylobacter* infection would need to be aware of the low level of public recognition.

Repeated consumer surveys (11) have indicated that consumers perceive foodborne illness as a minor sickness characterized by gastrointestinal upset without fever that comes on within a day of eating a contaminated food, most likely meat from a restaurant. Lacking from this perception is a recognition of the range and seriousness of possible symptoms, the long latency period associated with some pathogens, and the possibility that such infections can be acquired in the home.

The same survey (11) identified a subgroup of consumers who thought they had experienced a foodborne illness as those most likely to be interested in food safety measures. This group consisted of people under the age of 40 who had some college education. Paradoxically, the same subgroup was also more likely to practice risky eating behaviors, such as the consumption of raw foods of animal origin.

There also appear to be conspicuous gaps between knowledge of food handling principles and the practice of safe food hygiene, especially in relation to cross-contamination. In one survey of 1620 adults, 86% of respondents noted hand washing in general as important in decreasing the risk of food poisoning, but only 66% claimed to wash their hands after handling raw meat and poultry (1). Similarly, 80% of respondents knew that putting cooked steak back onto a plate that had held the raw steak increased the risk of food poisoning, but only 67% cleaned a cutting board after contact with raw meat or poultry. On the other hand, this disparity between knowledge and practice was not observed with respect to adequate cooking of meat and poultry.

Much of the consumer survey data suggests that for consumers to successfully receive and act on information on the nature and prevention of foodborne illness, they need some knowledge of basic microbiology (1, 21, 22). Simple and repeated lists of the do's and don'ts of safe food practice will not make any lasting impact on consumer behavior unless consumers have some understanding of the underlying principles involved. For instance, many consumers are not aware that raw food is probably the main source of bacterial contamination in the kitchen (22). Thus, the importance of many of the essential instructions regarding storage, temperature control, cooking, and the prevention of cross-contamination is not appreciated. In addition, many people do not realize that their bodies can be host to pathogenic microbes, and thus a source of contamination to food during handling and preparation, even when an individual is not ill or has recently recovered from gastroenteritis. As a result, the need for such measures of personal hygiene as hand washing before handling and preparing food is not fully understood (22).

C. The Potential for Inappropriate Food Practices at Home

Although the national data from various countries suggest that 50% or more of foodborne outbreaks occur within the home, there are few data at this time to indicate what proportion is attributable to poor food practice. While at least some of the home-based outbreaks are likely to have resulted from food prepared elsewhere and brought into the home, there are indications that inappropriate practices in the domestic kitchen may also be significant (26). Bryan (5) listed the use of contaminated foods, inadequate heat processing, and obtaining food from an unsafe source as the three leading factors contributing to foodborne illness in the home. While errors in time–temperature control probably constitute the most common contributory factor for food poisoning (5, 20), there is growing evidence that cross-contamination may also be an important factor. According to recent data from England, cross-contamination was a contributory factor in 36% of general outbreaks (9).

Various workers have reported on the potential for cross-contamination in the domestic kitchen. Several studies report on cross-contamination involving kitchen surfaces, hands, utensils, and dishrags (10, 24, 25). The potential for cross-contamination with *Salmonella* and *Campylobacter* organisms during the preparation of various foods in the domestic kitchen has also been investigated. These studies have surveyed the preparation of chicken (7, 8) and shell eggs (14), and the cross-contamination of cooked beef and melon from improperly cleaned surfaces (4).

II. PRACTICAL ADVICE AND INFORMATION FOR CONSUMERS

The following practical advice and information on safe handling of food covers the whole consumer food chain, beginning with shopping for food and transporting it home, through storing food, preparation and cooking, serving, and disposal of leftovers or tainted food. This information is similar to that offered to employees in restaurants and brings home food preparation more into line with food preparation in the foodservice sector. The information should be offered to consumers in conjunction with the supporting microbiology where appropriate.

In addition, consumers should be advised that the practice of good food hygiene at home may be compromised by the multifunctional role of the domestic kitchen. Often, the home kitchen is not solely a place for food preparation but also serves as the central hub of domestic life. However, such activities as the cleaning of pet tanks at the kitchen sink and emptying diaper pails at the kitchen sink, and the presence of pets on kitchen counters can result in a heavy inoculation of pathogens directly into the kitchen environment.

A. Shopping for Food and Transporting it Home

Safe food handling begins at the store, where food is selected and purchased. Usually foods are of high quality, but some are perishable and may contain pathogens, especially produce and raw foods of animal origin. After purchase, what happens to the food is largely determined by the consumer. Consumers should be encouraged to inform the store manager (and the local food inspector, if necessary) about any improper practices or spoiled foods they observe as they shop.

Food stores should look clean and well maintained, and there should be no spoiled food around, either inside the store or outside around the trash area. The store should be free from flies, other insects, and of course rodents. Apart from guide dogs, all pets should be kept out of food stores. The stores should be well lit and air-conditioned in hot climates. All food should be kept off the floor.

Staff should be clean and tidy in appearance, with clean uniforms. In some areas of the store such as the delicatessen (deli) counter and the bakery, workers may be required to wear a hat and disposable gloves as well as aprons, to protect the food from contamination. Staff should not have open wounds or sores on the hands or face—all wounds should be covered. In most jurisdictions, smoking in a food store is prohibited.

1. Fresh Produce

Unlikely as it may seem, even fresh produce such as melons, lettuce, and tomatoes may be contaminated with foodborne pathogens and may cause illness. Wrap loose produce in a bag to prevent any possible contamination from migrating to other foods in the basket or cart.

2. Fresh Meat

Usually fresh meat is either purchased ready-weighed and wrapped in plastic food wrap on a polystyrene tray from refrigerated shelves or is selected from a refrigerated meat counter and weighed and wrapped by a meat clerk. Only meat that has been kept refrigerated (or frozen) at less than 40°F (5°C) should be purchased—there should be a thermometer visible along the shelves.

When buying prepackaged meats, look for the date stamp. Always buy within the "sell-by" or "best before" date range (these terms differ by country); ground beef should be ground on the day of purchase. Meat packaging often leaks, and it is important to remember that the juices are likely to be contaminated. Therefore, it is best to avoid getting the meat juice onto hands or onto other foods in the basket or cart.

In addition to the quality of the packaging and the "sell by" or "best before" date, appearance and smell are important indicators. Do not purchase meat that has an unpleasant odor. Roasts of beef should be a bright cherry red, and although the outer surfaces of ground beef and steaks are often brown, the inside should be pink to red. Lamb should be light red, and pork should be pink with white fat, with a firm texture. Reject any meat that is brown or greenish or has brown, green, or purple blotches or black, white, or green spots. Slimy or sticky texture is another cause for rejection. Poultry should have no discoloration and should have a firm texture, with no stickiness under wings and around joints. Bad odor and discoloration on meat and poultry are all signs that the food is spoiling, either because it is not fresh or because it has been poorly handled, packaged, and/or stored. These signs do not in themselves indicate the presence of food poisoning organisms, but they certainly indicate that the store is not providing a wholesome product, and any pathogens present could have multiplied.

3. Seafood

Seafood can cause illness if it is contaminated with pathogens from polluted waters or seafood toxins. Fresh fish, shellfish such as clams, mussels, and oysters, and fresh Crustacea such as lobsters and shrimp should be displayed at cool temperatures [< 40°F (5°C)]. Seafood is highly perishable and is usually displayed packed with ice to keep it as fresh as possible.

Fresh fish should have no fishy odor; the eyes should be bright, clear, and full, and the texture of the flesh and belly firm. The shells of live molluscan shellfish should be closed, and the shells of lobster and shrimp hard. Fresh shellfish and Crustacea do not have a strong, unpleasant smell.

Risks of illness from seafood toxins cannot be predicted from the appearance of the fish, but the manager should be knowledgeable about the source of the fish and shellfish, and consumers should ask for this information if they have any concerns.

4. The Deli Counter

Although many of the foods at the deli counter are ready-cooked and/or ready-prepared, most are classified as high risk because they will not be cooked again and they support the growth of bacteria. Foods such as cooked cold meats and ready-prepared potato and pasta salads are very vulnerable to contamination from food handlers themselves. Food at the deli counter should be (a) displayed in a refrigerated cabinet [≤ 40°F (5°C)], (b) protected by a glass screen, and (c) completely separated from any raw foods, which could contaminate the cooked foods. The deli clerks should use disposable gloves and serving spoons instead of handling the foods directly.

Many deli counters offer freshly cooked, spit-roasted chicken for sale. Such chicken should either be eaten or refrigerated within 2 hours of being taken off the spit. Some stores label the cooked chicken with a 2-hour time stamp.

5. Salad Bars

Salad bars contain many of the high risk foods found at the deli counter but with the added risk of a self-service system. Self-serve systems in stores are vulnerable to misuse, and it is almost impossible for the store to maintain control over what happens. The food in the salad bar should be maintained at a refrigerated temperature of 40°F (5°C) or lower. Any food that is piled high above the container will probably be at ambient temperature, and only the food within the walls of the container will remain cold. There should always be a plastic sneeze guard or shield in a direct line between a customer's face and the food. Tongs or a long-handled ladle should be available for each item so that customers do not touch the food with their hands.

6. Bakery

Breads and baked cookies, cakes, pastries, and pies are generally to be considered low risk foods because the high temperature cooking process destroys all vegetative cells. If problems do occur with these products, it is usually as a result of something going wrong after the product has been baked, such as cross-contamination during handling. Some problems have also occurred because raw egg glazes or meat fillings were not cooked thoroughly. There are also concerns for icings and cream fillings that are uncooked and for partially cooked meringues. Self-service bins containing items such as unwrapped bread, rolls, and bagels are a popular feature in many stores. Just as at the salad bar, self-service may be a potential problem because customers are able to handle unwrapped foods and may contaminate the remaining foods by direct contact with their hands, or by coughing and sneezing. The best practice is for the bins to be so high off the ground that people will not be bending over them, and each bin should have its own tongs, which must be inserted into the bin through a hole just large enough for the bread item to pass through.

7. Refrigerated Cabinets

These days refrigerated cabinets contain a huge variety of packaged processed foods ranging from milk, butter, yogurt, cheese, cream, and eggs to sliced meats and hotdogs, smoked salmon, tofu, and orange juice. These foods should be kept at refrigerated temperatures [$\leq$ 40°F (5°C)], and the cabinets should be kept clean and neatly stacked. Items with damaged or leaky packaging should not be purchased. All these foods have a date stamp, either a "sell by" date or a "use by" date. Egg cartons containing cracked or dirty eggs should be rejected. Eggs should be less than 2 weeks old and should show "use by" dates; see Chapter 3 for more details.

8. Freezer Cabinets

Reject any frozen items with damaged packaging, and check for the "sell by" or "use by" dates. Thawing and refreezing are major dangers for frozen foods because bacteria can multiply during thawing and survive the refreezing. Telltale signs are large ice crystals, solid areas of ice, discolored or dried-out food, and badly deformed cartons or packages.

9. Canned, Dried, and Bottled Foods

Foods in the canned section are typically well processed and are classed as low risk foods. Although spore formers may survive the processing, as long as the foods are stored correctly, microbes will not grow because the medium either is dry, contains preservatives, has a high acid or sugar content, or has been baked or subjected to a high temperature canning or bottling process. Even so, these foods require careful storage, and all will have a recommended "use by" date.

Any cans with swollen sides or ends, damaged seals or seams, rust, dents, leaks, or foamy or bad-smelling contents must be rejected; see Chapter 11 for more details.

Dried-packaged foods, such as cereals and other grain products, sugar, flour, and rice, dried fruits, and vegetables, should be in dry, undamaged packaging. Dampness or mold is a sign that the food is spoiling and should not be eaten. Holes or tears may be signs of damage caused by insects or rodents.

10. The Checkout

To avoid excessive warming between the store and home, especially in warm climates, it is advisable to have all frozen, refrigerated, and high risk items put straight into a cooler or insulated plastic grocery bags. Alternatively, in the absence of insulated boxes and bags, bagging all the cool and cold items together makes it is easier to put a priority on getting these items unpacked and into the refrigerator or freezer at home.

11. Transporting the Food Home

It is important to transport food home as quickly as possible, especially in the summer. On hot days the trunk of a car can reach temperatures in the danger zone, and the temperature of the food itself will quickly begin to climb if left in these conditions. Using a cooler will allow more time to get the food home safely.

B. Storing Food at Home

The first decisions that have to be made in the domestic kitchen concern the appropriate storage of food. Of priority is the storage of high risk foods in refrigerators and freezers. Many consumers wrongly believe that low temperatures kill microbes, and this mistake leads to confusion about how long food can be kept in the refrigerator, cross-contamination prevention in refrigerators, safe thawing, and other issues such as when to refreeze.

In the kitchen, food storage can be divided up into the following areas: dried and canned foods and bottled foods and drinks, chilled foods, frozen foods, fruits and vegetables, and breads and pastries.

1. Chilled and Frozen Foods

Newly bought frozen and chilled foods should be returned to a freezer or refrigerator as quickly as possible: Once these foods begin to thaw and warm, any bacteria that are present start to multiply rapidly. These bacteria will not necessarily be killed by any subsequent chilling or even freezing. Frozen foods that have completely thawed should never be refrozen until they have been thoroughly cooked. Foods such as ice cream and frozen desserts should not be refrozen if they thaw completely.

2. Freezer and Refrigerator Temperatures and Thermometers

Refrigerators should be maintained at 40°F (5°C) or less, and the only way to check the temperature is by using a refrigerator thermometer. Very few domestic refrigerators have a built-in thermometer, and the dial for adjusting the temperature to "cooler" or "less cool" usually does not have a readout showing a numerical value. The temperature inside a refrigerator can rapidly rise to almost room temperature under certain circumstances—for example, if the door is left open while the refrigerator is being loaded on a hot day. It may also take many hours for the temperature to cool down again. Normally, the coldest area in the refrigerator is at the back and the warmest area is by the door. Refrigerator thermometers and combined freezer/refrigerator thermometers can be obtained from some kitchen stores for about $12 and from restaurant supply stores for considerably less. Domestic freezers operate at a temperature of 0°F (−18°C).

3. Refrigerated Storage

Store cooked and ready-to-eat foods above raw foods to avoid cross-contamination from drips or by direct contact.

Overwrap packages of raw meat, poultry, or fish, or place them on a dish before refrigerating so that their juices will not drip and cause cross-contamination.

Check foods for expiration dates.

Never line the refrigerator shelves. This cuts down on air circulation necessary for proper cooling.

Freeze meat, poultry, and seafood if they are not to be used within 2–3 days.

Some foods such as jams, jellies, and mayonnaise containing low levels of preservative require refrigeration after they have been opened. Read the storage instructions carefully.

Store eggs in the main area of the refrigerator, where it is coldest, not in the door.

Store salads, green vegetables, and fruit in the salad boxes.

Place a refrigerator thermometer near the front, as this tends to be the warmest area. Temperature should be maintained at 40°F (5°C) or less.

4. Freezer Storage

Foods placed in the freezer should already be frozen or chilled.

Thawing and refreezing damages food quality and can cause a food poisoning hazard, since thawing gives microorganisms an opportunity to multiply to dangerous levels.

Overwrap foods such as breads in freezer bags to prevent drying and freezer damage.

5. Safe Thawing

There are three ways to safely thaw frozen food at home.

1. *In a refrigerator.* Place the frozen item on a tray or dish on the bottom shelf to prevent the juices from dripping or splashing. Allow a day or more to thoroughly thaw large items such as turkeys and roast.

2. *As part of the cooking process.* This method is used only for small items such as vegetables, shrimps, and pie shells. Allow a longer cooking time for the frozen items. This method is not safe for big items and is not recommended for hamburger patties, which may not be thoroughly cooked. (However, some store-bought items (e.g., frozen turkeys) can be safely cooked in the oven if the instructions are followed.)
3. *In a microwave.* Food should be thawed in the microwave only if it is going to be cooked immediately. This method is not safe for big items (see Sec. II, D. 1. c).

6. Dried, Canned, and Bottled Foods

Dried, canned, and bottled food and drink can be stored for long periods at room temperature and out of sunlight.

Canned foods and many bottled foods need to be refrigerated once opened. Look for storage advice on the label.

Foods should not be left in unlined open cans, since oxidation of the metal can cause food taint and possible tin poisoning.

Keep dried fruits and vegetables, cereals, pastas and other grain products, sugar, flour rice, cookies, and snack foods in their own packaging or transfer to clean, dry, airtight containers.

Keep foods dry. Water and high humidity may cause food spoilage by allowing bacteria and mold growth.

Regularly check the packaging for expiration dates.

7. Fruits and Vegetables

Fresh fruit and vegetables have a limited shelf life, but proper storage will help to keep them in good condition.

Avoid warm, moist storage conditions and condensation, which will encourage bacterial and mold spoilage. Open plastic wraps to allow air circulation.

Store potatoes in a cool, but unrefrigerated dry place and away from light.

Most fruits and vegetables can be stored in the refrigerator.

Tropical fruits such as bananas and pineapples are best stored in a cool place [50°F (10°C)] but not at refrigerator temperatures.

8. Breads and Pastries

Fresh bread, pastries, and confectioneries have a very short shelf life before they become stale.

All should be stored in a cool, dry place.
Condensation inside the protective wrapping will quickly cause bread to become moldy.
Any items containing fresh or synthetic cream, custard, egg custard, or any meat products should be refrigerated and consumed within 2–3 days.

9. Advice on When to Reject Food

Sometimes decisions have to be made on whether to reject a food. Many foods can become spoiled for a number of reasons, such as damage to the packaging, poor storage conditions, and aging. Spoiled foods can be easily identified by signs such as discoloration, the presence of mold, which may have a slimy or hairy/fuzzy appearance, and/or bad or "off" odors and fermentation (gas bubbles and a "yeasty" smell). Reject and discard any damaged containers (especially canned and bottled foods, where damage and spoilage may be an indication of the presence of *Clostridium botulinum*) and any foods that show signs of spoilage. People often simply remove the layer of mold from a cheese or a jam and eat the rest of the food. This practice is not recommended because toxins from the mold can spread into the food beyond the area of visible mold. Unfortunately, food that contains pathogenic organisms does not necessarily show any signs of their presence, and there is not necessarily any relationship between the presence of food spoilage organisms and pathogens. If there is any doubt, the safest action is to discard the food without tasting it.

C. Food Handling and Preparation

Safe food handling and preparation is another very important part of reducing the risk of causing or acquiring food poisoning. With the knowledge that so many raw foods coming into the kitchen can be contaminated with pathogenic microbes, it is essential to prevent these organisms from getting onto other foods (cross-contamination), or onto hands and directly into mouths. Consumers need to understand that raw foods are likely to be contaminated and that cross-contamination is a contributory factor in foodborne illness.

Greatest care should be taken with raw foods of animal origin: that is, all meat, poultry, fish, and eggs, and including meat and fish juices. The safest way to deal with these foods is to assume that they are contaminated and always take the same precautions.

Hands should be thoroughly washed before and after handling raw foods. Choose single-use disposable paper towels instead of multiple-use hand towels.

Food handlers should not smoke, drink, or eat while preparing food, since by so doing they run the risk of transmitting pathogens directly to their mouths.

Avoid tasting even small quantities of raw foods of animal origin during preparation, including cake and cookie mixes and batters containing raw egg.

Use pasteurized eggs for recipes in which the eggs will not be cooked thoroughly: for example, in meringues, mousses, Caesar salad dressing, hollandaise and béarnaise sauces, eggnog, and mayonnaise.

Refrigerate prepared foods, especially those of animal origins, including egg batters, if there is a delay between preparation and cooking.

Refrigerate foods that are being marinated before cooking.

1. Preventing Cross-Contamination

Ensure that all surfaces, utensils, and cooking equipment are thoroughly clean.

Wash all kitchen items in hot soapy water or in the dishwasher after they have been in contact with raw foods.

Avoid splashing eggs or egg whites during preparation, as this can cause cross-contamination. Clean and sanitize any splashes.

Clean and sanitize all kitchen surfaces that have been in contact with raw foods. Use disposable paper towels for cleaning and sanitizing surfaces whenever possible to reduce the risk of cross-contamination via dishcloths and sponges.

If paper towels are not available and multiple-use sponges and dishrags are used instead, these need to be disinfected regularly, and always immediately after contact with raw foods and their juices, either by putting them through the dishwasher cycle, microwaving them for one minute (19), or sanitizing them with a bleach solution or a registered kitchen antibacterial cleaning product.

Wash fruits and vegetables under a cold running tap at the sink before preparation.

Keep the preparation of raw foods separate from other food preparation.

Use three separate cutting boards: one for raw meat, poultry, and fish, one for vegetables, salads, and fruits, and one for nonperishable items such as bread.

2. Cutting Boards

There has been much discussion in recent years on the use of wooden cutting boards versus plastic boards. It has been the practice to recommend the use of hard plastic boards and to avoid the use of wooden cutting boards. This advice is based on the fact that plastic boards are less easily scratched and damaged and more easily sanitized in a dishwasher than wooden boards. Deep scratches and gouges in any type of board can become contaminated with bacteria and food particles. Plastic boards are used almost exclusively in the commercial sector. However, research data do not support the common notion that wood surfaces are more likely than plastics to produce cross-contamination of foods (19). In addition, it was found that wooden boards can be effectively disinfected by microwaving at high setting for 4 minutes, whereas plastic, which is relatively inert to microwaving, cannot be disinfected successfully in the microwave. Undoubtedly these findings have led to some confusion and uncertainty. Perhaps the best advice is still that any boards used should be free from deep scratches; separate boards should be used for different foodstuffs; and all boards should be regularly sanitized, especially after contact with raw foodstuffs (16). Disinfection methods include the use of bleach solutions and registered antibacterial kitchen cleaners, the dishwasher for plastic boards, and microwaving for wooden boards.

D. Cooking and Serving

Cooking food to the right temperature is the best method of ensuring that any pathogens that might be present are destroyed, or at least reduced to such low numbers that the food is safe to eat. However, if toxins are already present in the food, cooking may not destroy them. Following cooking, it is important to ensure that the food is handled and served safely to prevent it from becoming contaminated again. Sources of pathogens for cooked foods include dirty utensils and preparation surfaces, unwashed hands, coughs and sneezes, pets, and pests such as flies. As a general rule, it is best to handle food as little as possible once it has been cooked; that is, cooked food should be either served quickly or cooled rapidly and refrigerated.

1. Cooking

Consumer cookbooks and some food safety guides offer a confusing range of cooking temperatures, and I believe that it is important to rationalize and simplify crucial information such as temperatures so that they can be easily remembered.

The only way to know the internal temperature of food is by using a thermometer. Instant-reading thermometers are the most accurate and can be used to check the temperature of anything that is cooked, either in the oven, on the stove top, on the grill or in the microwave. In addition, because they are calibrated to record temperatures between 0 and 220°F (32–104°C), the same instrument can also be used to check the temperatures of cold food.

Unlike the less accurate "meat thermometers," which are left in a roast as it cooks, the instant-reading thermometer is plunged into the thickest part of whatever food is being cooked until it hits the pan or a bone, and then withdrawn slightly. After 15 seconds the dial should register an accurate reading. Similar thermometer are now being recommended for cooking hamburger patties. The thermometer must be removed as soon as a reading has been obtained, or it will be damaged. After use, the thermometer should always be sanitized to prevent cross-contamination from one food to another via the thermometer. This can be achieved by using an alcohol wipe (available from pharmacies) or by wiping the stem with an antibacterial cleaner or diluted bleach and rinsing under the tap. These thermometers can be bought from kitchen stores for about $12; again, lower prices may be found at restaurant supply stores.

a. Internal Temperatures. The following foods should be cooked to internal temperatures of at least 160°F (70°C): **meat and poultry, stuffings containing meat products, leftovers, and stuffed meats and pasta.**

Fish should flake with a fork.
Eggs and egg dishes should be cooked until the yolks and whites are firm, never runny.
Cook stuffings separately from meat and poultry, and then stuff the cooked food with the cooked stuffing. Stuffing meat and poultry before cooking can insulate the internal surfaces and prevent thorough cooking.
Batters and breadings can also insulate the food they cover and prevent complete cooking. Items need to be thoroughly cooked.
Dispose of any leftover batter.

b. Cooking on the Grill. Cooking outdoors on the grill or barbecue is a very popular summertime activity and is often the center of a social event for families and friends. However, many cases of food poisoning have been traced to food hygiene errors occurring around grilling. Perhaps the very fact that barbecuing is a social event involving cooking outdoors leads people to forget that the same hygiene rules apply to this type of cooking activity as to all others. One of the major problems with grilling is that food is often overcooked on the outside but undercooked on the inside. Undercooked grilled meat, especially items such as chicken and hamburgers, are a food poisoning hazard, especially to children and other vulnerable people. The following guidelines for grilling and barbecuing will help to reduce the risks:

Marinate all food items in the refrigerator.
Keep food in the refrigerator until it is time to cook it on the grill.
Cook foods thoroughly—cut into pieces of meat or fish to check that they are cooked through and are not pink. Use an instant-reading thermometer to check the internal temperature of meat on the grill.
Thaw frozen foods before grilling, especially items such as beef patties.
Use clean utensils.

Put cooked items onto a *clean serving dish.* Do not put them back onto the dish of raw foods.

Do not allow leftover marinade from the raw foods to come into contact with cooked foods.

Teach children to ask for "well-cooked" burgers and to reject any that are pink inside.

c. *Microwave Cooking.* In microwave cooking, food molecules are caused to vibrate by the microwaves, producing friction and heat. The food cooks itself and continues cooking even when the power is off. Like conventional ovens, microwaves have hot and cold spots that may cause uneven heating, and this is why the food needs redistributing by turning or stirring if there is no built-in turntable. It is vital to ensure that the food is heated throughout to a high enough temperature to kill food-poisoning germs.

Microwaves work on varying power levels depending on the wattage of the oven. The most common power level is 650 W. The different power settings on most microwaves—high (100%), medium (50%), defrost (30%), and low (10%)—are analog representations of the amount of energy produced in bursts in a given time. This means that high is power coming on 100% of the cooking time, medium at about 50%, defrost 30%, and low at 10%.

Add a minimum of 25°F (14°C) to conventional internal cooking temperatures.

Rotate or stir midway through cooking to help spread the heat,

Let stand for 2 minutes after cooking so that all parts of the food heat to the required internal temperature (see above).

For roasting meat or poultry in the microwave, use a microwave-proof instant-reading thermometer; do not use a conventional metal meat thermometer. Combination ovens often have a built-in special temperature guide probe.

For cook–chill prepared meals and food, always follow instructions and timings on the packet to ensure that the food attains the required internal temperature. The same applies to reheating home-cooked foods.

Defrost power is about 30% output, but certain thinly sliced meats and fish may need low (10%) power to prevent the edges from starting cook to before the insides are fully thawed.

Small chickens and roasts thaw easily, but first place them on a dish with sheets of paper toweling to absorb the juices. Check inside poultry to see that it is free from ice before cooking. Discard the paper, and thoroughly wash the dish and your hands after handling the raw meat.

Continue with the cooking process as soon as the food has been defrosted.

Keep the internal surfaces of the microwave clean and sanitized.

2. Serving Food Safely

Once foods have been prepared, there are still precautions that should be taken to keep the food safe until it is eaten. The time between preparation and consumption is important because hygiene failures that occur at this stage will not be corrected by any further processing such as cooking.

Always wash your hands between preparation and serving.
Use only clean serving dishes and utensils for cooked/prepared foods to avoid cross-contamination.
All foods should be served within 2 hours of cooking/preparation; otherwise they should be cooled and refrigerated.
Protect food from being contaminated by contact with pets or pests.
Avoid touching the food; use serving utensils wherever possible.

a. Buffets. The serving of food buffet style brings a unique set of challenges to food safety. Hot food must be kept hot [≥ 140°F (60°C)] and cold food kept chilled [≤ 40°F (5°C)] throughout the duration of the buffet. In addition, the food must be protected from contamination resulting from poor service practices. Any combination of failures due to time–temperature abuse and contamination by the users may lead to contaminated food being held for a number of hours at warm temperatures, allowing massive proliferation of microorganisms. The maintenance of either hot or cold temperatures is essential for safe buffet-style serving.

Keep hot foods hot by using chafing dishes or warming trays. The minimum internal temperature should be 140°F (60°C). Use an instant-reading thermometer to check on this.
Do not keep food at 140°F (60°C) for longer than 2 hours. After this time, the food should be rapidly chilled and refrigerated or, if there was any doubt about whether the food had been maintained at a high enough temperature, any leftover should be discarded. It may seem wasteful to throw away apparently "good" food, but food that has been through a preparation and cooking process and then held at temperatures of less than 140°F (60°C) for 2 hours in a buffet (especially a self-serve buffet) may already be on the limits of safety for microbial contamination. Any further storage will allow the contamination levels to increase and make the food unsafe.
Cold food should be kept in dishes surrounded by ice. Replace the ice as it melts; otherwise the food will start to warm up. Alternatively, keep cold foods in the refrigerator until serving.
Keep the food covered and protected as much as possible.
For self-serve buffets, ensure that there are enough clean serving utensils

available so that people are not tempted to contaminate the food by serving it with their own used eating utensils.

b. Picnics. Picnics and packed lunches served away from home can be hazardous, especially if they involve high risk foods held at warm temperatures. The following advice can be offered for picnic preparation.

Chill all ingredients in the refrigerator before preparation.

Prepare foods as close to the time of consumption as possible, and chill foods once prepared.

Limit or avoid the use of high risk foods such as shellfish, cooked meats, chicken, egg dishes, raw eggs, and cream. Alternatives are spreads, peanut butter, cheeses, washed fruit, vegetables and salads, and dried fruit and nuts.

Wrap all items in food wrap or foil to protect them from contamination.

Pack picnics in insulated cool bags or boxes. Use ice or a freezer pack to maintain a low temperature.

Frozen juice boxes can be used as a freezer pack in children's lunch bags.

3. Cooling and Storing Cooked Foods

All cooked foods either should be eaten immediately or kept hot [140°F (60°C)] as in a buffet or they should be cooled and refrigerated within 2 hours of cooking. This ensures that any micorbial contaminants that survived the cooking process or were introduced after cooking do not have time to multiply to unsafe levels.

It can be quite difficult to cool foods rapidly in the home without the kind of equipment available in commercial kitchens. Cooling procedures that may be common but should be avoided include putting hot food directly into the refrigerator (causing everything else to heat up), leaving hot food cooling uncovered at an open window (food liable to contamination from sources such as pets and pests), and leaving hot food sitting in a pan on top of the stove or in a turned-off oven for hours (the food cools so slowly that any bacteria and their spores that survived the cooking process have time to recover and regrow).

The key to cooling food rapidly is to remember that the smaller the volume of the food and the larger its surface area, the quicker it will cool.

The aim is to reduce the temperature to 40°F (5°C) or less. Check temperatures with an instant-reading thermometer.

Soups and thick foods such as stew and chili can be cooled by pouring them into large, clean shallow containers surrounded by ice. Stir the food as it cools so that no warm spots are left.

Cut up large pieces of hot food into smaller pieces and divide large pans full of hot food into smaller, clean containers surrounded by ice.

Once the food has cooled, loosely cover the containers and place them on a shelf in the refrigerator so that air can circulate around them for further cooling.

a. Reheating Cooked Food. Consumers are often confused on the subject of reheating food and are not certain whether this is a safe practice. The safe practice is to reheat food once only and to ensure that it reaches a minimum internal temperature of 160°F (70°C) within a 2-hour cooking period. Food reheated in a microwave should reach a temperature of 185°F (85°C) and should be allowed to stand for 2 minutes after reheating, to permit the heat to spread evenly through the food. Ideally, any food that is not consumed after reheating once should be discarded.

b. Disposal of Unwanted Food. Food that is to be disposed of should be placed in the trash quickly so that it is not eaten by mistake. Food that is not fit for human consumption often is not good for pets either, and if it makes them sick, they may excrete the causative agent and transmit it to family members as a result of direct contact with the pet or with its vomit and feces. Food should be sealed in pest-proof trash containers to avoid attracting pests to the area. Kitchen trash should be emptied frequently.

III. THE FUTURE FOR CONSUMER EDUCATION

It is clear that the subject of food safety in the home is complex and much more than simply a matter of "common sense." In addition, changing and increasingly busy lifestyles, the advent of new and unfamiliar food technologies, and a growing population of vulnerable individuals who require "even safer" food combine to put further pressure on the domestic food preparer. While public health initiatives are to be commended, one of the most effective means of educating the community on issues of food hygiene is to be found at the primary and secondary school level. With the erosion or loss of home economics courses from many national educational programs, there may be no formal means of offering this subject in schools. In addition, practical advice on food preparation is often no longer passed along to children at home. If this trend is not reversed, the situation could become disastrous. With increasing concern in many countries about the future for food safety, as well as the huge national economic burdens of foodborne illness, it should be obvious that the most effective means of promoting food hygiene is via hygiene education programs in schools. The benefits of food hygiene education could include not only a reduction in the occurrence of foodborne illness at home, but also a workforce better equipped to apply hygiene principles and practice in the huge food industry and foodservice sectors of the national economies.

REFERENCES

1. Altekruse, S. F., D. A. Street, S. B. Fein, and Alan S. Levy. 1995. Consumer knowledge of foodborne microbial hazards and food-handling practices. J. Food Prot. 59:287–294.
2. Anonymous. 1996. Bacteria are Australian consumers' greatest food fear. World Food Chem. News, Nov. 27, p. 5.
3. Bean, N. H., and P. M. Griffin. 1990. Foodborne disease outbreaks in the United States, 1973–1987: Pathogens, vehicles and trends. J. Food Prot. 53:804–817.
4. Bradford, M. A., T. J. Humphrey, and H. M. Lappin-Scott. 1997. The cross-contamination and survival of *Salmonella enteritidis* PT4 on sterile and non-sterile foodstuffs. Lett. Appl. Microbiol. 24:261–264.
5. Bryan, F. L. 1988. Risks of practices, procedures and processes that lead to outbreaks of foodborne diseases. J. Food Prot. 51:498–508.
6. Council for Agricultural Science and Technology. 1994. Foodborne Pathogens: Risks and Consequences. Task Force Report 122. CASTS Ames, IA.
7. De Boer, E., and M. Hahne. 1990. Cross-contamination with *Campylobacter jejuni* and *Salmonella* spp. from raw chicken products during food preparation. J. Food Prot. 53:1067–1068.
8. De Wit, J. C., G. Broekhuizen, and E. H. Kampelmacher. 1979. Cross-contamination during the preparation of frozen chickens. J. Hyg. (Cambridge) 82:27–32.
9. Djuretic, T., P. G. Wall, M. J. Ryan, H. S. Evans, G. K. Adak, and J. M. Cowden. 1996. General outbreaks of infectious intestinal disease in England and Wales 1992–1994. CDR Rev. 4:57–64.
10. Enriquez, C. E., R. Enriquez-Gordillo, D. I. Kennedy, and C. P. Gerba. 1997. Bacteriological survey of used cellulose sponges and cotton dishcloths from domestic kitchens. Dairy, Food Environ. Sanit. 17:20–24.
11. Fein, Sara B., C.-T. Jordan Lin, and Alan S. Levy. 1995. Foodborne illness: Perceptions, experience, and preventative behaviors in the United States. J. Food Prot. 58: 1405–1411.
12. Guiguet, M., B. Hubert, and A, Lepoutre. 1992. Results of a one year surveillance of acute diarrhoea by general practitioners, in Proceedings of the Third World Congress on Foodborne Infections and Intoxications. Robert von Ostertag Institute, Berlin, pp. 72–75.
13. Hoogenboom-Verdegaal, A. M. M., and C. A. Postema. 1990. Voedselinfecties. Practitioner 5:549–554.
14. Humphrey, T. J., K. W. Martin, and A. Whitehead. 1994. Contamination of hands and work surfaces with *Salmonella enteritidis* PT4 during the preparation of egg dishes. Epidemiol. Infect. 113:403–409.
15. Kusch, O. W. D., and H. J. Klare. 1992. Possibilities and limitations of explanation of foodborne infections through investigations of food samples, in Proceedings of the Third World Congress on Foodborne infections and Intoxications. Robert von Ostertag Institute, Berlin, pp. 76–80.
16. Miller, A. J., T. Brown, and J. E. Call. 1996. Comparison of wooden and polyethylene cutting boards: Potential for the attachment and removal of bacteria from ground beef. J. Food Prot. 59:854–858.

16a. Mead, P. A. Finelli, L., Lambert-Fair, M. A., Champ, D., Townes, J., Hutwagner, L., Barrett, T., Spitalny, K. and Mintz, E. 1997. Risk factors for sporadic infection with *Escherichia coli* O157:H7. Archives of Internal Medicine. 157:204–208.
17. Ministry of Agriculture Fisheries and Food. 1988. Food Hygiene: Report on a Consumer Survey. Her Majesty's Stationery Office, London.
18. Mintel. 1990. Food Safety, Special Report. Mintel Publications, London.
19. Park, P. K., and D. O. Cliver. 1996. Disinfection of household cutting boards with a microwave oven. J. Food Prot. 59:1049–1054.
20. Roberts, D. 1986. Factors contributing to outbreaks of foodborne infection and intoxication in England and Wales 1970–1982, in Proceedings of the Second World Congress on Foodborne Infections and Intoxication. Robert von Stertag. Institute, Berlin, pp. 157–159.
21. Scott, E. 1983. Hygiene in the home—An appraisal of attitudes. Home Econ. 2: 67–70.
22. Scott, E. 1992. Food hygiene information and education—Is the consumer receiving the correct message? in Proceedings of the Third World Congress on Foodborne Infections and Intoxications. Robert von Ostertag Institute, pp. 949–950.
23. Scott, E. 1996. A review of foodborne disease and other hygiene issues in the home. J. Appl Bacteriol. 80:5–9.
24. Scott, E., and S. F. Bloomfield. 1990. The survival and transfer of microbial contamination via cloths, hands and utensils. J. Appl. Bacteriol. 68:271–278.
25. Scott, E., S. F. Bloomfield, and C. J. Barlow. 1982. An investigation of microbial contamination in the home. J. Hyg. (Cambridge) 89:279–293
26. Sockett, P. N. 1993. Foodborne statistics: Europe and North America, in Encyclopaedia of Food Science, Food Technology and Nutrition, Robert Macrae, ed. Academic Press, London, U.K., pp. 2023–2031.
27. Sockett, P. N., and J. A. Roberts. 1991. The social and economic impact of salmonellosis: A report of a national survey in England and Wales. Epidemiol. Infect. 107: 335–347.
27a. Scuderi, G., Fantasia, M., Filetici, E. and Anastasio, M. P. 1996. Foodborne outbreaks caused by *Salmonella* in Italy, 1991–4 Epidemiol. Infec. 116:257–265.
28. Sockett, P. N., J. M. Cowden, S. Le Baigue, D. Ross, G. K. Adak, and H. Evans. 1993. Foodborne disease surveillance in England and Wales: 1989–1991. CDR Rev. 3:159–173.
29. Todd, E. C. D. 1989. Preliminary estimates of costs of foodborne disease in Canada and costs to reduce salmonellosis. J. Food. Prot. 52:586–594.
30. Todd, E. C. D., and P. Chapman. 1997. Foodborne and Waterborne Disease in Canada. Annual Summaries 1990–1991 Health Protection Branch, Health Canada. Polyscience Publications, Quebec, pp. 28–31.
31. WHO. 1992. Surveillance Programme for Control of Foodborne Infections and Intoxications in Europe. Fifth Report, 1985–1989. FAO/WHO Collaborating Centre, Berlin.
32. Williamson, D. M. 1991. Home food preparation practice: Results of a national consumer survey. M. S. thesis, Cornell University, Ithaca, NY.

11
Canned Food Safety

Ashton Hughes
Health Canada, Ottawa, Ontario, Canada

I. INTRODUCTION

This chapter on canned food safety provides basic information on food safety issues relating to canned foods. Although it can be used by students of food microbiology, food science, food technology, and related disciplines, no previous knowledge of food microbiology is required. The chapter is intended to provide valuable information to alert the home canner and others involved in food preservation at the institutional level to the factors involved in food canning, with the goal of assuring a safe canned food product with minimal damage to organoleptic quality and nutritive value. The choice here is not to burden the text too much with theoretical principles, but to provide a more practical approach to the principles and practices required to preserve foods in hermetically sealed containers. The chapter should be a useful adjunct to the various publications on home canning, for although no recipes are published, the underlying safety principles that govern the safety of the various classes of canned foods are discussed.

"Canning" is a general term applied to the preservation of food through the application of heat. The canning process serves two objectives: to inactivate enzymes in the food that can degrade the food, with resulting loss of quality, and to eliminate pathogenic and food spoilage organisms. Traditionally, this involved the use of containers of two types, namely "tin cans" (in reality tin-coated steel) and glass jars. However, increasing use is being made of containers made of plastic composite material, so we no longer speak of canned food but of thermally processed food in hermetically sealed containers. Presently three types of hermetically sealed containers are recognized by the food industry: rigid, semirigid, and flexible containers (7). This chapter focuses primarily on home food preservation, and so only rigid containers (i.e., the three-piece metal can and glass jar) are discussed in detail. There are no reports to date of semirigid or flexible containers being used in home food processing, probably because this type of technology is not only expensive, but also technically very demanding. However, semirigid and flexible containers are briefly discussed. From the various publications on home food preservation, it would appear that the containers used are primarily glass jars, but since a few persons still preserve food in metal cans, containers of both types are discussed. General information on can construction and glass jar construction is presented next, since it is important that when cans or glass jars are being selected, only containers in good condition be used, to ensure that after double seaming of the cans or sealing of the glass jar, a hermetic container has been obtained. Further information on assessment of containers for defects is provided later (Sec. XI).

II. TIN PLATE CANS

The name "tin can" is really a misnomer. It was started shortly after such containers were first made in England, where they were originally called tin cannisters. Actually, the can is made of steel with only a very light coating of tin, and some modern cans contain no tin at all, hence often are referred to as tin-free steel cans. Tin plate is an ideal material for containers. Although tin is not completely inert to all foods, container corrosion and any product changes are slight when the correct type of coating is used.

A. Types of Can

Three types of tin can are used in home canning based on the type of tin plate used (19): plain tin, C-enamel (corn enamel), and R-enamel (sanitary or standard enamel). For most products plain tin cans are satisfactory. Enameled cans are recommended for certain fruits and vegetables to prevent discoloration of food, but the coating is not necessary to ensure the safety of the food. The three types of cans are recommended for different foods as follows:

Type of can	Recommended for
C-enamel	Corn, hominy
R-enamel	Beets, red berries, red or black cherries, plums, pumpkin, rhubarb, winter squash
Plain	All other fruits, vegetables, meat, and fish

B. Size of Can

Normally the tin plate can used in the home is cylindrical and is of the same type used in the food industry. In North America cans are designated by name, such as No. 2 can, or, in the case of cylindrical cans, by dimensions, such as 307 × 409. The diameter of a cylindrical can is given first, and the height of the can follows the "×." In the statement of each dimension, the first digit gives the number of whole inches, and the second and third digit give the fraction expressed in sixteenths of an inch. Thus 307 × 409 means that the can is 3 7/16 in. in diameter and 4 9/16 in. high. For home use, most fruits, vegetables, meat and fish are canned in No. 2 (307 × 409) or No. 2 1/2 (401 × 411) cans. A No. 2 can holds about 2 1/2 cups, and a No. 2 1/2 can about 3 1/2 cups (19).

The majority of cans used in the home today are still the cylindrical three-piece variety constructed from three basic parts: cylindrical body, bottom, and top. The cylindrical body is formed by overlapping the two ends of the body cylinder and either soldering or welding the resulting seam.

III. TWO-PIECE CANS

There are two basic parts to a two-piece can, the can body and the top. This is an advanced type of technology with the advantage that the can has no side seam, as found in the three-piece can, and there is only one end to be attached by the process of double seaming. Whether two- or three-piece cans are used, the most important factor is the formation of a reliable (leakproof) seal between the can body and the lid (top or bottom) through a process called double seaming, which involves the mechanical interlocking of the can lid (bottom or top) to the can body to form a hermetic seal (see Ref. 7, section on definitions).

IV. GLASS JARS

Glass containers of various types are used in the food industry, and when combined with the appropriate closure, the glass jar provides an inert, hermetic, durable packaging medium for a wide variety of foods. In addition, resealability and transparency make the glass jar a popular packaging medium for home food preservation. Home canning enjoys its popularity today as a result of the efforts of a tinsmith named John Landis Mason (3). This clever inventor patented a glass canning jar with a threaded opening that could be sealed with a metal cap and a rubber gasket. Up to that time, glass bottles and earthenware jugs were sealed with cork stoppers and wax, or tin containers were sealed with solder. This invention of Mason's greatly simplified home canning, making it easy and economical; this is what led to the popularity of home canning. While the term "Mason" was once a trade name, the patent on the original jar expired long ago, and "mason jar" is now a generic term.

The basic idea developed by John Mason is still used in several variations, but the only system recommended today is the "rounded square" or cylindrical jar and closure consisting of a two-piece system: a vacuum lid and a cap (3). The type of closure set for glass jars consists of a flat metal lid with a flanged edge, the underside of which has a rubberlike sealing compound (the vacuum lid). The cap is a threaded metal screw band that fits over the rim of the jar to hold the lid in place. The glass jars come in a variety of sizes and styles and are carefully made so that the closures will seal well; that is, when the lid is placed over the mouth of the jar, the sealing compound lines up directly over the rim of the jar. The glass in the jar is tempered to withstand the heat of the steam pressure canner or the subzero temperatures of the food freezer. The jar sizes in use today include, 236 mL or 1 U.S. half-pint, 473 mL or 1 U.S. pint, and 946 mL or 1 U.S. quart (2 pints). Also to be found are 500 mL, 1 L, and 1.5 L "rounded square" jars and 125 mL, 250 mL, 500 mL, and 1 L sizes cylinder-shaped jars. As in the case of metal containers, it is imperative that glass containers and closures be routinely inspected for defects, which could result in failure of the finished package. The glass and cap suppliers carry out their own sampling and inspection prior to shipment, but this does not alleviate the need for careful examination of closures and jars for defects. Examination for defects is covered in Sec. XI.

V. FLEXIBLE CONTAINERS

Flexible packaging for heat-processed foods has been an alternative to metal cans and glass jars since at least the mid-1950s (9, 14). Because of the complex requirements for filling, sealing, and processing, this technology is restricted at the present time to commercial applications and has no appeal to those engaged in home food preservation. Information presented here is primarily for general information for those interested in the technology, and details of the technology may be obtained elsewhere (6). Commercial application of this technology has been applied to acid foods (tomato-based sauces) as well as to low acid foods. Low acid foods packaged in flexible containers are heat-processed in specialized pressure vessels called retorts, and these pouch-shaped flexible containers are referred to as retort pouches.

A. Pouch Material

The retort pouch is a flexible, laminated, rectangular food package. The original pouch was a three-ply laminate but presently there are pouches of five- and even seven-ply structure. However, most retort pouches in common use today are constructed with a four-ply laminate, consisting of an outer layer of polyester bonded to an inside layer of nylon. Internal to the nylon is aluminum foil, with the final layer on the inside being polypropylene (6). Each component performs a specific function that is critical to shelf stability of the enclosed product and package integrity.

1. *Polyester:* high temperature resistance, toughness, and printability
2. *Nylon:* abrasion resistance
3. *Foil:* barrier to light, gases, microorganisms, and odors; extends shelf life; and offers stiffness, which allows for the tear notch
4. *Polypropylene:* heat seal integrity (good heat seal surface), flexibility, strength, and food compatibility

B. Pouch Fabrication

The retort pouch may be either a preformed (sealed on three sides) unit or one that is formed as an in-line operation with the filling and sealing combined in a pouch packager. The pouches are formed from roll stock, by folding a single roll along its center line, or by bringing two separate rolls together and heat-sealing the sides. The tubular material is automatically cut to length and the bottom sealed just prior to the product filling operation.

C. Sealing Operation

In the retort pouch, a hermetic seal is achieved by the fusion to each other of two heat-sealable layers of materials, such as polypropylene. Factors that can affect the proper formation of the seal include temperature of fusion, the pressure created by the sealing tool to hold the pieces together, sealing material compatibility, seal area contamination, and the condition of the sealing surface.

D. Thermal Processing

While retort pouch products can be thermally processed in a manner similar to that for glass or canned products, special attention must be given to the need for precise control of the pouch thickness to ensure that the pouch is adequately processed. Furthermore, the heat seal strengths of the pouch are reduced considerably at processing temperatures in the range of 230–250°F (110–121°C). To protect the seals and prevent rupturing of the pouch during thermal processing and cooling, air overpressure is required.

Flexible containers can be heat-processed in several different heating media (e.g., steam/air, water with superimposed air pressure, pure steam). During processing, gases (residual air and steam) expand in the pouch, causing it to bulge. This bulging will tend to stress both the container and all the seals. The use of air overpressure during processing will help control some of the bulging. For systems using steam/air, care must be taken to ensure that the mixture of air and steam is homogeneous, to avoid pockets of air resulting in cold spots in the processing vessel. Bulging can also be controlled by physical constraints. Often dividers are built into the racks suspending the pouches, so that the pouches can be restrained to the maximum thickness allowed as determined at the time the process was developed. More critical factors are involved with thermal processing of flexible pouches than of rigid containers. The additional factors (pouch thickness, residual gas, and pouch size) are due to the ability of the flexible pouch to change geometry and to the special racking required to restrain pouches during retorting. Since a flexible pouch cannot support a significant pressure gradient (pressure differential between the outside of the pouch and the inside) without jeopardizing seal integrity during thermal processing, overpressures are generally required.

E. Container Geometry

The intended geometry of pouches and other flexible containers is usually that of a slab. This shape offers the advantage of relatively rapid heat penetration into the food owing to the thin profile and high surface area to volume ratio of the pouch. As a result, the most significant factor affecting product heating in flexible containers is normally container thickness. Since the pouch is not rigid, thickness cannot be controlled by the container, and so it is desirable to physically define container thickness during processing by means of specially designed racks, trays, or cassettes.

F. Processing System

The processing system utilized for flexible containers should provide overpressure to ensure thickness conformity and seal integrity. Overpressure can be obtained by retorting in a mixture of steam and air, or in water with pressure applied by compressed air, or steam in the headspace above the water.

Regardless of the design and operational features, the processing system must be capable of providing every container with uniform temperature exposure to ensure proper sterilization. The flow rate and circulation pattern of the heating medium are critical to the performance of the processing system. Circulation of the heating medium is achieved by such means as fans in steam/air systems and bubbling air or circulation pumps in water systems.

The racking system used to process flexible containers should be designed to be compatible with the heating medium circulation pattern. Ideally, the circulation of the heating medium should be parallel to the container length or width. In designing a racking system, the following factors should be considered: separation between all container layers, incorporation of a circulation "channel" between layers of containers, provision for positive restraint at a designed maximum container thickness and suitable construction to maintain the original design specifications.

G. Postprocess Pouch Handling

The safe preservation of food in retort pouches is dependent on the ability of the sealed pouch to prevent reinfection by microorganisms by means of leakage through the seals or pouch body after the product has been heat-processed. Microbial contamination of the external surfaces of the pouches can be minimized by adhering to recognized sanitation requirements and by drying the pouches and enclosing them as quickly as possible in a protective outer wrap.

H. Advantages and Disadvantages of the Retort Pouch

The retort pouch has many advantages over canned and frozen food packages for both the consumer and the manufacturer (12). They are as follows:

1. Because the profile of the retort pouch is thinner than that of the metal can, heat penetration is faster in the retort pouch, and as a result less time is needed to reach lethal temperatures at the center of the food in the pouch than in cans. Consequently, there is less overcooking of the product in the pouch, especially around the periphery (as with cans), and the quality is better in terms of a truer color, fresher flavor, firmer texture, and less nutrient loss.
2. The pouched product can be reheated simply by heating the pouch in boiling water for 5 minutes.
3. The pouch is easily and safely opened by tearing across the top at the notch in the side seal or by cutting with scissors.
4. Because of the flat profile, pouches require less storage and disposal space than canned food packages.
5. Pouches can be packaged in a flat carton, providing a broader surface for product labeling and identification on the shelf than is possible on cans.
6. The pouch also offers the opportunity to market multipacks (e.g., entrée in one pouch and dessert in another).
7. Because of the thin profile, less brine is required in institutional size pouches than in cans of equivalent size when the same thermal process is delivered. In as much as more product and less brine can be placed in the institutional pouch, this makes the retort pouch desirable for use in institutional feeding.

In spite of its advantages, the pouch also has several disadvantages:

1. Some sort of secondary packaging, such as a carton, is required for protecting the pouch under normal conditions of marketing.
2. It is limited in size in terms of the advantage in thermal processing. The larger the pouch, the thicker the profile; and when the pouch becomes too thick, heat penetration becomes significantly slower and the sterilizing time gets longer. This results in overcooking and loss of quality advantages over canned foods.
3. The pouch is hampered by the lack of high speed filling equipment, resulting in greatly reduced output compared to packing speeds attainable with glass jars and cans.

VI. PHYSICAL HAZARDS

The integrity of the food product is always at risk from contamination by objects and materials in the environment. There are many possible sources of contamination, but these can be grouped according to type of contaminant.

A. Metal

Metal contaminants include nuts, bolts, washers, rivets, and other items that often fall in from manufacturing equipment. Containers to be used for packaging food should never be used as receptacles for these items, to prevent metal pieces from accidentally ending up in a finished product. Another point of consideration is that bits of metal may remain in the can after can fabrication or as a result of damage incurred during transportation and storage. A further source of contamination is wear and tear on equipment with metal parts. Therefore, before food is processed in any equipment, the device should be examined carefully for evidence of wear and any worm parts should be replaced. All cans should be washed thoroughly before filling.

B. Wood

Cutting and chopping of food on wood surfaces as well as the use of wooden utensils such as spoons normally results in the formation of wooden fragments. To minimize this type of contamination, wooden utensils should be avoided and plastic utensils used instead, since plastic utensils tend to be more durable and are easier to clean and disinfect.

Another source of contamination is twigs, stalks, bark, and general extraneous vegetable matter collected during the harvesting of raw materials. The first step in canning is to manually remove surface dirt and extraneous matter from fruits and vegetables. This is followed with washing by hand with frequent changes of water, depending on the type of contamination. It must be noted that leafy vegetables that have been bulk-stored are more difficult to wash because of layering (15). The leaves tend to become stuck to each other, forming a mat. Therefore, it is more desirable to use freshly picked leafy vegetables. Washing is followed by inspection and sorting by hand, during which time any remaining extraneous material and unwholesome product should be removed.

C. Glass

Fragments of glass may remain in the jars as a result of damage during transit or handling or because of imperfections during manufacturing. All glass jars with damaged tops as well as those with score marks on the body of the jar should be discarded. Any broken glass should be removed from the canning area.

VII. CHEMICAL HAZARDS

Gastroenteritis of varying degrees of severity, sometimes fatal, may be caused by the ingestion of toxic chemicals present in the food we prepare. Toxic chemicals may be ingested in several ways: (a) by mistaken use of a toxic chemical in the preparation of a food because of a close physical resemblance to a commonly used dietary product, (b) by the use of non-food-grade utensils for the cooking or storage of food or drink from which soluble toxic chemicals may leach out; (c) by the ingestion of raw fruits and vegetables contaminated on the surface with too high a level of residual pesticide, and (d) by the pollution of a drinking water source with fertilizer or pesticides.

The incubation period of chemical poisoning is usually very short, ranging from several minutes to an hour. In most cases the gastroenteritis manifests itself in the form of vomiting, abdominal cramps, and diarrhea. These reactions, though disagreeable, serve a useful therapeutic purpose, for the vomiting helps the stomach get rid of the food or fluid with the irritating chemical. Cramps are an indication that increased peristalsis is hurrying the toxic material along the intestinal tract, and the diarrhea evacuates it completely out of the digestive tract (16). However, even these functions cannot prevent the development of some serious conditions if too much of the toxic chemical is consumed.

A. Sodium Fluoride

Sodium fluoride, an ingredient of household insecticides, is a chemical that has on occasion been used in the preparation of food because it was mistaken for flour, sugar, baking soda, or salt. Symptoms of sodium fluoride poisoning are vomiting, abdominal cramps, and diarrhea, and in extreme cases, cold sweats, convulsions, and collapse (16).

B. Sodium Nitrite and Sodium Nitrate

Nitrites, used in the curing and preservation of meats such as bacon, are sometimes mistaken for sodium chloride owing to similar crystalline appearance. Ingestion of food contaminated with nitrates or nitrites results in cyanosis, a bluish discoloration of the skin due to deficient oxygenation of the blood (16).

C. Heavy Metals

The cooking of acid foods such as apples or the storage of acid liquids such as lemonade and orange juice in certain types of food container may result in metallic poisoning. If non-food-grade enamelware is used, antimony present in the enamel may leach out and cause gastroenteritis. Acid foods cooked in copper or galvanized pots may dissolve out a sufficient quantity of metal (copper or zinc) to cause vomiting and other intestinal illness. When vessels containing lead are used as containers for acid foods, lead leaches into the food, potentially causing gastroenteritis (16).

D. Pesticides

Pesticides, whether insecticides or herbicides, have become a necessity in the production of wholesome food. These chemicals are lethal to insects and plant life but are also harmful to humans if accidentally swallowed. Fruits and vegetables that have been sprayed with pesticide may still retain a residue when purchased, especially leafy vegetables that were sprayed with insecticides to protect them from attack by insects. It is therefore necessary to wash these foods thoroughly before processing. Food or utensils accidentally contaminated with pesticides may cause headache, nausea, dizziness, confusion, and convulsions within half an hour after ingestion (16).

E. Prevention of Chemicals from Entering the Food

The accidental ingestion of toxic chemicals may be prevented by exercising care in their storage and use. The use of kitchen utensils, pots, trays, and containers of questionable quality should be avoided. Stainless steelware, although more expensive, is the most desirable for it is the material of choice in the food industry, does not react even with the most acidic or alkaline foods, and is easy to clean and sanitize.

Fruits that are not not normally peeled before consumption and all leafy vegetables should be thoroughly washed before processing to remove all residual insecticide on the surface.

VIII. MICROBIOLOGICAL HAZARDS

Microbiological hazards are manifested in the form of foodborne disease that occurs after ingestion of food containing pathogenic microorganisms or their toxins. The illnesses can be either infections or intoxications.

Food infections occur when bacteria present in the food at the time of eating grow in the host and cause disease. In food intoxications, bacteria grow in a food, producing in it a substance that is toxic to humans. In cases of intoxication, the bacteria grow in the food and under normal circumstances do not grow in the host; it is during the growth of the bacteria in the food that the poisonous substances or toxins are produced. The poisonous materials, in this case, are already present in the foods when they are eaten. Canned foods have been responsible for both types of illness, but food that has been properly canned will keep indefinitely and is safe to eat as long as the seal remains intact. However, for best eating quality, use within one year is recommended. If the contents of a can are spoiled, the reason may be either insufficient processing (underprocessing) or a leak in the container through the double seam (cans), or lid (glass jars), giving access to microorganisms after processing (postprocess leakage).

A. Underprocessing

To ensure that an adequate process is delivered, it is of paramount importance to have the proper equipment, properly functioning. Heat treatments given canned foods are generally thought of as sterilization processes even though canned foods of various types are processed at quite different temperatures. In common use today are two terms, pasteurization and botulinum cook, which describe the type of heat treatment required to render food microbiologically safe. "Pasteurization" is the term most often applied to relatively mild heat treatments [process temperature usually that of boiling water, which is about 212°F (100°C)] given to some foods. Canned acid foods can be pasteurized because their acidity will not support growth of the more heat-resistant organisms, which require higher process temperatures. "Botulinum cook" is the term most often applied to more severe heat treatments, which are required for the control of the spores of the organism *Clostridium botulinum;* these severe heat treatments are in excess of 212°F (100°C) and can be only obtained through the use of pressurized vessels.

For home food preservation there are only two recommended and accepted methods for processing foods: the boiling water bath method (for delivering pasteurizing processes) and the pressure canner method (for delivering a botulinum cook). The selection of either process is based on the acidity of the food, which in turn will influence the type of microorganism capable of growing in the product. The most hazardous organism associated with food is the spore-forming organism *C. botulinum,* and laboratory research has shown that this organism will not grow at pH values of 4.6 or less. Foods at pH values of 4.6 or more are described as low acid foods, and failure to adequately process low acid foods will result in botulism, a potentially fatal form of food poisoning. Botulism is an intoxication caused by the *C. botulinum* neurotoxin. This organism lives in and comes from the soil in all parts of the world. It does not grow in the presence of air but requires an anaerobic environment (such as in canned foods) to grow and form its exotoxin, one of the most deadly neuroparalytic toxins known to man. Throughout the world a number of deaths occur each year from botulism contracted through the consumption of improperly home-canned foods. Because spores of *C. botulinum* are killed by heat, the most likely culprit in home canning is underprocessing, either by not using a high enough temperature, by processing for too short a time, or by a combination of the two. Meat, fish, poultry, dairy products, and vegetables other than tomatoes generally fall into the low acid food category. Spores of *C. botulinum* are classified as heat resistant and so are very difficult to destroy at the temperature of boiling water. The higher the processing temperature, the more quickly the spores are destroyed. Therefore, all low acid foods should be processed at temperatures of 240–250°F (115.5–121°C), attainable with pressure canners operated at 10–15 psig (68–102kPa) as measured by the gauge attached to the pressure canner.

Acid foods that are naturally acidic, such as most fruits, have pH values below 4.6. Foods that are artificially made acidic such as by pickling (cucumber pickles) or fermentation (sauerkraut) also have pH values below 4.6. Other acidic foods include jams, jellies, marmalades, and fruit butters. Although tomatoes, botanically classified as fruits, are usually considered an acid food, some varieties are known to have pH values slightly above 4.6. Figs, as well, have pH values slightly above 4.6. Therefore, if these two products are to be canned as acid foods, they must be acidified to a pH of 4.6 or lower with vinegar, lemon juice, or citric acid. Properly acidified, sweet tomatoes and figs are acid foods and can be safely processed in a boiling water canner. Spores of *C. botulinum* may be present in acid/acidified foods, but they are incapable of growing and forming toxin because of the acidic conditions.

The organisms capable of growth in acid foods and resulting in spoilage are yeasts, molds, and lactic acid bacteria. These organisms are heat sensitive and can be destroyed at temperatures obtained when boiling water is used. Therefore, the boiling water bath method is the method of choice for processing acid foods.

B. Process Adjustments for High Altitudes

The process time for canning food at sea level is not the same as at altitudes of 1000 ft or more. As altitude increases, atmospheric pressure decreases, and as a result water boils and steam is produced at a lower temperature. Such lower boiling temperatures may result in underprocessing if they are not compensated for by longer times. Increasing the process time or canner pressure compensates for lower boiling temperatures. Therefore, when following canning directions, always select the proper processing time or canner pressure for the altitude at which you live. The height above sea level can be obtained from the local meteorological department or department of the environment. Tables 1 and 2 give the adjustments necessary for boiling or pressure canning at altitudes from sea level to 10,000 ft.

IX. HOME CANNING EQUIPMENT PERFORMANCE AND MAINTENANCE

A. Boiling Water Bath Method

Acid foods can be processed in a boiling water canner (sometimes referred to as a boiling water bath), which can be any large pot with a rack on the bottom and a well-fitting lid, as long as the large container is deep enough to allow at least 2 in.

Table 1 Attitude Adjustment for Boiling Water Processing Method

Altitude		
ft	m	Processing time increase (min)
0–1000	0–305	No increase required
1001–3000	306–915	5
3001–6000	916–1830	10
6001–8000	1831–2440	15
8001–10,000	2441–3050	20

Source: Adapted from Ref. 4.

Table 2 Pressure Canning Altitude Adjustment[a]

Altitude		Dial gauge		Weighted gauge	
ft	m	psig	kPa	psig	kPa
0–1000	0–305	11	75	10	68
1000–2000	305–610	11	75	15	102
2000–4000	610–1200	12	82	15	102
4000–6000	1220–1830	13	89	15	102
6000–8000	1830–2440	14	95	15	102

[a]The processing time does not change if the pressure is increased.
Source: Adapted from Ref. 4.

of boiling water over the top of the smaller containers. For a rack, one can also use screw bands (part of the closure system for mason jars) tied together with string or wire. Boiling water bath canners are also commercially available and in addition to having a well-fitting lid contain a metal basket to hold the containers off the bottom of the kettle and to separate them from each other. The containers must be held off the bottom of the canner to allow the boiling water to freely circulate around them; they must be kept separated as well, so that they will not bump into each other or tip over in the boiling water. The containers must be covered by about 2 in. of water when it is briskly boiling to allow the heat to thoroughly penetrate the food at the top of the jar. An additional inch or two of air space should be allowed between the top of the boiling water and the top of the canner. The boiling water bath method is recommended for canning fruits, tomatoes, foods with vinegar added, and other acid foods. Butters, conserves, marmalades, jams, jellies, and preserves can also be processed in a boiling water bath at the temperature of boiling water [212°F (100°C)] for 5–20 minute (3).

B. Steam Pressure Method

Low acid foods must be processed in a steam pressure canner, which is a thick-walled kettle with a lid that can be clamped or locked down to make a steam-tight seal. The lid is fitted with a safety valve, a vent, and a pressure gauge. Two types of steam pressure canner are available commercially; those with dial gauges and those with weighted gauges.

1. Dial Gauges

Dial gauges, usually circular devices with an indicating needle that gives a direct reading of the internal pressure of the canner, must be checked periodically for accuracy according to the manufacturer's instructions, usually prior to use each year. The gauge should read zero at the prevailing atmospheric pressure. The scale should have a range such that the safe working pressure of the canner is approximately two-thirds of the full scale and should be graduated into divisions not greater than 7 kPa (1 psig). The dial gauge should be large enough to be easily and accurately read (2). If the gauge is incorrect, the processing of canned goods will not be accurate and the processing temperature may not be high enough to destroy the spores of *C. botulinum.* It is more convenient to replace the gauge every year than to try to have it checked for accuracy. The manufacturer of the pressure canner can provide information on checking of the gauges, as well as on replacing them.

2. Weighted Gauges

Weighted gauges, sometimes referred to as pressure regulators, control the amount of pressure that can be built up in the pressure canner. The pressure regulator consists of the pressure regulator body, two weight rings, a lock ring, and a knob. Completely assembled, it will maintain 15 psig. When one weight ring is removed, the pressure regulator will maintain 10 psig, the pressure used for processing low acid vegetables, meats, poultry, fish, game, and soups. When both weight rings are removed, the pressure regulator body will maintain 5 psig and can be used for processing fruits and tomatoes (i.e., acid and acidified foods).

The vent (small pipe on which the pressure regulator sits) on the pressure canner allows steam to escape under controlled pressure. Because the steam inside the kettle is pressurized, its temperature exceeds that of boiling water. At 10 psig, the temperature will reach 240°F (115.6°C), which is hot enough for the destruction of bacterial spores. If the vent should inadvertently become plugged, the safety valve or automatic vent is designed to pop, releasing pressure and preventing the kettle from exploding. When purchasing a pressure canner that uses a weighted gauge, it is advisable to ensure that there is a safety valve or automatic vent. For canners equipped with a dial gauge, the gauge should be of the type that indicates "CAUTION" when the pressure is too great.

3. Important Considerations for Steam Pressure Devices

a. Excess Pressure Inside the Canner. If the dial gauge is reading in the "caution area" or if the safety valve or automatic vent is released on a canner equipped with a weighted gauge, excessive pressure has built up in the canner and the following should be done immediately:

1. Turn off the heat source. **Do not attempt to remove the pressure cooker!**
2. Allow the cooker to stand until cold to the touch (this may take an hour or more). Only then, when the gauge reads "0," is it safe to remove the cover.
3. To clean the vent pipe, draw a clean piece of pipe cleaner or a small brush through the opening.

b. Keep Safety Openings Clear. It is important to regularly inspect the openings for the vent (on which the pressure regulator sits), the dial gauge, and the automatic vent and to ensure that steam can pass through these openings, located on the underside of the cover. The openings can become blocked by food from dirty and overfilled containers or by mineral deposits in the water. If any of the openings are blocked, inadequate venting or a buildup of pressure may occur.

c. Accuracy of Steam Pressure Gauge. The steam pressure gauge (dial gauge) should be replaced every year before the canning season begins. In most cases, it is cheaper and less bothersome to replace the old gauge than to have it checked for accuracy.

d. Water in Bottom of Pressure Canner. The bottom of the pressure canner must always be covered with water during processing. Follow the manufacturer's instructions for the volume of water to be added for processing. If the canner boils dry, damage will result to the container.

e. Filling the Pressure Canner. It is important to follow the manufacturer's instructions for filling the pressure canner with water, as this allows steam to circulate completely around the containers being processed. Do not add more water than is specified by the manufacturer; however, there must be enough liquid in the canner throughout the length of the processing time.

f. Dropping the Pressure Canner. Dropping the canner may cause damage, and so a canner that has been dropped must be examined carefully for any defects. Dropping on a hard surface, such as tile or concrete, may cause a crack to develop. Any cracks in the unit make it dangerous to use, even hairline cracks, not visible to the naked eye. Therefore, should the unit be dropped on a hard surface, it should be returned to the manufacturer or legal agent by a person able to explain the situation and request that the unit be thoroughly checked.

g. Opening the Pressure Canner Safely. At the end of processing, turn off the heating element or remove the pressure canner from the heat source and let the pressure in the canner drop on its own. There should be no steam pressure in the canner when the metal plunger in the safety valve has dropped or the dial gauge reads "0." Remove the pressure regulator from the vent pipe slowly and let the canner cool for a minute or two. Attempting to speed the cooling of the canner by placing wet cloths on the cover, or by setting the canner in water or in a draft or on a cold surface is not recommended. Such measures reduce the pressure in the canner more rapidly than in the containers, resulting in a higher pressure inside the container than in the canner, which in turn may cause jars to break or cans to buckle. Follow the manufacturer's instructions for opening the canner.

C. Home Canning Pressure Canner Maintainance

1. First-Time Use of a Pressure Canner

Before using the pressure canner for the first time, remove the sealing ring, which is usually a gasket made of some type of rubber material. Wash the sealing ring, groove, and automatic air vent (all located in the lid of the canner) to remove manufacturing oils. Clean the inside of the canner carefully with a good nonalkali aluminum cleaner or a steel wool cleaning pad. After the canner and rack have been washed and rinsed with warm water, the canner is ready for use (2).

2. Cleaning the Pressure Canner

Pressure canners as well as boiling water canners frequently acquire a buildup of hard water scale on the inside. The scale can be removed by soaking the vessels for several hours in a solution of 1 cup (250 ML) vinegar to 16 cups (4 L) of water. The vinegar concentration can be increased if the scale buildup is heavy (4). To clean the vent pipe of the pressure canner, draw a clean piece of pipe cleaner or string through the opening. If there is a sealing ring, periodically remove it and scrub the groove with a brush. Wash the sealing ring in hot soapy water, rinse in warm water, allow to air-dry, and replace.

3. Storing the Pressure Canner

The proper way to store the pressure canner is either to store the cover separately from the cooker or to place the cover carefully, upside down, on the cooker. This will allow air circulation, prevent odors from developing in the canner, and minimize wear and tear on the sealing ring. Place a piece of towel material or paper between the cover and body to protect the finish. Be sure the canner is properly dry to protect it against pitting and corrosion before putting it away. Store it in a clean, dry place (2).

X. METHODS TO AVOID

Only the boiling water bath and the pressure canner methods are recommended for canning. In the United States, home-canned food has been responsible for over 90% of the outbreaks of botulism over the past 70 years (11). In 1974, 15 outbreaks that resulted in 6 deaths were traced directly to home canning. These statistics indicate that it is very important for home canners to read instructions properly and to be sure that their instructions and equipment are up to date. Unfortunately, many people think that they can process any food in a boiling water bath and even continue to use the open-kettle and oven methods for low-acid foods (11).

A. Open Kettle

Open-kettle canning is not safe for the canning of low acid foods, nor is it recommended for home food preservation of acid foods. In the open-kettle method, food is first cooked in an open cooking vessel and then put into containers. The containers are filled and capped one at a time, and the lids are quickly put in place with the expectation that a proper seal will be achieved as the food cools (3). The open-kettle method is obviously *not safe for low acid foods* because it is impossible for the low acid food to reach temperatures necessary for the destruction of heat-resistant spores of *C. botulinum.* When this method is used for acid foods, there is a very good chance that postprocess contamination will take place while the food is being transferred from the kettle to the jar, and spoilage may result.

B. Microwave Oven

Canning should not be done in microwave ovens because heating the product in sealed cans or glass jars could result in uneven heating of the product. Underprocessing of the product would result, and any spores of *C. botulinum* present could possibly survive, especially in low acid foods. Spoilage organisms in acid foods could also survive.

C. Hot-Air Oven Canning

Oven canning is dangerous, for it is extremely difficult to obtain the required temperature in closed mason jars without the jars exploding. Hot air is very inefficient as a heat exchange medium for heating canned food, since canned food must be heated for a very long time to achieve the high temperatures required for processing, even for acid foods.

D. Steam Canning

Not to be confused with pressure canning, steam canning is conducted in a covered cooking vessel with a shallow bottom pan containing water. The water evaporates to form steam, which is allowed to circulate around the filled containers. Steam in this type of canner does not maintain a steady flow, and as a result the temperature of the food in the containers is uneven. Thus, it is virtually impossible to know whether the containers have been adequately heated (3).

XI. CONTAINERS: SELECTION

A. Three-Piece Metal Can

There are some basic rules for selecting cans to be used in food preservation.

1. *Use only cans in good condition.* Examine the can body and lid carefully. Discard dented or rusted cans or cans with damaged flanges. Discard lids with damaged curls or damaged gaskets (sealing compound). Keep lids in paper packing until ready for use. The paper protects the lids from dirt, moisture, and damage.
2. *Wash cans.* Just before use, wash cans in clean water and drain upside down. Do not wash lids with a detergent, for washing may damage the sealing compound. If lids are dusty or dirty, rinse them with clean water and allow to drain.
3. *Check the can sealer.* Make sure the sealer to be used is properly adjusted. Several types of machines are on the market for securing the lids onto the can body, a process called double seaming. The method of operation varies, so the instructions issued with the machine are specific for that particular type of double seamer and should be carefully followed to ensure that a proper double seam is

formed. The machine should be securely attached to a strong table or bench, either by the use of bolts or by means of a heavy-duty thumbscrew to prevent any wobbling movements of the machine.

The following general principle applies to most makes of home can-seaming machines. A baseplate, on which the can to be sealed is placed, is raised by means of a lever or screw until the can and the lid gently engage, and the unit is then pressed tightly against the chuck. This is specifically designed to fit into a recess in the lid so that the lid is fixed and held firmly in position on the can. The baseplate must be raised to its full extent so that the curl of the lid properly fits over the flange of the can body before the can is sealed, otherwise a faulty seam will result (13). The double-seaming procedure requires complete revolving of the can; this is accomplished by turning a handle. During this process, when the can is turning, two rollers are brought into operation, either automatically or by operating a lever.

The can double seam is generally formed in two operations, hence the name. In the first seaming operation, the curl of the end or lid is loosely interlocked (folded, clinched, or engaged) with the flange of the can body. The actual interlocking is performed by a specially contoured roll called the first operation roll. It is of utmost importance that the first operation seam be properly formed, and this can be done by following the manufacturer's direction for setting the first operation roll. There is no way to compensate for or correct a faulty first operation seam during the remaining steps. When the first operation seam is completed, the first operation roll is retracted and no longer contacts the can end or lid. The specially contoured second operation roll then moves into contact with the end to iron, smooth, and further compress the material in the seam. When the seam has been completed, the baseplate is lowered and the sealed can removed. The sealing compound and the mechanically interlocked can body and end work together to make the double seam a hermetic seal. Neither the sealing compound nor the interlocked can body and end by itself is able to hermetically seal the can. The components must complement each other.

B. Checking the Can Sealer

Make sure the sealer to be used is properly adjusted. To test, put a little cold water into a can and seal it. Grasp the can with a pair of tongs and, keeping the newly sealed end up, immerse it in boiling water for a minute or two. If air bubbles rise from around the double seam, the seam is not tight. Adjust the sealer, following the manufacturer's instructions.

C. Mason Jars and Lids

Discard any jars that are cracked or scratched, and any that show imperfections or nicks in the rim. With careful use and handling, mason jars may be reused many times, requiring only new lids for every canning project. Never attempt to reuse a lid that has already been used for processing. The sealing compound on the lid is softened when boiled prior to use, and once the lid sealing compound has molded itself to the contour of the rim of the jar, it will not seal properly a second time (4). The common self-sealing lid used with mason jars consists of a flat metal lid held in place by a metal screw band during processing. The edge of the flat lid is curled to form a trough, which is filled with a colored gasket compound. When jars are processed, the lid gasket is already soft from being preheated as mentioned above, and it flows slightly to assume the contour of the rim of the jar, yet allows air to escape from the jar. The gasket then forms an airtight seal as the jar cools. According to the manufacturers of mason jars, gaskets in unused lids work well for at least 5 years from the date of manufacture, but the gasket compound in older unused lids may fail to seal on jars (18). It is better to buy just the quantity of lids you will use in a year. Old, dented, or deformed lids, or lids with gaps or other defects in the sealing gasket should not be used.

D. Containers to Avoid

Commercial pint- and quart-size mayonnaise or salad dressing jars have been used with new two-piece lids for canning acid foods. Single-use jars, like those obtained from commercially prepared products (e.g., mayonaise jars, peanut butter jars, baby food jars, or jam/jelly jars) should not be used for home canning, however. These jars are made to serve a single purpose only—as a nonreusable package for a commercial food product. Single-use jars are not meant to be reused for home processing, and they are often made for a food product that is packed cold and, therefore, may not be able to withstand the temperatures used in home canning (3). If single-use are nevertheless used, one should expect more seal failures and jar breakage because the jars are not designed to be reused. Some of the commercial jars have mouths and threads that appear to be the same as canning jars. In fact, canning lids and caps seem to fit the threads of commercial jars. However, the commercial jar mouth and sealing surface may vary in width, and the rim may have dips. This can prevent the home canning lids from sealing properly, resulting in leakage (3). Because metal knives are commonly used to scrape the last bit of peanut butter or mayonnaise from a commercial jar, these jars are often scratched on the inside. This weakens the jars, and they may break when exposed to heat during canning.

Another disadvantage of using jars designed for commercial products is that commercial jars often do not come in exact pint or quart sizes but are in odd-ounce measures. If the jars are not exactly the size specified in the recipe, then the processing time given will be inaccurate (3). Like commercial jars, commercial lids are designed for a single use only. Some of them are made for dry products, while others have treated paper gaskets that are not intended to be airtight. These lids may not seal when used for home canning, resulting in postprocess leakage and spoilage of the food. Only the containers and closures made and sold especially for home canning should be used for that purpose.

XII. GLASS JARS

A. Cleaning of Jars

Immediately before use, wash the empty jars in hot water with detergent and rinse well by hand, or wash in a dishwasher. Detergent residues may cause unnatural flavors and colors. Stubborn stains due to scale or hard water films on jars are easily removed by soaking jars for several hours in a solution containing 1 cup of vinegar (5% acidity) per gallon (4.5 L) of water (18). Note that these washing methods do not sterilize jars.

B. Sterilization of Containers

Pickled products, acid foods, and all jams, jellies, and related products processed less than 10 minutes should be filled into sterile containers and covered with sterile lids (see Table 1 and Ref. 4). Containers to be used for food to be processed by the pressure canning method do not need to be sterilized prior to filling. To sterilize containers, put them right side up on rack in a boiling water bath and fill the bath and jars with hot (not boiling) water to 1 in. above the tops of the containers. Allow the water to boil for 10 minutes at altitudes of less than 1000 ft. At higher elevations, boil water for 1 additional minute for each additional 1000 ft of elevation (1). Remove and drain hot sterilized containers one at a time immediately before use. The hot water can be saved for processing filled containers (18). Dishwashers and ovens are unsuitable for sterilizing containers. Even on a sterilizing cycle, temperatures may not be high enough or the time period long enough to adequately sterilize the jars (4).

C. Filling of the Jars

Basically there are two ways of filling containers. Food may be placed into containers while the food is hot or cold. Many fresh foods, particularly plant material, contain from 10% to more than 30% air. The ability of canned food to retain high quality depends on how much air is removed from the food before the containers are sealed (18). One of the goals of canning is to remove air from the containers, to prevent adverse chemical reactions promoted between air and the preserved food. The less air in the jar at the beginning of processing, the more efficient is the air removal from the container. Fruit and vegetables contain very porous tissues that contain desirable juices, but they also contain air, which if not exhausted from the food can cause discoloration as well as floating. Hot packing helps exhaust this air. In hot packing, the freshly prepared food is added to the boiling syrup or liquid. The mixture is brought back to a boil and simmered for a few minutes; then the containers are loosely filled with the boiled food.

"Cold pack," or "raw pack" as it is sometimes called, is the practice of filling jars tightly with fresh prepared, but unheated food. Some foods that are delicate after they are cooked, such as whole peaches, are usually easier to handle if they are raw packed. The food is placed into the containers while it is raw. It should be packed firmly but not to the extent of crushing it. Some shrinkage will usually occur when the food is processed, and some foods, especially fruit, will float in the containers (3). The entrapped air in and around the food (through a chemical reaction) may cause discoloration within 2–3 months of storage. Raw packing is more suitable for food processed in a pressure canner (1), but whether the boiling water bath or pressure canner method is used, the packed jars should not be put into boiling water in the canner: the jars are still cold and might break as a result of cold shock (3).

Hot packing is considered to be the best way to remove air and is the preferred style for foods processed in a boiling water canner. Whether hot-packed or raw-packed processing is used, the juice, syrup, or water to be added to the foods should also be heated to boiling before it is added to the containers. This practice helps to remove air from food tissues and, in so doing, shrinks the food and helps to keep the food from floating in the containers. The hot liquid not only contains less air but also displaces air in the container, resulting in an increase in the vacuum in the sealed containers and an improvement in shelf life. Preshrinking food permits more food to be filled into each jar (1).

In the boiling water bath method, food that is hot-packed requires less processing time than raw-packed food because it is already hot when it goes into the canner and the initial heating is credited as part of the process. With the steam pressure canner, however, there is no difference in processing time because by the time the pressure canner reaches the required pressure (e.g., 10 psig, at which time you begin counting process time), the raw-packed food has become as hot as it would have been if it had been packed hot at the beginning (3).

Most raw-packed fruits and vegetables should be packed tightly into the container because they shrink during processing. There are a few exceptions: corn, lima beans, and peas should be packed loosely because they expand. Hot-packed food should be packed fairly loosely and should be at or near the cooking temperature when packed. There should be enough syrup, water, or juice not only to fill in around the solid food in the container but also to completely cover the food. Food at the top of the container tends to darken if not covered with liquid, as a result of chemical reactions between the food and residual air in the headspace (19).

D. Controlling the Headspace

The unfilled space between the inside of the lid and the top of the food or liquid is termed the headspace. This space is needed for expansion of food, entrapped air, and moisture as the containers are processed, and for forming a vacuum in the cooled containers. The extent of expansion of the contents of the container is determined by the air content in the food and by the processing temperature. Air expands greatly when heated to high temperatures, and moisture in the food is converted to steam. The solid part of the food expands less than air when heated (18). The amount of headspace to leave at the top of the container is determined by the type of food being canned. If containers are overfilled (i.e., too little headspace), the contents may boil out during processing. When this happens, food material may be deposited under the sealing compound or on the rim of the jar, and this condition may prevent an airtight seal from forming. When too much headspace is left at the top of the jar, processing time may not be long enough to drive out all the extra air and may result in the formation of a poor vacuum and discoloration of the food at the top of the container owing to chemical reactions between the food and residual air present in the headspace. Follow the directions for headspace in the recipe being used. However, when specific directions for headspace are unavailable, the following general directions can be used: for jams and jellies, leave 1/4 in. (6 mm) headspace; leave 1/2 in. (12 mm) for fruits and tomatoes to be processed in boiling water, and 1–1 1/4 in. (2.5–3.2 cm) for low acid foods to be processed in a pressure canner.

XIII. SEALING

A. Metal Cans

Metal cans are sealed before processing, once they have been filled, and the correct headspace allowed for. The temperature in the cans should be at least 170°F (77°C) when the cans are sealed (19). Food is heated to this temperature to drive out air so that there will be a good vacuum in the can after processing and cooling. Food packed raw must be heated in cans (exhausted) before the cans are sealed. Food packed hot may be sealed without further heating if precautions are taken to ensure that the filling temperature of the food is at least 170°F (77°C). To make sure, test with a thermometer, preferably a food thermometer, by placing the sensing area at the center of the can after its contents have been thoroughly mixed. If the thermometer registers lower than 170°F (77°C), or if you prefer not to make this test, exhaust the cans (19).

B. Exhausting Metal Cans

Exhausting is the procedure of removing air from the headspace of canned foods to prevent oxidation of the food and to promote the development of a good vacuum after processing and sealing. To exhaust, place open, filled cans on a rack in a kettle in which there is enough boiling water to come to about 2 in. below the tops of the cans. Ensure that the level of the boiling water is not high enough to spill into the cans. Cover the kettle and bring the water back to a boil. Boil until a thermometer inserted into the liquid in the center of a can registers 170°F (77°C)—or for the length of time given in the directions for the food you are canning (19). Remove cans from the water one at a time, and check the headspace. Boiling packing liquid (brine) or water may have to be added to bring the headspace back to the required level. Place a clean lid on the filled can and seal at once before removing the next can from the kettle.

C. Glass Jars

After the jars have been filled, remove the trapped air bubbles by sliding a spatula (rubber or plastic, but not metal) between the jar and food. Slowly turning the jar and moving the spatula up and down encourages the bubbles to escape (1). After the air bubbles have been removed, be sure to readjust the headspace to the required level with packing liquid. Wipe the jar rim, preferably with a hot, clean, damp cloth to remove any stickiness, which could result in seal failures. Put the lid on with a sealing compound next to the rim of the glass jar and screw the metal band down by hand. Apply the screw bands just until they are fingertip tight. Fingertip tightness allows some "give" between the lid and the jar. This allows rising steam created inside the jar during heat processing, as well as air, to be exhausted.

XIV. PROCESSING

A. Boiling Water Canner Method

Jams, jellies, marmalades, conserves, fruit butters, fruit and fruit sauces, tomatoes with added acid, pickles, relishes, chutneys, and condiments can be all processed in a boiling water canner. Although pressure canners are used for processing acid foods, boiling water canners are recommended for the purpose just named because they are faster. Processing a load of containers in a pressure canner could take from anywhere from 55 to 100 minutes, while the total time for processing most acid foods in a boiling water canner varies from 25 to 60 minutes. Processing in a boiling water canner loaded with filled jars involves heating the water to boiling before timing of the process. A full canner load would require about 20–30 minutes of heating before the water begins to boil.

A loaded pressure canner requires about 12–15 minutes of heating before steam is generated and the unit begins to vent, 10 minutes to vent or completely remove the air from the canner, 5 minutes to achieve the processing pressure required, 8–10 minutes to process the acid food, and another 20–60 minutes to cool and depressurize the canner before the containers can be removed (18). The following steps have been published for successful processing using the boiling water canning method (18):

1. Fill the canner halfway with water.
2. Preheat the water to 140°F (60°C) for raw-packed foods in glass jars and to 180°F (82°C) for hot-packed foods in cans or glass jars.
3. Load filled containers, fitted with lids, into the canner rack and use handles to lower the rack into the water; alternatively, fill the canner, one jar at a time, with a jar lifter.
4. Add more boiling water, if needed, so that the water level is at least 1 in. (2.5 cm) above the tops of the containers.
5. Turn on the heat to its highest position until the water boils vigorously.
6. Set a timer for the minutes required for processing the food. Timing starts when the water starts to boil vigorously.
7. Cover the canner lid and lower the heat setting to maintain a gentle boil throughout the process schedule.
8. Add more boiling water, if needed, to keep the water level above the containers.
9. When the containers have been processed for the recommended time, turn off the heat, remove the canner lid, and immediately remove the containers.
10. Using a jar lifter, remove the containers. Place jars on a towel and cans on a clean sanitized surface, leaving at least an inch between the containers during cooling.

B. Pressure Canning Method

Low acid foods must be processed in a manner that will eliminate the risk of botulism. Microorganisms are destroyed during processing based on time and temperature of the process, and equivalent time–temperature combinations have been calculated for processing low acid foods. These time–temperature combinations have the same quantitative destructive effect on bacteria and vary from low temperature/long time to high temperature/short time; however, the most reliable way to produce microbiologically safe food of good nutritional quality is the use of high temperatures and correspondingly short times. These temperatures, usually around 240°F (116°C), are higher than that of boiling water and can be achieved only in a pressure canner.

Pressure does not destroy microorganisms, but based on the laws of physics, an increase in pressure results in an increase in temperature. It is this increased temperature that is effective in destroying bacteria. The effective destruction of all microorganisms capable of growing in canned food is based on the temperature obtained in pure steam, free of air. It is extremely important that the steam used in processing in a pressure canner be free of air and that this important step be taken care of during venting. Air trapped in a pressure canner (steam/air mixtures) lowers the temperature of pure steam and results in underprocessing; this is why it is extremely important to respect the venting time specified for the particular pressure canner.

The highest volume of air trapped in a pressure canner occurs in processing raw-packed foods in dial gauge canners, and once these canners have been pressurized to the required pressure after venting, they do not allow the slow escape of steam during processing (18). Home canners can choose between pressure canners containing a dial gauge and those with a weighted gauge. Always refer to the manufacturer's instructions for specific instructions on operating a pressure canner. These basic principles apply to all pressure canners.

To ensure the complete removal of air, pressure canners of all types must be vented completely before being pressurized. When venting a pressure canner, ensure that the vent port is not blocked, then leave the vent port uncovered on newer models or manually open petcocks on some older models (18). When a canner containing water and containers of food has its lid locked into place and is heated, the water boils and generates steam that escapes through the petcock or vent port. The first gassy cloud to escape is a mixture of steam and air and must be purged through the process called venting. One must set the timer for the required vent time, after which the petcock is closed or the counterweight gauge placed over the vent port to pressurize the canner.

1. Weighted Gauge Pressure Canners

Weighted gauges regulate 5, 10, or 15 psig (34, 68, or 102 kPa) pressure only; pressures between these levels cannot be achieved. The 10 psig weighted gauge is used for pressure canning at sea level and up to 1000 ft (305 m). At altitudes above 1000 ft the 15 psi weight must be used, while the 5 psi weight may be used for processing acid foods (not low acid foods). Models of pressure canners with weighted gauges exhaust tiny amounts of air and steam each time the gauge rocks or jiggles during processing. They control pressure precisely, and there is no need for constant attention during processing to ensure that the correct pressure is being maintained. The hissing sound of escaping steam and the sound of the weight rocking or jiggling indicates that the canner is maintaining the recommended pressure.

2. Dial Gauge Pressure Canners

The pressure inside the canner is shown on an exterior dial called a pressure gauge. With dial gauge canners, the pressure and/or temperature inside the canner can be measured and controlled more precisely than is possible with the weighted gauge canners. This offers an advantage at high altitudes (4). At sea level and up to 2000 ft (610 m), 11 psi (75 kPa) pressure on the dial gauge is needed to safely preserve low acid foods (4). For every additional 2000 ft, the pressure must be increased by 1 psig (6.8 kPa) (4).

During processing with dial gauge canners, constant attention must be paid to the dial gauge to ensure that the recommended pressure is being maintained. The gauge must also be regularly and professionally checked to ensure that it is measuring pressure accurately. In some cases, it may be easier to replace the dial gauge before the canning season starts rather than trying to have it professionally checked.

3. Pressure Canning of Low Acid Foods

The following steps have been published as a guide for successful pressure canning of low acid foods (18).

1. Put 2–3 in. (5–8 cm) of hot water in the canner. Using a jar lifter, place filled containers on the rack. Fasten canner lid securely.
2. Leave the weight off the vent port or open the petcock. Heat at the highest setting until steam flows from the petcock or vent port.
3. Maintain high heat setting, exhaust steam for 10 minutes, and then place weight on vent port or close petcock. The canner will pressurize during the next 3–5 minutes.
4. Start timing the process when the pressure reading on the dial gauge indicates that the recommended pressure has been reached, or when the weighted gauge begins to jiggle or rock.
5. Regulate the heat under the canner to maintain a steady pressure at or slightly above the correct gauge pressure. Quick and large pressure variations during processing may cause unnecessary liquid losses from jars. Weighted gauges either jiggle (about two or three times per minute), or rock slowly throughout the process.
6. When the time process is completed, turn off the heat, remove the canner from the heat source, and let the canner depressurize. **Do not force-cool the canner.** Since the normal cooling time was considered to be part of the processing time when the process was developed, forced cooling may result in food spoilage. Cooling the canner with cold running water or opening the vent port before the canner is fully depressurized will cause loss of liquid from jars, seal failures, or buckling of cans. Force-cooling may also warp the canner lid of older model canners, causing steam leaks. Depressurization of older models should be timed. Standard-size heavy-walled canners require about 30 minutes when loaded with pints and 45 minutes when loaded with quarts. Newer thin-walled canners cool more rapidly and are equipped

with vent locks. These canners are depressurized when their vent lock piston drops to a normal position.

7. When the canner is depressurized, remove the weight from the vent port or open the petcock. Wait 2 minutes, unfasten the lid, and remove it carefully. Lift the lid away from you carefully so that the steam does not scald your face.
8. Remove containers with a lifter, and place them on a towel or cooling rack, if desired.

XV. POSTPROCESS HANDLING AND STORAGE

A. Glass Jars

1. Postprocess Handling

Before pressure-canned glass jars are stored, they must be cooled and tested for integrity of seals according to the following steps.

1. After jars have been removed from the pressure canner, place them upright and well separated on a wire rack, folded dry cloth, board, or newspapers.
2. Avoid cold surfaces and drafts, which may cause cracking due to cold shock; do not cover jars with cloth, as this retards cooling.
3. Sometimes there is an increase in the headspace in the jar due to shrinkage of food or loss of liquid during processing. Never open a jar after processing to fill up space: as this may expose the food to post-process contamination and could result in spoilage of the food.
4. Leave jars upright during cooling. That is, it is not a good practice to turn jars upside down, as this might break a seal that has already formed.
5. Do not tighten bands after cooling: as retightening of hot lids may cut through the gasket and cause seal failures (18).
6. After jars have cooled for for 12–24 hours, remove the screw bands carefully. If a band sticks, covering it for a moment with a hot, damp cloth may help loosen it. Wash the bands and store them in a dry place. Test the seals with one of the following options (18):
   a. Press the middle of the lid with a finger or thumb. If the lid springs up when you release your finger, the lid is unsealed. A properly sealed lid is securely held in place by vacuum and does not yield to finger pressure.
   b. Tap the lid with the bottom of a teaspoon. If it makes a dull sound, the lid is not sealed. Food in contact with the underside of the lid

also will cause a dull sound. If the jar is sealed correctly, the teaspoon will make a ringing, high-pitched sound.

c. Hold the jar at eye level and look across the lid. The lid should be concave (curved down slightly in the center). If the center of the lid is either flat or bulging, it may not be sealed.

2. Reprocessing Unsealed Jars

Reprocessing is not recommended, since food would be badly overcooked and the quality poor. Should it be considered, reprocessing is practical only if a number of jars are to be so treated. Remove the lid and check the jar-sealing surface for defects. If necessary, change the jar, add a new, properly prepared lid, and reprocess within 24 hours using the same processing time (18). Alternatively, the headspace in unsealed jars may be adjusted to 1.5 in. (3.8 cm) and the jars frozen instead of reprocessed. Foods in single unsealed jars may be stored in the refrigerator and consumed within several days (18).

B. Food Processed in Cans

After canning in tin cans, turn off the heat source and release the pressure slowly and as safely as possible. This can be done by removing the weighted gauge carefully or opening the petcock slowly, using oven mits in both cases. Use a jar holder to remove the cans from the canner and plunge cans directly into cold clean water immediately to cool them. Keep the water cold by changing it regularly or by leaving the cold tap water running slowly. Using a jar holder, turn the cans over several times so that food near the center will cool faster (1). Cool the cans in water until the cans feel warm, so that there is enough residual heat for the cans to dry in air. This can be done by removing the cans with a jar holder and touching the can body with the hand. **Do not touch the double seam or side seam with bare hands.** After processing, the sealing compound is hot and sets to form a proper seal on cooling. During this time the vacuum is being formed and the sealing compound has not yet set, and thus the can is highly susceptible to postprocess contamination. Moisture is needed for transfer of bacteria into the can through the double seam. Manual handling of the double seam while it is still wet presents a risk of contamination from bacteria, which could be transferred from the hands onto the container. Do not overstack cans, but rather stack them in such a way that air can freely circulate around them.

When cans are cool and dry, usually 24 hours after removal from the cooling container, they should be checked to ensure that there is a proper seal, as indicated by a good vacuum. A simple test for vacuum in processed cans and jars consists of tapping the lid with a hard object such as a spoon and listening for the resulting tone. When containers with a high vacuum are tapped with a suitable object, a high-pitched tone results. Any container with a low vacuum will give an easily discernible lower tone, while a "thud" indicates an overfilled can. A dull note does not always mean a poor seal, so it is advisable to carry out more than one of the tests as mentioned for options a–c in Sec. XV. A. 1. Cans in which the vacuum is suspect can be reprocessed, but as mentioned earlier for glass jars, this is not recommended. It would be preferable to use the canned food right away or refrigerate and use within 1 week.

XVI. LABELING AND STORING OF CONTAINERS

Wipe containers with a damp cloth, dry thoroughly, and label with a permanent marker. Store them in a clean, cool, dark, dry place and avoid areas where freezing of the canned food may occur. If the storage place is not dark, the mason jars may be wrapped in paper or stored in cardboard boxes to prevent light from affecting the color of the food. Canned food stored in warm places [> 86°F (> 30°C)] could lose some of the desirable quality attributes in a few weeks or months, depending on the temperature. Low acid foods also suffer from a type of degradation called thermophilic spoilage. This occurs when these foods are stored at temperature above 110°F (93°C) and spoil because extremely heat-resistant spore-forming thermophilic bacteria are present. These bacteria grow normally at temperatures between 131 and 158°F (55–70°C), but some can grow at lower temperatures. Thermophilic bacteria are not pathogenic and occur naturally in agricultural soils. Occasionally, canned low acid foods contain low numbers of certain thermophilic spores that will not cause spoilage unless the food is stored at temperatures above 110°F. Thermophilic spoilage of canned food is more often found when the canned product is cooled inadequately or is stored at high temperatures.

XVII. PLANT PRODUCTS PACKED IN OIL

Various fresh, dried, or rehydrated fruits, vegetables, mushrooms, herbs, and spices have historically been stored in oil to extend their shelf life and/or to flavor the oil. Such products include sun-dried tomatoes, lupin beans, hot peppers, mushrooms, artichoke hearts, eggplant, garlic, onion, spices, and herbs. Food poisoning outbreaks associated with vegetable-in-oil products have been reported in Canada and various other countries around the world. There is now increased concern over the safety of plant products packed in oil. The hazard of concern associated with plant products packed in oil is a microbiological problem, specifically botulism.

The organism that causes botulism, *C. botulinum,* is normally present in soil and marine environments and is expected to be present in soil-borne vegetables. It may also be present on stems, leaves, fruits, and vegetables however. In the preparation of flavored oils, as well as fruits and vegetables in oils, it is impractical to remove the contaminating botulinum spores, and thus they are expected to be in the plant-in-oil preparation. Therefore, in the preparation of plant-in-oil products, conditions that would allow the botulinum spores to grow and form toxin must not be permitted. One of the main factors governing the growth of bacteria (including botulinum spores) is available moisture, to which the term "water activity" is applied. Not all the moisture present in a food, in this case the oil as well as the plant material, is available for microbial growth because some of the moisture is physically and chemically bound. Only the free or unbound moisture is available for microbial growth. Pure vegetable oil does contain some small amounts of moisture, but not enough to support the growth of botulinum spores. Botulism becomes a problem only when the moisture content is too high because of moisture present in the plant material, and then precautions must be taken to prevent the germination of the botulinum spores in the plant-in-oil preparation.

A. Homemade Flavored Oils

Flavored oils are used to enhance the flavor of cooked as well as raw foods such as salads. These oils can contain spices, herbs, nuts, seeds, and other plant parts. The microbiological safety of homemade flavored oils is dependent on the method of preparation, which then determines the method of storage and handling.

1. Homemade Flavored Oils Requiring Refrigeration

Fresh or raw plant material including seeds can be added to oil to make flavored oil preparations. However, fresh plant material is high in moisture, and this results in high and available moisture in the oil. This increase in moisture content along with the absence of air, results in conditions suitable for growth of botulinum spores, and this type of preparation requires continuous refrigeration from the time of preparation until use. Botulinum spores associated with plant material are inhibited from growth under refrigerated conditions. Practices such as adding herbs, seeds, and other ingredients to oil and then placing the container in a sunny window for weeks to develop the flavor are considered to be very dangerous. Homemade flavored oils made from fresh or raw plant materials must always be kept under refrigeration and used within 1–2 weeks. Such preparations, if left any longer than 2 hours between 4 and 60°C, could support growth of botulism spores with the formation of botulinum toxin, without any evidence of spoilage such as "off" odor, taste, or appearance (8). Should such a preparation be exposed to overheating as just described, it should be discarded, preferably by burying the contents in the ground. Failing this, the container should be sealed in a heavy-duty plastic bag and disposed of as explained later (Sec. XVIII. B).

It is recommended that home-flavored oils be made in small batches so that they can be used up quickly. However, to eliminate the potential for food poisoning, this type of product should be made fresh and used immediately. Leftover quantities should be discarded (17). Commercial flavored oil products made from fresh plant material and not requiring refrigeration can be obtained from the grocery store. Either these commercial products contain preservatives such as salt or their acidity (normally listed on the label) is increased sufficiently to prevent the growth of botulinum spores. Even though these products are sold at room temperature on grocery shelves, once purchased and opened, they should be refrigerated.

2. Homemade Flavored Oils Not Requiring Refrigeration

Flavored oils can also be made through a process called infusion. The easiest and most reliable way to make an oil infusion is to heat the oil for a given length of time at a low oven temperature [300°F (150°C)] with the desired plant material. This allows the oil to absorb the flavor, after which the infusion is cooled and then strained to remove the burnt plant residue. This flavored oil preparation can be stored at room temperature, but it is desirable to make small batches at a time because they are highly flavored and a little goes a long way (17).

B. Plant Material Stored in Oil

Plant materials such as whole or chopped garlic or eggplant are sometimes processed in oil. In this case, it is the plant material that is used in preparing the food, not the oil. The use of fresh or raw plant material stored in oil and prepared in the home is discouraged because the high available moisture could foster the growth of botulism spores. Furthermore, refrigeration of this type of home-prepared product should not even be considered because of the potential for temperature abuse. Dried plant material such as tomatoes, chili peppers, and peppercorns in oil can be safely stored at room temperature, because the amount of moisture remaining in the dried plant material is too low to support the growth of bacteria. Therefore, if the plant material is being dried in the home, steps must be taken to ensure that the product is adequately dried.

Home food preservation requires close attention to detail to produce safe and wholesome food. The advantages of home food preservation are lost when one starts with poor quality raw material. Moreover, when containers fail to seal properly, food spoils, and flavors, texture, color, and nutrients deteriorate rapidly owing to poor processing techniques or process application. However, successful home food preservation is a rewarding hobby in that one is able to prepare safe, wholesome, and nutritious food specifically to suit one's taste. This gives a sense of accomplishment and a feeling of pride, as well as the enjoyment of the economic savings of preserving food bought when it is plentiful and less expensive.

XVIII. PROBLEMS ASSOCIATED WITH HOME FOOD PRESERVATION

A. Safe Food Starts with Proper Home Canning

If the processing equipment and containers were in good condition and every canning step was followed correctly, both acid and low acid foods should be safe to eat. In as much as spoilage and toxin formation can occur without visible signs, particularly in canned low acid foods, it is extremely important to follow the published prescribed steps. As a precaution, some published documents have advised consumers to boil home-canned low acid foods before eating them. The concern is that botulinum toxin could be present in food that shows no signs of spoilage. Because the spores of *C. botulinum* are killed by heat, their survival in canned low acid foods is an indication that understerilization has occurred, whether because the temperature used was not high enough, the processing time was too short, or a combination of these two conditions. This is why it is extremely important to correctly follow published scientifically proven procedures that result in safe and wholesome food, rather than depending on boiling of low acid canned food before consumption as a means of protection against process failure.

B. Warning Signs That Food May Be Unsafe to Eat

Before consuming any home-canned or commercially prepared canned product, one should examine the container carefully for signs of spoilage. Hold the can or jar upright at eye level, and rotate it while examining the outside of the container for streaks of dried-food material originating from the seal area. Upon opening the container, smell for unnatural odors and look for rising air bubbles, unnatural color, bubbly liquid, or mold growth on the surface of the food or underside of the lid. Mold growth generally has a white cotton-like appearance, but it may be blue, black, or green. Do not taste food from any can that looks swollen, rusted, or dented. The same precaution applies to food inside glass jars, or cans in which the food inside smells or looks different from the way it normally does. The obvious signs of microbiological spoilage include putrid odors, cloudy brine, sediment in brine, gassiness or small bubbles in the food, seepage around the seal, mold around the seal, and shriveled or spongy-looking food. Tasting even a small portion of such food can be extremely dangerous. Before opening the container, check the end (can) or top (glass jar) for any abnormal appearance such as swells or stains on the label. Eat food only from containers that have slightly depressed (concave) ends (cans) or tops (glass jars), which is an indication of the presence of a vacuum. Upon opening the container, you should hear air being drawn into the container. A bulged container end or top is a reliable sign that the food inside is unsafe to eat. Spoiled low acid canned foods may not exhibit any signs of spoilage. Therefore, all canned low acid foods made with tomato and tomato-based products that appear suspicious should be assumed to contain **botulinum** toxin and handled carefully. This means not only not tasting the food but avoiding the spread of aerosols upon opening the swollen containers.

If a swollen can or suspect glass jar is still sealed, it should be placed in a heavy-duty, leak-proof plastic bag or similar receptacle, to contain the contents, in the event of rupture. The plastic bag or equivalent should then be placed in a heavy-duty garbage and disposed of with the regular garbage (18). If the suspect can or glass jar has been opened or is leaking, it should be detoxified before disposal.

1. Detoxification Process

Since boiling will destroy botulinum toxin, food and its container can be placed in a boiling water bath. Using disposable rubber gloves, carefully place the suspect containers and lids on their sides in a container suitable for boiling water and large enough to accommodate the cans, glass jars, and water, which will be boiling vigorously. Place the pot on the heating appliance and carefully add water to the pot to completely cover the food containers with a minimum of a 1 in. (2.5 cm) level above the containers. Cover the pot, heat to boiling, and boil for approximately 30 minutes. Allow all items to cool to room temperature and empty food material into a heavy-duty, leakproof plastic bag or some similar container for liquid food, and then place in a heavy-duty garbage bag. Do not try to detoxify containers with swollen lids or sides, as they may explode. Food material can also be buried in the ground, and the empty container disposed of with the regular garbage (18).

2. Decontamination of Equipment and Clothing

Prepare a saturated solution of baking soda in warm water, about 1 part baking soda to 10 parts warm water (25°C) and thoroughly scrub all countertops, containers, and equipment, including all utensils and hands that may have contacted the food or containers (5). Discard any sponges or washcloths used in the cleanup. These can be placed in a plastic bag, which can be placed into a garbage bag and disposed of with the regular garbage. Contaminated clothing can be soaked in the baking soda solution for a few minutes and the clothing washed thoroughly.

XIX. QUESTIONS AND ANSWERS

Q. Is it necessary to use published recipes and procedures, or can I use information passed along over the years?

A. Use only recipes and procedures that have been published in scientifically reputable publications. These recipes and procedures have been developed by food preservation experts based on the most up-to-date scientific information.

Q. How much can I modify these published recipes and procedures for my own use?

A. The recipes and procedures should never be modified.

Q. Why do processing times and temperature differ for home canning recipes?

A. Texture and consistency are the two main factors that influence the time the food takes to achieve the correct temperature. Differences in these factors cause variations in the rate at which heat penetrates each type of food.

Q. What do I need to know about basic food types before I start canning?

A. For canning purposes, foods are divided into two groups: acid and low acid. The canning of foods in these two groups requires specific methods that are not interchangeable.

Q. What types of food are classified as acid foods?

A. Acid foods are foods with pH values less than 4.6. These foods naturally contain acid and generally include all fruits and some vegetables, (e.g., tomatoes and rhubarb are acid foods). Consult any book on canning to determine which foods are acidic.

Q. What types of food are classified as low acid foods?

A. Low acid foods are foods with pH values exceeding 4.6. All meats, poultry, seafoods, soups, and most vegetables (e.g., peas) are low acid foods. These foods must be processed in a pressure canner. Consult any book on canning to determine which foods are low acid.

Q. Can I use a boiling water bath cannner for processing all foods?

A. A boiling water bath canner must never be used for processing low acid foods. The high temperature required to process low acid foods cannot be achieved in a boiling water bath canner.

Q. What causes seal failure with lids on mason jars?

A. There are various reasons for seal failure, such as incomplete removal of air from the headspace due to insufficient process time, defective lids or jars with defective rims, and food contamination between the jar and lid. Too large or too small a headspace can result in the development of insufficient vacuum.

Q. What should be done when seal failure occurs?

A. This applies only to lids checked 12–24 hours after sealing. Check the rim for defects such as tiny nicks. If necessary, change the jar and lid and reprocess within 24 hours for the full process time. Alternatively, the food may be refrigerated in the jar and consumed within several days, or the food stored frozen in the jar after the headspace has been reduced to about 1.5 in. (3.75 cm) (18).

Q. Why is it necessary to keep jars in hot water until ready for use?

A. This exercise is to minimize the chances of jar breakage. Sudden temperature changes may impart thermal shock to the glass and can result in jar breakage. When the boiling water bath canner is used, the water in the canner should be hot but not boiling when the jars are added to the canner. This stage serves to equilibrate the temperature inside and outside of the glass container, and only after this stage should the water be brought to a rolling boil.

Q. Is it important to have the correct amount of headspace?

A. It is important to have the correct amount of headspace for both cans and glass jars. Too much headspace may prevent the formation of an adequate vacuum within the container and may be insufficient to hold the lid in place on the rim of the jar. Not enough headspace could also result in insufficient vacuum, and in extreme cases the food may be forced out of the jar (or cause the can to bulge) because there is not enough room for the food to expand during processing. Such a situation, which results from overpacking the container with food, could lead to underprocessing, with the possible survival of spoilage and food-poisoning bacteria.

Q. Should liquid lost during processing be replaced?

A. No, never. Reopening a container, whether it be a glass jar or metal can, to top it up with liquid encourages postprocess contamination, necessitating the reprocessing of the container. Loss of liquid could result in poor sealing of the lid; containers that have leaked during processing should be refrigerated and used within a few days.

Q. Is it important to remove air bubbles from the food container before sealing and processing?

A. Yes. Air bubbles would rise to the surface during processing and increase the headspace volume. This situation could lead to the development of an insufficient vacuum.

Q. Can acid and low acid foods be safely processed in a oven?

A. No. This practice is considered dangerous. Use of high oven temperatures leads to jar breakage due to thermal shock. Use of lower oven temperatures results in inadequate processing and in spoilage, or the potential for botulism in the case of low acid foods.

Q. Why is the open-kettle method of canning not recommended for acid foods?

A. In the open-kettle method of canning, the food to be processed is cooked in an ordinary cooking utensil, then packed into hot jars and sealed without processing. For whole or cut fruits, the temperatures obtained in the open-kettle method are not high enough to destroy the spoilage organisms. For tomato sauce and puréed fruits, the processing temperature may be high enough, but postprocess contamination could still occur during transfer of the food from the kettle to the jar.

Q. In place of a water bath canner, can a pressure canner be used for processing acid foods?

A. Yes. Some publications provide home canning recipes for use with a pressure canner. Should this procedure be chosen, the instructions should be followed exactly.

Q. Should glass jars and lids as well as metal cans and lids be sterilized by boiling before canning?

A. No. Not when the pressure canner method is used. However, acid foods processed for less than 10 minutes should be filled into sterile containers and covered with sterile lids (4).

Q. Should chemicals be added to help preserve foods being canned?

A. No. Chemicals such as aspirin are not needed to make canned food safe. Some chemicals could even be harmful. Good quality containers and lids, along with proper processing, are all that is needed for canning food that is safe to eat.

Q. Should citric acid or any other acid be added when canning tomatoes?

A. Yes. This is really a safety precaution. Recently developed varieties of tomatoes as well as overripe tomatoes have been found to be not acidic enough to be safely canned in the boiling water canner.

XX. DEFINITIONS

Hermetically sealed container: a container designed and intended to be secure against entry of microorganisms including spores.

Rigid: a filled and sealed container whose shape or contours are neither affected by the enclosed product nor deformed by an external mechanical pressure of up to 70 kPa (10 psig; i.e., normal finger pressure).

Semirigid: a filled, sealed container whose shape or contours are not affected by the enclosed product under normal atmospheric temperature and pressure but can be deformed by an external mechanical pressure of less than 70 kPa (10 psig, i.e., normal finger pressure).

Flexible: a filled, sealed container whose shape or contour are affected by the enclosed product.

Double seam: the closure formed by interlocking and compressing the curl of the end and the flange of the body to form hermetic seal. After double seaming, the flange is referred to as the body hook and the curl forms the cover hook or end hook. This form of closure is called a double seam because it is commonly produced in two operations. The first roll operation forms the metal to produce five thickness or folds, and the second operation roll flattens these to produce the hermetic seal.

Processing: the heating of filled containers of food to a specified temperature for a specific time.

REFERENCES

1. Agriculture Canada. 1975. Canning Canadian Fruits and Vegetables. Publication 1560, Ottawa, Ont., Canada.
2. Anonymous. 1985. Pressure Canning and Cooking, "Presto." Supreme Aluminium Industries Ltd. Scarborough, Ont., Canada.
3. Anonymous. 1994. Ball Blue Book, The Guide to Home Canning and Freezing. Alltrista Corporation, Consumer Products Company, Consumer Affairs Department, P.O. Box 2729, Muncie, IN 47307-0729.
4. Anonymous. 1995. Guide to Home Preserving. Bernadin Ltd., Toronto, Ont., Canada M8Z 5V5.
5. Austin, J., and B. Blanchfield. 1997. MFHPB-16. Detection of *Clostridium botulinum* and its toxins in suspect foods and clinical specimens, in: Compendium of Analytical Methods, Vol. 2., Polyscience Publications, Laval, Quebec, Canada.
6. Canadian Food Inspection Agency, Fish/Seafood & Production Division. 1994. Flexible Pouch Defects. Identification and Classification. CFIA, Ottawa, Ont., Canada.
7. Health Canada. 1990. Recommended Canadian Code of Hygienic Practice for Low-Acid and Acidified Low-Acid Foods in Hermetically Sealed Containers (Canned foods). Health Canada, Ottawa.
8. Health Canada. 1994. Garlic- and- Oil Products. Issues. Health Protection Branch, Ottawa.
9. Hu, K. H., A. I. Nelson, R. R. Legault, and M. P. Steinberg. 1955. Feasibility of using plastic film packages for heat-processed foods. Food Technol. 9: 236–240.
10. Labuza, T. P., 1977. Food and Your Well-Being. West Publishing, New York.
11. McDaniel M. R., R. Diamant, E. R. Loewen, and D. H. Berg. 1981. Dangerous canning practices in Manitoba. Can. J. Public Health 72:58–62.
12. Mermelstein, N. H. 1976. The retort pouch in the U.S.—An overview. Food Technol. 30(2): 28–37.
13. Ministry of Agriculture, Fisheries, and Food, 1982. Home Preservation of Fruit and Vegetables. Her Majesty's Stationery Office, London.
14. Nelson, A. J., K. H. Hu, and M. P. Steinberg. 1956. Heat processible food films. Mod. Packag. 29: 173–179, 248, 250–251.
15. Rutledge, P. 1991. Preparation procedures, in: Vegetable Processing. D. Arthey and C. Dennis (eds.), Blackie, VCH Publishers, New York. pp. 42–68.
16. Tartakow, I. J., and J. H. Vorperian. 1981. Foodborne and Waterborne Diseases: Their Epidemiologic Characteristics. AVI Publishing., Westport, CT.
17. Topp, E., and M. Howard, 1997. Put a lid on it. Small batch preserving for every season. Macmillan Canada, Toronto, Ont., Canada.
18. U.S. Department of Agriculture. 1989. Complete Guide to Home Canning, Information Bulletin 539. Pennsylvania State University, University Park.
19. U.S. Department of Agriculture, Extension Service. 1983. Home Canning of Fruits and Vegetables. Home and Garden Bulletin 8. USDA, Washington, DC.

12
Safe Handling of Ethnic Foods

Gloria I. Swick
Perry County Health Department, New Lexington, Ohio

I. INTRODUCTION

This chapter is designed to expose you to some common violations and situations found during ethnic restaurant inspections. Perhaps you will gain some insight on what to look for and why things are the way they are, and even discover alternative ways to address issues. Every culture and region has its culinary specialties, such as haggis in Scotland and chitterlings (pig intestines) in the southern United States, and each of these culinary specialties presents certain health concerns. For instance, a documented outbreak of *Yersinia enterocolitica* occurred in infants and children in Atlanta; the bacteria had been transmitted from raw chitterlings to the children via the hands of the food handlers (7). This chapter is not intended to be an exhaustive review of all types of ethnic restaurants and foods, but rather an overview of the most common situations that I have encountered as an inspector in Japanese, Chinese, Mexican, Korean, Vietnamese, and Greek restaurants in the United States.

II. JAPANESE

Although the Chinese may have invented *sashimi* and *sushi* (2), *sushi* in Japanese literature is recorded as far back as to the *Heian* period (A.D. 794–1185), and in the present style dates back to the middle of the *Edo* period (A.D. 1700) (14). Many Japanese restaurants today serve *sushi,* which contains vinegared rice, called *shari, sushi rice,* or simply *sushi,* and also *sashimi,* which is fish, octopus, eel, squid, clams, scallops, cockles, shrimp, and other seafood cut up into small pieces, usually served raw. Sashimi, however, can also be made of other meats, such as raw beef.

Each piece of sushi, no matter what shape, is usually a two-or three-bite size (12). It is rice mixed with a vinegared dressing, garnished with seafood and vegetables, and shaped into many different forms (12). There are different kinds of *sushi,* including *oshi-zushi,* which is made by pressing the rice and often other ingredients into little molds; *chirashi-zushi,* which is pieces of cooked or raw seafood and vegetables on a bed of rice; and *maki-zushi,* which is rice and seafood or vegetables rolled in seaweed. The sushi is rolled in seaweed inside a bamboo mat called a *sudare,* which cannot be sanitized and is usually wiped clean by hand. The rice is prepared in a very large wooden cedar bowl called a *handai,* which is usually too large to fit through a dishwasher. Rather, the bowl is simply wiped with a cloth moistened with a vinegar solution, as are the wooden molds that are used to press sushi. Imported Japanese cooking utensils may not meet the food safety requirements in some U.S. jurisdictions.

There are a number of food safety questions concerning sushi. What is the pH of that particular rice steamer of sushi rice? Is the inside of the rice grain, where *Bacillus cereus* may be found, the same as the outside of the grain, where more of the vinegar is found? Can the chef guarantee that each batch of sushi rice is prepared according to the exact same recipe? The final pH of the product is supposed to be approximately 4.2, which would be sufficient to inhibit bacterial growth (11). Is it possible, however, that someone could become ill from eating these foods prepared by these methods? Also, are the conditions of preparation permitted under the existing health codes?

One recipe for sushi calls for ¼ cup of rice vinegar or 3 tablespoons of mild white vinegar (different vinegars may have different pH values.), 3½ tablespoons of sugar, 2½ teaspoons of salt, 1½ of tablespoons *merin* (sweet *saki*) or 1 tablespoon of pale dry sherry, ½ teaspoon of monosodium glutamate (MSG), 2 cups of Japanese or unconverted white rice cooked in 2½ cups of water until it is all absorbed, and a 2 in. square of *kombu.* Even if the recipe was followed exactly each time, there is no way of knowing what the pH would be (16). Another recipe for sushi requires 3 cups of white rice cooked according to the directions on the package and excess water drained off, ½ cup of rice vinegar, ⅓ of cup sugar, and 1 tablespoon of salt (12). As the recipes, chefs, and types of vinegar differ, so will the pH of the end product.

The process of wrapping the fish inside the rice ball is another concern, because the sushi chef dips his fingers in a little bowl of water and works the water into the rice to make it stick together. Some chefs will reuse the same water bowl for an entire shift, and it may sit out at room temperature for many hours. The rice is usually maintained in a rice steamer that has been lined with a cotton or linen cloth. Sushi chefs say that the rice, which is usually maintained at 115 to 120°F (46–49°C) is safe because sushi rice contains vinegar, which lowers the pH, and therefore makes it safe to keep at a temperature lower than is required by most food codes. Sushi chefs say that if the rice is maintained above 140°F (60°C), it will partially cook the fish, which not only would spoil the flavor, but also would destroy the character of the sushimi, or raw product. Shimizu (14) states that room temperature vinegared rice unites the flavors magically with the various types of fresh fish topping or whatever you choose.

Nori, composed of dried sheets of laver, a thin type of seaweed, is used to cover rice balls, for garnishes, and as a skin for *nori makizushi* (12). If seaweed or any other greenery is used as a food or garnish, it should be harvested from non-contaminated water, washed thoroughly, and dipped in a chlorine solution. Some health departments are now insisting that vegetables and garnishes be washed in a mild chlorine solution, and other departments require washing in a strong chlorine solution followed by a potable rinse. "Other ingredients frequently used in the preparation of vegetables, seafood, meat, and fowl include: a green horseradish powder called *wasabi; daikon,* the giant white radish that resembles the white turnip in taste; the strips of dried gourd called *kanpyo; gobo,* or burdock root; and *shirataki,* a preparation made from a yam-like tuber." (16).

Miso is a fermented soybean paste made from rice, soybeans, water, and salt. There are two kinds: red, or *aka miso;* and white, or *shiro miso* (12). An inexperienced inspector, seeing the product in different colors, may think that the color variations are due to spoilage or molds. An inspector who does not know should ask the manager or workers to explain what an ingredient is, how long it has been sitting out at room temperature, or what is in the mixture. Communication is a very important part of any inspection.

Often, *sashimi* is wrapped inside or over a ball of rice, which is called the *sushi* or *sushi rice,* and the display of these products raises concern. The seafood is often stored in refrigerated display cases maintained at about 55°F (13°C), which is considered in the danger zone for rapid bacterial growth, and, furthermore, the seafood has been contaminated by hands, greenery, and seashells. Sushi chefs tell me that if the fish is maintained below 41°F (5°C), it has less or a different flavor.

Sandler (12) notes that *sashimi* is prepared with the freshest fish fillets and shellfish and artfully presented on shredded vegetables with dipping sauces. The fish itself must be of a saltwater variety, since freshwater fish are capable of transmitting parasites to humans. They must be fresh, not 3 days old, and never frozen, since freezing would adversely affect the flavor and texture of the *sashimi.* There is a possibility that the fish may contain *Anisakis simplex,* commonly known as herringworms, (1–4 cm long, off-white organisms, difficult to see with the unaided eye) or *Pseudoterranova decipiens,* or codworms (about 2.5–5 cm long, ranging from white to yellow to brownish or reddish, and about the diameter of a human hair). These parasites may be difficult to see without a lighted preparation surface. In an attempt to prevent the sale of parasitized fish, seafood processing plants "candle" species of fish that are commonly infected. Candling is done by placing the fish fillets on a brightly backlit table and removing any worms that are visible.

The disease anisakiasis is produced when the *Anisakis* roundworms get stuck in the process of trying to attach to or burrow through the stomach or intestinal wall. Anisakiasis, produced by herringworms, is characterized by sporadic abdominal pain, nausea, vomiting, fever, diarrhea, and bloody stools within 1–5 days after eating the contaminated seafood. These worms, if ingested, usually pass through the gastrointestinal tract and are expelled in the human feces; they live no longer than 7–10 days in human digestive tracts. Ingestion of codworms generally results in less severe symptoms or no symptoms other than a tingling throat before coughing up the worms (2). These parasites can be killed if the product is cooked thoroughly to an internal temperature of 140°F (60°C), or if the raw seafood is frozen at −31°F (−35°C) for at least 15 hours or at −10°F (−23°C) for at least 7 days (12). Unfortunately, seafood that has been subjected to those procedures is no longer fresh, and there is a discernible difference in taste and texture. Connoisseurs prefer fresh seafood, which has never been frozen. Frozen fish cannot be satisfactorily shredded when thawed; therefore, fresh fish is used in seafood salad preparations, etc., in many restaurants.

Shellfish have been linked to a number of diseases, including those caused by bacteria such as vibrios, a variety of viruses, including hepatitis A and Norwalk agent, and those caused by toxins. Foodborne infections caused by bacteria and viruses in the shellfish can be prevented by cooking seafoods thoroughly, storing them properly, and protecting them from contamination after cooking (19). (See Chapter 5 for further details.) Steaming clams does not kill all the bacteria and viruses. Cooking shellfish that contains toxins (e.g., paralytic shellfish poison, neurotoxic shellfish poison, diarrhetic shellfish poison, amnesic shellfish poison) does not destroy the toxins, even though the shellfish may look and taste normal. Shellfish screening programs test shellfish, monitor the safety of the harvest beds, and will not approve the harvest of shellfish from sewage-contaminated waters (20). Inspectors should verify that all shellfish in a restaurant comes from approved waters, is properly labeled, and remains in the original container until it is sold. The shellfish tags with harvest and shucking information should be retained for a period of at least 90 days, in case of a foodborne outbreak.

Satterwhite (13) states that in some restaurants it is not unusual to see live shrimp jumping around on the preparation counter, or buckets of slithering eels being carted through the kitchen. Sights such as these are rare in the United States but may be fairly usual in some parts of the world. These situations, which could result in the cross-contamination of all the food contact surfaces, tend to occur among workers who are trained in such kitchens and do not see cross-contamination as a violation. Rather, they see all the items as food, and it will all be eaten. Exotic meats such as boar, venison, and even horsemeat are often cooked *nabe-*

mono-style (i.e., in hot grease at the table) (13). In Ohio, the meat inspector is in the slaughter room and inspects the animal as it walks in to be killed. When small cans of venison or boar meat appear on the shelves of an ethnic grocery, the inspector must assume that the animals were slaughtered under the same strict inspection conditions required in the United States, rather than being sickly animals that fell behind and were killed and gutted in the bush during a hunt. However, it is safe to say that canned meat is probably less hazardous than fresh or frozen meat.

In Japanese restaurants, the popularity of egg dishes has been growing. The Japanese version of the omelet consists of layer upon paper-thin layer of egg, each layer cooked in an oblong pan for a few seconds and then rolled around the previous layer, and served either hot or at room temperature and garnished (16). Many of the omelet pans are metal, and the sharp squared corners of the finished dish are produced when the metal is folded much like bedsheets folded in "hospital corners." Problems arise because the crack in the fold cannot be properly cleaned and sanitized to remove particles of food that may lodge there. Since many of these pans, as well as other cooking utensils used for ethnic cuisine, are produced in foreign countries, they do not meet all the necessary standards for local health jurisdictions. Also, since the utensils are mainly produced abroad, they are difficult and expensive to replace when worn. Often they are used far beyond their life expectancy and remain in service despite being pitted, cracked, and broken.

According to Steinberg (16), the Japanese insist that *Shincha* tea be brewed for 1 or 2 minutes in water between 140 and 160°F (60–71°C), and the cups warmed to the same temperature beforehand. Ordinary tea requires about 180°F (82°C), and only the cheapest grades, *bancha,* are exposed to water near the boiling point. Tea leaves are often harvested from fields far away from sanitary facilities, and the workers cannot properly wash their hands after relieving themselves. Often donkeys are used to pull the carts in the fields at harvest time. The tea leaves may be dropped and stepped on before being returned to the drying racks, or exposed to rodents and insects in warehouses or ships. In Ohio there have been suspected cases of foodborne illness due to *E. coli* in iced tea; therefore, brewed tea is now required to reach a temperature of at least 170°F (77°C) to kill any bacteria that might be present.

In Japanese cuisine, the presentation of food is every bit as important as the taste, and serving food attractively is an art that has been practiced for centuries (13). For instance, *mukimono* means the art of vegetable peeling (6). The atmosphere and appearance is a major part of the *sushi* experience. Often the little pieces of fish, eel, octopus, etc., which have been cut into fancy shapes, and the balls of sushi rice, are arranged artfully with greenery on a lacquered wooden tray, multitiered tray, or plate. According to Steinberg (16), the Japanese are adept at imparting additional excitement to food through the use of exquisite dishes and bowls selected for their harmony with particular foods. This dedication to visual appeal actually enhances a meal. The palate is captivated by tastes calculated to complement each other as the eye is intrigued by contrasts of color, shapes, and texture. Many of these imported food contact surfaces cannot be sanitized effectively and are not approved by the National Sanitation Foundation (NSF) or an equivalent body, as is required in some health districts. Other decorative items may be included as well, such as old seashells in which to put the food or vegetation. Many health districts prohibit the reuse of real seashells because they cannot be effectively sanitized. There are artificial shells produced for this purpose that can be sanitized; however, they are considerably more expensive.

In Japanese restaurants, there usually are cloth half-curtains at the top of the doorways from the kitchen into the serving area. Each time cooks, servers, or other employees walk through the door with food, they must move the curtain with their hands, and soon the curtain becomes soiled and stiff from the starch in the food, carried on the workers' hands. Also, it is important to notice whether chefs are wiping their hands on their aprons or multi use rags, or are properly washing them according to the prevailing health code of that the appropriate jurisdiction. While the infractions mentioned above do not actually cause disease, they do facilitate the transmission of bacteria and other disease-causing agents and, therefore, are violations of most health codes.

During an inspection, if health violations are found, the inspector should write the violation properly on the legal inspection form, citing the appropriate health code number, describing what was found that is in violation, and writing down what should be done to correct the problem. The inspector is required to document critical violations and to document the time frame in which violations are to be corrected. Violations such as improper temperatures and poor personal hygiene, which are considered to be critical in most health jurisdictions, usually must be corrected immediately. There is usually a reinspection of the foodservice operation if critical violations exist. What happens next depends on departmental policy. Recurring critical violations or defiant behavior may warrant a hearing or suspension or revocation of license. Many departments tend to allow a certain amount of violations in ethnic restaurants owing to the cultural differences and lack of knowledge on the part of directors, commissioners, and boards of the departments.

III. CHINESE

Several years ago, employees of the Ohio Department of Health walked to a nearby Chinese restaurant for lunch. Later, they developed symptoms commonly associated with *Bacillus cereus* gastroenteritis. This type of illness probably affects people frequently, but since the symptoms tend not to be recognized as foodborne, the illness is almost never reported. However, in the situation just described, the victims recognized the symptoms. Thus they reported the outbreak to the local health department, which investigated the incident, inspected the restaurant's method of handling rice to make fried rice, and educated the restaurant workers on proper food handling techniques. It is not uncommon to enter a Chinese restaurant when it opens in the morning and find large tubs of rice that have been out at room temperature all night cooling, to make fried rice the next day.

Chinese restaurants vary in size from very small to those with large elegant dining areas. Despite these size differences, the establishments usually have many things in common. For example, Asians are seen serving the food and seating customers. Often, however, the food in the back is prepared by persons of other ethnic groups. For example, in central Ohio you might find that your *war su gai* was prepared by Mexicans. Jobs for foodservice workers are traditionally low paying, and people who have just arrived in the country can readily find employment despite having little or no experience and no local references. Often, Chinese restauranteurs struggle with the English language at best and speak few, if any words, in Spanish. Communication is a real problem and, as a result, the inspector is faced with two sets of ethnic food handling customs. It may be necessary to label food containers in multiple languages.

Also, in most Chinese restaurants the kitchen is usually very small in relation to all the activities going on. Most restaurants are designed to eliminate all unnecessary expense by designing the kitchen as small as possible, which may suit some types of cuisine, such as those that buy portioned frozen meat, frozen precut fries, and pastries baked elsewhere. However, most Chinese restaurants order whole, raw product and cut, dice, chop, slice, and roll everything by hand. In a Chinese restaurant that does a high volume of business, many people work to prepare and serve the food and soon fill up the kitchen. Limited space in the kitchen, however, often results in health violations. For example, food may be found too close to dirty dish areas, mop sinks, or hand-washing sinks. Sometimes employees will prepare food on a tray laid on top of a garbage can. Cooked noodles often are drained into large colanders that just fit inside the top of the trash or garbage can.

Oriental and Mexican restaurants purchase large quantities of rice, and finding adequate storage space is often a problem. The storage room becomes crowded and close examination of corners becomes difficult. Many health codes do not explain exactly what the word *clean* means and an inspector's interpretation may differ from that of a restaurant owner or employee. Were food particles in corners deposited today, or has the dirt been there for weeks? In addition, each inspector has different expectations. Some health officials think that floors, walls, and ceilings have little bearing on foodborne illness, while others see them as indicators of how particular the cleaning people are and how well they maintain the restaurant. Also, if a food source is available, rodent and cockroach populations will soon become established and they can and do cause foodborne illnesses.

Persons from countries in which mice may be seen in restaurants do not consider evidence of mice to be a health violation. With this in mind, the corners of shelves and the backs of drawers must be checked for mouse excreta. Mice and cockroaches thrive in secluded areas with an ample food source. Mice seem to like the hard, crunchy noodles served in ethnic restaurants and will feast upon any food left out for the Buddha in Chinese restaurants. It is not uncommon to find unwrapped Twinkies, cookies, fruit, and egg rolls set out as offerings. Experienced inspectors will examine such shrines well for excreta, dead mice, and cockroaches. If contamination of this type is found, the area should be thoroughly cleaned. After extermination efforts, if more excreta, dead insects or rodents appear, restaurant managers will know their attempts to alleviate the problem were not successful. Restaurants should be encouraged to obtain maintenance agreements with extermination companies who know local requirements and standards, as well as the safe and proper use of pesticides.

In the preparation of Peking duck, the skin is made crispy by overnight hanging and airing (8): Some restaurants hang the ducks on a clothesline, which is strung from the back door to the dumpster a practice that should be discouraged.

Often, the back door of a restaurant is propped open a critical violation in many health jurisdiction. In warm weather, when the kitchen becomes hot and crowded, employees may cut up chickens on a table outside, as is done in their native country.

Usually chickens are delivered once or twice per week. If the restaurant is busy, cases of whole chickens may be left out at room temperature until they can be processed. The birds are taken out of the boxes, cut up, deboned, and processed, and then everything is put into the cooler at the same time. Thus some of the birds may have been at room temperature for many hours. The meat is frozen on trays and then sliced thin to be cooked in the wok. The bones are put in buckets or the original delivery boxes, until they are cooked to make broth. Nothing is thrown out. After the chickens have been prepared, there are trays, boxes, bowls, and piles of chicken and chicken parts everywhere—on the floor, on tables, drain boards, garbage cans, etc. Sometimes, the trays are stacked one on top of the other and refrigerated. If the bottom of one tray touches chicken parts, it will contaminate other surfaces.

An experienced inspector carefully evaluates the cleaning of equipment, such as the slicer, because employees often wipe it down with a cloth instead of washing, rinsing, and sanitizing it between uses. As in most restaurants, the can opener is often a source of contamination as well. Asking employees how they clean the equipment may be the only way to know how it is done; however, the response may be that the employee does not understand the question.

Chinese foods, as well as other ethnic foods such as Japanese and Korean, are very labor intensive. The Chinese eat with chopsticks, which means that food must be prepared in bite-sized pieces. Chinese, restaurant employees make egg rolls, won tons, etc., by hand in the back room or kitchen during the nonpeak hours. It is not uncommon to find waiters, cooks, waitresses, and managers sitting around tables eating their lunch while they prepare food. Even in fast-paced restaurants, employees will eat and drink while they work. There is hand-to-mouth contact, as well as the food handling involved, which leads to food contamination.

Proper hand-washing practices often are neglected in all types of food establishments, including ethnic restaurants. Often you will find that to save time people wash their hands not at the hand sink, but in the most convenient sink, even if it is a food preparation or utensil sink. In some countries, running water is available at only one sink and that is considered a luxury. Instead of thorough washing with soap, people will often rinse their fingertips in cold water and wipe them on their aprons or a multiuse towel. A Vietnamese chef remarked that in his country one does not cut down trees, grind them up, and make paper out of them so people can wipe their hands once and throw the paper away! Single-use hand towels are a requirement in most food services. Again, laws may differ depending on the health jurisdiction.

Multiuse wiping cloths, used in most ethnic establishments, are often found lying around preparation and cooking areas. In most health districts, sanitized multiuse cloths are acceptable for wiping equipments only if they are rinsed and stored in a proper solution of sanitizer water. They are not approved for wiping hands or faces. If hands are dirty enough to require wiping, they should be washed at the hand sink using the proper method.

In Asian restaurants, it is common to find woks rather than conventional cooking stoves. There is probably a connection between the shape of the wok and the design of the helmet used by the invading Mongols during the Bronze Age (15). Food could be cooked in the helmet over the fire, and since heat is concentrated in the bottom of the pan, or helmet, the method also represents a very efficient use of fuel. A good carbon steel wok, well seasoned, is essential for cooking Chinese food. Although the wok is one of the oldest kitchen devices, still in use, it is difficult to see how it could be improved. However, woks are always a question in the minds of the inspectors. Should they be on wheels or stationary? Should they have quick-attaching connectors for gas and water or not? Should they have 2 in. (5 cm) air gaps or be plumbed direct into a floor drain?

In Ohio, the wastewater from a wok must be plumbed through a grease interceptor (4). During a busy period, woks are always untidy in appearance. The extreme heat kills bacteria. All bacteria? Obviously not, because the heat may not be evenly distributed throughout the product, and the temperature may not be equal over the entire surface of the wok area. Woks are not washed, rinsed, and sanitized between uses. Instead, restaurant employees use a metal scourer in the wok, stir hot water around the vessel rapidly, and toss all waste contents aside, where the water runs off down the drain. Sometimes, metal bits come off during the scouring and get into the food.

All Oriental restaurants have a supply of cupped-shaped plates or shallow bowls, usually metal, that the workers use to measure out ingredients by hand for each order of food they prepare. They put the contents of the bowl into the wok and cook it rapidly at a high heat before serving it. Most restaurants reuse the metal bowls throughout the shift, sometimes throughout the day, rather than washing, rinsing, and sanitizing them between uses. Between uses, the bowls are left out at ambient temperature, or by the wok where it is warmer. This provides ample opportunity for bacteria to multiply on bowl surfaces. If the food is heated to high enough temperatures for an acceptable period of time and is not recontaminated, the food is safe because harmful bacteria are killed. However, if the food is recontaminated, or maintained at improper temperatures for long periods of time, perhaps while being transported in a take-home container, bacteria can multiply to dangerous levels.

Ingredients are not always refrigerated properly. Most items other than spices and peppers commonly found in the preparation area or near the wok should be refrigerated. Cornstarch mixtures, chopped vegetables, cooked noodles, and meats usually require temperatures to be maintained below 41°F (5°C). Refrigeration units are expensive, and if the food is refrigerated, there is often moisture on the surface, which pops and spatters when it hits the hot grease in the wok. In addition, it takes longer for the internal temperature of the food to reach cooking or serving temperatures if, at the outset, the temperature is below 41°F (5°C), rather than 75°F (24°C) or room temperature. By the time the internal temperature of the food is high enough, the breading on the outside may be cooked too much.

In Oriental restaurants, chefs often prepare the meats and sauces in advance. When they deep–fry meat in batter for sweet and sour pork or *war su gai,* they will drain the excess fat after frying. Corrugated cardboard is often used to absorb excess fat from the food as it sits at room temperature. Cardboard works well but is not approved as a food contact surface because it cannot be sanitized. The cardboard that comes into a foodservice operation has usually been in warehouses, on docks, and often shipped halfway around the world. It is far from clean, let alone sanitized. Sometimes insects, or insect eggs can be found inside the corrugated compartments. Even the cardboard lids of raw chicken boxes are reused to drain grease from cooked foods. Without a doubt, many of these boxes are contaminated with salmonellae and will cross-contaminate the cooked foods.

A more acceptable use for cardboard boxes is to lay them on the floor in front of the wok area to prevent the floor from becoming dangerously slick from the grease that often ends up on the floor. Many health codes require that the floors be smooth and easily cleanable. Some inspectors will allow cardboard on the floor if it is clean in appearance and changed regularly.

In Oriental restaurants, especially the full-service or plush, expensive ones, there is a violation often found regarding the use of tablecloths to cover work surfaces, even surfaces that are smooth, impervious laminants and stainless steel. It is difficult to persuade employees to clean and sanitize work surfaces once they have developed the habit of using tablecloths. They use cloth napkins to cover food and cloth towels to line rice containers. Most health codes do not allow multiple uses of fabrics in this manner, even if the cloths appear to be clean and changed regularly. Some establishments will simply accept a written violation as long as they are not forced to change their procedure. Some inspectors will write the violation, which is not usually considered critical, and debit the points to protect themselves, but will not force the restaurant owner to change the procedure unless the fabrics are used for too long a time or do not appear to be relatively clean.

Uncovered food in storage is a regular violation in ethnic as well as other types of restaurants. Trays of uncovered food (in the refrigerator, freezer, walk-ins, storage room, under the prep table, etc.) are not uncommon. At times, napkins are used, but often there is no covering at all. This may relate to reluctance to use expensive food-grade paper or plastic that must ultimately be thrown away. When a manufactured lid is used the product cannot cool as rapidly and holds in moisture that will make the battered dips soggy and peel away from the meat. In addition, the lids are an added expense and are breakable.

"What has never failed to astonish Westerners are [sic] the strange ingredients that find their way onto the table in China: shark's fin, birds' nest soup, bamboo shoots and sea cucumber salad, not to mention mouse, snake, dog and cat meat. It is true that the Chinese are one of the few races that have no taboos with regard to food" (3). For example, one of the great dishes is *birds' nest soup,* which is made from the mucus from the salivary glands of the small salagane. The mucus is collected by boiling the nests of these birds (17). In Hong Kong restaurants, you could order looed duck tongues moderately priced or any type of dried seafood (15); *bear's paws,* another famous dish, is becoming extremely rare (17). Again, inspector who finds exotic products on the shelves of ethnic grocery stores may assume that these items are from an approved source and were carefully inspected during processing and/or packaging. An inspector who has doubts about the contents or sanitation of an imported product should make them known to regulatory authorities and ask for a written response for the files.

Tea eggs are hard-boiled eggs that are boiled again in tea, with salt added. The shell is somewhat cracked, but not detached, during the second boiling (8). "*Hundred-year-old eggs*" are not over 100 years old but have been slowly cooked by chemical reaction brought about by immersing them in a lime bath (17). "Thousand-year-old eggs" (*songhua dan*) are not exactly antiques; they are really only a couple of months old and have been pickled and buried so that when shelled they have the appearance of jade laced with transparent veins (3,8). Normally obtainable from Chinese markets, these are duck eggs preserved in lime, pine, ash, and salt and coated with dried mud and husks, which gives them the appearance of aged jade. When the mud has been washed off and the shell removed, the egg is greenish-black, with a yellowish-green yolk. It has a pungent, cheesy taste (8). According to the Ohio Food Service Regulations, eggs are considered potentially hazardous and must be maintained below 45°F (7.2°C) or above 140°F (60°C). Picking mud off of greenish-black eggs and serving them to the public is not an option in many health jurisdictions.

IV. KOREAN

Koreans are night people. After dark, the streets bustle, especially in Seoul and Pusan.

> It is in the early hours of the evening that the main streets in the cities vibrate with activity, as the street-food people emerge with their offerings served from small carts with heating units (gas), or with simple baskets of a regional specialty. The women vendors bring out their bags of tricks—specialties that are popular and easily transported. . . . Beef strips and rice rolls in a tomato and hot chili sauce are sizzling on gas griddles. Side dishes of several types and their soy sauce dips are standing by to be taken on a help-yourself basis. Competitors selling the same foods are neck and neck, sitting on the narrow road closed to vehicular traffic. Very little meat is seen except the roasted, bronze-colored pigs' feet, which look wonderful but are as hard as rocks (9).

Korean dishes incorporate such unconventional mixtures as pine nut porridge, octopus porridge, fermented cucumber kimchi, bracken, taro stems, 30 varieties of pickled fish, and pickled fish throat. Other delicacies include seafood soup, beef entrails soup, carp and chicken soup, grilled eel, fried laver, seasoned raw skate, fish eggs, pressed and salted fish roe, and pickled crabs and oysters. The black fungus known as tree ears or cloud ears (*Auricularia polytricha*), which grows on decaying trees, is used in Korean cooking; pieces of it are sold dried in plastic bags in Asian groceries (9).

V. VIETNAMESE

Many of the same problematic food handling practices that are found in Chinese foodservices are also found in Vietnamese foodservices. Like the Chinese, Vietnamese feature rice in many of their dishes and go out of their way to prevent waste.

Fish sauce is to Vietnamese cooking what salt is to Western and soy sauce is to Chinese cooking, and it is included in practically all recipes. It is prepared from fresh anchovies and salt, layered in huge wooden barrels, and fermented for 6 months (10).

Santa Clara County, California, has been considered for years as one of the "melting pots" of the United States. Many immigrants have settled in this region, bringing their customs and cuisine with them. In the 1980s, there was a major influx of people from southeast Asia to Santa Clara County, and the newcomers introduced previously unknown foods and preparation methods. The inspectors there had little experience with those foods and were unprepared to deal with the cultural differences, which resulted in a task force being established to assure that all foods being served from food establishments were safe and wholesome regardless of ethnic origin.

The task force had nine objectives (1). These were (a) to identify ethnic communities which, by tradition may utilize uncommon methods of food storage and preparation in food establishments; (b) to institute a means of educating and continually updating inspectors in the identification, preparation, and handling of ethnic specialty foods, (c) to study the perishability of various ethnic food products; (d) to seek cooperation and assistance of other appropriate agencies, institutions, and community organizations in studies of ethnic specialty foods; (e) to provide a standardized approach to food establishment inspection; (f) to educate various ethnic communities in proper food handling and storage practices; (g) to implement a means of disseminating new information to staff and other interested parties; (h) to foster communication, education, and cooperation; and (i) to encourage food establishment owners and operators to maintain an acceptable level of sanitation while continually striving for improvement.

The following foods are not considered to be potentially hazardous: air-cooled hard-boiled eggs, foods with a water activity (a_w) of 0.85 or less, foods with a hydrogen ion concentration (pH) of 4.6 or below when measured at 75°F (24°C), foods that have been adequately commercially processed and remain in their unopened hermetically sealed containers, and foods for which a variance has been granted by the regulatory authority (20). Variance are given based on laboratory evidence demonstrating that rapid and progressive growth of infectious and toxigenic microorganisms, including growth and/or toxin production of *Clostridium botulinum,* cannot occur in the foodstuffs in question.

Working in cooperation with the University of California, Davis, Department of Food Science and Technology, the task force conducted food histories of 33 different Vietnamese-prepared food products. Water activity and pH measurements of food samples were taken in an effort to determine the potential hazard of each food item (5). To prepare a Chinese-style roasted whole pig, the body cavity is marinated with a mixture of soy sauce, which contains sugar, vinegar, and spices. Then the pig is hung and roasted at 700–750°F (317–399°C) until the internal temperature reaches 165–175°F (74–79.4°C). The conclusion from microbial tests performed on the various parts of the Chinese-style roasted whole pig showed that there was no growth on either the skin or the glazed surfaces, even though the pig had been maintained at temperatures favorable for microbial growth. As a result of the reduced a_w of the skin, the only microbial growth occurred on the cut surfaces of the meat. According to this research, the Chinese-style roasted whole pig is not potentially hazardous until it is cut, and thus it can be maintained whole at ambient temperatures [72°–110°F (22–43°C)] during the first 8 hours or until it is cut up. Inspectors should consult with their superiors in their health jurisdiction to determine whether this should be noted as a violation under the local laws. Some of the results of other ethnic recipes that were tested by the Task Force and the University of California, Davis, are listed in Table 1.

In summary, all the ethnic food recipes that were tested by the task force and the University of California, Davis, were found to have a pH favorable to microbial growth. All except one of the foods tested had substantial moisture to support the growth of microorganisms that could cause foodborne illness. Under favorable conditions for microbial growth, these foods could be hazardous.

VI. GREEK

Gyro meat, or donairs, cooked on a large rotisserie, is found on display in Greek as well as other restaurants. It is usually a lamb or mutton comminuted (finely ground and reshaped) meat product in the form of a large spool; however, ground beef and chicken slices are sometimes used alone or in combination with each other or with lamb (18). The meat is slowly turned on a spit and the outside is seared or brazed while the inside of the meat is in the danger zone for hours. If business is slow, the spit is turned down or off, or the exterior of the spool is trimmed and the meat stored so that it does not overcook. Storage is usually below 140°F (60°C) so that the meat will not dry out or overcook. At the end of the day, the meat remaining on the spool is placed in the cooler or left on the rotisserie until the next day.

According to research data, cooking temperatures in typical donairs may fail to reach 140°F (60°C); however, the aerobic colony counts (ACCs) still decreased in one example from 10^7 aerobic colony counts per gram (ACC/g) in the raw product to 10^5 ACC/g in the cooked product. *S. aureus, B. cereus,* and *E. coli* were present only in low numbers in the uncooked product, however *C. perfringens* can survive the cooking and freezing processes to become a potential problem if the donair is left at room temperature for a lengthy period of time. The data also showed that some foodservice operators failed to properly maintain temperatures during cooking, reheating, and cooling. Additionally the spits are sometimes manually turned, which results in uneven heating of the donairs (18).

Smaller establishments with a relatively low volume of business, might cut the spool of meat in half and put only one piece on the rotisserie at a time. The meat that has been in the center of the spool in the danger zone for hours should be disposed of, not reheated and served the following day. If the spool of meat has been shaved off and sold in a rather short period (i.e., 2–4 hours) and the center portion remains, it could be shaved or sliced off and cooled in individual portions.

Table 1 The Composition, pH, and Water Activity (a_w) of Various Ethnic Recipes

Food type	Components	pH	a_w
Xoi bap (hominy corn sweet rice)	Sweet rice, hominy, sugar, and canned coconut milk	5.8	0.93
Xoi dau phong (peanut sweet rice)	Sweet rice, peanuts, sugar, and canned coconut milk	5.8	0.93
Banh xu kem (choux ala Creme)	Pastry shell made of egg flour and water and filled with milk and instant custard pudding	6.2	0.97
Banh hoi (rice vermicelli)	Rice flour, water, soy oil, and salt	5.9	0.87
Pâté chaud	Frozen French puffed pastry dough filled with ground beef, onions, French bread, milk, sugar, salt, MSG, egg yolks, and spices	6.0	0.96
Cha bo (fried beef patty)	Ground beef, soy oil, dill, anchovy fish sauce, flour, sugar, salt, baking powder, and MSG	6.1	0.96
Banh xu xe (green sushi)	Starch, dried coconut, mung bean squares, water, sugar, green FD&C food color	5.8	0.86
Banh boa	Pork filling and bun	Dough, 6.1 Filling, 6.6	Dough, 0.87 Filling, 0.95
Banh ich	Mung bean, sugar, and rice flour folded in banana leaves	6.0	0.95
Banh ich	Rice flour, mung bean, and coconut milk flour in folded banana leaves	6.0	0.94
Banh it nhan dau	Rice flour, mung bean, and coconut milk flour folded in banana leaves with flaked coconut	6.1	0.95
Cha chien	Fried pork roll	6.2–6.6	0.84–0.85
Cha lua	Steamed pork roll	6.2–6.5	0.96
Che bap	Sweet corn dessert	6.1	0.93
Che xoi nuoc	Mung bean, sugar, and flour dessert cup	5.8	0.94
Che khoai mon	Sweet potato (taro) dessert cup	6.0	0.91
Banh day	Rice dough and pork patty sandwich	Rice dough, 5.9 Pork patty, 6.4	Rice dough, 0.88 Pork patty, 0.94
It tran	Rice flour, pork, and shrimp	6.4	0.94
Banh chung	Square rice cake with mung bean and pork in banana leaves	6.4	0.95

Before reserving, it should reach a temperature above 165°F (74°C). Again, any temperature violations found during an inspection must be documented.

VII. ETHNIC RESTAURANT INSPECTION CONCERNS

The Food and Drug Administration (FDA) is one of the oldest consumer protection agencies in the United States, and, first and foremost, it is a public health agency charged with protecting U.S. consumers. Its approximately 9000 employees monitor the manufacture, import, transport, storage, and sale of about $1 trillion worth of products each year. There are 1100 investigators and inspectors who cover 95,000 FDA-related businesses in the United States, located in offices in 157 cities across the country (21). The FDA inspectors are taught standardized inspection methods, so that an inspector in a foreign country will enforce the same regulations in the same manner and expect the same standards of cleanliness, as an inspector in the United States when inspecting similar plants or foodstuffs. It is difficult to standardize inspectors in the same health jurisdiction when they are inspecting identical chain establishments, let alone ethnic food items throughout the world.

The volume of imported goods regulated by the FDA has tripled from an estimated half-million shipments in 1971 to about 1.5 million in 1992. Most meat and poultry are regulated by the U.S. Department of Agriculture; however, all other food is subject to examination by the FDA upon importation. These food products must meet the same standards as domestic goods and must be pure, wholesome, safe to eat, and produced under sanitary conditions; all labeling and packaging must be informative and truthful (22).

The FDA has entered into agreements called memoranda of understanding (MOUs) with foreign governments to ensure that foreign products are manufactured under sanitary conditions and sampled to ensure compliance with U.S. quality standards. Under the authority of the MOUs, inspectors from the FDA can inspect manufacturing plants in other countries to verify that their practices meet U.S. quality standards. The FDA has nine MOUs with countries that export shellfish to the United States; these MOUs help ensure that the shellfish are raised, processed, packaged, and shipped properly (22).

The FDA automatically detains products that are consistently in violation of FDA standards or are known or suspected health hazards. Shipments of products under automatic detention alerts are denied entry without further inspection. Examples of products currently under automatic detention include ceramic ware (because of possible lead contamination), swordfish (because it contains high levels of mercury), and canned mushrooms (because they have caused several outbreaks of staphylococcal food poisoning) (22).

There are small, convenient, and inexpensive lead test kits available that can be used in restaurants to test for lead in dishes and other food contact surfaces. These test kits contain small tubes with cotton tips that turn pink when touched to the surface of lead contaminated products. Often foreign-manufactured ceramics are coated with glazes that contain lead because this treatment produces brighter colors at less expense. Since the lead can be leached out into the food or beverage, some ceramics (e.g., those manufactured in the United States) no longer contain lead.

At times, large shipments of dried fish or barrels of sauce are imported into a city, such as Chicago or Toronto, and from there are divided into smaller lots and delivered (in the backs of vans or cars) to the final destinations. The inspector finds the dried or frozen fish, which may or may not be eviscerated, in a plastic bag or simply in a box or freezer with no identification. The food must come from an approved source, however, and in most health jurisdictions the burden of proof is with the owner of the restaurant or market. Asking the owner for the bill or receipt may be the only way the inspector has to locate the source. All food in the operation or establishment must be covered or properly protected with wrapping to prevent contamination. All food must be labeled to identify the contents and should be dated. If any employees speak only one language, all languages spoken at that restaurant must be on the labels to permit read identification of the contents. An inspector who has any questions about the contents or the integrity of any product should contact the FDA and ask for assistance.

Often dining areas in restaurants are dimly lit. This practice may give a romantic impression, but it also may mask the general overall cleanliness of the restaurant. Available lighting may be a problem in restaurants because of the expense of electricity and lightbulbs. Sometimes, there are inadequate foot-candles (fc) of light for cleaning, and the staff is unable to do a thorough cleaning job because they are unable to see the dirt. The inspector should always carry a flashlight to examine the corners thoroughly.

Most health codes have lighting requirements. In the kitchen, preparation, and dishwashing areas, it is not uncommon to find a requirement for 40 fc of light, 20 fc may be required in the walk-in cooler, walk-in freezer, and other food storage areas. Often there is no lighting requirement for dining areas unless they are being cleaned, at which time it must be possible to turn the lights up to a specified level. While examining the lights, the inspector should look for light shields or coated bulbs, which would satisfy the local health requirements. Shielded bulbs are often required in food preparation, cooking, dishwashing, and storage areas, as well as in refrigeration units and freezers.

Plumbing in ethnic restaurants should be evaluated as carefully as in establishments of other types. Hiring a professional plumber may be expensive, but many ethnic food restaurant managers do not know local plumbing requirements. Some countries have no plumbing code, and people do their own work. They may attach PVC to copper or ABS pipes, or they may use duct tape or hose clamps. Drain pipes and hoses may run uphill, and they may not be safety-wasted [i.e., there may be no 2 in. (5 cm) air gap between the drain pipe and the floor drain]. Hot water heaters, too, may be improperly self-installed, which could be deadly, and necessary grease traps may be omitted. If there are questions concerning the validity or installation of plumbing, a qualified plumbing inspector should be consulted. Serious foodborne illness can result from improper plumbing. For example, a drain hose from an ice machine in a large hotel in Ohio was stuck down into the drain and not air-gapped. During a large banquet, ice in the bottom of the bin was used, to meet unusually high requirements. The ice had been contaminated with wastewater that had backed up through the drains, and it made several hundred people ill.

Many people around the world take care of their own children, rather than seeking child care facilities. Therefore, another commonality among Asian and Mexican restaurants is the presence of children in the kitchen or back room. This practice violates many health codes on three counts. (a) there must be a definite separation of foodservice operation and living quarters; (b) unnecessary persons are not permitted in the food preparation, cooking, or dish areas; and (c) hands must be properly washed prior to food handling or the handling of sanitized utensils. The inspector who observes the situation closely will often find baby beds, toys, diapers, etc., in the foodservice operation and the children touching food contact surfaces or helping themselves to food when they are hungry. Also, it is not uncommon to find young children, who are not properly schooled in proper food handling procedures and handwashing techniques, helping in the restaurant. Parents sometimes will wipe a baby's nose, feed him, or change his diaper and return to their kitchen duties without proper hand washing because they do not view their baby as a source of contamination.

Problems with the storage of personal items often surface in Asian restaurants. Many Asian people are at work so many hours throughout the day that they basically live at the restaurant, even if they sleep elsewhere. They soon have numerous pairs of shoes, jackets, books, newspapers, and other possessions that accumulate and become cluttered in with the utensils or canned goods. These areas are perfect hiding places for insects and rodents.

If the restaurant is not on a public sewer system, the inspector should inquire about the waste removal arrangement, be it a septic tank and leaching system or aeration system (the small domestic type or extended aeration "package" plant). Often, proper cleaning and maintenance are not emphasized. Many people who have never had a private or semiprivate sewage system are not aware that the units must be pumped regularly and that dumping bleach, yeast, or enzymes into the system does not replace thorough cleaning when solids build up in the tank. Maintenance may include checking, and perhaps replacing, the baffles or tees in the septic tank, or oiling the motor, cleaning the lines, checking the paddles/weirs, and checking the effluent in an aeration system. A properly functioning sewage system has very little odor. Some restaurants build additions onto an existing structure and cover up the entire septic tank. Others pour concrete over the tank and then, years later, when a problem occurs, find that there is no way to clean the tank.

One Mexican restaurant in a small midwestern village has a cleanout for the septic tank in the storage room of the restaurant adjacent to and under the walk-in cooler. Unprotected sewage lines cross over the walk-in cooler from the apartments upstairs. The walk-in cooler, as well as the reach-in freezer, sit on top of the septic tank, which is a concern because of the age of the existing septic tank and the flooring.

For restaurants that are not supplied by a municipal or public water supply, regular water samples should be submitted to a reputable, qualified laboratory for analysis to test for total coliform organisms.

In Korea, prior to 1980, many government officials were hated and despised by the people because health inspectors and other government workers would take bribes and expect pay offs. The health inspectors would only criticize the restaurant owners and would not help them or educate them. In the past, in Shanghai, China, the health inspector was respected, but in recent years corruption has crept in. The salary was low but inspectors received graft in the form of free meals, cash, and liquor. In all countries, there are still some inspectors who expect free meals and see their position as one of power.

For the Chinese, food is not a *part* of living, it is living (15). The sharing of a meal is what seems to hold the culture together, and a common greeting, "*Chi fan le mei you?*" translates as "Have you eaten yet?" An inspector, however, should not accept meals, or anything else of value . When a restaurant owner asks an inspector if he or she would like a cup of tea, that is a way of making the inspector feel welcome in the restaurant and of being polite. Tea is offered only to someone who is respected, and it may be considered impolite to refuse this hospitality. Some inspectors will accept a glass of water. The inspector should not accept the tea or water if it is prohibited by the department standards and should explain that to the restaurant owner.

The health inspector's job is to evaluate, enforce, and educate. Evaluation is at times difficult because the food is different and is prepared differently with unfamiliar equipment. Enforcement can be difficult as well because people will usually do what they have been taught throughout the years. If the inspector is persistent about changing a procedure, restaurant employees will make the change when the inspector is watching, or for a brief period of time following the inspection. The best way to modify behavior or procedures permanently is to educate the owners and employees. The inspector should invite the owner or manager to be present during the inspection, explaining faults and expected procedures as the occasions arise. If the owner or manager does not speak English, the inspector should ask for a translator or interpreter. The owner or manager should be given a copy of the current codes, or laws, for that health jurisdiction in a language that can be understood.

Inspectors should set a good example for the employees by displaying immaculate appearance and always washing their hands thoroughly with soap in the approved method at the hand-washing sink upon entering the food preparation area. Another way to educate restaurant workers is to develop signs to put above the hand sink and three-compartment sink to give instructions in the languages spoken in that establishment. In addition, signs regarding good personal hygiene, temperature requirements, how to properly cool large quantities of rice, and so on may be helpful. For employees who cannot read the signs in any language, pictures may be helpful. Some health departments provide classes for food handlers at no charge to explain the local health code, demonstrate the proper way to mix and test sanitizers, how to properly use and calibrate a thermometer, etc. Having the assistance of a bilingual employee or translator is beneficial if there are employees who do not understand the common language for that area.

Monosodium glutamate (MSG) is a powder that is a flavor enhancer (16), but it has been has been identified as the cause of Chinese restaurant syndrome.

> After eating in Chinese restaurants, which use MSG more extensively than do Japanese, some diners have on occasion complained of burning sensations, a feeling of pressure in the chest, and facial tightness, or numbness. . . . The cause appears to be the use of ¼ teaspoon or more of MSG per serving (5).

The symptoms, which may include migraine headaches as well, are transient and often go unidentified by physicians. Most people do not relate numbness and tingling of their hands and feet or tightness in the chest that makes taking a deep breath impossible as a foodborne illness.

Trash removal is very important and often a neglected area. Some cultures equate throwing things out with waste, and children are taught not to be wasteful from a very early age. To complicate matters, trash hauling service is becoming expensive. Refuse dumpsters should always have the lids closed when not in use, and the hole in the bottom, which is just the right size for a rat to come and go, should be plugged.

VIII. CONCLUSION

Some common violations and situations found during ethnic restaurant inspections have been explained. It is hoped that the reader has gained some insight and knows what to look for in ethnic restaurants and why things are the way they are in these dining establishments. Perhaps alternative ways to address issues have been suggested.

While it is not necessary for everyone to be the same, enjoy the same things, and eat the same foods, it is good for people to know what they are eating, and contracting foodborne illness should not be part of anyone's restaurant experience. Education is very important in furthering this goal. Inspectors, as well as the restaurant operators and foodservice workers, should be educated, and in order to educate, one must communicate.

REFERENCES

1. Anonymous. 1992. Handout 2. Santa Barbara County Health Department, Department of Environmental Health, Santa Barbara, CA.
2. Anonymous. 1994. Nematode Parasites of Fish. New England Fisheries Development Group, Boston.
3. Anonymous. 1994. Baedeker's China. Prentice-Hall Travel, New York.
4. Anonymous. 1998. Article 4101:2-63-02 of the Ohio Administrative Code. Banks-Baldwin, Cleveland.
5. Gans, J. H. 1996. Ethnic foods and cross-cultural communication. Proceedings of the Annual Meeting of the International Association of Milk, Food, and Environmental Sanitarians, Seattle, WA.
6. Haydock, Y., and R. Haydock. 1980. Japanese Garnishes: The Ancient Art of Mukimono. Holt, Rinehart & Winston, New York, p. 1.
7. Lee, L. A., A. R. Gerber, D. R. Lonsway, J. D. Smith, G. P. Carter, N. D. Puhr, C. M. Parrish, R. K. Sikes, R. J. Finton, and R. V. Tauxe. 1990. *Yersinia enterocolitica.* New Engl. J. Med. 322:984–987.
8. Lo, K. 1979. The Encyclopaedia of Chinese Cooking. Bristol Books by William Collins and Co., New York.
9. Marks, C. 1993. The Korean Kitchen: Classic Recipes from the Land of Morning Calm. Chronicle Books, San Francisco.
10. Ngo, B., and G. Zimmerman. 1979. The Classic Cuisine of Vietnam. Barron's Educational Series, Woodbury, NY.
11. Pong, L. W. *Anisakis* Parasites and Sushi: Some Food for Thought. San Francisco.
12. Sandler, S. T. 1974. The American Book of Japanese Cooking. Stackpole Books, Harrisburg, PA.
13. Satterwhite, R. 1998. What's What in Japanese Restaurants—A Guide to Ordering, Eating, and Enjoying. Kodanska International/USA through Harper & Row, New York.
14. Shimizu, K. (ed.). 1988. Sushi at Home. Shufunotomo/Japan Publications, New York.
15. Smith, J. 1989. The Frugal Gourmet Cooks: Three Ancient Cuisines: China-Greece-Rome. William Morrow, New York.
16. Steinberg, R. 1976. Recipes: The Cooking of Japan. Time-Life Books, Alexandria, VA.
17. Summerfield, J. 1994. Fodor's China: The Complete Guide with Scenic Tours and Business Travel Tips. Fodor's Travel Publications, New York.
18. Todd, E. C. D., R. Szabo, and F. Spiring. 1986. Donairs (gyros)—Potential hazards and control. J. Food Prot. 49:369–377.
19. U.S. Centers for Disease Control and Prevention. 1997. Food and Waterborne bacterial Diseases.
    http://www.cdc.gov/ncidod/diseases/bacter/shelfish.htm

20. U.S. Food and Drug Administration. 1995. Food Code. U. S. Public Health Service, Department of Health and Human Services, Washington, DC.
21. U.S. Food and Drug Administration. 1997. The Food and Drug Administration: An overview. U. S. Public Health Service, Department of Health and Human Services, Washington, DC.
http://www.fda.gov/opacom/hpview.html
22. U.S. Food and Drug Administration. 1997. Imports and FDA. U. S. Public Health Service, Department of Health and Human Services, Washington, DC.
http://vm.cfsan.fda.gov/~1rd/imports.html

13

Food Safety Information and Advice in Developing Countries

Frank L. Bryan
Food Safety Consultation and Training, Lithonia, Georgia

I. INTRODUCTION

Diarrhea is common among citizens as well as tourists in developing countries, and frequently it is the primary cause of death, particularly among children, in these countries. Foodborne disease surveillance is often nonexistent. Therefore, most available data are reports of cases of diarrheal illnesses by physicians or clinics, anecdotal articles in newspapers or complaints by tourists; seldom are the reports laboratory-confirmed. Because of limited knowledge of the foodborne diseases on the part of investigators (if, indeed, investigations are made) or because of political pressures to not casting blame on an industry, an establishment, or an individual, little is learned to serve as bases for subsequent preventive and control measures.

Food safety information and advice can be given to health authorities and citizens of developing countries, but implementation is difficult because of many barriers. The lack of financial resources is the primary barrier. Political considerations and the need to sell agricultural commodities and provide employment for citizens of agrarian cultures may be counter to improved food safety efforts. Administrative structure and protocol in developing countries often are slow and sometimes prevent either rapid or effective action. The lack or limited utilization of food microbiology, food technology, and food safety training, or the provision of outdated information on these subjects, can confuse officials about priorities or the best courses of action to take. Deficiencies in health education of the public may prevent citizens from knowing what to do to prevent foodborne illnesses among their family members or the public to which they cater foods. Part of the reason for inaction is not knowing to what extent a foodborne disease problem exists or indeed, whether it exists at all. Furthermore, foodborne diseases are often considered as low in priority compared with apparently more severe social and economic problems (e.g., poverty, malaria, yellow fever).

Although it is often assumed that there is a high risk of acquisition of foodborne illness from foods prepared and/or sold on streets and in small food shops in developing countries, there are only scant epidemiological data to support this hypothesis (8). Nevertheless, certain epidemiological data from other sources can give information on which to base estimates of risk. Of particular value among these are data on vehicles and factors that contribute to the causation of foodborne disease outbreaks.

II. HAZARDS AND HIGH RISK FACTORS

A hazard is an unacceptable contamination of a microbiological, chemical, or physical nature, and/or unacceptable survival or persistence, and/or unacceptable growth or increase of the contaminant. The assessment of hazards must be done in reference to the severity of outcome if they persist or influence the final product and to the probability of their occurrence during food production, processing, preparation, and vending.

A. Factors That Contribute to Outbreaks

Preparation and storage practices that have contributed to the causation of outbreaks of foodborne diseases indicate high risks and direct attention to operations that are apt to be designated as critical control points. A critical control point is an operation at which preventive or control measures can be taken that will either eliminate, prevent, or significantly reduce, minimize, or delay one or several hazards. For example, contributory factors in the United States (2, 5) and in England and Wales (34, 35) are classified into vital and trivial categories and listed in Table 1. Similar data not included in the table have been tabulated in Australia (25) and Canada (41). The factors occurring in each country are remarkably similar. The same factors occur elsewhere even though epidemiological data may be sparse and not yet collected or tabulated in such detail. Differences in incidence of foodborne diseases between cultures will primarily be affected by agents present in or on the raw foods and by food handling and storage practices. In all situations resulting in foodborne disease, there must be contamination, then survival or contamination after heat processing, and often situations that allow proliferation of pathogens for typical foodborne illnesses to occur. It can be assumed that such situations occur in developing countries also, and probably the risk of their occurrence is high.

Table 1 Factors That Contributed to the Occurrence of 1918 Outbreaks of Foodborne Disease in the United States (1961–1982) and 1479 Outbreaks Foodborne Diseases in England and Wales (1970–1982)

	United States		England/Wales		Total	
Contributory factor	Number	Percent[a]	Number	Percent	Number	Percent
Vital factors						
Improper cooling	839	43.7	1034	69.9	1873	55.1
Storage at ambient temperature	405	21.1	566	38.3	971	28.6
Inadequate cooling	378	19.7	468	31.6	846	24.9
Lapse of 12 or more hours between preparation and serving	434	22.6	844	57.1	1278	37.6
Inadequate reheating	203	10.6	391	26.4	594	17.5
Inadequate cooking/canning/heat processing	298	15.5	223	15.8	521	15.3
Colonized person handled implicated food	348	18.1	65	4.4	413	12.2
Incorporating contaminated raw food/ingredient into foods that received no further cooking	303	15.8	93	6.3	396	11.7
Improper hot holding	255	13.3	77	5.2	332	9.8
Contaminated processed food (source unidentified)	246	16.6			246	7.2
Cross-contamination	104	5.4	94	6.4	198	5.8
Obtaining food from unsafe source	192	10.0			192	5.7
Intermediate factors						
Use of leftovers[b]	66	3.3	62	4.2	128	3.8
Improper cleaning of equipment/utensils	103	5.4			103	3.0
Inadequate/improper thawing	7	0.4	95	6.4	102	3.0
Toxic containers/pipelines	61	3.2			61	1.8
Extra large quantities prepared			48	3.2	48	1.4
Intentional additives	46	2.4			46	1.4
Mistaken for edible varieties	33	1.7			33	1.0
Improper fermentation	25	1.3			25	0.7
Incidental additives	24	1.3			24	0.7
Trivial factors						
Inadequate acidification	5	0.3			5	0.1
Poor dry storage practices	5	0.3			5	0.1
Contaminated water	4	0.2			4	0.1
Postprocessing contamination	3	0.2			3	0.1
Slow/inadequate drying	2	0.1			2	0.06
Misbranding	2	0.1			2	0.06
Faulty sealing	1	0.05			1	0.03
Soaking time too short	1	0.05			1	0.03
Growth during seed germination	1	0.05			1	0.03
Improper preservation	1	0.05			1	0.03
Inadequate dishwashing (contamination afterward)	1	0.05			1	0.03
Contamination by fertilizer or soil	1	0.05			1	0.03
Flies on foods	1	0.05			1	0.03

[a]Percentage exceeds 100 because multiple factors contribute to single outbreaks.
[b]Also lapse of 12 or more hours.
Source: Refs. 3–5.

B. Vehicles

Data on foods that are commonly implicated as vehicles of foodborne pathogens are available in national surveillance reports in some developed countries. If such data are unavailable on a local or regional basis, and related operations (e.g., food-service) are carried out, similar risks ought to occur when the same types of foods are prepared in developing countries. For example, Mexican-style (particularly beans and ground or shredded meats) (9) and Chinese (particularly fried rice) (10) foods are common vehicles of outbreaks of foodborne diseases in the United States (6). These or similar ethnic foods, originally and commonly prepared in homes, small food shops, and restaurants, and by street vendors in one region, are now often produced for more diverse populations in many regions throughout the world. Hence, risks are implied regardless of whether epidemiological data exist to confirm them. Other foods (e.g., gyros or donairs) that are prepared and sold in some countries are implicated occasionally (19). Many other foods because of their composition and preparation practices are potential (if not actual but thus far undetected) vehicles of foodborne illness. Others, however, are quite shelf stable and present a low to negligible risk. (See Ref. 3 for further explanation and classification of risks.)

Whether pathogens reach foods depends on raw ingredients, handling, and preparation procedures. Whether any pathogens present survive, depends on the type (e.g., vegetative cells or spore formers) and quantity of contaminants and on the extent of heating, acidification, or other operations intended to inactivate the pathogens or their toxins. Whether survivors or newly acquired contaminants propagate depends on (a) time–temperature exposures, (b) atmosphere (E_h) surrounding the food, (c) characteristics of the pathogen, (d) ratio of total microbial flora to pathogen, (e) and characteristics [e.g., nutrients, pH, water activity (a_w), (E_h), and natural or added inhibitory substances] of the food.

III. HAZARD ANALYSES

Despite the limitations associated with the lack of epidemiologic data on diarrheal diseases in developing countries, information of potential contributory factors can be obtained by conducting hazard analysis at typical food processing, preparation, and vending operations. This will, however, require knowledge of the ecology of the common agents that cause foodborne illness and an awareness of other potential etiologic agents on the part of the analyst(s). Thus, either authorities from outside the country or key persons from the ministry of health or others responsible for food safety need to be trained and equipped to do this task. This requires financial resources and commitment to understand the factors that contribute actually or potentially to outbreaks of foodborne diseases.

A few hazard analyses have been done in homes, at street vending sites, and in food stands and shops in a few countries in widely scattered geological regions

(14–18, 21–28, 30–33, 36–40). At these locations, a multiplicity of foods have been prepared in ways traditional to ethnic groups residing in the regions. The hazard analyses have discovered situations that have demonstrated sources and modes of contamination, explained the manner by which the contaminants survived cooking or other potentially lethal processes, and/or demonstrated the way pathogens propagated to the populations or generated the toxins that can cause illness. The foods analyzed in developing countries in several continents vary considerably in composition and methods of preparation, but they are exposed to similar sources of contamination wherever grown, harvested, processed, or prepared. Despite variation in foods prepared in these places, hazards observed or otherwise identified have been remarkably similar (14–18, 21–28, 30–33, 36–40).

Hazard analyses done in developing countries can detect on-site hazards and assess related risks where foodborne disease surveillance is either underdeveloped or underutilized. Such evaluations are part of the hazard analysis critical control point (HACCP) approach to food safety. This approach consists of the following successive, interrelated activities: (a) conduct hazard analysis, assess severity of outcomes if hazards are not prevented or controlled, and estimate risks of occurrences of the hazards; (b) determine critical control points; (c) select effective preventive or control measures and set appropriate criteria (or critical limits); (d) monitor critical control points; (e) take prompt corrective actions when results of monitoring show that a hazard exists or that control has been or is being lost; and (f) verify that monitoring is being done effectively and the HACCP system is in place and maintained (34–36). Although the HACCP concept was initially developed for use in food processing plants, it, or at least part of it, is applicable for preparing, holding, and vending foods.

Hazard analyses include (a) determining the extent of contamination of raw foods and ingredients; (b) watching preparation, handling, and holding practices; (c) measuring, as appropriate, time–temperature exposures during heating and holding, pH, and/or water activity of certain foods; (d) sampling and testing foods at appropriate stages of preparation for contaminants of concern, as applicable to confirm hypotheses about sources of contamination, survival, and growth/concentration/attenuation; and (e) conducting challenge studies, if necessary, to provide confirmation of hazards (7, 11, 29). Such studies have demonstrated that hazards are readily detectable and risk predictable.

Any one or a combination of the following situations can contribute to high populations of microorganisms on or in raw foods that are purchased in developing countries. These and other hazardous operational procedures can be observed, measured, or tested for during hazard analyses.

A. Raw Foods

Hazards associated with foodborne diseases begin at the place of growing, harvesting, or production of foods and continue as the foods are transported, processed, and prepared. At the place of production or harvesting, contaminants may come from (a) the soil or mud in which crops are grown; (b) water in which the animal or vegetable develops, water that contacts vegetables during their development, or water that is ingested by food source animals or farm workers; (c) infected or colonized animals; (d) human or animal fecal waste that directly or indirectly reaches the food; (e) persons who handle animals or pick the produce coupled with poor hygienic practices; (f) insanitary practices that spread contaminants or allow survival of the contaminants during processing; and (g) long durations of storage at temperatures that are conducive to microbial growth. For example, soil and mud introduce spore formers (e.g., *Clostridium botulinum, Clostridium perfringens, and Bacillus cereus*) and other soil-associated bacteria (e.g., *Listeria monocytogenes*) to the foods. The water used to irrigate or spray pesticides, growth promoters, or fungicides onto crops may introduce contaminants and/or provide an environment or time for parasites to develop to infectious stages. Sewage discharges into water courses are obvious sources of human intestinal pathogens for fish, shell fish, and water-grown vegetables. Additionally, some pathogens (e.g., *Vibrio parahaemolyticus, Vibrio vulnificus*) occur naturally in seawater and readily contaminate marine life. Fertilization with human feces (night soil) or animal manure is an obvious source of pathogenic enteric microorganisms. The habitual environments and contacts of plants or animals include other sources of contamination that are unavoidable, and contamination is an expected result. After harvesting of crops, water (such as that obtained from rivers, lakes, or canals) used to wash produce or to cool heat-processed foods may introduce additional contaminants. Carts, truck beds, and railcars that previously held live animals, raw foods, or toxic substances and come in contact with raw foods or foods packaged in permeable bags or other containers may be an additional source of contamination.

Consequently, raw foods are frequently contaminated with foodborne pathogens. For example, in homes and at street vending stands in a mountain town in Pakistan, salmonellae were isolated from raw ground meat, raw chicken flesh, egg shells, and raw buffalo milk (22, 24). Greater than 10^5 cfu/g coliform bacteria were isolated from raw milk, ice cream mixes and products, and pulse patty mixes. Raw foods were further contaminated by bare hands of persons cutting, chopping, mixing, or otherwise handling them and by unclean or improperly cleaned utensils and equipment surfaces. They also were subjected to time–temperature conditions conducive to bacterial growth.

B. Cooking

Hazard analyses, with few exceptions, have revealed that foods are usually thoroughly cooked in developing countries (9, 21–24, 30, 39). Hence, vegetative forms of pathogenic bacteria ought to have been killed at least on surfaces if not in the interior of foods during cooking. Bacterial spores, however, would survive and germinate later when temperatures become conducive for bacterial growth. During cooking of Greek or Middle Eastern *gyros (shawarma, dona kebabs, donairs),* temperatures lethal to vegetative pathogenic bacteria were attained at the meat surface and in a thin layer beneath the surface, but nowhere else (19). Only meat sliced from the surface, however, is normally served.

The major hazards for cooked foods commence after cooking. These are fourfold: (a) handling cooked foods with bare hands; (b) preparing cooked foods on cutting boards, on tables, and/or with utensils previously used for raw foods (i.e., resulting in cross-contamination); (c) holding foods at outdoor or room ambient temperatures for several hours (sometimes with the aid of charcoal or heating devices); and (d) insufficient reheating, if indeed the foods are reheated. All these situations have led to contamination, survival, or growth of foodborne pathogens during food preparation and storage (2, 5, 25, 34, 41).

C. Handling Foods After Cooking

Foods sold commercially are frequently handled after heating and then held for many hours on display. In homes, the degree of handling is often less, but the holding time is often as long as in the commercial setting. In Pakistan, for example, staphylococci reached cooked potatoes while they were being peeled, cut, and otherwise handled (23). These and other bacteria were also transferred to products on display during shaping and garnishing. Staphylococci increased ($\leq 10^5$ cfu/g) and elaborated enterotoxins while the contaminated foods were held for several hours on display. Large numbers (usually $> 10^5$ cfu/g) of coliform bacteria and aerobic mesophilic microorganisms (10^6–10^9 cfu/g) were isolated from all foods after handling and then holding for several hours. Furthermore, salmonellae were isolated from wooden (often heavily soiled) cutting boards used often for both raw and cooked foods by street vendors (23). In Chinese-style restaurants, cooked ducks and *char siu* (marinated baked cuts or strips of pork) were subjected to cross-contamination during cutting and other handling after cooking (12, 20).

Confectioneries are often sold on streets or in small shops. In Pakistan, for example, several confectioneries are made from milk products. These include, *khoa* (a concentrated milk having an a_w value of approximately 0.97) and a cheese made by a renin process (40). The *khoa* as received by the candy maker was contaminated with *Staphylococcus aureus* and contained enterotoxin. The *khoa*-based confectionery was subsequently cooked to temperatures that would be expected to be lethal to staphylococci but insufficient to inactivate staphylococcal enterotoxins. Nevertheless, high populations of staphylococci were often found in the finished products because of additional contamination during the handling that followed the heating. Furthermore, *khoa*-filled confectionery and confectionery made from cheese were contaminated by salmonellae. These bacteria reached the products either during cooling in water or during handling after cooking. Multiplication occurred in the warm environment at the place of manufacture and could continue in products that have a sufficiently high a_w during transport, at vending sites, and within retail outlets.

D. Holding Foods at Room or Outdoor Ambient Temperatures and Leftover Foods

Allowing foods to remain at room or outdoor temperatures for several hours is the most frequently occurring factor that contributes to foodborne illness (2, 5, 25, 41). Hence, leftover foods are a high risk of becoming vehicles of foodborne disease agents. Rice, chick peas, cornmeal cereal, and beans are often held at ambient outdoor temperatures while on display on vendors' stands in many parts of the world. In Egypt, for example, foods held in hotels, restaurants, and small food shops and by street vendors had a lower mesophilic aerobic colony counts and a lower prevalence of *B. cereus* when held at temperatures above 54.4°C than when held below this temperature. Food temperatures after cooking decreased with increased storage time until they reached the ambient room or outdoor temperature with accompanying large bacterial populations (27, 35). *Bento* (Japanese-style box lunches) are kept at room temperature while on display by vendors or in shops; microorganisms multiply as time passes (13). In homes and at street vending operations in Pakistan, large populations (10^4–10^7 cfu/g) of *C. perfringens* were isolated from samples of cooked pulses, ground meat dishes, and chickpeas on display for sale, 8–10 hours after cooking (21–23). Populations of *B. cereus* ($\leq 10^5$ cfu/g) were isolated from cooked foods after a 6-hours or longer holding period. Holding stacks of pulse patties on the cooler edge of a griddle for several hours would have allowed germination and growth of bacterial spores. Aerobic colony counts were also high in these and other foods that were held for several hours, unless kept hot at temperatures above 55°C throughout the holding period or periodic reheating practiced (which was done by only a few vendors).

In the Dominican Republic, large populations of aerobic mesophilic organisms, but not always associated with pathogens, were found in fried foods (e.g., pork, fish, chicken, *yuca,* a root vegetable) held for several hours at outdoor ambient temperatures (15). Many of these were prepared early in the morning and displayed at ambient temperatures throughout the day until sold. Those not sold were held unrefrigerated overnight and often not reheated the next day. To the contrary, in Zambia, *nshima* (boiled and whipped maize meal) was held at high temperatures in pans over glowing charcoal in which steam was generated throughout the entire holding period during the day, preventing bacterial growth. However, large (> 10^5 cfu/g) populations of *B. cereus,* aerobic mesophilic colonies, thermotolerant coliforms, and/or *Escherichia coli* were found in the *nshima,* porridge, and rice after the foods were held unrefrigerated overnight (14, 30, 39). Additionally, salmonellae were isolated from a sample of leftover *kapenta* (cooked dried minnows).

E. Cooling

In developing countries, cooked leftover foods often are not cooled. This is partly due to the lack of refrigerated units, but the practice often prevails where refrigerators are used to store raw foods and to keep beverages cold. Even when foods were refrigerated in foodservice facilities, they did not always cool rapidly because of improper storage practices (e.g., attempting to cool foods in large containers, using tight-fitting lids that prevent removal of heat from the foods, stacking of pans on top of others) (9, 10).

F. Reheating

Temperatures attained during reheating have been quite variable. It has been observed that although foods are usually thoroughly heated during cooking, they often fail, during reheating, to attain temperatures that would be lethal to surviving foodborne pathogens, or those that multiplied during improper holding (9, 21).

G. Summary of Risks

Foods (e.g., Chinese, Dominican, Egyptian, Greek, Japanese, Mexican, Pakistani, Peruvian, and Thai) have been shown to have high risks whether prepared in homes, in foodservice establishments, or by street vendors in any of the cultures studied (4). The risks depend on (a) microorganisms that are likely to reach the foods, (b) preparation and holding practices, and (c) the prevailing understanding of ways in which preparers can handle and hold foods to reduce contamination, kill pathogens, and prevent or slow bacterial growth. Risks are evaluated on the basis of operations that contributed to contamination, survival, and growth of etiologic agents based on observations and measurements.

IV. FOOD SAFETY MEASURES

Critical control points must be determined from hazard analyses. They become the focus of preventive actions at the place of production, processing, or preparation, during official inspections, and through educational efforts. Practical monitoring procedures must be devised by health authorities and applied scientists and implemented by preparers and vendors of foods at commercial operations. Health officials must verify that foods are indeed handled in a safe manner and that monitoring is being done, and done effectively. Critical control points for many foods include source of ingredients, formulation, cooking, manipulation of foods after cooking, holding cooked foods, cooling, and reheating. Simple, but effective, monitoring procedures must be taught to food workers and vendors and to those who verify their operations.

A. Source of Ingredients

Obtaining and receiving incoming foods may be a theoretical critical control point, but for practical reasons, monitoring is often limited to obtaining foods from sources as safe as is practicable or observing signs of decomposition or the state of being frozen, if applicable. Quality may be suspect for many foods purchased. In practice, in developing countries, the food is usually accepted as is at the time of purchase, and the contaminants it harbors must be dealt with during subsequent preparation and holding.

B. Formulation

In developing countries where climatic temperatures are often high, many traditional foods have frequently depended on a formulation that prevents bacterial growth, or at least rapid bacterial growth. Formulation of acidified foods can be a critical control point if sufficient quantities of high acid ingredients are added, and there are provisions to ensure adequate mixing and time for marinating. Formulation can also be a critical control point for heavily salted products (e.g., salted fish), highly sugared foods (e.g., confectioneries), or dried foods (e.g., certain dried seafoods). If sufficient amounts of water are either tied up or removed, pathogenic bacteria cannot multiply, but they can survive for long durations. Amount of high acid ingredients, thorough mixing, time of marinating, characteristic sourness, amount of moisture, and percentage salt and/or sugar potentially can be monitored by preparers and vendors, and public health personnel can perform "verification" by means of pH and water activity meters. Formulation as a critical control point, however, is limited in application by knowledge of the characteristics of the food in question and applicable monitoring and verification procedures using calibrated and applicable monitoring equipment.

C. Cooking

Cooking is a critical control point for most cooked foods. To be effective in attaining its microbiologic goals—to kill parasites, viruses, and vegetative forms of pathogenic bacteria that are initially present in raw foods or ingredients or that reach foods during preparation—temperatures must be sufficiently high for a sufficiently long duration to result in lethality to pathogens. For moist foods, temperatures of 55°C or more can produce lethal effects on vegetative forms of pathogenic bacteria in moist foods if exposure at these temperatures is long enough (i.e., up to 2 h at 55°C). As temperatures increase, time for lethality decreases. For instance, an internal temperature of 74°C will inactivate large numbers of these microorganisms in a few seconds. Foods of decreased a_w require greater time–temperature exposures to attain similar microbial lethality. Cooking, however, is not a critical control point for spore-laden foods because these time–temperature combinations are insufficient to kill spores. In fact, they may heat-shock the spores into rapid germination as the food cools.

Subjective monitoring may be done by observing boiling, by observing change of color of interior portions or juices, or by detecting changes in texture. Effective (objective) monitoring, however, can be done only with a thermometer, thermocouple, or similar temperature measuring devises accompanied by time pieces. Health authorities charged with verification must use such instruments.

D. Manipulation of Foods After Cooking

Manipulation of foods after cooking is a critical control point. Touching cooked foods is a commonly identified practice that can lead, and has led, to outbreaks of staphylococcal food poisoning, typhoid fever, shigellosis, streptococcal septic sore throat, hepatitis A, and gastroenteritis due to small round structured viruses. When the contaminants from the hands are bacterial pathogens and the foods are to be held within a temperature range that is conducive to bacterial growth, minimal infective or toxigenic doses can be attained. Minimal infective doses for viral and some virulent bacterial pathogens (e.g., *Campylobacter, Escherichia coli* O157:H7, *Shigella*) do not need the subsequent time–temperature exposure to cause illness. To minimize chances of contamination, handling must be such that pathogens are not acquired from bare hands of food workers (e.g., workers must use clean utensils rather than bare hands). Equipment surfaces that have contacted raw foods of animal origin are usually contaminated with pathogens, so they must be cleaned between such usages. Monitoring and verification are by observation. For these checks to be accomplished effectively, food workers must be aware of food safety hazards and practice self-discipline.

E. Holding Cooked Foods

Holding foods after cooking at improper temperatures is the greatest hazard and calls for a critical control point. Foods that are not held hot (i.e., above maximum temperature for multiplication of pathogenic bacteria) are often near the optimal temperatures for microbial growth. Hence, to remain safe, foods must either be held for only a short duration, or holding temperatures must be at or above those at which spores cannot germinate and resulting cells and newly acquired vegetative pathogens cannot multiply. Temperatures exceeding 55°C should suffice, but the regulatory criterion is often 60°C. Effective monitoring can be done only with a temperature measuring device and/or a timepiece.

F. Cooling

Cooling, when it is done, is a critical control point. The easiest solution is to eat foods promptly after cooking so that they are not held long, but this typically is not done in many homes, at most street vending operations, and in many foodservice establishments. If foods are left over or prepared several hours ahead of serving, they should be put into shallow containers to a depth not exceeding 7.5 cm and cooled rapidly in refrigerators or by ice. However, this stricture is applicable only if cooling facilities are available and within the economic resources of the preparer or purchaser of the foods. Monitoring is done by observing practices, measuring depths of foods being stored, and measuring temperatures of food and cooling units.

G. Reheating

When leftovers or previously cooked and held foods are reheated, the step can be a critical control point. It is often the last line of defense. As with cooking, time–temperature exposures need to be sufficient to inactivate large numbers of infectious microorganisms or heat-labile toxins. If there has been time–temperature abuse during storage, larger quantities of pathogens often must be killed than were present during the initial cooking. Periodic reheating (e.g., every 4–6 h) could eliminate cells that generated from spores during intervening intervals at which bacterial growth could occur. Monitoring and verification must be done with devices that measure time and temperature.

Heat-stable toxins (e.g., staphylococcal enterotoxins and *B. cereus* emetic toxin), however, will not be inactivated during reheating. Prevention of exotoxin-mediated illnesses rests with preventing their formation by (a) eating foods before the toxins can be elaborated, (b) cooling foods rapidly and holding them cold, or (c) holding foods at temperatures above or below which toxins are formed.

H. Strategy for Prevention

The Pareto principle states that only a few problem situations (e.g., hazards) occur commonly (hence are referred to as the vital few); but many other problems occur either less frequently or rarely (hence are referred to as the trivial many). Priorities for attention should address the vital few hazards, which may represent the 10–20% that cause 90–80% of the harm. For example, situations related to aesthetics (e.g., dust blowing or settling on foods) fall into the trivial or low priority category. On the other hand, certain operations foods undergo (such as handling cooked foods with bare hands and holding these foods within a temperature range that is conducive to growth of bacterial foodborne pathogens) fall into the vital or high priority category. Data on which to make such classifications come from either epidemiological studies or on-site observations and measurements, with rational interpretations being based on scientific information about the microbial ecology of foodborne pathogens in foodstuffs. (See Table 1.)

The Pareto principle must be kept in the forefront of decision making to ensure that attention is focused on high risk operations (i.e., critical control points), not on matters of either minor public health consequences or aesthetics. Hence, holding of cooked foods at outdoor ambient or warm temperatures for several hours is a matter of major concern (one of the vital few or a critical control point) that must be given high priority by health agencies.

Health agency officials in developing countries, food industry personnel, and the public need to become aware of the hazards already described and of appropriate preventive measures. Control actions, training agenda, and educational campaigns ought to be focused on these critical control points.

Management of public health activities for protecting consumers of restaurant and street foods should be based on the HACCP approach rather than on (a) traditional inspection, (b) prepared-product (end product) sampling, or (c) action for correction of facilities deficient merely in aesthetic considerations; otherwise, food safety issues may either go undetected or seem overwhelming. This will require, particularly in developing countries, a change of attitudes of many persons responsible for food protection, equipment to assess hazards and to monitor and to verify operations critical to food safety, and skills in making hazard analyses and applying the HACCP concept to processing, preparation, and vending of foods. Food safety activities must concentrate on informing those who handle, prepare, process, and store foods about specific hazards and means by which control can be applied at critical control points.

A strategy to implement these actions is to first alert and train public health officials (e.g., epidemiologists, food microbiologists, sanitarians, and nutritionists) so that they can focus attention on preparation practices that are hazardous. As hazards are identified by either epidemiologic investigations, hazard analyses, or scientific studies (or hypotheses of likely hazards confirmed by on-site observations and measurements or challenge studies) and probability of occurrence (risks) determined, preventive measures that are practical under prevailing circumstances must be chosen. If no such measures are available, they must be devised. These measures should be demonstrated to operators and action taken to implement them. Health officials must verify that appropriate preventive and control measures are implemented and maintained in all food operations.

V. INFORMATION FOR TRAVELERS TO DEVELOPING COUNTRIES

Travelers to developing countries who eat and drink local food and water are at higher risk than those who ingest the same items in industrialized countries (see also Chapter 15). Some precautions can be taken, but complete assurance of food safety is difficult to attain under present circumstances.

A. Water

Even when chlorinated or otherwise treated, community water supplies in developing countries often do not meet standards applied in many developed countries. Even chlorination does not provide protection against parasites such as *Giardia lamblia, Cryptosporidium,* or *Cyclospora.* Water treatment plant sand filtration is required to remove these pathogens. Even if operations at the treatment plant meet international standards, contamination can occur during distribution. Unbeknownst to water and health authorities cross-connections may exist between the potable and nonpotable supplies. Back-siphonage can occur, bringing polluted water into the main supply when water outlets are submersed and negative line pressure occurs following shutting off the water supply or excessive nearby uses. Power failures are not uncommon in some developing countries, and periodically power is cut off for load shedding. Pump failure can shut down water distribution, other than that from higher elevations than the supply, and this situation can be conducive to back-siphonage where submersed inlets exist. In more rural settings, the potential for occurrence of all these situations is even higher, and the water treatment may be inadequate, or there may be no treatment at all. Furthermore, the water may come from unprotected sources close to sites of fecal pollution. Additional hazards can come from ice in drinks and by brushing teeth with potentially contaminated water.

Consequently, travelers should be advised to drink beverages made from boiled water (such as coffee or tea), or canned or bottled carbonated beverages, beer, or wine (which are acidified by acid ingredients, carbonation or fermentation). Use of water from the hot tap (which receives some heat treatment and holding of the heated water for an undetermined interval) offers some degree of protection when brushing teeth or drinking water in hotels. Ice should be avoided because it may be made from contaminated water and is a known vehicle of pathogens that survive freezing. Bottled water can be ingested, but some local brands have been contaminated with pathogens (1). Some room attendants at hotels collect used bottles and fill them with nonpotable water, either for resale or as a means of obtaining tips. Water can be made safe by boiling, but this is often not feasible for travelers. Small immersion heating elements can be used, however, or the traveler can disinfect water with iodine or chlorine. Water-disinfecting iodine tablets are available from pharmacies and sporting good stores. Five drops of a 2% tincture of iodine solution for clear water and twice the dose for cloudy or cold water can be put into a quart/liter. Chlorine is available in 5% liquid (e.g., household bleach) and in up to 70% powdered (e.g., swimming pool disinfectants) forms, and when sufficiently diluted can be used for water disinfection. A 1 ppm (1 mg/L) solution should be safe. Many parasites, however, survive these treatments. Small-size water filters are sold, but there is insufficient scientific evidence to recommend their use with any assurance of safety.

B. Foods

Contrary to previously held beliefs, foods pose far greater risks of enteric diseases for tourists than water. This is partly due to the practice by many travelers of not drinking community water because of anecdotes about its potential as a vehicle of enteric pathogens. Of greater consequences, however, is that personal action to improve the safety of purchased foods are more difficult than seeking tap water substitutes or disinfecting small quantities of water. All raw foods (e.g., milk, meat, produce) are subject to contamination. Therefore, salads or other dishes made from them pose risks. Milk is often not pasteurized. Hence, products such as cheese made from it would also be potentially hazardous. Eggs may contain pathogens in their yolks, or their shells may be contaminated. Meat and meat products, particularly those prepared in continental restaurants, may be cooked insufficiently to kill pathogens in an effort to provide the style of meats preferred by many American and European customers. Certain species of fish are toxic, particularly if harvested from tropical and subtropical insular waters. Raw seafoods are common vehicles of *Vibrio cholerae, Vibrio parahaemolyticus,* hepatitis A virus, and small, round structured viruses.

Food workers may not practice good personal hygiene, and facilities to enhance such practices are often lacking or deficient. Cooked foods may be prepared in advance at foodservice establishments and by caterers, and held in a manner that is conducive to the multiplication of pathogenic bacteria that survive cooking; such foods may become contaminated by handling after cooking, as well. This time–temperature abuse can occur while the foods are left at room or outdoor ambient temperatures, held warm but not hot in steam tables or warming cabinets, or cooled in large masses and either not reheated or reheated inadequately. Time–temperature abuse is the greatest risk factor leading to foodborne illness in developed countries, and it must be an even higher risk in developing countries.

Under any of these circumstances, personal protection is limited. Raw fruits and vegetables, if purchased for personal use, can be peeled with clean implements in a manner to avoid cross-contamination of the peeled surfaces from the outer surfaces; and hands can be washed afterward. The types of potentially hazardous foods just cited should not be selected or ordered while in undeveloped countries, if this is at all practicable. Hot, recently cooked foods should be selected, if available, or chosen from a menu if the preparation history can be anticipated. Menu descriptions and even observations of food on display, however, can be deceiving. Foods are often prepared in advance and held until ordered in foodservice establishments and at street vending stands. Foods that are supposed to be hot but look or taste cold can be rejected, or reheating can be requested, if practical, but such actions may result in spiteful recontamination on the part of disgruntled kitchen staff. Most of the food protection that is available is limited to the quality control policies and daily operating procedures practiced by the staff of the place at which food is obtained, and sporadic supervision by the responsible regulatory agency. Reputations of establishments are usually unknown by tourists, and the aesthetic appearance of a dining room may be misleading. Therefore, there are risks in large hotels and expensive restaurants as well as with small food shops. Risks are increased as large volumes of foods are prepared and held for several hours or overnight before serving, which are practices that are more likely to occur in large operations. Hence, protective measures are limited, but by understanding them before purchasing or ordering foods in developing countries, one can minimize the risks.

VI. CONCLUSIONS

Food safety in developing countries has not yet been attained. Foodborne disease surveillance and control activities are at various stages of development in different countries. Nevertheless, risks of acquiring foodborne illness for both residents and visitors are high under prevailing circumstances. Reduction of the incidence of detected and undetected illnesses poses a challenge to homemakers, the food industry, and public health and food regulatory agencies. Some of the hazards have been described in this chapter, others will be detected after (a) increased epidemiologic surveillance of foodborne diseases, (b) hazard analyses and observations of food preparation, processing, and storage techniques, and (c) laboratory and challenge studies of foods that become either identified or suspected as being vehicles of foodborne pathogens. A decrease of risk of foodborne illness in developing countries will come only after action is taken to minimize and reduce contamination of foods, kill pathogens that reach foods, and prevent or substantially slow growth of foodborne pathogenic bacteria and toxigenic molds. Education of the public and training of food industry personnel and of food regulatory officials will be necessary to attain these goals. The education and training activities should be based on hazard studies done within the countries as well as on presently accepted food safety techniques.

REFERENCES

1. Blake, P. A., M. L. Rosenberg, J. H. Costa, P. S. Ferreira, C. L. Guimaraes, and E. J. Gangarosa. 1974. Cholera in Portugal. I. Modes of transmission. Am. J. Epidemiol. 105:337–343.
2. Bryan, F. L. 1978. Factors that contribute to outbreaks of foodborne disease. J. Food Prot. 41:816–827.
3. Bryan, F. L. 1982. Foodborne disease risk assessment of foodservice establishments in a community. J. Food Prot. 45:93–100.

4. Bryan, F. L. 1988. Safety of ethnic foods through the application of the hazard analysis critical control point approach. Dairy Food Sanit. 8:654–660.
5. Bryan, F. L. 1988. Risks of practices, procedures and processes that lead to outbreaks of foodborne diseases. J. Food Prot. 51:663–673.
6. Bryan, F. L. 1988. Risk associated with vehicles of foodborne pathogens and toxins. J. Food Prot. 51:498–508.
7. Bryan, F. L. 1992. Hazard Analysis Critical Control Point Evaluations. A Guide to Identifying Hazards and Assessing Risks Associated with Food Preparation and Storage. World Health Organization, Geneva.
8. Bryan, F. L. 1996. Hazard analysis: The link between epidemiology and microbiology. J. Food Prot. 59:102–107.
9. Bryan, F. L., and C. A. Bartleson. 1985. Mexican-style foodservice operations; hazard analyses, critical control points and monitoring. J. Food Prot. 48:509–524.
10. Bryan, F. L., C. A. Bartleson, and N. Christopherson. 1981. Hazard analysis, in reference to *Bacillus cereus,* of boiled and fried rice in Cantonese-style restaurants. J. Food Prot. 44:500–512.
11. Bryan, F. L., C. A. Bartleson, C. D. Cook, P. Fisher, J. Guzewich, B. Humm, R. C. Swanson, and E. C. D. Todd. 1991. Procedures to Implement the Hazard Analysis Critical Control Point (HACCP) System. International Association of Milk, Food and Environmental Sanitarians, Ames, IA.
12. Bryan, F. L., C. A. Bartleson, M. Sugi, B. Sakai, I. Miyashiro, S. Tsutsumi, and C. Chun. 1982. Hazard analyses of *char sui* and roast pork in Chinese restaurants and markets. J. Food Prot. 45:422–429, 434.
13. Bryan, F. L., I. Fukunaga, S. Tsutsumi, L. Miyashiro, D. Kagawa, B. Sakai, H. Matsuura, and M. Oramura. 1991. Hazard analysis of Japanese boxed lunches (*bento*), J. Environ. Health 54:29–32.
14. Bryan, F. L., M. Jermini, R., Schmitt, E. N. Chilufya, M. Michael, A. Matoba, E. Mfume, and H. Chibiya. 1997. Hazards associated with holding and reheating foods at vending sites in a small town in Zambia. J. Food Prot. 60:391–198.
15. Bryan, F. L., S. Michanie, P. Alvarez, and A. Paniagua. 1988. Critical control points of street-vended foods in the Dominican Republic. J. Food Prot. 51:373–383.
16. Bryan, F. L., S. Michanie, N. Mendoza Fernandez, M. Moscoso Vizcarra, D. Taboada, P. S. Navarro, O. A. Bravo Alonso, and E. Guerra Requejo. 1988. Hazard analyses of foods prepared by migrants living in a new settlement at the outskirts in Lima, Peru. J. Food Prot. 51:314–323.
17. Bryan, F. L., S. Michanie, M. Moscoso Vizcarra, S. Navarro, D. Taboada, N. Mendoza Fernandez, E. Guerra Requejo, and B. Perez Muñoz. 1988. Hazard analyses of foods prepared by inhabitants near Lake Titicaca in the Peruvian Sierra. J. Food Prot. 51:412–418.
18. Bryan, F. L., B. Phithakpol, W. Varanyanond, C. Wongkhalaung, and P. Auttaviboonkul. 1986. Phase II: Food handling: Hazard analysis critical control point evaluations of foods prepared in households in a rice-farming village in Thailand. Food and Agriculture Organization, Rome.
19. Bryan, F. L., S. R. Standley, and C. Henderson. 1980. Time–temperature conditions of *gyros*. J. Food Prot. 43:346–353.

20. Bryan, F. L., M. Sugi, L. Miyashiro, S. Tsutsumi, and C. A. Bartleson. 1982. Hazard analyses of duck in Chinese restaurants. J. Food Prot. 45:445–449.
21. Bryan, F. L., P. Teufel, S. Riaz, S. Roohi, F. Qadar, and Z. Malik. 1992. Hazards and critical control points of vending operations at a railway station and a bus station in Pakistan. J. Food Prot. 55:534–341.
22. Bryan, F. L., P. Teufel, S. Riaz, S., Roohi, F. Qadar, and Z. Malik. 1992. Hazards and critical control points of street-vending operations in a mountain resort town in Pakistan. J. Food Prot. 55:701–707.
23. Bryan, F. L., P. Teufel, S. Roohi, F. Qadar, S. Riaz, and Z. Malik. 1992. Hazards and critical control points of street-vended *chat,* a regionally-popular food in Pakistan. J. Food Prot. 55:708–713.
24. Bryan, F. L., P. Teufel, S. Roohi, F. Qadar, S. Riaz, and Z. Malik. 1992. Hazards and critical control points of foods preparation and storage in homes in a village and a town in Pakistan. J. Food Prot. 55:714–721.
25. Davey, G. R. 1985. Food poisoning in New South Wales: 1977–1984. Food Technol. Aus. 37:453–456.
26. El-Sherbeeny, M. R., M. F. Saddik, H. E.-S. Aly, and F. L. Bryan. 1985. Microbiological profile and storage temperatures of Egyptian rice dishes. J. Food Prot. 48: 39–43.
27. El-Sherbeeny, M. R., M. F. Saddik, and F. L. Bryan. 1985. Microbiological profiles of foods sold by street vendors in Egypt. Int. J. Food Microbiol. 2:355–364.
28. Hartog, B. J. 1992. Application of the HACCP concept to improve the safety of street foods. Food Lab. News 8:23–39.
29. International Commission on Microbiological Specifications for Foods. 1988. Microorganisms in Foods, Vol. 4. Application of the Hazard Analysis Critical Control Point (HACCP) System to Ensure Microbiological Safety and Quality. Blackwell Scientific Publications, Oxford.
30. Jermini, M., F. L. Bryan, R Schmitt, C. Mwandwe, J. Mwenya, M. H. Zyuulu, E. N. Chilufya, A. Matoba, A. T. Hakalima, and M. Michael. 1997. Hazards and critical control points of food vending operations in a large city in Zambia. J. Food Prot. 60: 288–299.
31. Michanie, S., F. L. Bryan, P. Alvarez, and A. Barros Olivo. 1987. Critical control points for foods prepared in households in which babies had salmonellosis. J. Food Microbiol. 5:337–354.
32. Michanie, S., F. L. Bryan, P. Alvarez, A. Barros Olivo, and A. Paniagua. 1988. Critical control points for foods prepared in households whose members had either alleged typhoid fever or diarrhea. Int. J. Food Microbiol. 7:123–134.
33. Michanie, S., F. L. Bryan, N. Mendoza Fernandez, M. Moscoso Vizcarra, D. Taboada, P. S. Navarro, O. A. Bravo Alonso, and L. Santillan. 1988. Hazard analyses of foods prepared by inhabitants along the Peruvian Amazon River. J. Food Prot. 51:293–302.
34. Roberts, D. 1982. Factors contributing to outbreaks of food poisoning in England and Wales 1970–1979. J. Hyg. 89:491–498.
35. Roberts, D. 1989. Factors contributing to outbreaks of foodborne infection and intoxication in England and Wales 1970–1982, in Proceedings of the Second World Congress on Foodborne Infections and Intoxications, Vol. 1. Institute of Veterinary Medicine–Robert von Ostertag Institute, Berlin, pp. 157 ff.

36. Saddik, M. F., M. R. El-Sherbeeny, and F. L. Bryan. 1985. Microbiological profiles of Egyptian raw vegetables and salads. J. Food Prot. 48:883–886.
37. Saddik, M. F., M. R. El-Sherbeeny, and F. L. Bryan. 1996. Microbiological status of cooked rice, salads and cooked vegetables served at Cairo and Alexandria hotels, in: Proceedings of the Second World Congress on Foodborne Infections and Intoxications, Vol. 2. Institute of Veterinary Medicine, Berlin.
38. Saddik, M. F., M. R. El-Sherbeeny, B. M. Mousa, A. El-Akkad, and F. L. Bryan. 1985. Microbiological profiles and storage temperatures of Egyptian fish and other seafoods. J. Food Prot. 48:403–406.
39. Schmitt, R., F. L. Bryan, M. Jermini, E. N. Chilufya, A. T. Hakalima, M. Zyuulu, E. Mfume, C Mwandwe, W. Mullungushi, and D. Lubasi. 1997. Hazards and critical control points of food preparation in homes in which persons had diarrhea in Zambia. J. Food Prot. 60:161–171.
40. Teufel, P., F. L. Bryan, F. Qadar, S. Riaz, S. Roohi, and Z. Malik. 1992. Risks of salmonellosis and staphylococcal food poisoning from Pakistani milk-based confectioneries products in Pakistan. J. Food Prot. 55:588–594.
41. Todd, E. D. C. 1983. Factors that contribute to foodborne disease in Canada, 1973–1977. J. Food Prot. 46:737–747.

14

Food Safety Information for Those in Recreational Activities or Hazardous Occupations or Situations

Ewen C. D. Todd
Health Canada, Ottawa, Ontario, Canada

I. INTRODUCTION

This chapter describes situations that may call for specific hygienic requirements because people work in contact with animals (e.g., hunting, trapping, fishing) or in leisure time activities (recreational camping, exploring, mountaineering). Those continually exposed to infected animals should limit their direct skin contact, but this is usually not done. For some pathogens, protection may come from a buildup of immunity. Where preparation, cooking, and storage of food outdoors is concerned, often general good sense and some practical tips will serve to prevent foodborne illness from occurring. In other cases, those who receive the bare minimum of nutrition and live in close proximity to one another, especially under conditions of poor sanitation, are vulnerable to the spread of diarrheal infections. Such persons include workers in recreational camps, refugee camps, or concentration camps, where food safety is only one of many issues facing the operators of the facilities. Military personnel and astronauts often work where access to good quality food is limited for operational reasons, but the possibilities of foodborne and waterborne disease causing serious damage to missions is recognized. In both these areas, considerable efforts have been made to overcome these risks. The most extreme concern is for space travelers, where there is no outside help available, and extra care is taken to prepare and serve safe food and potable water.

The extent and impact of natural disasters, such as floods, hurricanes, and earthquakes, however, are difficult to predict. These events can destroy sanitary systems and prevent the transport from unaffected areas of potable water and nutritious, safe food. Under these conditions, enteric infections (e.g., cholera and shigellosis), can result. In regions where these are prone to occur, preparation is essential.

One area in which it is difficult to mount a defense is in the deliberate use of infectious agents to immobilize or kill people to promote a particular political position. These biological substances have been considered to be agents of terror, either in national warfare or in the hands of small dissident organizations. Their effectiveness on a large scale has yet to be demonstrated, although deliberate contamination of food in rare isolated incidents has been documented.

II. SPORTING AND RECREATIONAL ACTIVITIES

A. Explorers and Mountaineers

Those who explore and climb mountains often do so under extreme conditions of endurance and climate, and if illnesses or deaths occur, it is usually because of injury or exposure. Hodgdon et al. (42) showed that cold weather did not affect troops' cognitive or physical performance, provided they had adequate clothing, nutrition, and sleep. However, if there is something wrong with the food, disastrous situations may occur, as with the Franklin expedition of 1845 to explore the North West Passage. From analysis done on the bones of some of the crew and physical evidence of the direction taken after their ships sank, it is now believed that everyone perished because of bad judgment of the leaders, which was due partly to lead deposited in their bodies following consumption of canned meat that had been sealed with a lead-based solder (9). A less dramatic event occurred in northern Ontario when medical and paramedical personnel were on a survival course and had to spend two days and one night with minimum facilities, including a plastic sheet as a shelter and dehydrated soups, tea, and honey as food (107). Six of the 16 participants complained of gastroenteritis. It was later found that the dehydrated soups contained 47–232 ng/g of staphylococcal enterotoxin C. Under the conditions of limited nutrition and shelter, the relatively small amounts of toxin were apparently enough to cause illness. Therefore, it is possible that on some expeditions, insufficient nutrition coupled with exposure to the elements have made explorers more vulnerable to infectious diseases or intoxications.

Another example illustrates problems associated with outdoor walking. In British Columbia, two young hikers took a cloth-covered metal canteen filled with orange-flavored crystals dissolved in water. One and a half hours after each had consumed half a cup, the hikers developed severe abdominal cramps, bloating, and vomiting (106). An examination of the canteen showed that it was made of zinc, which the acidic liquid had partly dissolved; it was estimated that each victim had consumed 950 mg of zinc. The manufacturer subsequently labeled the canteens, cautioning against use for beverages other than water.

Advice for mountaineers, trekkers, and hill walkers consists mainly of advocating such commonsense practices as wearing or carrying enough proper clothing, bringing extra food and water, knowing the different ways of starting a fire, telling others of the planned trip, and knowing personal limitations. Perishable food should not be taken on any trips lasting more than a few hours. If game meat is prepared or fish caught on the trip, the food should be thoroughly cooked to prevent bacterial or parasitic infections. Giardiasis can result from ingestion of apparently pristine water sources in remote areas; these may originate from infected campers or from animal reservoirs. It is also useful to know that altitude sickness can mimic gastroenteritis, with headaches and vomiting, and the action taken for this condition is clearly different from the requirements for dealing with an infection. Getting patients down to lower altitudes to receive medical attention applies in both cases, however. For outdoor events in general, surface water supply may be contaminated and a source of potable water should be carried or obtained en route.

B. Camping and Recreational Camps

Recreational camps are popular in many parts of the world for people to explore new locations or as a retreat. These can vary from cottages with meals prepared in a central kitchen to backpacking in remote areas with limited facilities. Sporadic cases of gastroenteritis among campers may be considered to be some version of travelers' diarrhea because of an unknown "bug" or water supplies with an unfamiliar chemical composition. Therefore, it is not surprising that individual cases of foodborne or waterborne illness involving campers are not usually reported. Outbreaks, however, may cause sufficient concern (especially if children are involved) to encourage health authorities to investigate them. Some examples may be cited.

Two Cub Packs went to a campsite in Alberta, Canada, in two separate vehicles but carrying ham and potato salad from the same source in their trunks (1). The first car, whose food was in coolers, arrived uneventfully at the site and waited for the second car, which was delayed by a mechanical breakdown and took 6–8 hours for the trip. No one was ill after eating the refrigerated food. Unfortunately, the second car had no cooler and the temperature was over 30°C; those who ate the food from this vehicle developed acute gastroenteritis 3–4 hours later, and 17 were hospitalized with *S. aureus* intoxication. This event was almost a case-control study to demonstrate the importance of refrigeration of potentially hazardous foods on a camping trip.

Thirty school children and four adults went away on an overnight field trip in British Columbia in April 1975 (107). Preprepared hamburger stew had been taken to the camp unrefrigerated in two large pots and was simmered for 5 hours before consumption. Diarrhea and abdominal cramps were experienced by 28 of the children and 2 adults 7–12 hours later; over 10^7 of *Clostridium perfringens*/g was found in the leftover stew, which looked and smelled normal. The "simmering" had allowed the pathogen to grow, instead of destroying it.

Sixty-nine of 95 attendees at a Bible camp in Ontario in 1980 were infected with *Salmonella agona* after consuming leftover roast pork slices (111). Temperature abuse of the pork was likely, since several of the campers had noticed an off-taste.

An outbreak of foodborne illness occurred among participants at an Australian youth camp attended by 820 people on the long Easter weekend in April 1994 (46). The young people and their leaders were predominantly between the ages of 10 and 29 years old. On the second day of the camp, 230 people were ill with gastroenteritis; 118 were treated in emergency departments and 13 were transported by bus and ambulance to a hospital in the Sydney metropolitan area. A total of 43 work days was lost. This was the largest reported outbreak of *Clostridium perfringens* foodborne outbreak in Australia, and the improper handling and storing of chicken was identified as the cause.

E. coli O157:H7 was identified as the agent in an outbreak of bloody diarrhea at a summer camp in Virginia in 1994 (31). The median age of those affected was 12 years, with a duration of illness of 3–10 days. Of the 18 who were ill, many had eaten rare ground beef, especially at meals cooked over a campfire on an overnight trip. No other factors, including consumption of well-cooked ground beef, handling uncooked ground beef, sharing water bottles, or swimming frequency were associated with illness. In 1994 there were three reports of *E. coli* O157:H7 outbreaks at U.S. summer camps. In another episode in 1995, 30 of 109 persons who attended a Boy Scout retreat in Maine suffered from *E. coli* O157:H7 infection after eating lettuce that may have been contaminated from ground beef (66).

Salmonella enteritidis PT 6B caused illness in 46 of 49 members of a Boys and Girls Brigade camping group in North Wales in 1994, and 33 were hospitalized (15). The campers brought all their food, both cooked and uncooked, from their hometown 300 miles away without refrigeration facilities. A lemon meringue pie was the vehicle of infection. Fresh shell eggs, stored after purchase for 2 days in high ambient temperatures, appear to be the most likely source of infection, with multiplication during preparation and subsequent storage of the pie a significant contributory factor. Another *Salmonella enteritidis* outbreak in the same year affected 14 of 28 Scouts at a camp in northern England who had eaten mince prepared with raw shell eggs (23). The authors recognize that campers may be at greater risk than others and should consider the use of portable coolers for the transport and storage of eggs, and avoid the preparation of lightly cooked egg products under typical outdoor camping conditions.

Gastroenteritis may be documented in camps but the cause never identified, as occurred in British Columbia in 1976 (108). Five separate outbreaks were reported between March and June, despite inspection by the public health authorities after the second and third incidents, with no risk factors being identified. After the fifth episode, the camp was closed for the summer.

Mushroom poisoning occurs with regular frequency in persons outdoors who unknowingly pick toxic varieties (110), or in some cases deliberately seek hallucinogenic experiences (109). In Hungary, Russia, and Ukraine, mushroom poisoning is a major cause of foodborne morbidity and mortality (94, 114).

Campers and those traveling on rivers or lakes may become infected with pathogens even though the water appears to be pure. For instance, 21 campers developed campylobacteriosis after drinking untreated stream water in the Rocky Mountains in Wyoming (103). In 1989, 195 of 600 persons at a Scout camp in Sweden were infected with *Aeromonas hydrophila* present in the camp's water supply, which had not been chlorinated because the chlorinator was out of order (34). At a school camp in Christchurch, New Zealand, 67 students and leaders developed campylobacteriosis over 2 weeks; probably the camp water supply was the source of the microbe (13).

It is not possible to travel without water, and access to a potable supply is not always possible. Bottled water for drinking or mixing with food can be obtained but is not practical for backpacking to remote areas. Stream and river waters should be considered unsafe to drink because of bacterial pathogens such as *Campylobacter* and *Shigella.* Illnesses can also occur from ingestion of *Giardia* protozoa, which can come from water contaminated with feces from humans and wild animals; Frost et al. (32) found that 6.3–19% of beavers and 35.2–42.6% of muskrats trapped were infected. Giardiasis is normally contracted through consumption of contaminated water supplies, but also occasionally food accidentally contaminated with water containing feces or through an infected food preparer. Standard chlorination or commercial purification tablets do not destroy parasitic cysts. Filtration is the only method that is certain of removing them. If these treatments are not available to the camper, stream or lake water should not be consumed unless it is boiled. Ongerth et al. (73) determined that heating water to at least 70°C kills cysts and that 8-hour contact with iodine-based chemicals is more effective than treatment with chlorine compounds. Portable filtration devices were also examined, and only two of the five filter kits available for backpackers had a pore size low enough to remove *Giardia* cysts. Potable water should be used not only for drinking but also for preparing food, washing dishes and pots, and brushing teeth. If the water supply is limited, those working with food can use disposable wipes to clean their hands.

In areas where wild rodent or sylvatic plague and tularemia are endemic, campers, backpackers, and hunters are at risk from these diseases if they walk in areas where infected fleas are found (e.g., a prairie dog colony) or if they handle animals in preparation for food. In 1996 five cases of plague, two of which were fatal, were reported from western U.S. states (104). Recommended control measures include warning signs for campers and backpackers in epizootic areas (e.g., at trail heads and campgrounds); avoiding sick or dead animals in these areas; using repellents, insecticides, and protective clothing during potential exposure to rodent fleas; restricting cats and dogs from going into areas with wild rodent plague; removing sources of rodent food and harborage from residences or campsites; and using gloves when handling animals killed by hunting or trapping. Tularemia can be contracted not only through biting fleas but also through direct infection of the skin, or conjunctiva, inhalation of contaminated dust or dried feces, or consumption of water polluted by infected animals or undercooked flesh of rodents, rabbits, or groundhogs. Those at most risk are hunters and trappers who skin and cut up the infected animals. In North America in the past, there have been several instances of trappers being infected at the same time, particularly after handling muskrat. Laboratory workers are also at high risk because of the danger of aerosols.

Keeping perishable food cool is difficult on camping trips. Polystyrene coolers are lightweight, low in cost, and retain the cold well, but they can be expected to have a limited lifetime. Plastic, fiberglass, or steel coolers are more durable but, once filled, can weigh up to 20 kg. Blocks of ice keep longer than packs of ice cubes, but are not as easy to find. At the start of the trip it is possible to use clean, empty milk cartons to prefreeze blocks of ice or use frozen gel packs. Also, food itself, (e.g., bags of milk) can be frozen and used as a coolant, then consumed when unthawed. Unfortunately, care must be taken to avoid freezing in a way that may expand the bags sufficiently to cause small punctures, with resulting leaks.

When one is filling the cooler with frozen or refrigerated foods, the first foods to be packed should be the last to be used. The foods should be wrapped in waterproof packages to prevent contamination by melting ice and leaked or spilled food. Some rearrangement of the foods is usually necessary after a few meals, especially to replace the ice, to consolidate some of the food supplies as they are eaten, and to add newly purchased food. To save space, foods should be obtained in the smallest size needed based on consumption rate and potential for spoilage or staleness.

The ambient temperature will determine how well a cooler performs. Unless a vehicle is air-conditioned, sunshine will heat an automobile to considerable temperatures, whether it is being driven or parked at the campsite. If the ice is likely to run out overnight, it is preferable to obtain a new supply. The trunk of a car can act as an incubator, since the hot air has no chance to escape. Insulation of the cooler with a covering such as a blanket or tarpaulin may prevent the direct rays of the sun from striking the cooler. Because of wild animals often in the vicinity of campsites, it is often imprudent or even illegal to leave foods, including coolers, on the ground. All foods must be left in the vehicles or suspended from trees, out of reach of racoons, bears, etc.

For the backpacker or wilderness camper, ice may not be available or a cooler too bulky and heavy to transport. Nonperishable foods may be the only option. Apart from some baked products and peanut butter, which have reasonable shelf life as long as conditions are not too hot and humid, there are some products that are suitable for traveling. These include beef jerky and other dried meats, dried noodles and soups, dried fruits and nuts, rice, powdered milk, and flavored drink crystals. Canned foods, bottled drinks, sterilized or UHT milk, and concentrated juice boxes are excellent but may be too heavy to carry far. Dehydrated foods or retortable pouches are lightweight with a reasonable choice of menu, but can be expensive. Often too much in the way of choice is carried into wilderness situations. If the most exotic or perishable products will not be consumed first, they should be left off the list altogether.

Supplementing a food supply from the wild is sometimes possible. If such activities are permitted, hunting, trapping, or fishing can bring in meat, fish, and shellfish. There are risks to health in doing these, however (see also under Sec. III. A). Also, harvesting mushrooms and wild plants for salads, fruits, and teas may result in poisonings unless the local botanical knowledge is adequate.

At some sites cooking is permitted on a stove or open fire. Take only the required number of pots and pack them so that they nest in each other. Aluminum foil wrap can be used effectively to roast potatoes and bake biscuits, as well as to broil fish and meat. Experience in this style of cooking is valuable in determining the length of time required to adequately cook food for the party involved.

Many camping areas prohibit campfires, at least in dry weather. So, a stove usually has to be taken. Stoves can take a long time to cook food if the pressure is not high enough, and usually only two items can be heated at once. Keeping foods warm while other portions or different menu items are cooked may not be possible, and dishes should be eaten as soon as prepared. This is one reason for keeping meals simple.

On an open fire, the heated food can either be quickly burnt or hardly warmed depending on how well the fire was constructed or the fuel burns. A mature fire with red-hot coals cooks food more evenly, provided there are means for placing the pots over the heat (e.g., metal grills or a barbecue pit in a recognized campground; carefully arranged branches over the fire or the pots directly on the coals in remote areas). Since it is easy for outdoor roasts or barbecues to cook food unevenly and incompletely, raw food of animal origin, particularly ground meat and poultry, should be cooked until the juices run clear before it is served.

Leftover food should be burned or buried, not dumped. Dish and pot washwater should be disposed of either at specific sites or away from lakesides and other sources of fresh water. What cannot be consumed or burnt must be taken back in garbage bags for dumping at appropriate disposal sites or at home. This requirement must be considered when heavy items like canned products are hiked in—empty cans have to come out, too.

C. Sporting and Recreational Activities

Nine of 26 white water rafters contracted leptospirosis on a trip on flooded waters in Costa Rica in 1996 (86). They drank the water or fell from the rafts and involuntarily swallowed some water. *Leptospira* causes zoonoses in many wild and domestic animals and is excreted though the urine. If it is in the soil, it can be released into water when the ground is saturated or flooded, and infections have been associated with canoeing, wading, and swimming (86). About 200 participants at a Finnish national game shooting competition who attended an event one summer evening in 1997 and ate roasted whole pigs and drank beer became ill with salmonellosis (92). The pigs had been kept in a cool room for 2 days after slaughtering until July 26, when they were transported to the site, roasted, and served.

A similar event occurred at an outdoor pursuit center in southern England in April and May, 1998. Forty persons who attended paintball games were unwell, and at least 10 were infected with *Salmonella java* (16). In a paintball event two teams dress in combat gear and run through woodland shooting paint capsules at each other. Of the three potential risk factors identified during the investigation—eating spit-roasted pork (hog roast), drinking water (perhaps untreated), and accidental ingestion of the paintballs themselves (made from gelatin and animal fat and stained with food coloring), the first was the most likely. At one meal at the pursuit center, the pig was stored at room temperature for up to 38 hours before being cooked, and the apparatus used for roasting likely did not heat the inner parts of the hog to a safe temperature.

In Sweden, several boys on a soccer team had gastroenteritis after drinking water in a neighboring town. *S. typhimurium* DT124, a rare phage type in Finland, was cultured from stools. A nearby creek had been connected to the drinking water system (94). At an outdoor center in Scotland, a party of schoolchildren contracted campylobacteriosis 2–3 days after arrival at the center (5). A failure in chlorination of the water supply was the probable cause. In the spring of 1997, a similar event at the center stemming from a failure in the water filtration system caused another *Campylobacter* outbreak. In a more atypical recreational activity, more than half of 32 people returning in June 1997 from an archaeological dig in Carthage, near Tunis, had *Campylobacter* isolated from their stools. Members of the dig, which was organized through the University of Alberta, were from six Canadian provinces, the United States, and the United Kingdom (Alberta Health; Red Deer Community Health Centre; Laboratory Centre for Disease Control, Health Protection Branch, Health Canada, 1997). Although no cause was stated, contaminated water or possibly food was the most likely source of the infection.

Swimming pools and other communal water play areas have been implicated in causing enteric infections such as *E. coli* O157 hemorrhagic colitis (14), giardiasis (37, 40, 84), cryptosporidiosis (52, 97), Norwalk agent gastroenteritis (55), hepatitis A (64), and enterovirus-like illness (59). In 1991 visitors to a relatively small and stagnant Oregon park lake experienced *E. coli* O157:H7 and *Shigella* infections (21 and 38 cases, respectively) (56). Those who were ill tended to be swimming in the water a longer time than controls and to have swallowed some water.

Similar types of recreation-associated illness have arisen from *Shigella* infections in California, Kansas, Michigan, and Oregon (56). Even a collapsible paddling pool used by relatively few children has been implicated (14). Although the cause is not always established, person-to-person transmission from fecal contamination among those in the water has been suspected in most outbreaks. One investigation of 67 cryptosporidiosis cases in England found a connection between a learner pool and the main sewer, and those who immersed their heads in the water were more likely to have become infected (52). In a *Cryptosporidium* outbreak in Los Angeles County, an unintentional defecation in the pool preceded the infection in the 44 cases in different groups using the pool (97). In 1998, 26 young children who were playing in a kiddie pool at a water recreational park in Georgia were infected with *E. coli* O157:H7 as a result of one or more children, ill with diarrhea, defecating in the water on three occasions (65). The infections occurred even though the pool was chlorinated. These episodes should reinforce the need to prevent children in diapers or who are not toilet-trained from entering recreational waters used by those other than the immediate family. Diapers should not be changed near pools, and fecally-contaminated hands should not be rinsed in pool water. The impact of accidentally excreted stools can be reduced but not prevented by having toddlers wear close-fitting swimsuits (48). Most enteric pathogens can survive for several days in recreational water. Over 42 strains of *Aeromonas* (mainly *A. hydrophila*) were isolated from a rowing and fishing area in a river in Italy (18). Therefore, there are risks in playing or swimming in water that is not treated and filtered, especially if the water is still. This is in contrast with water with a high turnover—for example, in large lakes or rivers with some current, and the sea with tidal movement of water away from the beach.

Even if there are some means to remove pathogens from pools, closure of pools may be necessary until sufficient time has passed for filters or chlorine to be effective. Drainage and replacement of the water may be necessary in sites of persistent contamination. Such a situation occurred in Riverton, Utah, in 1998 after a 3-million-gallon reservoir was contaminated with *E. coli*—more than half the city's population was forced to either boil the water or buy bottled water (6). When the reservoir was drained, the carcass of a large raccoon was found at the bottom of the reservoir.

Apart from gastroenteritis, campers and cottage dwellers who bathe or wade may be exposed to skin infections from organisms such as *Vibrio* spp., which invade wounds to produce ulcerous lesions, and cercariae of specific parasites that penetrate the skin to cause dermatitis through an allergic-type reaction (swimmer's itch). Vibrios occur naturally in salty or brackish water, and the schistosomes live in freshwater snails as intermediate hosts and are excreted into the water as cercariae, normally destined to infect birds or wild mammals.

III. WORK AND SUBSISTENCE ACTIVITIES

A. Hunters, Trappers, and Fishers, and Scavengers of Dead Carcasses

Hunting, trapping, scavenging, and gathering of food from the wild is as old as the human race. However, we know that even today these activities are not without risks from infection or poisoning. Hunters may contract parasitic diseases after eating game meat such as bear and wild boar. One particularly large outbreak occurred in 1973 in Manitoba, Canada, when a number of Indian families in a tribal band ate fried, boiled, smoked, and dried bear meat (106). In most cases the meat was insufficiently cooked or cured to eliminate the *Trichinella spiralis* present. Of the 82 persons eating the bear, 53 were infected, 5 were hospitalized, and one died. In Korea in 1994, three men who ate raw viscera and meat from a boar they had hunted experienced gradual loss of vision (20). They were diagnosed as having unilateral toxoplasmic chorioretinitis and had high antibody titers to *Toxoplasma gondii*. The infection was assumed to have arisen from *T. gondii* in the boar meat.

Botulism arising from consumption of fermented seal or whale meat and blubber is a continual risk for Inuit hunters in North America and Siberia. Inuit hunters in Canada and Greenland have also contracted trichinosis from eating walrus, polar bear, and whale (89). A number of those who hunt and consume caribou have serological evidence of infection with echinococcosis (83). In the Lebanon–Israel area of the Middle East, *Echinococcus* is more often associated with hunting wild boar, home slaughter of sheep, and keeping of dogs, particularly in the Druze population (119). Hunters and trappers of rodents may contract tularemia while handling the dead animals (see also under Sec. II.B).

Other diseases that potentially can be contracted by hunters are tuberculosis and brucellosis. *Mycobacterium bovis* has been reported in white-tailed deer, mule deer, elk, and moose (88). In recent years, the source of the infection was probably infected cattle, captive bison, or elk herds. In 1995 between 0 and 1% of elk sampled in three areas in Montana had serological evidence of *Brucella abortus* (88). Even if the meat is not eaten, touching a carcass or fecal matter may result in an enteric infection. Rangiferine brucellosis is enzootic in Siberia, Alaska, and the Northwest Territories of Canada (29). Sera from caribou in Baffin Island, Northwest Territories (now Iqaluit), showed that 15–43% of animals had evidence of brucellosis and that 60% of the caribou with lesions consistent with the disease had *Brucella suis* biovar 4 isolated from them (29). The average incidence of brucellosis in Baffin Island is 34 in 100,000, the highest rate in the Northwest Territories. Since the consumption rate for caribou by the Inuit in Baffin Island is 2.2 per person per year, and each hunter kills an average of 27 animals each year, there is a risk of hunters contracting the disease through handling infected animal parts. However, since most cases occur in individuals between 6 and 19 years of age, touching carcasses is not the only factor. Consumption of raw or incompletely cooked meat is another source of infection. In another situation, a farmer cleaning up feces from two horses with diarrhea subsequently suffered from bloody diarrhea himself, and both the horses' and the farmer's stools contained the same strain of *E. coli* O157 (19).

Anthrax is enzootic in Texas, with infrequent, sporadic cases in livestock. In July 1997, because of heavy rainfall in the spring and summer, anthrax was confirmed in cattle, goats, and white-tailed deer (Martin Hugh-Jones, personal communication). Anthrax is endemic in the deer population, especially among deer kept specifically for hunting. Because of the recent introduction of Texas and Louisiana cattle through the Plains states into Canada, one strain of *B. anthracis* has been found in cattle all the way from Texas to Alberta. No doubt deer throughout this area are also occasionally infected, hence present risks to hunters. However, unless the animals die from the disease and the meat is consumed, foodborne anthrax is unlikely. This is not necessarily the case in developing countries.

In Africa, wild animals such as hippopotamus and elephant succumb to anthrax and carcasses may be utilized as human food. In 1976, 155 villagers in Uganda ate a zebu cow that was later found to have died from anthrax septicemia; 143 contracted the disease, and 9 children died (70). In Ghana, in 1997, 30 died after eating anthrax-infected beef. In Vietnam, where anthrax-infected cattle are used for meat, hundreds of persons develop the disease and three to seven die each year (71). Another example of illness arising from scavenged dead animals occurred in Sri Lanka. A monkey that had been found dead was made into a curry: nine persons were infected with *S. enteritidis,* and one died (58). Since other monkeys were found dead, it was thought that a troop had suffered from salmonella septicemia.

Fur trappers may be exposed to hazards during handling of live animals or carcasses. *Giardia* is present in beaver and muskrat (32), and *Toxoplasma* is in river otters, mink, fishers, martens, nutria, beaver, and muskrat (105). Plague, normally contracted through handling of wild animals, is also a risk in endemic areas (see also under Sec. II.B).

Those consuming fish and shellfish are also at risk from parasitic infections (e.g., *Diphyllobothrium* tapeworms in freshwater fish, *Anisakis* and *Pseudoterranova* from herring, cod, and other marine fish). The Asian freshwater clam has the ability to adapt to areas in which there is agricultural and industrial pollution and urban waste (36). It has the highest filtration rate of all freshwater bivalves and can develop dense beds in rivers passing through cities and fields. Since a single clam can retain over a million *Cryptosporidium* oocysts, people who eat raw clams probably are at risk for cryptosporidiosis. Green and other species of turtles also carry the parasite, probably from contact with human sewage entering the sea (35). Other wild animals may harbor this or other parasites and either should be avoided or thoroughly cooked.

Certain species of fish, particularly those in the Scombrideae family, such as tuna, mackerel, bonito, and also mahi-mahi and marlin, can cause histamine poisoning (scombroid poisoning) if allowed to spoil after capture. Tropical fish may contain ciguatoxin (see also under Sec. IV.A), which affects many Pacific islanders and those living in the Caribbean region. For many the risk is acceptable because fish is their main source of protein and fishing is a traditional way of life. Paralytic shellfish poison, diarrhetic shellfish poison, neurotoxic shellfish poison, amnesic shellfish poison, and tetramine poisoning have all been documented as causing illness in various parts of the world, and all are heat stable; that is, cooking will not destroy the toxins. Cessation of harvesting at appropriate times through local knowledge and the posted results of government testing of shellfish in endemic areas (mainly for paralytic shellfish poison) is the only certain way of preventing illnesses.

B. Workplaces Under Isolated or Exposed Conditions

The workplace environment was a frequent location for contracting foodborne disease, mainly in factory canteens in eastern European countries. In some countries, work camps are established where foreign workers stay for short periods of time and then return home. If hygienic conditions in these camps are inadequate, illnesses may occur. Workers of Indian or southeast Asian origin were the groups most frequently affected by outbreaks through mishandling of food in their work camps in Saudi Arabia. Insufficient cooking and improper storage of food were the main factors contributing to the incidents. In one particular outbreak in 1985, 168 of 419 Filipino workers at a workers' camp in Damman, Saudi Arabia, contracted salmonellosis (2). A rice dish with meat and vegetables was the suspected food, and one food handler and 57 ill workers were positive for *Salmonella minnesota.* The source of the bacterium was thought to be raw meat.

Migrant camps (e.g., crop pickers from Central America living in the United States) are also at risk from developing diarrhea in their small communities and perhaps contaminating the fruits and vegetables they harvest. These workers are often mobile, following the harvest season northward. Despite the apparent risks of infection and lack of adequate hygienic facilities in the field, it has been difficult to link these workers to outbreaks associated with the products they pick.

A small but increasing number of highly paid people live and work on oil rigs for extended periods of time. Although high quality food and sanitary conditions are expected, these are not always provided, and illnesses have been reported. Between 1985 and 1989, eight outbreaks caused by *C. perfringens* were documented on oil rigs off Scotland's coast (34,85), but similar occurrences there do not appear to have been reported recently. However, in 1996 in the Gulf of Mexico, an oyster-associated outbreak was attributed to a malfunctioning sewage disposal system on an oil rig on which some workers had been ill with Norwalk-like gastroenteritis (28).

The risk to divers from enteric pathogens present in water is relatively high (49). Twenty-six diving operations in the United States, Russia, and Ukraine were studied from 1989 to 1992. More than 70% of the divers were exposed to specific pathogens. The frequency of the isolation from divers or water was as follows: *Pseudomonas aeruginosa,* 64.5%; *Aeromonas hydrophila,* 6.9%; *Vibrio cholerae* O1, 2.8%; *V. cholerae* non-O1, 2.8%; *V. vulnificus,* 1.3%.

IV. MILITARY-RELATED ACTIVITIES

A. Military Operations

It has long been known that contaminated food and water can be major contributors to death in military operations. From a Roman army manual, it is clear that disease is related to unhygienic water supplies. Vegetius (24) advises:

> Do not allow the army to use water that is unwholesome or marshy, as drinking bad water, just like poison, causes illnesses in men. . . . If a large number of troops remains for considerable amount of time in summer or autumn in the same place, this can cause very unwholesome diseases from the contamination of the waters and the foulness of the smell itself, as the drinking water is tainted and the air infected. The only way to prevent this is frequent changes of encampment.

Roman commanders in the early centuries A.D. were careful in their sanitary arrangements (24). In permanent locations such as forts, aqueducts, wells, and springs provided fresh water for drinking and washing. Drains were built to discharge into rivers downstream of the watering area for animals, or septic tanks were built. Bath houses were constructed and latrines were continuously flushed to remove the sewage far from the forts. Hospitals were also available, isolated from other fort activities, with separate bathing and kitchen facilities. In campaigns, troops lived in tents and there were even field hospitals. Treatment of non-wound infections was, however, limited to herbal cures, such as plantain and certain types of wine against dysentery or diarrhea.

Exercise and diet were other important elements for keeping soldiers healthy. Troops were exercised every day, indoors if the weather was inclement. Soldiers also received a good menu selection, including corn, lard, cheese, wine, meat (ox, sheep, pig, poultry), beans, lentils, fresh fruit and vegetables, and nuts.

Some of these were grown locally and others imported from around the empire. On campaigns soldiers carried bacon, hardtack, and sour wine. Bacon fat could substitute for oil, the corn ration would be included in the hardtack, and water would be added to the wine. Hunting could add venison and wild boar to the table, and shellfish were considered a delicacy. It is interesting to note that meat for the troops had to be boiled or roasted. Occasionally on expeditions, meat was not prepared properly, as in 151 B.C. when soldiers became ill from eating excessive quantities of meat cooked without salt. Also, those who were most sick received easily digested items, such as soft-boiled eggs and oysters. There is no indication, however, that these soldiers developed infections from *Salmonella enteritidis, Vibrio* spp., or enteric viruses!

However, foodborne disease occurred. An officer in the Egyptian fleet suffered a "violent and dreadful attack of fish-poisoning . . . and for the next five days I was unable to drop you a line, not to speak of going up to meet you. Not one of us was even able to pass through the camp gate" (24). Apparently the fish tasted delicious. Thus this may have been an example of scombroid or ciguatera poisoning. The medical staff were well occupied in treating this kind of complaint, as well as typical ailments and wounds suffered in peacetime policing. Because of their importance to the well-being of the army, military doctors were granted special rights and privileges.

It would be many centuries before military commanders recognized the need for prevention of enteric disease to be successful in the field, at least in long campaigns. For instance, the French lost 75,000 (78%) to disease out of 96, 615 men who died during the Crimean War (1854–1856). And in the American Civil War (1861–1865), 97,000 died in battle and 184,000 from disease (65%) (25). Army Regulations in the United States in 1825 indicate that control over spoilage or foodborne disease agents at that time was limited. Article 40 recommends that fresh meat be hung out at the back window on hooks, but not in the sun, and Article 27 states that "fresh meat issued to soldiers in advance, in hot weather, may be preserved by half boiling it; or, if there be not time for that operation, the meat may be kept some 24 hours, by previously exposing it for a few minutes to a very thick smoke" (62).

By 1856 one J. R. Martin recognized that a total knowledge of the environment

> respecting the surface and elevation of the ground, the stratification and composition of the soil, the supply and quality of water, the extent of marshes and wet ground, the progress of drainage; the nature and amount of the products of the land; the condition, increase and decrease, and prevalent diseases of animals maintained thereon; together with periodical reports of the temperature, pressure, humidity, motion, and electricity of the atmosphere

was necessary for understanding diseases affecting British troops and the local population in India (7). Some of Martin's thinking reflects the miasma theory, namely, that something in the atmosphere (heavy rainfall, temperature changes, rotting vegetation, etc.) played a role in cholera transmission. He also recognized, however, that human excrement was a major factor in the spread of cholera, dysentery, and diarrhea, diseases common to the army at the time. Such diseases were "accompanied by profuse discharges, with which the air, water, bedding, linen, closets, walls of hospitals and barracks become more or less infected." (7).

Cholera was of particular concern because its spread in a population was often rapid but not in a predictable way. There were at least 15 million deaths in India from 1817 to 1865, and a further 23 million between 1865 and 1947. Large-scale movements of infected troops and Hindu and Muslim religious pilgrimages could carry the disease over vast areas of India and beyond. It was not until 1884, when Koch discovered *Vibrio cholerae* in a Calcutta water tank, that water was recognized as playing a key role in cholera transmission (7). Only in the last few decades was food also identified as an important vehicle.

Although deaths from infectious disease were less significant in the twentieth century, they still had a major impact. In World War II, 30% of Rommel's North African forces were immobilized by diarrhea because of the lack of adequate sanitation, and 73% of hospital admissions in Vietnam were caused by disease. In Operation Desert Shield in 1990, the chief risks to soldiers were dehydration, diarrhea, and other intestinal ailments (25).

Today, however, ration packs are not as likely to cause infections as the water supply or local food. In the United States and many other countries, manufacturing establishments serving the armed forces have to show that they have HACCP or quality assurance systems, and their products are accepted on a risk-based inspection system (25). The frequency and depth of inspection is based on the degree of risk to the consumer. Traditionally, potentially hazardous foods like milk, meat, and seafood have higher risks than breads, soft drinks, and fresh fruits. However, recent outbreak data indicate that fresh fruits and vegetables and nonpasteurized juices are of more concern than they used to be in industrialized countries, and there are pathogens present that are difficult to eliminate without thorough cooking.

A case for irradiating such products, at least for the armed forces, may soon be made. However, U.S. military personnel have a low awareness of what food irradiation is and have a high concern about consuming products so treated (95). Although there is more willingness to eat such food in the field than in the dining mess, introduction of irradiated foods into the military should be attempted only with a supporting educational campaign. It has been long known that canned and retort pouch items are of particular risk for botulism if they are underprocessed or if they become punctured in the field. Since these are used extensively in military rations, much work has been devoted to ensuring their safety, and no botulism outbreaks have occurred from these products in rations. In the U.S. Army Veterinary Service, apart from an HACCP approach to food inspection, a limited microbiological testing program is carried out for ground beef and ice cream and for *Listeria monocytogenes* in potentially hazardous foods (25).

The types of food served to personnel in the field varies with different armies. For example, canned foods, dried bread, tea, and sugar were given to Warsaw Pact troops away from field kitchens a decade ago (101). In China during the same time period combat rations included compressed food bars, dehydrated rice, noodles, canned meats, and seasoning. Retort pouches with meat, vegetables, boiled rice, and peaches in syrup were offered for special operations (101). During the Falkland Islands campaign, British troops were given fresh or frozen food on the long voyage to the Islands, according to typical navy menus (30). On landing they were given basic 24-hour rations. These contained breakfast, main meal, and a snack consisting of small cans and packets weighing 3.5 lb (4000 kcal). An Arctic pack that was similar but designed for individual heating, with dehydrated and powdered food and required water, weighed 3 lb (4500 kcal) (43). Later the soldiers indicated that for extensive traveling on foot, these 24-hour rations were bulky and they would have preferred to have more snacks and drinks. However, these operational packs were consumed for more than 30 days, with most meals eaten cold without detriment to the soldiers.

The ration packs were improved after the campaign based on the use of the retort pouch with breakfast, midday snack, and an evening meal incorporated into seven different menus, as well as nine drinks. There were absolutely no facilities on the Islands to support foodservice operations, and everything had to be imported at high cost. Once bases were established, a port built, an airfield constructed for long-range cargo planes, a bakery assembled and fresh vegetables grown locally in a hydroponic solution, and a variety of frozen products available, the quality of food served was high, with imaginative menu items available every day.

Unlike the Falkland Islands campaign, where little could be anticipated for provisioning such an event, peacekeeping forces may be given more time for foodservice preparation. When about 5000 U.S. troops were stationed in Tuzla, Bosnia, they had a Force Provider system developed from the experience of Operation Desert Storm in 1991 (60). This is a containerized, highly deployable bare-base system engineered to provide climate-controlled sleeping and dining areas, showers and latrines, and laundry and recreational facilities. Before this was erected, the area had to be cleared of mines and gravel put down. Portable restaurants provided by a British company for both U.S. and U.K. forces in Bosnia (93) were rapidly erected so that troops could quickly come off their combat rations.

No foodborne illness has been reported for any peacekeeping missions, but it must occasionally occur, especially if troops consume local food. For instance, in Sweden, 30 soldiers suffered from campylobacteriosis during an exercise in 1985 after chickens they had prepared themselves were eaten partially raw. Three years later 72 soldiers contracted salmonellosis after eating in a military canteen; although four of the kitchen staff were infected, no specific food was incriminated (34). In another outbreak the same year in a camp, 45 soldiers in one company fell ill after eating leftover rice.

Specific studies have been done to monitor diseases in the U.S. and U.K. armed forces. From 1966 to 1984 declines were observed in some infectious diseases, including diarrhea, in enlisted Navy men and women (44). In naval personnel between 1975 and 1983, seaman and hospital corpsmen were at greater risk from parasitic diseases (68), which included amebiasis and cestode- and helminth-caused infections. The specific rates per 100,000 for three of these that could be foodborne or waterborne were amebiasis, 1.9; toxoplasmosis, 0.7; and trichinosis, 0.1. Cultural dietary practices among some ethnic groups, such as Filipinos, indicate that certain people are at higher risk for certain diseases. From 1975 to 1984 the rates per 10,000 for initial hospitalizations for various infectious and parasitic diseases in the U.S. Navy were as follows: shigellosis, 0.3; typhoid/paratyphoid, salmonellosis, amebiosis, 0.2; and toxoplasmosis, 0.1 (79). The rate for ill-defined infections was much higher, at 8.8. However, all initial hospital admissions for specific infectious and parasitic diseases declined over the 10-year period. These diseases were sometimes related to the ports ships had called on in the Pacific and Mediterranean, and there were more diarrheal diseases in the Atlantic Fleet than in the Pacific (115).

Disease and nonbattle injury rates were computed for ships of the British Royal Navy that were deployed during wartime and peacetime operations. The wartime sick list admission rates were lower aboard carriers, battleships, and cruisers than for their counterparts deployed in peacetime (12). The illness rate differences for battleships and cruisers were statistically significant. Several categories of disease also yielded significant differences in the wartime/peacetime contrasts. Illness rates varied by ship type, with the lowest rates evidenced aboard carriers. One reason for a decrease in infectious or parasitic disorders was less exposure to people and conditions in foreign ports. Generally, there was a lower hospitalization rate for submariners than for crew of surface ships (17). However, the rate for diarrheal diseases was greater in submarines (74.0 vs. 45.3). The disease and nonbattle injuries to marines in Okinawa and Korea was higher during battles, especially if they were intense (10). Apparently, combat stress lowers the body's resistance to disease. Marines in Vietnam suffered from severe febrile illness at a rate of 0.09 of 1000 persons per day (11). Diseases of helminthic origin, diarrhea, and staphylococcal food poisoning were identified with rates of 0.016, 0.009, and 0.00007, respectively.

From these data, it is difficult to estimate the impact of foodborne disease compared with other diseases and injuries in battle or at other times, since specific foodborne diseases are rarely specified in studies. However, diarrhea has a major impact on effectiveness of personnel, and the likelihood of contracting it depends on the troops' location or the ports that ships' crews enter. Experience gained since World War II has been valuable in reducing the number of hospitalizations.

Recent arrivals in a combat or training areas are prone to disease (27). American soldiers posted to South Korea often developed cramps and diarrhea in the first few weeks after arrival (55% in the first 6 weeks in one study). No bacterial, viral, or parasitic pathogen was isolated in connection with these symptoms. These soldiers ate at central eating facilities where coliforms were found on lettuce, cutting boards, waitresses' and cooks' fingers, and eating utensils, and the temperature of the dishwashing rinse water was lower than specified in the guidelines, all indicating that sanitary conditions were far from ideal. This result confirmed an earlier (1976) study in which 694 marines with travelers' diarrhea in South Korea were examined, although no causative agent was found (26). With today's greater knowledge about pathogens (e.g., *Cryptosporidium, Cyclospora*), it is possible that the causative agents could now be identified.

Studies in Turkey indicated that, soldiers in Ankara garrisons were exposed to unacceptable levels of *Aeromonas hydrophila* if well water was used as an emergency supply (38). No specific illnesses were associated with the well water, but the source was considered to be too risky to use. This situation may well occur in operations where troops are exposed to water of lower quality than from municipal supplies, and diarrhea can result. This may have occurred in 1988, when 41 soldiers in one company developed severe diarrhea during military exercises in California (100). Water taken into the field was nonchlorinated, and ice was made with nonpotable water. However, raw meats may well also have been involved, as they were left for 12–24 hours unrefrigerated in the extreme heat of the desert before being prepared for serving. Fecally contaminated water was probably the cause of a hepatitis E outbreak affecting 32 of 692 Nepalese soldiers at an isolated training camp (21). Water was obtained from various surface sources such as springs and creeks, generally downhill from training, dining, and sleeping quarters. Similar hepatitis E outbreaks occurred in military camps in Chad, Ethiopia, and India after troops had ingested fecally contaminated drinking water (21).

Another issue is that of a combination of infections. For instance, schistosomiasis is contracted through skin contact with the cercariae of the intravascular parasite *Schistosoma.* Thousands of cases in troops were documented in the Philippines and China (72). Rash, fever, pneumonitis, and spinal cord lesions may occur, as well as chronic salmonellosis. The schistosomes can be rapidly colonized with salmonellae, and the bacteria present in the adult worms can colonize the gut. Prolonged recurrent bacteremia can persist for up to 1 year. Successful therapy requires treatment of both the parasitic and bacterial infections; otherwise relapses occur.

In Operation Desert Shield in 1990, 57% of 2022 surveyed U.S. troops in Saudi Arabia had at least one episode of diarrhea, and 20% reported that they were temporarily unable to carry out their duties because of their symptoms (50). Enterotoxigenic *E. coli* (ETEC) and *Shigella sonnei* were the most frequently isolated pathogens, and most of the isolates were resistant to at least one antimicrobial drug. During the same period, 26% of stool specimens from 181 British troops hospitalized with diarrhea in Saudi Arabia contained ETEC; most of these belonged to a new serotype of O159:H? (117). This was different from the serotype (O148:H28) detected in British soldiers in Aden over two decades earlier (91). ETEC was also detected in troops with diarrhea deployed to Thailand, Indonesia, and the Philippines in 1995 (96).

In 1995 a joint military exercise between U.S. and Egyptian armed forces was conducted near Alexandria, Egypt, and lasted 30 days. Two percent of the 1200 U.S. troops involved reported sick for treatment of diarrhea (74). ETEC and enteroadherent *E. coli* were present in 42% of diarrheal cases; *Entamoeba histolytica, Cryptosporidium,* other protozoan parasites and enteropathogenic *E. coli* were isolated to a much lesser extent (5–11%). No *Campylobacter, Salmonella, Shigella,* or *Vibrio* spp. were found. The relatively low rate of diarrhea was attributed to the work of an advance team, which had prepared the way for the troops, warned them of the risks of eating in certain local restaurants, and given them a general education on diarrheal diseases. Interestingly enough, acute diarrheal cases were common among the advance party personnel who lived and ate in local establishments.

As indicated in the study of Oyofo et al. (74), an important function of any military medical service during combat operations is the prevention of infections and parasitic disease. Surveillance requires careful monitoring and a thorough understanding of the trends in incidence and distribution of known endemic agents in the areas troops are deployed. A field medical surveillance system with appropriate laboratory support could be set up for the prompt detection and prevention of illnesses that may occur during foreign assignments or conflicts. Occasionally, a foodborne outbreak is documented, such as the one that occurred at the U.S. Naval Hospital in Bethesda, Maryland, in 1983 (61). Acute dysentery was experienced by 6% of active-duty hospital staff and 12 others. *Shigella dysenteriae* type 2, unusual in the United States, was isolated from the patients. Raw vegetables from the salad bar in the cafeteria were implicated, and the extent of the outbreak was limited by the exclusion of civilians from the cafeteria and by the practice of preparing inpatient food separately.

Ciguatoxin and other fish toxins are hazards to military personnel overseas in tropical areas. Ciguatera poisoning has been documented from historic times until today. Endemic areas for ciguatoxin include the north Caribbean Sea and the Pacific and Indian Oceans, where reef fishes are found. The largest known outbreak of probable ciguatera poisoning occurred in 1748 when British Admiral Edward Boscawen, attempting to invade Mauritius in the Indian Ocean, "lost upwards of fifteen hundred men, which occasioned the expedition to fail against the Isle de France." The sailors had consumed veille fish, now known as a grouper (39). Later, because French troops were experiencing intoxications in tropical seas, a document on poisonous fishes was published in 1877; it was the first military survival manual on this problem. The Russian Navy published a similar illustrated book in 1886. Japanese military personnel in World War II also produced an illustrated guide to toxic fish, having experienced enough illnesses and deaths in the Pacific to warrant the issuing of such advice. In the same year, 1943, an Australian publication was found to be useful by the Allied forces. However, these guides did not prevent illnesses from occurring in Americans (e.g., 71 sailors in the Mariana Islands in two episodes in 1945–1946 and 13 soldiers in Puerto Rico in 1945). Establishment of military bases with reef destruction and dredging during the war and postwar period may have contributed to an increase in the toxin source benthic dinoflagellate *Gambierdiscus toxicus,* which grows on surfaces including dead coral and man-made structures.

These illnesses continue up to the present day. In 1995 six U.S. solders serving with the Multinational Force in Haiti ate a large amberjack purchased from a local fisherman. All six presented to the aid station 5–8 hours later with typical ciguatera symptoms, and all recovered within 1–3 months (82). The Caribbean ciguatoxin C-CTX-1 was recovered from the leftover portion of the cooked amberjack. It is not easy to predict the occurrence of ciguatoxic fish in endemic areas, especially without local knowledge, and avoidance of large tropical fish known to be toxic is the safest course of action for military personnel.

B. Concentration and Refugee Camps

Concentration camps are less frequent today than during the international conflicts earlier in this century. A few of these have been used for testing of agents used in biological warfare, with serious illness and deaths occurring (see also Sec IV, C). However, these camps are also notorious for inadequate hygienic facilities and poor nutrition, and they rarely allow inspection by international authorities to check on the well-being of the prisoners. Perhaps to a lesser extent, the same can be said for many refugee camps, which typically hold more people than they were originally designed for. In Africa, several cholera outbreaks have occurred since the seventh pandemic began in 1970. In particular, those in refugee camps in Malawi, Somalia, Ethiopia, and the Sudan have experienced severe morbidity and mortality (67). In one camp heavy rains destroyed latrines 15 days before the outbreak began, and *V. cholerae* from the fecal waste probably contaminated the groundwater supply. In 1992, 772 cases of abdominal cramps and bloody diarrhea were documented in a camp housing 60,000 Mozambicans. Most of these appeared to be caused by *E. coli* O157:H7 and *Shigella dysenteriae* type 1 (75), and the mortality rate was 4.7%.

Refugees are becoming more numerous in many parts of the world, particularly in Africa, where there are many civil wars and much national unrest. This leads to many camps having cholera and other diarrheal diseases. For example, in April 1997, a cholera outbreak occurred among 90,000 Rwandan refugees residing in three temporary camps in the Democratic Republic of Congo. This outbreak had a higher death rate than usual for refugee camps. The daily crude mortality rate ranged from 7 to 14 per 10,000 population (63). Most patients with cholera were severely malnourished and suffered from other health problems (e.g., malaria or acute respiratory illnesses). Cholera also occurred among health care workers at the cholera treatment center. Cholera control interventions included filtration and chlorination of the camps' water systems, health education, and construction and maintenance of latrines.

Three factors accounted for the high mortality among the refugees in this outbreak: (a) the refugees had been without adequate food, shelter, or access to health care during the preceding 5 months; (b) the camps were located far from the nearest villages, and relief personnel had limited time for patient care; and (c) camps were moved during the outbreak, requiring relocation of ill persons, rebuilding of cholera treatment facilities, and delaying the proper construction of water treatment and sanitation facilities.

In August 1998, in another situation where food was the source of shigellosis, more than 500 Palestinians from a refugee camp in the West Bank town of Nablus came down with diarrhea after eating a variety of foods such as falafel and hummus sold from food stands (87). Food samples taken contained shigellae and showed evidence of spoilage.

C. Biological Warfare and Terrorist Activity

Biological warfare has been practiced from at least medieval times, when plague victims and manure were used as ammunition in siege campaigns and, later, smallpox-infected blankets were given to native peoples in North America (22). The impact of these specific interventions compared with spread through populations by personal contact, water, or food is not known, but it probably was limited. An unsuccessful attempt to immobilize soldiers in the Naples campaign occurred in 1495 when Spanish troops gave French soldiers wine contaminated with blood from leprosy patients (90). There were plans in World War I to use anthrax against reindeer in Norway, sheep in Romania, and various livestock in Argentina to disrupt the Allied forces, but these were not carried out. Between 1932 and 1945 biological weapons were tested by the Japanese in Manchuria and China, and 10,000 prisoners used in experiments died (22). At least 11 Chinese cities were attacked through release of plague-infected fleas, and cultures of *B. anthracis, Salmonella* spp., *Shigella* spp., and *V. cholerae* were either dropped from aircraft or used to contaminate food and water. Illnesses and deaths occurred, not only in the target populations but among the invading Japanese troops, who also had no adequate defense against the pathogens. Cholera seemed to cause most of the casualties. During World War II the Germans developed biological weapons, but the only recorded attack was the contamination of a reservoir in Bohemia with raw sewage in May 1945 (22).

No recent attempts at biological warfare have been successfully waged by nations. However, in 1979, at least 77 cases and 66 deaths resulted from an accidental release of anthrax spores from a military microbiology facility in Russia. In Iraq stockpiles of anthrax slurry, botulinum toxin solution, and aflatoxin were made between 1985 and 1991 and incorporated into bombs (120). None was used in action, but unless they and their means of delivery are destroyed they could be used against any number of nations in the Middle East during a time of conflict. The most likely risk today for use of biological agents, however, is through small groups intent on carrying out terrorist activities by means of infective or toxigenic agents, which are relatively easy to prepare at low cost. The most effective way is through aerosols, but the agents most likely to be used are also associated with foodborne disease, namely *Bacillus anthracis* spores and *Clostridium botulinum* toxin (98). The Japanese Aum Shinriyko cult, which used sarin nerve agent in an attack that killed 12 people and injured 5000 in the Tokyo subway system, had prepared the biological agents just named and had tried unsuccessfully to disseminate them earlier (54, 99).

Although food as a vehicle for dissemination is not as effective for mass casualties, biological agents deliberately added to food have been used for specific targets. These range from mushroom poisoning in Roman times to enteric infections in this century. A few examples are described.

Over 26 days in November and December, 1961, 23 cases of hepatitis A occurred among personnel at a U.S. Navy base. The infection was traced to potato salad eaten by officers on October 26 and 27 (53). The preparer of the salad had suffered from loss of appetite and diarrhea and possibly had anicteric hepatitis, for about 2 weeks after October 10. This individual had a personality disorder with episodes of wandering around in a daze, and a history of urinating on objects. It is probable that he urinated in the potato salad as an act of defiance to the officers, but did not know that by so doing he would be transmitting the virus. From December 1964 to March 1966, over 200 persons in Japan were ill with typhoid fever and dysentery in several outbreaks in two hospitals and homes in four different towns (3). A 33-year-old physician and research bacteriologist, who had deliberately contaminated food that was used as a vehicle for transmitting the pathogens, was arrested in April 1966. The culprit had acted because he had a deep antagonism to the seniority system that prevailed in medical circles at that time. However, he infected not only medical personnel (four of whom died), but his own relatives. A member of his family committed suicide over this event. The physician was judged to be abnormal but not insane. The administrators of the hospital at which he worked suspected him to be the source of the illnesses but did not remove him from the laboratory and kept quiet about the event.

In 1970, four Canadian university students were infected through food deliberately contaminated with embryonated *Ascaris suum* ova. This parasite causes ringworm in pigs and the students developed pulmonary infiltrates, asthma and eosinophilia (80).

In September and October 1984, in The Dalles, Oregon, 751 persons were infected with *Salmonella typhimurium* associated with salad bars in 10 restaurants (113). Illnesses occurred in two peaks separated by about 10 days. No specific risk factors were identified, and a criminal investigation implied that there had been deliberate contamination of salad bars by members of a religious commune. The followers of Bhagawan Shree Rajneesh had purchased a large ranch in the area, and there was local conflict over cultural values and land use. It was later admitted that cultures of the *Salmonella* were poured over salad bar items and into coffee creamers to incapacitate residents from voting in upcoming local elections, the results of which might have affected the commune adversely. The culturing had been done at the clinical laboratory in the commune with an American Type Culture Collection strain, which was found during the investigation and was identical to the outbreak strain. In 1986 two members of the commune were sentenced to 4.5 years in prison for violating the federal antitampering act.

In 1989–1990 in Scotland, 30 people contracted giardiasis from water that had been deliberately contaminated with fecal matter (33).

In 1996, 12 laboratory workers in a large medical center in Texas experienced severe gastrointestinal illness after eating blueberry muffins and assorted doughnuts anonymously left in their break room between the night and morning shifts of October 29 (57). An unsigned invitation to eat these was posted on laboratory computer screens. *S. dysenteriae* type 2 was isolated from stool cultures. This strain, which is a rare cause of disease, was identical to that in the laboratory's own stock culture collection for the last two years. No research was currently being done with the organism, so accidental contamination seems unlikely. Although the motive and method of contamination remain unknown, it was probably the act of someone who knew how to culture bacteria from beads, had access to the freezer, could send e-mail messages through the supervisor's computer, and was able to enter the locked break room. All this points to one of the medical center staff, although no one was implicated during the investigation.

Many experts say it is no longer a question of whether a major bioterrorist attack will occur, but when; it's just a matter of time (102). The threat today is greater than before because a precedent has been set by the Aum Shinriyko cult, and the techniques for growth and dispersal of pathogens are better known. On a clear, calm night, a light plane dispensing 100 kg of anthrax spores over a city could kill up to 3 million people. Such an act would require considerable effort to develop a dispersal mechanism, but the Japanese cult had over 40,000 members, hundreds of scientists, and $1 billion in assets, and some work had progressed on botulism, anthrax, and Ebola virus as bioterrorist agents.

Unlike aerosolization of pathogens as a terrorist attack procedure, malicious contamination of food to cause illness is a low-tech activity, although it appears to be rare and the motives are often unclear. Therefore, it is not easy to predict the occurrence of such incidents or to prepare counter measures, especially if the perpetrators are coworkers or relations. However, guidelines should be established for secure storage and close surveillance of laboratory stock cultures.

The impact of terrorism and warfare using biological weapons is not restricted to those who are ill or die. The psychological effect can be even greater, since whole populations may take unnecessary evasive actions and disrupt lifestyles. The sarin gas attack that killed 12 in the Tokyo subway system also caused more than 5000 injuries in the resulting panic (8). Overwhelming of the health care system, general demoralization, and blaming of governments for insufficient preparation are the types of problem that can follow an attack or even the threat of one (47). Contingency plans for a major attack need to be prepared at the national level, and these should ensure the availability of antibiotics and antisera against toxins or organisms, and to minimize the horrific effect of a mass contamination of the air, water, or food. Microbiologists need to be a part of the group that creates policies protecting against biological weapons but also allowing necessary research to go on (8).

V. PREPARATION FOR DISASTERS

Preparation for a disaster is important because basic services may not be available after an earthquake, flood, hurricane, winter storm, or other catastrophic event. Gas, water, electricity, and telephone service may be cut off for days, and relief workers may not arrive on the scene immediately. Authorities have the prime responsibility to prevent or at least reduce the impact of disasters, especially if they impact large communities. In countries that have very little in the way of a relief program, however, the military may have to be called in to assist. All these arrangements take time in a situation where survival is at stake. In areas prone to natural disasters, it is prudent to be prepared to supply potable water and safe food to a stricken population. The following information is based on "Before Disaster Strikes," a booklet prepared in 1997 by the International Association of Milk, Food, and Environmental Sanitarians (51).

Foods that have come in contact with flood waters or water from broken pipes must be considered to be contaminated. Where power is not available, refrigerators and freezers can keep foods cold for only a limited time, depending on the ambient temperature. These units should be opened as infrequently as possible to retain the cold, and the most perishable foods should be eaten first. Raw foods of animal origin probably should be discarded immediately if there are no means to cook them. However, there may be plenty of debris to build a fire in the open air, and the meat could be cooked thoroughly in the same way as in a remote camping area (see also Sec. II B). This should not be attempted if there is any likelihood of escaping natural gas or in the presence of flammable material including gasoline. An indoor fire usually is not desirable if the occupied building has sustained structural damage.

Other perishable foods that should be consumed within 2 hours of the temperature rising above 5°C are milk, cream, cooked pasta, rice, custard, puddings, cheese pies, casseroles, soups, cookie dough, cream-filled pastries, salads, and egg meals. Some foods, like yogurt, soft cheese, cooked whole potatoes, pastries without uncooked dairy or egg ingredients, lettuce, and other greens, may be kept for several hours longer, but only if they have not been contaminated. The time depends on the ambient temperature, but should be no longer than 6 hours. Margarine, butter, fresh fruits and other vegetables, peanut butter, jams, jellies, commercially prepared mayonnaise, dressings and sauces, condiments, and hard or processed cheeses can be kept for many days, but once containers are opened, spoilage by mold, yeast, or nonpathogenic bacteria can occur. Contamination by chemicals, dust, smoke, and glass shards is also possible if food was not protected during the disaster event.

Survival is difficult in extreme weather situations, such as those encountered in desert sand storms, avalanches, and plane crashes in remote areas, or on sinking ships. Here, conservation of food and even water may be less important than resisting the elements until help arrives, and there may be little time to make life-and-death decisions. For example, it is better to grab a lifebelt than a sandwich if the prospect of being dumped into stormy water is imminent. It is most important for those in potential disaster situations (explorers, mountaineers, and wilderness trekkers by foot, ski, bike, or boat, as well as sailors, air crew, and rescue teams) to plan ahead and practice survival procedures.

In one recent example, the ice storm that affected northeastern United States and eastern Canada in January 1998, electrical power supplies were knocked out by the breaking of utility poles or wires, and there was loss of power to 600,000 persons. Some of those without power in Maine were followed up to find out how well they were coping (45). Warmth was supplied by propane or kerosene heaters or gasoline generators, which could lead to carbon monoxide exposures. Drinking water was available, but sump pumps could not work without power. Although telephone lines were down, most residents had access to radio or television. However, physicians' offices were without power. Luckily, illnesses or injuries were only slightly higher than usual. However, there was a slight increase in the number of persons with vomiting and diarrhea. This type of study helps prepare a plan for future disasters.

Individuals can be ready by assembling an emergency supply kit (51). This can include a large clean, covered container (such as a large plastic trash can) to store items; chlorine bleach solution, a plastic resealable container such as a soft drink bottle to store water; a metal stem thermometer, a can opener, a permanent ink pen, measuring spoons, and eating utensils. At least a 3-day supply of nonperishable food should be obtained. The items selected should require little or no water to prepare and should be edible without cooking, if necessary. Examples are commercially canned ready-to-eat meats, fish, fruits, and vegetables, ready-to-drink juices, milk, soup, and soft drinks; high energy foods (e.g., peanut butter, jelly, crackers, granola bars, trail mix, chocolate bars); rice, flour, cookies, hard candy, sweetened cereals, instant coffee, tea, sugar, salt, and pepper. For certain people (young and aged) specific items such as vitamins, infant formula, and special diets should be considered. Specially packed foods, such as retort pouches for military rations or wilderness explorers, can be added to this supply, as these will keep well and cannot be damaged by water or heat. If food has become wet, it may not be edible and will have to be discarded. That is why hermetically sealed foods with a long shelf life are highly desirable inclusions in an emergency supply kit. Water-contaminated intact cans and pouches should be disinfected with a strong detergent solution followed by immersion in dilute chlorine bleach before they are opened.

The water supply must be potable if it is to be used for drinking, brushing teeth, washing fruits and vegetables, cooking, and cleaning. Consider all tap water unsafe following a flood, earthquake, or large fire. If cold weather has frozen water pipes and lowered the water pressure, the water supply may be contaminated by sewage. Public announcements may be made on the safety of the water supply with a recommendation to boil the water. If fuel is not available, water purification tablets or addition of small amounts of bleach can substitute, but these measures may not destroy parasitic cysts, such as *Cryptosporidium* or *Giardia,* which have caused many waterborne outbreaks of disease around the world.

Kitchen counters and other areas where food is to be stored and prepared should be thoroughly sanitized. If there is no functioning existing waste disposal system (toilets may be blocked or overflowing, or sewage backing up into floor drains), an alternate location must be organized. A pit latrine can be dug outside if conditions allow this. Disinfected water will then have to be available for washing hands after using the privy. Packages of wet wipes can be substituted if water is in short supply.

Apart from food and water, the emergency container should contain toilet paper, disposable rags or paper towels for disinfecting surfaces, a supply of light (candles, flashlights), waterproof matches, flares (to attract attention), rope, knives, and a first aid kit. Waterproof and warm clothing and footwear are also important in many disaster situations to prevent hypothermia and should be readily available if the need is anticipated. Finally, common sense and knowledge of how to cope with potential specific disaster situations and development of leadership skills to take charge in such a crisis can be invaluable to prevent panic and may save lives. In most disasters, foodborne or waterborne disease is not the most immediate concern, but outbreaks can severely affect individuals who have been weakened through injury, exposure, or malnutrition. Therefore, it is in the days following a catastrophe that good leadership and preparation will prevent these diseases from occurring.

WHO has published a book, Management of Environmental Health in Emergencies and Disaster, A Practical Guide, which includes a chapter on food safety (116). This volume should also be consulted for practical tips.

VI. FOOD SAFETY IN SPACE

A. Early Space Flights

When the National Aeronautics and Space Administration (NASA) initiated space flights in the 1960s, one criterion for foods was the absence of pathogens or toxins. Also, the compressed time schedules for flights made finished product testing difficult to accomplish. This led one of the contract firms supplying food, Pillsbury, to develop the HACCP system in which critical control points were identified in the process of each food, from ingredients to the meals eaten by the flight crew, to prevent the incorporation of hazards (pathogens or toxins) into the process or to eliminate or reduce them if they were present (118). In the early days of space flight (Mercury, Gemini, and Apollo programs) the amount of food was limited, although the Apollo craft had facilities for heating. Crew on Apollo capsules could use retort pouches for thermostabilized foods. Following rehydration of the contents, a pressure-type plastic zipper was opened and the food removed with a spoon. The moisture content in the food enabled it to cling to the spoon, making eating in zero gravity a more normal experience.

B. The U.S. Space Shuttle Program

On the U.S. Space Shuttle program, the orbiter provides a sufficient amount of potable water generated by fuel cells at an hourly rate of 25 lb (11.4 kg) (69). This water is passed through a microbial filter before it is consumed by the crew or used to rehydrate dried food. Preparation is started by a crew member 30–60 minutes before meal time. A full meal for a crew of four can be set up in about 5 minutes, with an additional 20–30 minutes for any heating and rehydration of the food. In the rehydration station in the galley, a water dispenser needle penetrates the septum and the appropriate amount of water is added. Mixing and heating follow, if required. Beverages are drunk through straws and solid foods handled with knives, forks, and spoons. Condiments can also be added. Heating is done in an oven between 145 and 185°F (70–96°C). Three 1-hour meal periods are scheduled for eating and cleanup each day. Foods to be eaten on a flight are chosen ahead by each crew member. Because sinuses tend to be congested on space flights and food taste is diminished, most astronauts ask for spicy foods. Menus include scrambled eggs, beef steak, shrimp cocktail, meatballs with spicy tomato sauce, tortillas, broccoli and cheese, and chocolate pudding. There are no freezing nor refrigeration facilities on the orbiter, unlike the situation in Skylab, where frozen filet mignon, vanilla ice cream, and chilled fruits and beverages were available.

The following types of packaged foods are available for the Space Shuttle program:

1. Heat-processed foods in aluminum or bimetallic tins and retort pouches

2. Irradiated foods packed in flexible foil-laminated pouches
3. Intermediate moisture products: dried foods with a low moisture content such as dried apricots, packed in flexible pouches
4. Freeze-dried products, some of which may be eaten as-is (e.g., fruits), whereas others require the addition of hot or cold water before consumption
5. Rehydratable dried foods and cereals packed in semirigid plastic container with a septum for water injection
6. Foods such as nuts, crunch bars, and cookies, packed in flexible plastic pouches
7. Dry beverage powder mixes packed in rehydratable containers

To reduce their exposure to infectious agents, astronauts are quarantined 7 days before a mission and are not allowed to eat in public, although they may eat any foods they like. Only foodborne or waterborne diseases with long incubation periods such as hepatitis A or typhoid fever would not be apparent during this time and could become manifest on the flight. This has not yet happened. In fact, no foodborne or waterborne disease problem has been documented on any of the space flights to date. However, as space flights become more frequent, especially with the development and use of the International Space Station, the probability of illness increases. An emergency medical kit is available on the orbiter and includes oral medications that could reduce the amount of diarrhea and vomiting if gastroenteritis were to occur. A microbiological test kit is available to detect some bacterial infections, but it is unlikely to be comprehensive enough to identify many foodborne disease agents. *S. aureus* is considered to be a pathogen that could spread from person to person in crowded facilities on a spacecraft and possibly through food.

Pierson et al. (81) examined *S. aureus* isolates from nose, throat, urine, and feces of crew members and from air and surfaces in the orbiter over a 2-year period in the Space Shuttle program over 11 flights after previous studies with Apollo, Apollo–Soyuz, and Skylab spacecraft showed that *S. aureus* was transferred between astronauts. Pulsed-field gel electrophoresis of isolates showed that the shuttle crew members had unique strains from the beginning to the end of the flights, and there was only one instance of transfer between the crew.

C. Long-Term Space Missions

Further studies are recommended for long-duration missions such as on the International Space Station or on interplanetary voyages. Irradiated foods is one option at present. NASA's future ingredient list includes fresh vegetables, frozen and refrigerated foods, handheld foods like tortillas, flavors and spices, aromas (to counteract the loss of smell under zero gravity), condiments, nutraceuticals, and micronutrients (112). A U.S. Commercial Space Center for Food Technology is being planned by NASA and the National Food Processors Association to develop new foods for both long-term space travel and commercial applications on earth (4). Part of the challenge will be to have a self-sustaining food production and processing system rather than relying on prepackaged food prepared in advance of flights. Crops being considered for advanced life support in space include cereals, beans, salad greens, tomatoes, carrots, and strawberries. Some of these could be contaminated with bacterial or parasitic pathogens present in the crew on earth, possibly in a nonvirulent form, if such microorganisms were to mutate during the duration of the total space trip (there and back), or to be trapped in the seeds used to grow the crops.

One area where space travel may benefit from military input is in studies in isolation. Personnel who winter-over in the Antarctic are almost as isolated as those in a spacecraft or space station, such as that contemplated on Mars (78). Over a 15-year period, illness, especially hospitalization, was not as frequent in those who over-wintered as in a control group consisting of people who were accepted for winter duty but did not go (76). Psychological changes, however, may be apparent in members of small base stations. Foodborne disease in the Antarctic has not been documented, probably because, as in space, most food is commercially sterile or frozen and there are a limited number of menu items. However, gastrointestinal disorders are frequently noted among other complaints, and people's attitudes to types of food and feeding habits are important to those in close confinement (e.g., sloppy eaters are not easily tolerated by those who are more fastidious) (77, 78).

D. Waste Disposal and Use

Every spacecraft must have an integrated multifunctional system to collect and process biological wastes from the crew (69). It consists of a commode that is operated like a normal toilet, although there are straps to assist the evacuation process in zero gravity. Also, urine is delivered to a wastewater tank through funnels designed for both men and women. Feces and paper tissues are deposited into a bag, and then dried into solid waste. Air returning to the cabin is deodorized and filtered. Wastewater is dumped overboard into space. If this system fails, there is a backup system with a contingency urine collection device, and defecation can be carried out in Apollo fecal bags, which can be stored. The commode seat and urinal should be cleaned once a day with a biocidal detergent by spraying and wiping. Each astronaut has a personal hygiene kit (toothbrushing, hair care, shaving, nail care), pressure-packed agents for cleaning the hands, face, and body, and tissue towels. In addition, two washcloths and one towel per crew member per day are provided for each of the 7 days. Hand washing after defecation is desirable but may not always happen.

As more experience in spaceflight is gained, food systems for space travelers will continue to improve. New foods and their containers are being developed and evaluated. Fresh fruits and vegetables are being considered for space station menus, especially foods that can be prepared under modified atmosphere packaging to retain their shelf life (41). However, these are not necessarily free from pathogens and may have to be irradiated to guarantee safety. As missions become longer and crews larger, storing or resupplying food, water, oxygen and other life-support materials becomes more difficult and expensive, requiring, as well, increased storage space.

Crew members' daily needs without resupply consist of several elements: thermal control, air generation and purification, and food, water, and solid waste management. Ongoing research is exploring the use of chemical processes to convert carbon dioxide, wastewater, and solid wastes to breathable air, potable water, and food, respectively. Water should be relatively easy to sterilize or filter. However, pathogens in fecal matter may be transferred to food grown in the processed manure, and if the food were to be eaten fresh and not cooked, it might cause illness. Eventually, research that has developed space food will end up generating a series of food products for the general population, including backpackers and terrestrial explorers, and people in hospitals and nursing homes for whom extended bed rest is prescribed.

VII. CONCLUSION

In some unusual or even unique activities and occupations, food safety is not necessarily a high priority. Explorers, mountaineers, and backpackers are greatly concerned with how much they can carry. Therefore, light items are favored over cans. Fortunately, there are increasing numbers of products that are vacuum-packed or sold in retort pouches, but they are more expensive.

Those living off the land, such as hunters or fishers, can run risks of illness either through handling carcasses or through eating poorly prepared or stored meat or fish. Even some toxic plants can be mistaken for edible varieties. Recreational campers in more permanent locations with wells for water intake and lakes or septic tanks for waste disposal also have risks of foodborne disease, more typically with agents that cause disease in mass-catered facilities. When refugees or prisoners are herded together in very crowded conditions, person-to-person spread of disease is common; moreover, it is difficult to ensure that adequate food is quickly served, and opportunities for bacterial growth no doubt frequently occur. People under these conditions, who usually are undernourished and prone to infections from opportunistic pathogens, must be considered to be susceptible populations.

In the past, very occasionally, prisoners have been deliberately infected to observe the effects of specific agents. The military is very much aware of the possibility of biological warfare for disabling or killing persons in areas that an enemy wishes to neutralize. Even if attack is not the object, knowledge of appropriate detection and neutralizing mechanisms is required to protect troops and civilians from illness or death. Troops on the ground and naval staff can also experience enteric diseases during operations or in training; these can have a serious effect on fighting capabilities. Some of these diseases are transmitted through nonpotable water and contaminated food. If medical staff are aware of the potential diseases that may affect military personnel and prepare for them ahead of operations, the impact of such outbreaks will be greatly diminished.

VIII. SUMMARY OF RECOMMENDATIONS

A. Mountaineers, Backpackers, Trekkers, Outdoor Sports Enthusiasts, Campers, Hunters, and Trappers

Anticipate problems and bring extra food and water.

Ensure that coolers have enough ice at all times.

Throw out unwrapped, perishable leftovers that have had contact with ice water or raw food.

Use dehydrated foods or retort pouches if canned or other processed food is too heavy.

Avoid bringing perishable food, or eat such items early in the trip.

Thoroughly cooked all game meat (mammals, birds, and reptiles) or fish and shellfish. Camp gas stoves tend to take a long time to cook food, especially at high altitudes, and usually only a few items can be heated at once.

Harvest mushrooms and wild plants for salads, fruits, and teas only with local knowledge of how to distinguish poisonous from edible species.

Burn or bury leftover food.

Use local knowledge to avoid certain species of tropical fish that are prone to ciguatoxin, as well as shellfish that may be exposed to toxic plankton or enteric viruses.

Do not let fish remain at ambient temperature for more than a few hours; otherwise scombroid poisoning may result.

Consider all stream and river waters unsafe to drink. Water can be purified by boiling or filtration, chlorination, or other chemical treatments, although the last may not be effective against parasitic cysts. Portable filtration devices for the backpacker are available but need to be checked out before they are used to remove parasites.

Avoid drinking or swallowing recreational waters, since untreated water can carry pathogens from human and animal sources.

Do not allow children in diapers to enter recreational waters.

Be aware that the risks of skin infections from organisms in seawater, brackish water, or ponds are diminished if there are no cuts or scratches and the exposure time is not too long.

Avoid exposure to sick or dead animals.

Use gloves when handling animals killed by hunting or trapping, since parasites and bacterial pathogens are often present in fur-bearing animals.

Remove sources of rodent food and harborage from residences or campsites.

Know how to take care of a person stricken by severe vomiting, cramps, or diarrhea in a remote area.

B. Camps for Refugees and Foreign Workers, and Those in Exposed or Isolated Work Environments

Expect problems, since many refugees are weak, malnourished, and exposed to crowding, or because foreign workers may be unused to the local food.

Isolated or exposed workplaces need to have food safety issues specifically addressed.

C. Military Operations

Observe the food safety precautions as for camping or trekking, except that cooking and eating may have to be done when troops are exposed to enemy action. However, freeze-dried or retort pouch rations that can be quickly prepared are readily available.

Avoid eating local food during operations through enemy territory, especially if action has interrupted electrical supplies. Use water or ice from a potable source.

Scout out the region where troops are to be deployed and learn about local food hygiene and endemic foodborne diseases.

D. Bioterrorism

Be aware that aerosolized pathogens or toxins can be dangerous not only from inhalation but also from their contamination of food and water supplies.

Have contingency plans that include supplies of antibiotics and antisera.

E. Emergency Situations for Disasters

Prepare an emergency supply kit and at least a 3-day supply of food (nonperishable).

Boil or disinfect water before drinking it.

F. Food Safety in Space

Use foods that are commercially sterile for short space flights. Irradiated foods and retort pouches are present options. Immediate needs are to provide more variety of foods for the astronauts.

New ways for producing safe food for long space voyages where crops and animals are required need to be developed.

REFERENCES

1. Adler, K. 1970. Staphylococcal food poisoning. Epidemiol. Bull. (Department of National Health and Welfare, Canada) 14:35.
2. Al-Ghamdi, M., S. Al-Sabty, A. Kannan, and B. Rowe. 1989. An outbreak of food poisoning in a workers' camp in Saudi Arabia caused by *Salmonella minnesota.* J. Diarrh. Dis. Res. 7:18–20.
3. Anonymous. 1966. Deliberate spreading of typhoid in Japan. Sci. J. 2:11–12.
4. Anonymous. 1997. NASA moves to set up commercial space center for food technology. Food Chem. News, September 22, pp. 11–12.
5. Anonymous. 1998. Outbreak of *Campylobacter* infection at an outdoor centre. SCIEH Wkly. Rep. 32(31).
6. Anonymous. 1998. Dead raccoon suspected in tainted Riverton water. Deseret News, October 6.
7. Arnold, D. 1993. Colonizing the Body: State Medicine and Epidemic Disease in Nineteenth Century India. University of California Press, Berkeley, pp. 1–354.
8. Atlas, R. M. 1998. Biological weapons pose challenge for microbiology community. ASM News 64:383–389.
9. Beattie, O., and J. Geiger. 1987. Frozen in Time: The Fate of the Franklin Expedition. Bloomsbury, London. pp. 1–180.
10. Blood, C. G., and E. D. Gauker. 1993. The relationship between battle intensity and disease rates among Marine Corps infantry units. Mil. Med. 158:340–344.
11. Blood, C. G., D. K. Griffith, and C. B. Nirona. 1989. Medical resource allocation: Injury and disease incidence among Marines in Vietnam. Report 89–36, Naval Health Research Center, San Diego, CA, pp. 1–145.
12. Blood, C. G., W. M. Pugh, E. D. Gauker, and D. M. Pearsall. 1992. Comparisons of wartime and peacetime disease and non-battle injury rates aboard ships of the British Royal Navy. Mil. Med. 12:641–644.
13. Bohmer, P. 1997. Outbreak of campylobacteriosis at a school camp linked to a water supply. N. Z. Public Health Rep. 4:58–59.
14. Brewster, D. H., M. I. Brown, D. Robertson, G. L. Houghton, J. Bimson, and J. C. M. Sharp. 1994. An outbreak of *Escherichia coli* O157 associated with a children's paddling pool. Epidemiol. Infect. 112:441–447.
15. Brugha, R. F., A. J. Howard, G. R. Thomas, R. Parry, L. R. Ward, and S. R. Palmer. 1995. Chaos under canvas: A *Salmonella enteritidis* PT 6B outbreak. Epidemiol. Infect. 115:513–517.

16. Brusin, S. 1998. An infectious hazard of playing soldiers: Outbreak of *Salmonella java* infection associated with a paintball event. Eurosurveillance Wkly. July 2.
17. Burr, R. G., Palinkas, L. A., and A. S. Pineda. 1989. Disease and non-battle injuries for U.S. Navy submarine personnel and surface-ship personnel by occupational group. Report 89–10, Naval Health Research Center, San Diego, CA, pp. 1–26.
18. Cenci, P., M. Vitaioli, A. Stefanati, and L. Prati. 1996. Isolation of *Aeromonas* spp. from water used for aquatic sports. Ig. Mod. 106:567–576.
19. Chalmers, R. M., R. L. Salmon, G. A. Willshaw, T. Cheasty, N. Looker, I. Davies, and C. Wray. 1997. Vero-cytotoxin-producing *Escherichia coli* O157 in a farmer handling horses. Lancet 349:1816.
20. Choi, W.-Y., H.-W. Nam, N.-H. Kwak, W. Huh, Y.-R. Kim, M.-W. Kang, S.-Y. Cho, and J. P. Dubey. 1997. Foodborne outbreaks of human toxoplasmosis. J. Infect. Dis. 175:1280–1282.
21. Clayson, E. T., D. W. Vaughn, B. L. Innis, M. P. Shrestha, R. Pandey, and D. B. Malla. 1998. Association of hepatitis E virus with an outbreak of hepatitis at a military training camp in Nepal. J. Med. Virol. 54:178–182.
22. Christopher, G. W., T. J. Cieslak, J. A. Pavlin, and E. M. Eitzen Jr. 1997. Biological warfare: A historical perspective. JAMA. 278:412–417.
23. Communicable Disease Report. 1995. *Salmonella* in humans, England and Wales: Quarterly report, 10 March. 5(10):47–50.
24. Davies, R. W. 1989. Service in the Roman Army. Columbia University Press, New York, Chapter IX, The Roman military diet, pp. 187–206; Chapter X, The Roman military medical service, pp. 209–236.
25. Derstine, H. W. 1994. The history of food safety in the military. Activities Report of the R & D Associates 46(1):29–38. [Research and Development Associates for Military Food and Packaging Systems, San Antonio, TX.]
26. Echeverria, P., F. A. Hodge, N. R. Blacklow, J. L. Vollet, G. Cukor, H. L. DuPont, and J. H. Cross. 1978. Travelers' diarrhea among United States Marines in South Korea. Am. J. Epidemiol. 108:68–73.
27. Echeverria, P., G. Ramirez, N. R. Blacklow, T. Ksiazek, G. Cukor, and J. H. Cross. 1979. Travelers' diarrhea among U.S. Army troops in South Korea. J. Infect. Dis. 139:215–219.
28. Farley, T. A., L. McFarland, M. Estes, and K. Schwab. 1997. Viral gastroenteritis associated with eating oysters—Louisiana, December 1996–January 1997. MMWR November 28. 46:1109–1112.
29. Ferguson, M. A. D. 1997. Rangiferine brucellosis on Baffin Island. J. Wildl. Dis. 33: 536–543.
30. Fleming, H. E. 1988. Effectiveness of British military foodservice lessons learned from the Falklands Campaign. Activities Report of the R & D Associates 40(2):20–24. [Research and Development Associates for Military Food and Packaging Systems, San Antonio, TX.]
31. Frost, B., C. Chaos, L. Ladaga, W. Day, M. Tenney, D. McWilliams, E. Barrett, L. Branch, S. Jenkins, M. Linn, E. Turf, D. Woolard, G. B. Miller Jr., S. Henderson, B. Campbell, M. Mismas, J. Dvorak, D. Patel, D. Peery, J. Morano, and K. Campbell. 1995. *Escherichia coli* O157:H7 outbreak at a summer camp—Virginia, 1994. MMWR 44(22):419–421.

32. Frost, F., B. Plan, and B. Liechty. 1980. *Giardia* prevalence in commercially trapped mammals. J. Environ. Health 42:245–249.
33. Gelinas, P. 1997. Handbook of Foodborne Microbial Pathogens. Polyscience Publications, Morin Heights, Quebec pp. 1–206.
34. Gerigk, K. 1992. WHO Surveillance Programme for Control of Foodborne Infections and Intoxications in Europe, 1985–1989, 5th Report. Institute of Veterinary Medicine, Berlin.
35. Graczyk, T. K., G. H. Balacs, T. Work, A. A. Aguirre, D. M. Ellis, S. K. K. Murakawa, and R. Morris. 1997. *Cryptosporidium* sp. infections in green turtles, *Chelonia mydas,* as a potential source of marine waterborne oocysts in the Hawaiian Islands. Appl. Environ. Microbiol. 63:2925–2927.
36. Graczyk, T. K., R. Fayer, M. R. Cranfield, and D. B. Conn. 1997. In vitro interactions of Asian freshwater clam (*Corbicula fluminea*) hemocytes and *Cryptosporidium parvum* oocysts. Appl. Environ. Microbiol. 63:2910–2912.
37. Greensmith, C. T., R. S. Stanwick, B. E. Elliot, and M. V. Fast. 1988. Giardiasis associated with the use of a water slide. Pediatr. Infect. Dis. J. 7:91–94.
38. Gursoy, T. K., H. T. Aktan, and A. Yurtyeri. 1994. Isolation of *Aeromonas* species from water sources of military units. Rev. Int. Serv. Santé Forces Armées 67:151–153.
39. Halstead, B. W. 1967. Poisonous and Venomous Marine Animals of the World, Vol. 2, Vertebrates. Government Printing Office, Washington, DC, pp. 1–1070.
40. Harter, L., F. Frost, G. Grunenfolder, K. Perkins-Jones, and J. Libby. 1984. Giardiasis in an infant and toddler swim class. Am. J. Public Health 74:155–156.
41. Hill, S. D. 1990. Space station freedom food system. Activities Report of the R & D Associates 42(1):119–122. [Research and Development Associates for Military Food and Packaging Systems, San Antonio, TX.]
42. Hodgdon, J. A., R. L. Hesslink, A. C. Hackney, R. R. Vickers, and R. P. Hilbert. 1991. Norwegian military field exercises in the Arctic: Cognitive and physical performance. Arct. Med. Res. 50(suppl. 6):132–136.
43. Hogg, M. D. 1983. Feeding plan for support of the British Forces in the recent Falklands operation. Activities Report of the R & D Associates 35(1):24–33. [Research and Development Associates for Military Food and Packaging Systems, San Antonio, TX.]
44. Hoiberg, A. 1987. Infectious disease trends in the U.S. Navy, 1966–1984. Report 87-40, Naval Health Research Center, San Diego, CA, pp. 1–22.
45. Holt, D., J. Even, W. W. Young Jr., P. E. Chalke, D. Stuchner, L. Covey, S. King, M. A. Johnson, M. Twomey, S. Steinkeler, P. Pelletier, P. Boucher, D. Mills, G. Becket, A. Hawkes, D. Shields, N. Sonnenfeld, R. Wolman, A. Smith, L. Crinion, C. Sloat, J. Sherman, P. Pabst, M. Bouchard, J. Matthews, J. Hardacker, D. Smith, A. Drake, and K. Gensheimer. 1998. Community needs assessment and morbidity surveillance following an ice storm—Maine, January 1998. MMWR 47:351–354.
46. Hook, D., B. Jalaludin, and G. Fitzsimmons. 1996. *Clostridium perfringens* foodborne outbreak: An epidemiological investigation. Austl. N. Z. J. Public Health 20:119–122.
47. Holloway, H., A. E. Norwood, C. S. Fullerton, C. C. Engel Jr., and R. J. Ursano.

1997. The threat of biological weapons: prophylaxis and mitigation of psychological and social consequences. JAMA 278:425–427.
48. Hosaka, T. 1998. *E. coli* rarely spreads in pools, officials say, officials offer reassurance after GA outbreak. Washington Post, June 27.
49. Huq, A., M. A. R. Chowdhury, J. A. K. Hasan, G. A. Losonsky, and R. R. Colwell. 1994. Detection of selected pathogens in divers and diving sites in the United States, presented at the 94th General Meeting of the American Society for Microbiology, Las Vegas, NV, May 23–27. 1994. Abstr. Q-211, p. 425.
50. Hyams, K. C., A. L. Bourgeois, B. R. Merrell, P. Rozmajzl, J. Escamilla, S. A. Thornton, G. M. Wasserman, A. Burke, P. Echeverria, K. Y. Green, A. Z. Kapikian, and J. N. Woody. 1991. Diarrheal disease during Operation Desert Shield. N. Engl. J. Med. 325:1423–1428.
51. IAMFES Food Sanitation Professional Development Group. 1997. Before Disaster Strikes . . . A Guide to Food Safety in the Home. International Association of Milk, Food, and Environmental Sanitarians, Des Moines, IA, pp. 1–6.
52. Joce, R. E., J. Bruce, D. Kiely, N. D. Noah, W. B. Dempster, R. Stalker, P. Gumsley, P. A. Chapman, P. Norman, J. Watkins, H. V. Smith, T. J. Price, and D. Watts. 1991. An outbreak of cryptosporidiosis associated with a swimming pool. Epidemiol. Infect. 107:497–508.
53. Joseph, P. R., J. D. Millar, and D. A. Henderson. 1965. An outbreak of hepatitis traced to food contamination. N. Engl. J. Med. 273:188–194.
54. Kadlec, R. P. 1997. Biological weapons control: Prospects and implications for the future. JAMA 278:351–356.
55. Kappus, K. D., J. S. Marks, R. C. Holman, J. K. Bryant, C. Baker, G. W. Gary, and H. B. Greenberg. 1982. An outbreak of Norwalk gastroenteritis associated with swimming in a pool and secondary person-to-person transmission. Am. J. Epidemiol. 116:834–839.
56. Keene, W. E., J. M. McAnulty, F. C. Hoesly, L. P. Williams, K. Hedberg, G. L. Oxman, T. J. Barrett, M. A. Pfaller, and D. W. Fleming. 1994. A swimming-associated outbreak of hemorrhagic colitis caused by *Escherichia coli* O157:H7 and *Shigella sonnei.* New Engl. J. Med. 331:579–584.
57. Kolavic, S. A., A. Kimura, S. L. Simons, L. Slutsker, S. Barth, and C. E. Haley. 1997. An outbreak of *Shigella dysenteriae* type 2 among laboratory workers due to intentional food contamination. JAMA 278:396–398.
58. Lamabadusuriya, S. P., C. Perea, I. V. Devasiri, U. K. Jayantha, and N. Chandrasiri. 1992. An outbreak of salmonellosis following consumption of monkey meat. J. Trop. Med. Hyg. 95:292–295.
59. Lenaway, D. D., R. Brockmann, G. J. Dolan, and F. Cruz-Uribe. 1989. An outbreak of an enterovirus-like illness at a community wading pool: Implications for public health inspection programs. Am. J. Public Health 79:889–890.
60. Lindsay, T. C., J. McLaughlin, and N. Bruneau. 1996. Force Provider deploys to Operation Joint Endeavor. Activities Report of the R & D Associates 48(2):26–40. [Research and Development Associates for Military Food and Packaging Systems, San Antonio, TX.]
61. Longfield, R., E. Strohmer, R. Newquist, J. Longfield, J. Coberly, G. Howell, and R. Thomas. 1983. Hospital-associated outbreak of *Shigella dystenteriae* type 2—Maryland. MMWR 32(19):250–252.

62. Luecke, B. K. 1990. Feeding the Frontier Army, 1775–1865. Grenadier Publications, Eagen MN, pp. 1–141.
63. Matthys, F., S. Malé, and Z. Labdi. 1998. Cholera outbreak among Rwandan refugees—Democratic Republic of Congo, April 1997. MMWR 47:389–391.
64. Mahoney, F. J., T. A. Farley, K. Y. Kelso, S. A. Wilson, J. M. Horan, and L. M. McFarland. 1992. An outbreak of hepatitis A associated with swimming in a public pool. J. Infect. Dis. 165:613–618.
65. McKenna, M. A. J., and V. Anderson, 1998. Cobb boy's infection puts total at 20. Atlanta Journal-Constitution, July 2.
66. Mermin, J., P. Mead, K. Gensheimer, and P. Griffin. 1996. Outbreak of *E. coli* O157:H7 infections among Boy Scouts in Maine. Abstr. Intersci. Conf. Antimicrob. Agents Chemother. 36:257.
67. Moren, A., S. Stefanaggi, D. Antona, D. Bitar, M. G. Etchegorry, M. Tchatchioka, and G. Lungu. 1991. Practical field epidemiology to investigate a cholera outbreak in a Mozambican refugee camp, 1988. J. Trop. Med. Hyg. 94:1–7.
68. Mueller, L. D., and F. C. Garland. 1989. Parasitic disease in the U.S. Navy. Report 89-30, Naval Health Research Center, San Diego, CA, pp. 1–9.
69. NASA. 1998. NSTS Shuttle Reference Manual. *www.ksc.nasa.gov/shuttle/technology*
70. Ndyabahinduka, D. G. K., I. H. Chu, A. H. Abdou, and J. K. Gaifaba. 1984. An outbreak of human gastrointestinal anthrax. Ann. First Super. Sanit. 20:205–208.
71. Nguyen, V. X. 1992. Epidemiology of foodborne infections and intoxications, in: Proceedings of the 3rd World Congress on Foodborne Infections and Intoxications. Institute of Veterinary Medicine, Berlin, p. 184.
72. Oldfield, E. C., M. A. Wallace, K. C. Hyams, A. A. Yousif, D. E. Lewis, and A. L. Bourgeois. 1991. Endemic infectious diseases of the Middle East. Rev. Infect. Dis. 13 (suppl. 3):S199–S217.
73. Ongerth, J. E., R. L. Johnson, S. C. Macdonald, F. Frost, and H. H. Stibbs. 1989. Backcountry water treatment to prevent giardiasis. Am. J. Public Health 79:1633–1637.
74. Oyofo, B. A., L. F. Peruski, T. F. Ismail, S. H. El-Etr, A. M. Churilla, M. O. Wasfy, B. P. Petruccelli, and M. E. Gabriel. 1997. Enteropathogens associated with diarrhea among military personnel during Operation Bright Star 96, in Alexandria, Egypt. Mil. Med. 162:396–400.
75. Paquet, C., W. Perea, F. Grimond, M. Collin, and M. Guillod. 1993. Aetiology of haemorrhagic colitis epidemic in Africa. Lancet 342:175.
76. Palinkas, L. A. 1987. A longitudinal study of disease incidence among Antarctic winter-over personnel. Aviat. Space Environ. Med. 58:1062–1065.
77. Palinkas, L. A. 1988. The human element in space: Lessons from Antarctica. Report 88-8, Naval Health Research Center, San Diego, CA, pp. 1–20.
78. Palinkas, L. A. 1989. Antarctica as a model for the human exploration of Mars, in: The Case for Mars III: Strategies for Exploration—General Interest and Overview, Vol. 74, Science and Technology Series, C. Stoker (ed.), American Astronautical Society–Univelt, San Diego, CA.
79. Palinkas, L. A., T. S. Pineda, R. G. Burr, and K. C. Hyams. 1989. Ten-year profile of infectious and parasitic disease hospitalizations. Report 89-4, Naval Health Research Center, San Diego, CA, pp. 1–52.

80. Phills, J. A., A. J. Harrold, G. V. Whiteman, and L. Perelmutter. 1972. Pulmonary infiltrates, asthma and eosinophilia due to *Ascaris suum* infestation in man. N. Engl. J. Med. 286:965–970.
81. Pierson, D. L., M. Chidambaram, J. D. Heath, L. Mallary, S. K. Mishra, B. Sharma, and G. M. Weinstock. 1996. Epidemiology of *Staphylococcus aureus* during space flight. FEMS Immunol. Med. Microbiol. 16:273–281.
82. Poli, M. A., R. J. Lewis, R. W. Dickey, S. M. Musser, C. A. Buckner, and L. G. Carpenter. 1997. Identification of Caribbean ciguatoxins as the cause of an outbreak of fish poisoning among U.S. soldiers in Haiti. Toxicon 35:733–741.
83. Poole, J. B., and R. A. Marcial-Rojas. 1975. Echinococcosis, in: Pathology of Protozoal and Helminthic Diseases with Clinical Correlation, R. A. Marcial-Rojas (ed.), Robert E. Krieger, New York, pp. 635–657.
84. Porter, J. D., H. P. Ragazzoni, J. D. Buchanon, H. A. Waskin, D. D. Juranek, and W. E. Parkin. 1988. Giardia transmission in a swimming pool. Am. J. Public Health 78: 659–662.
85. Reid, T. M. S., G. P. Sinton, R. J. Gilbert, and M. F. Stringer. 1985. *Clostridium perfringens* food poisoning on North Sea oil installations. Lancet i:272.
86. Reisberg, B. E., R. Wurtz, P. Dias, B. Francis, P. Zakowski, S. Fannin, D. Sesline, S. Waterman, R. Sanderson, T. McChesney, R. Boddie, M. Levy, G. Miller Jr., and G. Herrera. 1997. Outbreak of leptospirosis among white-water rafters—Costa Rica, 1996. MMWR 46:577–579.
87. Reuters. 1998. Food germ infects Palestinian refugees. August 31.
88. Rhyan, J. C., K. Aune, D. R. Ewalt, J. Marquardt, J. W. Mertins, J. B. Payeur, D. A. Saari, P. Schladweiler, E. J. Sheehan, and D. Worley. 1997. Survey of free-ranging elk from Wyoming and Montana for selected pathogens. J. Wild Dis. 33:290–298.
89. Ribas-Mujal, D. 1975. Trichinosis, in: Pathology of Protozoal and Helminthic Diseases with Clinical Correlation, R. A. Marcial-Rojas (ed.), Robert E. Krieger, New York, pp. 677–710.
90. Robertson, A. G. 1995. From asps to allegations: Biological warfare in history. Mil. Med. 160:369–373.
91. Rowe, B., J. Taylor, and K. A. Bettelheim. 1970. An investigation of travellers' diarrhoea. Lancet i:1–5.
92. Ruutu, P., M. Jahkola, and A. Siitonen. 1997. Outbreak of a rare phage type (DT 124) of *Salmonella typhimurium* in Finland. Eurosurveillance Wkly. August 7.
93. Schad, P. 1996. PKL Group (UK) Ltd., Specialists in Containerized Kitchen Technology. Activities Report of the R & D Associates 48(2):43–49. [Research and Development Associates for Military Food and Packaging Systems, San Antonio, TX.]
94. Schmidt, K. 1995. WHO Surveillance Programme for Control of Foodborne Infections and Intoxications in Europe, 1990–1992, Sixth Report, Institute of Veterinary Medicine, Berlin, pp. 1–335.
95. Schutz, H. G., and A. V. Cordello. 1997. Information effects on acceptance of irradiated foods in a military population. Dairy, Food Environ. Sanitat. 17:470–481.
96. Serichantalergs, O., W. Nirdnoy, A. Cravioto, C. LeBron, M. Wolf, A.-M. Svennerholm, D. Shlim, C. W. Hodge, and P. Echeverria. 1997. Coli surface antigens associated with enterotoxigenic *Escherichia coli* strains isolated from persons with traveler's diarrhea in Asia. J. Clin. Microbiol. 35:1639–1641.

97. Sorvillo, F. J., K. Fujioka, M. Tormey, R. Kebabjian, W. Tokushige, L. Mascola, S. Schweid, M. Hillario, and S. H. Waterman. 1990. Swimming-associated cryptosporidiosis—Los Angeles County. MMWR 39:343–345.
98. Steffen, R., J. Melling, J. P. Woodall, P. E. Rollin, R. H. Lang, R. Lüthy, and A. Waldvogel. 1997. Preparation for emergency relief after biological warfare. J. Infect. 34:127–132.
99. Stevenson, J. 1997. Pentagon-funded research takes aim at agents of biological warfare. JAMA 278:373–375.
100. Still, J. 1991. Safe food preparation in the desert. Army Log (July–August):36–37.
101. Sutphen, E. 1988. Field Service in the Warsaw Pact Countries and China. Activities Report of the R & D Associates 40(2):12–13. [Research and Development Associates for Military Food and Packaging Systems, San Antonio, TX.]
102. Taylor, R. 1996. All fall down. New Sci. 150 (2029), May 11.
103. Taylor, D. N., K. T. McDermott, J. R. Little, and M. J. Blaser. 1981. Diseases acquired from "pure" mountain streams, presented at the EIS (Epidemic Intelligence Service) Conference, Atlanta, April 20–24.
104. Tenborg, M., B. Davis, D. Smith, C. Levy, B. England, B. Koehler, D. Tanda, J. Pape, R. Hoffman, R. Fulgham, B. Joe, and J. Cheek. 1997. Fatal human plague—Arizona and Colorado, 1996. JAMA 278:380–381.
105. Tocidlowski, M. E., M. R. Lappin, P. W. Sumner, and M. K. Stoskopf. 1997. Serologic survey for toxoplasmosis in river otters. J. Wildl. Dis. 33:649–652.
106. Todd, E. C. D. 1976. The first annual summary of food-borne disease in Canada. J. Milk Food Technol. 39:426–431.
107. Todd, E. C. D. 1978. Foodborne and waterborne disease in Canada—1975 Annual summary. J. Food Prot. 41:910–918.
108. Todd, E. C. D. 1981. Foodborne and waterborne disease in Canada—1976 Annual summary. J. Food Prot. 44:787–795.
109. Todd, E. C. D. 1982. Foodborne and waterborne disease in Canada—1977 Annual summary. J. Food Prot. 45:865–873.
110. Todd, E. C. D. 1985. Foodborne and waterborne disease in Canada—1978 Annual summary. J. Food Prot. 48:990–996.
111. Todd, E. C. D. 1987. Foodborne and waterborne disease in Canada—1980 Annual summary. J. Food Prot. 50:420–428.
112. Toops, D. 1997. What's cooking in space? Food Process., December, 64, 67.
113. Török, T. J., R. V. Tauxe, R. P. Wise, J. R. Livengood, R. Sokolow, S. Mauvais, K. A. Birkness, M. R. Skeels, J. M. Horan, and L. R. Foster. 1997. A large community outbreak of salmonellosis caused by intentional contamination of restaurant salad bars. JAMA 278:389–395.
114. Warren, M. 1998. Mushroom and melon eaters dice with death. Electron. Telegr., Russia, July 29.
115. White, M. R., L. A. Hermansen, and E. K. Shaw. 1991. The effects of port visits on infective and parasitic diseases in U.S. Navy enlisted personnel. Report 91-11, Naval Health Research Center, San Diego, CA, pp. 1–10.
116. WHO. 1998. Management of Environmental Health in Emergencies and Disaster, A Practical Guide. World Health Organization, Geneva.
117. Willshaw, G. A., T. Cheasty, B. Rowe, H. R. Smith, D. N. Faithfull-Davies, and

T. G. J. Brooks. 1995. Isolation of enterotoxigenic *Escherichia coli* from British troops in Saudi Arabia. Epidemiol. Infect. 115:455–463.
118. Wooden, R. 1991. An overview of HACCP—History and impact. Activities Report of the R & D Associates 43(2):43–47. [Research and Development Associates for Military Food and Packaging Systems, San Antonio, TX.]
119. Yarrow, A., P. E. Slater, E. M. Gross, and C. Costin. 1991. The epidemiology of echinococcosis in Israel. J. Trop. Med. Hyg. 94:261–267.
120. Zilinkas, R. A. 1997. Iraq's biological weapons; the past as future? JAMA. 278: 418–424.

15
Food Safety Information and Advice for Travelers

O. Peter Snyder, Jr.
Hospitality Institute of Technology and Management, St. Paul, Minnesota

I. INTRODUCTION

Over 400 million travelers cross international boundaries each year (29). People often become ill upon traveling from one area of the world to another. Contributing factors include traveling in areas where there are lower standards of hygiene and going to regions that lack appropriate technology, equipment, and supplies for producing and providing a reliable supply of safe food and potable water. These sanitation concerns are usually due to unhygienic, inefficient, or nonexistent food and water safety control systems. This problem is a major concern for those who visit small villages and outlying areas where there is little or no choice of ready-to-eat food except what is offered at small, local restaurants or by street vendors. To assure the safety of food, correct hazard and control information must be supplied by government sources throughout the world to producers and preparers of food, who must be trained and motivated to use this information. Tourists must also become informed on how to make correct choices of food and beverages when traveling.

Foodborne disease surveillance programs must also be developed throughout the world (51). At present, most countries have some type of reporting for notifiable diseases but few have foodborne disease surveillance programs. Surveillance programs are necessary to identify and record outbreaks and to inform a population about contaminated food and water. Improvement of global surveillance is critically important as travel and trade increases between various areas of the world. Imported foods from one region of the world to another have also been implicated in disease transmission. For example, imported produce from Mexico, Central America, and other tropical and semitropical areas are important vehicles for the transmission of enteric pathogens in North America (40).

II. FOODBORNE ILLNESS: A GLOBAL PROBLEM

Foodborne illness is a global problem (1, 23, 33, 37, 50, 51, 53, 54). In developing regions of the world, a wide range of foodborne diseases prevail. These include cholera, typhoid and paratyphoid fevers, shigellosis, *Escherichia coli* gastroenteritis, brucellosis, and parasitic diseases such as amebiasis and taeniasis. In developed countries, foodborne illnesses are more likely to be due to *Clostridium perfringens, E. coli, Salmonella* spp., Norwalk virus, *Bacillus cereus,* and *Campylobacter jejuni* (29, 50).

Travelers to Third World countries run a higher risk of acquiring intestinal parasitic infections than travelers in developed countries. Studies of large numbers of returning travelers indicate that helminthic or worm infections most often are acquired in rural areas that are visited infrequently by tourists. In recent years, many foreign-born Americans have been returning to their homelands to visit family and friends. This group is exposed to parasitic infections when they stay for extended periods of time (>1 month) in rural communities with relatively primitive sanitary facilities. As the length of stay is increased, the risk of acquiring endemic parasitic infections is also increased. Intestinal protozoan infections found in returning travelers are more common than infections of helminths, because the former multiply in the human host, causing clinically relevant and symptomatic disease. Short-term travelers are unlikely to be infected with clinically important numbers of worms, which usually do not reproduce in the human host. Studies have shown that protozoan infections (from *Entamoeba histolytica, Giardia lamblia, Cryptosporidium spp., Blastocytis hominis, Balantidium coli, Dientamoeba fragilis,* and *Entamoeba polecki*) are more than 10 times as frequent as those caused by worms [*Ascaris lumbricoides, Trichuris trichura* (whipworm), hookworms, and tapeworms] (57).

Käferstein (28) of the World Health Organization (WHO) stated:

> Contaminated food is the cause of serious health problems in both developed and developing countries. It is one of the main etiological factors of malnutrition in developing countries due to its role in causing diarrhea. Foodborne diarrhea kills about 5 million children per year in Asia, Africa, and Latin America. Although the mortality rates of foodborne disease are negligible in industrialized countries, some experts believe that these diseases rank second only to respiratory diseases in morbidity.

Table 1 is a list of some documented incidents of foodborne illness throughout the world that have affected significant numbers of people. These reported in-

Table 1 Worldwide Reports of Foodborne and Waterborne Illness Outbreaks

Year	Location	Cause	Number of people affected
1981	Spain (44)	Toxic oil syndrome (TOS): edible oil that had been chemically adulterated	20,000 people affected, 300 deaths
1984	British Airways (flight between Europe and Canary Islands) (44)	*Salmonella in mayonnaise*	1000 passengers and personnel
1985	Illinois and surrounding states (44)	*Salmonella* in pasteurized milk	200,000 people
1985	California (44)	*Listeria monocytogenes* in white cheese	90 people; 29 deaths and several abortions (miscarriages)
1987	Canadian tourists at Caribbean resort (44)	Ciguatera toxin in fish casserole	57 people
1987	Canada (44)	Amnesiotoxicosis caused by marine toxin *Pseudonitzschia pugens* in blue mussels	150 people
1988	China (44)	Hepatitis A from contaminated clams	292,000 people; 32 fatalities
1988	Pennsylvania and Delaware (44)	Viral illness from packaged ice	5000 people
1989	Central America and Mexico (44)	Paralytic shellfish poisoning (PSP)	Several hundred people
1990	Latin America (44)	Shigellosis due to contaminated water supply	More than 2600 people
1990–91	Israel (44)	Niacin (nicotinic acid) intoxication through ground beef (added for color retention)	149 people
1991	Mediterranean country (44)	Botulism from salted fish	81 people; 18 fatalities
1991	Australia—air travel (44)	Norwalk-like virus, or small round structured virus (SRSV), from orange juice	3051 people
1992	Airline flight from Argentina and Peru to Los Angeles (15)	Cholera	31 culture confirmed cases: of 356 passengers and crew, 54 people reported diarrheal illness
1986–1993	Cruise ships—United States (31)	Various causes (undercooked seafood and eggs; polluted water; poor food handling practices)	31 investigated outbreaks of diarrheal illness involving 7626 passengers and 601 staff members
1994	Sweden (56)	*Shigella sonnei* from imported lettuce	55 people affected
1994	Tourists of southeast Asia (55)	*Vibrio cholerae* O139 Bengal infection	6 people from a cruise ship, after eating yellow rice in Bangkok
1995	London (19)	Scombrotoxic fish poisoning from tuna imported from Sri Lanka	27 people affected
1996	Travelers from Minnesota to Mexico (36)	*Salmonella enteritidis* (subtype 11)	14 confirmed cases
1997	Travelers to Canary Islands from England, Wales, and Finland (20)	*Escherichia coli* O157:H7 from nonpotable well water used in hotels	8 confirmed cases

Source: Adapted from Ref. 44.

cidents involved both residents and travelers. In addition, many illness incidents go unreported, hence are never recorded.

A. Traveler's Diarrhea

One of the most common concerns and complaints of travelers is diarrheal illness, even though incidents of bacterial, viral, and protozoan enteritis can also affect people in their home environment. Each year, millions of people travel from industrialized countries to developing countries, which tend to be in the tropics and subtropics. During the course of travel, 20–50% of all travelers become ill and experience diarrhea caused by a wide variety of enteric pathogens acquired through ingestion of contaminated food and/or water (21, 24). In many countries, crops are grown in soil fertilized with human excreta. Thus, food grown in this soil is contaminated with enteropathogens (21). These bacterial, viral, and parasitic pathogens are spread by water, particularly during rainy seasons. Diarrhea is more likely to occur in travelers exposed to these pathogens because people from developed countries have not acquired a resistance to some types of pathogen because their diet contains fewer pathogens.

The onset of traveler's diarrhea is usually within the first week of travel. Duration of the illness is usually short, lasting 24 hours or less in 20% of cases, and 2–7 days in 60% of individuals affected. Some travelers (8–15%) experience illness for a week or longer (39). Symptoms of traveler's diarrhea are usually characterized by the passage of at least 3, and up to 10, unformed stools in a 24-hour period. Other symptoms include nausea, vomiting, abdominal pain or cramps, fecal urgency, tenesmus, and/or the passage of bloody or mucoid stools (dysentery). Each of these symptoms occurs in 10–20% of all cases. The severity of the illness is usually dependent on the pathogen or pathogens responsible for illness and the immune status of individual affected. Approximately 20% of patients are confined to bed for 1–2 days (21, 39). Occurrence of traveler's diarrhea is slightly more common in young adults than in older persons. There seems to be no difference in incidence because of gender (16).

When the stools of the majority of these ill individuals are tested, one or more pathogens are usually found to be present. Enterotoxigenic *E. coli* spp. generally are the most frequently identified pathogens (30). For example, studies of travelers' diarrheal episodes have found that pathogenic strains of *E. coli* have been found to be responsible for 40% of travelers' illnesses in Latin America, 31–75% in Africa, and 20–30% in Asia (39). Other pathogens have also been implicated as causes of diarrhea in a smaller fraction of ill travelers. These include *Shigella* spp., *Salmonella* spp., *Campylobacter jejuni, Vibrio* spp., *Aeromonas hydrophila, Entamoeba histolytica, Giardia lamblia,* rotavirus, and Norwalk virus (Table 2).

Plesiomonas shigelloides, adenoviruses, and *Cryptosporidium* may also

Table 2 Pathogens Commonly Isolated in Cases of Traveler's Diarrhea

Organism[a]	Percentages of traveler's diarrhea				
	Asia	Middle East	Latin America	Africa	Worldwide
ETEC	20–30	57	40	31–75	36
EAEC	—	—	5	33	5
EIEC	3	—	6.5	2	3
Salmonella spp.	6–18	2–7	7	2–25	1
Shigella spp.	2–17	4–20	15	4–15	4
Campylobacter spp.	5–41	2	3	1–28	3
Aeromonas spp.	1	—	2	1–8	2
Vibrio non-O1	1–16	—	2	<1	2
Rotavirus	—	6	10	5	2
Entamoeba histolytica	—	—	<1	—	2
Giardia lamblia	1	—	4	—	2
Unknown	42	—	15	38–40	39

[a]ETEC, enterotoxigenic *E. coli;* EAEC, enteroadhesive *E. coli;* EIEC, enteroinvasive *E. coli.*
Source: Adapted from Ref. 39.

cause episodes of traveler's diarrhea. Mixed infections of two or more of these pathogens can also occur. Frequently, multiple pathogens are isolated from the same individual (7, 26, 30). The relative proportion of enteropathogens isolated varies with the population studied, season of the year, and site of travel (25, 26). Pathogens responsible for enteric travelers' illnesses vary somewhat according to region and time of the year (21, 52). For example, travelers to Mexico and Morocco in the rainy summer months of the year are more apt to become ill due to enterotoxigenic *E. coli, Salmonella* spp., and *Shigella* spp., while illness due to *C. jejuni* is more common during the drier winter months. *Giardia* and *Cryptosporidium* are important causes of illness in mountainous areas of North America and Russia (21). The threat of diarrhea and illness represents economic costs to travelers and the host country as well, through loss of business.

B. Other Factors Contributing to Illness

"Changes in the environment affect all people, coming and going" (18). Traveler's diarrhea and other foodborne diseases that affect travelers are usually of biological origin. However, other factors such as jet lag, fatigue, disruption in eating habits, change in climate, and low immunity to the microflora of the new environment decrease the resistance of travelers and contribute to their susceptibility to illness (29, 45).

C. Understanding Some of the Causes of the Problem

To reduce the risk of foodborne illness and travelers' diarrhea, travelers must understand what makes some foods unsafe. Then they can choose food and beverages based on preparation and processing methods that assure safety. Table 3 lists some of the sociocultural factors that affect food safety (23).

D. Street-Vended Food for Travelers (Tourists) as a Cause of Foodborne Illness

Travelers to foreign countries often wish to try some foods from the local area, or indeed may have no choice but to buy and consume street-vended food. However,

Table 3 Sociocultural Factors That Affect Food Safety

Positive widespread practices	*Negative widespread practices*
Thorough cooking of foods	Unsanitary marketing practices
Washing of fruits, vegetables (when water is safe)	Inadequate water supplies
Peeling uncooked fruits	Unsanitary waste disposal
Eating recently slaughtered, well-cooked meat	Poor personal hygiene of food handlers
Boiling of milk	Preparation of food long before serving
Breast feeding	Storage of food at ambient temperatures
	Uncontaminated food brought into contact with contaminated food or surfaces
	Unhygienic street vending of food
	Failure to protect cooked food from recontamination (e.g., through flies)
Positive culture-specific practices	*Negative culture-specific practices*
Cutting food, especially meat, into small pieces to effect maximum heat penetration (countries of the Orient)	Preference for uncooked/undercooked meat, fish
Buying only sufficient food for one meal at a time	Partial precooking of large pieces of meat
Correct fermentation processes, especially of milk (Asian nomads)	Aging of food without refrigeration to enhance flavor
Food is preserved safely through proper methods of salting and/or sun or air drying	Precooking of large quantities of rice (southeast Asia)
	Unsanitary preparation of infant formulas
	Taboo on hand washing (Latin America)
	Addition of commercial detergents to make whiskey foam (Thailand)
	Practice of drying and salting ungutted fish

Source: Adapted from Ref. 23.

travelers are at risk when they consume this type of food, since it is often a source of enteropathogens. For example, *Vibrio cholerae* O1 was isolated in 11% of samples of *ceviche* (a raw fish product seasoned with spices and vegetables in an acid marinade) taken from street vendors, and in 6% of *ceviche* samples obtained from restaurants in Guadalajara, Mexico (52). Research microbiologists (17) have also reported the high occurrence of shiga-like toxin-producing strains of diarrheagenic *E. coli* in raw ground beef samples in Rio de Janeiro, Brazil. Documented studies of the hazards associated with street-vended foods have been reported by Bryan and others (8–14, 27, 47).

An interesting analysis of street-vended foods in China (Yichang City and Puqi City) was reported by Liang and Yuan (34). The study showed that the number of coliform bacteria on cooled food (food not heated or heated a very short time before sale) was much higher than on cooked food (food heated sufficiently before sale). *Shigella* spp. were found in hot dry noodles, cooled noodles, stewed meat, rice, ice water, and steamed, stuffed buns. However, *Salmonella* spp. and pathogenic bacteria responsible for cholera and paratyphoid fever were not found in any of the 290 samples of 16 kinds of street-vended food. The unfixed street-vending units could not possibly assure the safety of their products because they had no safe drinking water supply, no netting for preventing contamination from flies and dust, and no cleaned and disinfected tableware. It was reported that tableware used by street vendors certainly contaminated food products. *Shigella* spp. were detected on the surface of 4.7% of tableware. The aerobic bacterial count on the surface of tableware averaged 4.7×10^4 cfu/cm^2, and the coliform count averaged 1.4×10^2 cfu/100 cm^2.

Bryan (8) summarized and listed hazards of street foods. High populations of microorganisms may be present on foods purchased from vendors. If tourists see any of the following practices, they should avoid the food.

1. Washing or freshening produce with polluted water
2. Unsanitary practices, spreading of contaminants, and survival of contaminants during processing:
   a. Handling cooked food with (unwashed) bare hands
   b. Preparing cooked foods on cutting boards, on tables, and/or with utensils previously used for other foods (cross-contamination)
   c. Holding foods at outdoor or, in some cases, indoor ambient temperatures
   d. Insufficient reheating, if foods are reheated at all
   e. Long duration of storage at temperatures that are conducive to microbial growth

Street-vending operations are not the only source of agents causing food-borne illness. Pathogenic bacteria have also been isolated in hotel dining rooms.

El-Sherbeeny et al. (22) reported isolating *Salmonella* in Oriental rice prepared in a five-star hotel in Egypt, and *Shigella* in boiled rice in a four-star hotel.

In developed nations such as the United States, Canada, and many northern European countries, illness due to growth of *C. perfringens* and *B. cereus* in improperly cooled or improperly refrigerated food products is a cause of foodborne illness (50). In areas of the world where refrigeration storage is lacking or non existent, inadequate pasteurization (heat insufficient to destroy vegetative pathogenic microorganisms) of food and time–temperature abuse of food after cooking (which allows the growth of pathogenic bacteria) are more likely to be the cause of illness due to unsafe food handling practices.

These reports stress the need for, and importance of, ensuring a supply of safe, wholesome, reasonably priced food in developing countries. Public health officials must inform and train all food preparers and food handlers to use preventive measures that are practical under prevailing circumstances. Training in the identification and control of hazards should be given to home food prepares and to those who work in large hotel dining rooms, as well as to street vendors. The measures can be demonstrated to people, but action must be taken to ensure implementation of the measures whenever possible. The following suggestions come from WHO authorities (2):

> Vendors should use neither grossly contaminated food, nor hazardous material or additives in the preparation of food. Places of preparation should be clean and used only for the purpose of preparing and selling food. Containers used for cooking, storage and display should be easy to clean and their surfaces should not release toxic substances into food. The point of sale, whether stationary or mobile, should be located where there are minimal risks of contamination from sewage, animal and human fecal matter and other hazardous materials. Vending operations should not interfere with vehicular or pedestrian traffic, nor place customers, particularly children, in danger from traffic or other hazards. Particular care is taken to see that drinking-water is safe and that water for washing utensils is clean and not reused.

III. PREVENTION OF FOODBORNE ILLNESS AND INJURY: UNDERSTANDING THE HAZARDS

The starting point for food safety is to identify the possible hazards in food. Then, appropriate controls can be assigned to each hazard. Biological, chemical, and physical hazards are causes of foodborne illness and injury for people in both developed and underdeveloped countries. Table 4 provides an overview of the hazards that can harm people anywhere in the world. Some of these hazards (e.g., allergic reactions to food) have not yet been considered in most government programs, because the focus of hazards in food is usually centered on microbial hazards (42).

Table 4 Hazards in the Food System

Chemical	Physical	Biological
Poisonous substances	Hard foreign objects	Microorganisms and their toxins
Toxic plant material	Glass	Bacteria: vegetative cells and spores
Intentional (GRAS) food additives[a]	Wood	Molds [mycotoxins (e.g., aflatoxin)]
Chemicals created by the process	Stones, sand, and dirt	Yeasts (*Candida albicans*)
Agricultural chemicals	Metal	Viruses and Rickettsiae
Antibiotic and other drug residues in meat, poultry, and dairy products	Packaging materials	Parasites
Unintentional additives	Bones	Marine animals as sources of toxic compounds
Sabotage	Building materials	Fish
Equipment material leaching	Filth from insects, rodents, and any other unwanted animal parts or excreta	Shellfish
Packaging material leaching	Personal effects	Pests as carriers of pathogens-insects, rodents, birds
Industrial pollutants	Functional hazards	
Heavy metals	Particle size deviation	
Radioactive isotopes	Packaging defects	
Adverse food reactions food sensitivity)	Sabotage	
Food allergies	Choking/food asphyxiation Hazards: Pieces of food	
Food intolerances	Thermal Hazards: Food so hot that it burns tissue	
Metabolic disorder–based reactions		
Pharmacological food reactions		
Idiosyncratic reactions to food		
Nutrition		
Excessive addition of nutrients		
Nutritional deficiencies and/or inaccurate formulation of synthesized formulas		
Antinutritional factors		
Destruction and unnecessary loss of nutrients during processing and storage		

[a]GRAS, generally recognized as safe.
Source: Adapted from Ref. 42.

IV. THE ROLE OF GOVERNMENTS IN THE PREVENTION OF FOODBORNE ILLNESS AND INJURY

Governments in every nation must require that people supplying and selling food become aware of the possible hazards in food and establish appropriate preventive measures that can be taken to assure food safety (3). Bryan (8) has cited critical control points for ensuring the safety of foods, whether in a home, hotel kitchen, or street-vending operation. When deciding what food to eat or where to eat, travelers should consider whether there is a reasonable likelihood that person(s) preparing and serving the food followed these guidelines; if it is suspected that the food has not been prepared and handled in this manner, the food should not be eaten.

1. *Source of ingredients:* ingredients for preparing foods must be obtained from safe sources of supply.
2. *Formulation:* food must be formulated so that it will not produce illness when it is consumed.
3. *Cooking:* food must be heated to temperatures necessary to destroy vegetative pathogenic microorganisms.
4. *Manipulation of foods after cooking:* food should be handled or manipulated as little as possible after cooking to avoid cross-contamination. Clean, sanitized utensils should be used to serve foods.
5. *Holding cooked foods:* cooked foods should be held at 140°F (60°C) or higher until consumed, or cooled to temperatures below 40°F (4.4°C) and held at these temperatures for times that are insufficient for the growth of pathogenic bacteria to hazardous levels.
6. *Reheating:* cooked foods should be heated to at least 165°F (73.9°C). Heat-stable toxins will not be inactivated by reheating, and prevention of associated illnesses is dependent on preventing their formation by eating foods before toxins can be formed, cooling foods rapidly, and holding foods at temperatures above or below those at which toxins are formed. Periodic reheating (e.g., every 4–6 h) of food can eliminate cells that have germinated from spores during the holding time at temperatures that allow bacterial growth to occur.
7. *Cooling appropriately:* if facilities are available, foods should be continuously cooled to below 40°F (4.4°C) as rapidly as possible.

People sometimes believe that preemployment physical examination of food employees reduces risk of disease transfer. However, Käferstein and Motarjemi (29) have stated that there is overwhelming evidence that preemployment and routine medical examination of food handlers to exclude carriers from handling food are of no value in the prevention of foodborne diseases. Therefore, WHO (53) recommends that this practice be discontinued and be replaced by education of food handlers and insistence on adherence to good personal hygiene practices.

V. THE WORLD FOOD SYSTEM AND THE TRAVELER

A traveler is anyone who is away from his or her regular environment and source of food. Travel can include being anywhere in the world and eating food at hotels, restaurants, and campsites; on cruise ships, trains, and airplanes; at street-vending operations, food stands at carnivals, and outdoor food markets. Food at an establishment is basically defined as safe if travelers who eat it do not get sick as a result.

Although all age groups are at risk of illness while traveling, incidence is the highest among the very young (40%). This is presumably because infants and children lack exposure to many pathogens, hence have no or limited immunity. In the 15- to 29-year age groups (36%), the higher incidence of such illness is thought to be due to adventurous travel style and the ingestion of high volumes of potentially contaminated food. Males and females are affected equally. If compliance with the traditional preventive dietary recommendation of "boil it, cook it, peel it, or forget it" is not followed by travelers, the frequency of illness is usually proportional to the number of dietary indiscretions (44).

The general hazards and controls for any food operation in the world can be seen in Table 5. Controls will be implemented, however, only when individual governments become more concerned about the health status of the general population and travelers, as affected by the safety of the food and water supply. Tourism is likely to increase in areas of the world where travelers can be more certain of the safety of food and water.

The information in Table 5 is based on the premise that the traveler is any individual more than 5 years of age. [It is assumed that babies less than 6 months old are breast fed or are given a liquid formula that has been processed for safety and formulated to meet the infant's nutritional requirements.] The traveler's immune system is normal, and he or she is not taking any immune-suppressant drugs. If the traveler is more than 65 years old, it is assumed that he or she is not taking antibiotics and has no known medical problems such as diabetes, ulcers, cholecytitis (inflammation of the gallbladder), or heart disease.

VI. METHODS AND AGENTS USED TO PREVENT TRAVELER'S DIARRHEA

Travelers going to high risk regions of the world (Asia, Africa, South America, and Mexico) may consider medical, pharmalogical, or biological methods to prevent or treat illness. Some of these agents have a beneficial effect in preventing traveler's diarrhea; some do not. Beneficial effects are dependent on the type and number of pathogenic microorganisms responsible for causing disease or illness and the immune status of the traveler. Even when travelers use these methods to prevent illness, they must still be cautious and carefully select all beverages and foods consumed.

A. Bismuth Subsalicylate

Bismuth subsalicylate is commonly known by its commercial name, Pepto-Bismol™. In 1900 a physician in New York developed a liquid preparation con-

Table 5 General Hazards and Controls for Any Food Operation in the World

Hazard source identification	Control
Consumer	
Consumer/traveler has no knowledge of causes of foodborne illnesses and disease.	Consumer/traveler Is educated with respect to the hazards and causes of foodborne illnesses and disease. Utilizes this knowledge to select food and places to eat where the risk of acquiring a foodborne illness or disease is low. Eats a variety of food in moderation so that antinutritional factors and other possibly toxic compounds in food are not a problem. Does not consume foods that have remained at ambient temperatures or are warm (refrigeration of foods in developing countries is marginal to nonexistent). Consumes only fruits and vegetables that can be peeled, if eaten fresh. Fresh fruits and vegetables that are cooked and maintained at ≥140°F (60°C) can also be consumed if a residual amount of agricultural chemicals is not a consideration. May consume commercially canned foods (fruits, vegetables, meats, poultry, fish, milk, etc.) if consumer knows or recognizes the brand name as a known reputable source. Never eats any meat, poultry, or fish that is rare (red) or raw. These products should be cooked until all parts of the products have reached 150°F (65°C) for 1 min.
Management and personnel	
Owner: If the person operating the food establishment has no knowledge of hazards, it must be assumed that controls are nonexistent, and the safety of the food depends on chance.	The operation has established a HACCP program. The owner: Demonstrates food safety leadership and provides resources that enable employees to strive for zero safety defects. Knows the risk of hazards in the food supply and uses validated controls to assure food safety. Can answer questions such as: What is the source of the food? Were the farmers' growing processes following an HACCP approach? If the food is plant material, how was it fertilized—with human or animal fecal material and/or with chemical fertilizers? How and when was the food harvested? Processed? Stored? What has been the temperature of the food since it was harvested, caught, or slaughtered? Were the fruits and vegetables washed? If so, in what kind of water?

Table 5 Continued

Hazard source identification	Control
Management and personnel (*continued*)	To what temperature was the food cooked? Has the food been reheated? If so, how many times and to what temperature? Trains all employees to prepare and serve food safely before assigning them to handle and prepare food. During operations, coaches employees to improve performance and measures actual performance of each process. Takes immediate action to find the cause of any product deviation and corrects the problem. Gains more education, training, and information each year to improve performance.
Environment	
Water is contaminated with hazardous chemicals and/or pathogenic microorganisms.	Consumer/traveler: Does not consume water in the food operation unless it is known that the water is from a safe or potable source. Will not drink bottled water unless it comes from a sealed container. If the water supply is not safe, drinks only pasteurized water [heated to 158°F (70°C) or above for 1 min in a covered container], hot beverages such as coffee and tea, canned or bottled carbonated beverages, beer and wine. Avoids consuming ice unless it is known that the ice is produced with potable water. Wipes water from the surface of a can or bottle so that the surface that touches the mouth is clean and dry, and potential levels of pathogenic microorganisms are removed. Does not brush teeth with tap water unless the city or community certifies the safety of the water. Is careful not to get water in mouth when shaving, washing, bathing, or showering if water supply is not potable. Knows that water can also be made safe by filtering it with filters designed to remove parasites and then chemically disinfecting it with either chlorine or iodine tablets, and allowing at least 20 min for the disinfection process.
Air contains hazardous pollutants. In a food operation or hotel with cooling towers, pathogens such as *Legionella pneumophila* and harmful molds will multiply and can cause serious illness.	Cooling towers in air-conditioned buildings are cleaned and are kept chemically disinfected on a regular basis to prevent accumulation of harmful pathogens.

(*continued*)

Table 5 Continued

Hazard source identification	Control
Environment (*continued*)	
Insects and rodents are present and cause contamination of food supplies. They carry pathogenic bacteria on their feet and bodies, and must not be allowed to contact ready-to-eat food.	At the food facility: There is no obvious indication of insect and rodent contamination. There is proper disposal of waste and garbage. Screens and doors prevent the entrance of pests. Food is displayed so that insects are excluded.
Facilities and equipment	
Facility is dirty (floors, table surfaces, trash areas, etc.).	The facility is clean and well maintained inside and outside.
Restroom and toilet areas are dirty. No soap and paper towels are provided for hand washing.	Toilet areas are clean, and there is a supply of soap and paper towels for washing and drying hands.
Table and glassware are not clean.	Tableware is clean and is stored properly.
Equipment is poorly maintained, does not function correctly.	Equipment functions properly and is maintained.
Food contact surfaces are not cleaned and disinfected. These surfaces (plastic, wood, metal) will all be a hazard when used for preparing raw meat, fish, and poultry if they are not washed before contacting ready-to-eat food.	Food contact surfaces (plastic, wood, metal) used for preparing raw meat, fish, and poultry are washed with clean, disinfected, flowing water before contacting ready-to-eat food.
Raw food. No precautions are taken when raw food is handled (e.g., washing, and preventing cross-contamination). It is assumed that all raw food (raw fruits and vegetables; raw meat, poultry, and fish) are contaminated with pathogenic microorganisms, chemicals, and particulate matter.	Food facility controls: Raw fruits and vegetables are washed twice with safe water: first in water in a clean sink, not used for meat, fish, or poultry; then with flowing water in a colander. All meat, fish, and poultry are pasteurized for a 7D *Salmonella* spp. kill. For example, heated until all parts of the food reach a temperature of 150°F (65.6°C) for 72 sec.
Ingredients and processes	
Fermented foods such as salami, cheese, yogurt, and sauerkraut can contain pathogens if not pasteurized, and if lactic acid cultures do not grow properly.	Only fermented products that are produced at a facility with an HACCP certification should be consumed.
Acidified foods such as mayonnaise, salad dressings, and hollandaise sauce are not sufficiently acidified and may be, contaminated with pathogenic bacteria from the ingredients, (raw eggs, spices, etc.).	Salad dressings and mayonnaise are prepared according to procedures that assure their safety. These products have a pH below 4.1 if made with raw ingredients and have been held at room temperature for more than 3 days to assure safety (destruction of *Salmonella* spp.).

Table 5 Continued

Hazard source identification	Control
Ingredients and processes (*continued*)	
Dry foods such as rice, beans, and dried peas are normally contaminated with pathogenic spores of *Bacillus cereus, Clostridium perfringens,* and *Clostridium botulinum.*	Rice, dried beans, and peas are prepared so that they achieve temperatures of at least 140°F (60°C) in less than 6 h, are pasteurized, and are kept at temperatures above at least 140°F. If not served immediately, these products are continuously cooled from or above to below 45°F (7.2°C) in less than 15 h.
Salted foods such as butter and some fermented sausages may contain pathogenic microorganisms such as *Staphylococcus aureus. S. aureus* can produce toxin in food with a salt concentration of 12%.	Food such as butter suspected as being an *S. aureus* threat, if displayed at room temperature, contains more than 12% salt in the water phase.
Spiced and herbed foods	
Spices and herbs are a major source of pathogenic microorganisms and toxins.	Food containing spices and herbs is eaten hot or within 2 h after the food containing the spice has been cooked.
Spices are used to mask or cover up spoilage of food.	When spices and herbs are added to food, the food is cooked or heated for at least 12 min at 140°F (60°C) or more to assure adequate pasteurization of the spice ingredients.
Container food such as dry infant formula, canned (sterilized) food, and eggs in the shell are susceptible to contamination whenever the product is opened, the seal is broken, or the shell is cracked.	Dry infant formula is reconstituted with disinfected water and is consumed within 2 h. Canned food, if not refrigerated, is consumed within 2 h. Raw eggs in a shell may be contaminated and must be pasteurized by heating until all parts of the egg have reached a temperature of 158°F (70°C) for 1 min.
Cooked–pasteurized food. All raw products (meat, poultry, fish, eggs, vegetables, etc.) must be assumed to be contaminated. Sauces, gravies, soups, stews, and many other combination food products are often made far ahead of being served and held at less than 140°F (60°C). Thus, growth of spores can occur.	For reliable safety, all raw food items should be consumed while they are still hot from the initial cooking. There should be no pink center in meat or poultry products. Fish is cooked until it flakes. Sauces, gravies, soups, stews, hot combination products, etc. are served hot [>150°F (65.6°C)]. Cooked hot food is never allowed to remain at room temperature for more than 2 h before it is consumed.
Food on display. Hot food is not kept at temperatures of at least 140°F (60°C). Thus, pathogenic spores germinate and vegetative cells multiply.	The temperature of hot food on display must be above 140°F (60°C), or it must be consumed within 2 h of the time it was cooked initially or removed from this heat source.

(*continued*)

Table 5 Continued

Hazard source identification	Control
Ingredients and processes (*continued*)	
Food left at ambient temperatures is not sufficiently acidified to prevent growth of pathogenic bacteria. The food can become dangerous if held for more than 9 h at 70°F (21.1°C).	It should be obvious that hot food has not been reheated. (Reheated leftovers may contain dangerous toxins.) Food that is lukewarm should not be consumed.
There is pathogenic contamination of cooked foods from handling. Multiplication of these pathogens occurs rapidly when foods are stored at temperatures above 50°F (10°C).	Food held at ambient temperatures is safe if it is sufficiently preserved with acids, salt, sugar, fermentation, or alcohol to retard the growth of harmful pathogenic microorganisms. Cooked food that is allowed to cool to ambient temperatures should be eaten within 2 h. Owners/food prepares must use correct hand-washing procedures and correct washing and sanitation procedures for all food contact surfaces. Cold foods contain as much acid, sugar, salt, or other preservatives as possible to be still palatable. Cold foods must be kept at temperatures below 50°F (10°C). Consumers/travelers should not consume cooked cold food such as hors d'oeuvres or sandwiches unless the food preparer or operator can prove they are safe.
Leftover food is stored at room temperature. It may or may not be reheated before it is sold. Leftover foods are always mixed in with freshly prepared food items so that the food is not wasted. These conditions allow cross-contamination of food products, pathogenic spore outgrowth, and vegetative cell multiplication.	Owners/food prepares should never mix leftover food items with freshly prepared items. If the temperature of the food exceeds 140°F (60°C), it will be safe, but the quality and nutritional value of the food is poor after 2 h. Consumers/travelers should eat only freshly prepared, well-done, hot food, or cold food (prepared under conditions that prevent food-borne hazards) that is kept below 50°F (10°C).

taining bismuth subsalicylate, zinc salts for their astringency, salol (phenylsalicylate) for its antiseptic ability, oil extracts of wintergreen for flavor, and a red dye to make the product pink, to enhance its appeal to children. This combination was called "*Mixture Cholera Infantum*" and was successful in the treatment of the illness. Discussions of the use of this preparation are given by Bierer (6), Dupont and Ericsson (21), Manhart (35), Okhuysen and Ericsson (39), and Steffen (48).

Bismuth subsalicylate (BSS) is a very insoluble salt of trivalent bismuth and salicylic acid, which is hydrolyzed in digestion to bismuth oxychloride and salicylate. When the BSS is dissociated, salicylate is absorbed. The bismuth ions travel to lower segments of the bowel where the ions, in turn, form several nonabsorbable salts that possess antimicrobial activity. These bismuth salts adhere to bacteria, causing leakage of ATP from the bacterial cell, and ultimately cell death. It is also believed that bismuth salts interfere with attachment of bacteria to specific host gut receptors (39).

Steffen et al. (48,49) reported that BSS in liquid form reduced the incidence of diarrhea in U.S. students living in Mexico and, when used in tablet form, it reduced the incidence of diarrhea in volunteers challenged by enterotoxigenic *E. coli.* In tourists visiting various developing countries, a randomized double-blind study was conducted in which 390 persons received either a total of 2.1 or 1.05 g of BSS daily or a placebo in tablet form in two doses. BSS reduced the incidence of diarrhea by 41% in the high dose group and by 35% in the low dose group, without causing adverse reactions.

Thus, some members of the medical profession recommend that BSS be used to prevent traveler's diarrhea. Sometimes BSS is also used to treat symptoms of traveler's diarrhea, and it has been demonstrated to decrease the number of unformed stools by 50%, possibly because of the antisecretory effect of salicylate in addition to its antibacterial and anti-inflammatory properties. It has also been found to decrease symptoms associated with viral gastroenteritis (21).

In general, BSS is safe for use by healthy adults and older children. However, it should not be taken by individuals with a history of hypersensitivity to aspirin (salicylates) or by those with diabetes, gout, or renal insufficiency. It should also not be taken by individuals on oral anticoagulants or hypoglycemic agents (39). Dupont and Ericsson (21) reported that encephalopathy may develop in AIDS patients who take excessive amounts of BSS. This medication should not be taken by anyone for more than 3 weeks.

B. Lactobacillus Preparations

Lactobacilli are bacteria that metabolize carbohydrates to lactic acid and other organic acids, thus reducing the pH in the gut and inhibiting the growth of some enteropathogens. The efficacy of lactobacilli is dependent on both the organisms in the cultured product (yogurt, acidophilus milk) and the pathogen responsible for the illness. Steffen et al. (49) reported that ingestion of *Lactobacillus acidophilus* preparations did not reduce the incidence of traveler's diarrhea due to virulent enterotoxigenic *E. coli* (ETEC).

C. Vaccinations

The risk of infection varies with the region of the world that is visited. Areas of high risk include the developing countries of Africa, the Middle East, and Latin

America. Individuals living in these areas often develop a *naturally acquired immunity* to certain diseases after exposure to bacterial pathogens and viruses that may or may not result in disease or illness. By being vaccinated for pathogenic infections before traveling to other regions of the world, travelers can gain *artificially acquired immunity.* Immunity that lasts a lifetime is called *active* immunity and can be acquired either naturally or artificially. Immunity that lasts only a few months or years is called *passive* immunity. Passive immunity to some diseases can also be acquired either naturally or artificially (43).

An effective vaccine developed for hepatitis A requires a booster shot every 6–12 months (this is an example of artificially acquired, passive immunity). Vaccines have also been developed for cholera that provides protection for a 6-month period and for typhoid fever that provides protection for 3–4 years (43). There have also been reports of the development of vaccines for other enteric pathogens, including *Giardia lamblia, Shigella flexneri,* enteropathogenic *Campylobacter* spp., and ETEC (32, 39, 46). However, at present, there is no single vaccine or medication capable of providing general protection against diarrhea or other foodborne disease. The best strategy for travelers is to become informed of the illness risks in the regions visited, to seek medical advice, and to make safe choices of food and beverage based on this information.

D. Antimicrobial Treatment

Examples of antibiotics used to prevent and treat travelers' diarrhea include trimethoprim, trimethoprim–sulfamethoxazole [TMP-SMZ], doxycycline, ampicillin, and tetracycline (46). However, resistance of enteric pathogens (e.g., some types of *Shigella, Salmonella,* and *Campylobacter*) to these agents may prevent the success of antibiotic prophylaxis and therapy (21, 24, 38, 39).

These drugs may also produce adverse reactions (24). Dupont and Ericsson (21) reported that the following side effects may occur with the use of these antibiotics: skin rashes, vaginal candidiasis, photosensitivity reactions, bone marrow hypoplasia or aplasia, dental staining, antibiotic colitis, and possibly anaphylaxis in a very small number of people. Thus, travelers choosing to take antibiotics to prevent traveler's diarrhea run the risk of a potentially fatal reaction to prevent an illness that is usually self-limiting.

VII. TREATMENT IN THE EVENT OF ILLNESS

Hoge et al. (26) state that "Travelers should continue to be instructed in common sense hygiene rules by avoiding theoretically contaminated foods such as leafy vegetables, untreated water and previously cooked foods that have not been thoroughly reheated." However, it seems to be just as important to inform travelers that disease may occur despite these efforts. Therefore, travelers should provide themselves with knowledge to treat themselves if they do indeed succumb to traveler's diarrhea or illness.

Usually, individuals want relief from two major complaints: abdominal cramps and diarrhea. The following preparations have been used in an attempt to alleviate these conditions.

Activated charcoal and *kaolin* and *pectin* are often taken for diarrhea. Activated charcoal is ineffective in the treatment of diarrhea. Kaolin and pectin appear to give stools more consistency but have not been shown to decrease cramps and frequency of stools or to shorten the course of infectious diarrhea (24).

Antimotility agents such as diphenoxylate and loperamide provide prompt symptomatic but temporary relief. However, these agents should not be used by patients who have high fever or blood in the stools. The use of antimotility agents should be discontinued if symptoms persist beyond 48 hours. Diphenoxylate and loperamide should not be given to children under 2 years of age (24).

Traveler's diarrhea is associated with a mild or moderately severe loss of fluid and electrolytes. Severe fluid deficits are encountered only rarely (4, 5, 21, 24). Significant morbidity occurs only in older adults, or in individuals with chronic intestinal diseases or other chronic diseases (cardiac, pulmonary, or renal diseases). Treatment of fluid and electrolyte deficits may be effectively achieved by rehydration with *oral rehydration solution* (41) or with a commercial solution of similar composition (e.g., Pedialyte™ (4) (Table 6)

It is usually recommended that adults and children be discouraged from fasting and that they resume their usual diet with small snacks. However, there are also reports that attempts to start feeding may prolong diarrhea when diarrhea is associated with a rotavirus infection. Drinks or foods containing caffeine or lactose should be avoided, since they prolong diarrhea. Fluid and electrolyte balance can be maintained by drinking potable fruit juices and caffeine-free hot or cold beverages and by eating salted crackers (4, 24). Consumption of alcoholic beverages should also be avoided.

Antibiotics may be required for individuals who develop diarrhea with three

Table 6 Components of Oral Rehydration Solution for Acute Diarrhea

Component	Weight (g)	Volume measurement
Sodium bicarbonate (baking soda)	2.4	1/2 teaspoon
NaCl (table salt)	3.5	1/2 teaspoon
KCl (potassium chloride)	1.5	1 1/4 teaspoons
Sucrose (table sugar)	40.0	3 tablespoons plus 1 teaspoon
Water	1000.0	1 liter

Source: Adapted from Ref. 41.

or more loose stools in an 8-hour period, especially if diarrhea is associated with nausea, vomiting, abdominal cramps, fever, or blood in the stools. Medical advice and treatment must be sought for these cases.

VIII. CONCLUSION

A review of travelers' illnesses leads to the following conclusions.

1. From 20% to as many as 50% of all travelers experience the occurrence of diarrheal illness when traveling.
2. Enterotoxigenic *E. coli* spp. are the most commonly identified enteropathogens in all regions of the world and are thought to be the cause of most cases of traveler's diarrhea.
3. Other microbial causes of traveler's illness include a long list of other bacterial, parasitic, and viral pathogens. Sometimes mixed infections can occur. Analysis of stool specimens can be used to determine the exact cause of the illness.
4. Travelers to high risk areas of the world may consider vaccination for known causes of endemic illness and the use of bismuth subsalicylate or other antimicrobial therapy to prevent and treat traveler's diarrhea. Before traveling to one of these areas of the world, individuals should consult a physician concerning the prevention and treatment of traveler's diarrhea and other possible illnesses.
5. Medical advice should be sought for severe cases of traveler's illness. It is important to prevent dehydration in individuals affected by this illness. Travelers should consume bottled water from a safe source of supply (e.g., boiled water or chemically -treated water). To maintain the body's electrolyte balance, an oral rehydration fluid can be prepared according to WHO instructions, or a commercial preparation of a fluid such as Pedialyte™ should be consumed.
6. As described by Todd (51), foodborne disease surveillance programs must be developed globally. At present, most countries have some type of reporting for notifiable diseases, but few have foodborne disease surveillance programs. Hence, little is known of foodborne disease in general on a worldwide basis. Improvement of surveillance is critically important with the development of increased travel and trade between various areas of the world. Surveillance programs will require cooperation between governments and industry. Currently, the World Health Organization provides information regarding international foodborne disease incidents and outbreaks. However, this information is generally limited to countries that have adequate surveillance programs and supply this information to the WHO Surveillance Program for Control of Foodborne Infections and Intoxications.
7. Travelers may be able to avoid some of these pathogens by paying careful attention to the foods they eat and the beverages they consume. However, a

major decrease in travelers' illness will occur only when potable sources of water are available every where and food handlers in foodservice establishments (restaurants, hotels, street vendors, catering operations, cruise ships, etc.) throughout the world are trained to use good personal hygiene practices and safe methods of preparing, storing, and serving food.

REFERENCES

1. Abdussalam, M., and D. Grossklaus. 1991. Foodborne illness: A growing problem. World Health, July–August.
2. Abdussalam, M., and F. K. Käferstein. 1993. Safety of street foods. World Health Forum 14:191–194.
3. Allen, R. J. L., and F. K. Käferstein. 1983. Foodborne disease, food hygiene and consumer education. Arch. Lebensmittelhy. 34:81–108.
4. Banwell, J. G. 1986. Treatment of travelers' diarrhea: Fluid and dietary management. Rev. Infect. Dis. 8 (suppl. 2):S182–S187.
5. Banwell, J. G. 1990. Pathophysiology of diarrheal disorders. Rev. Infect. Dis. 12 (suppl. 1):S30–S35.
6. Bierer, D. W. 1990. Bismuth subsalicylate: History, chemistry, and safety. Rev. Infect. Dis. 12 (suppl. 1):S3–S8.
7. Black, R. E. 1990. Epidemiology of travelers' diarrhea and relative importance of various pathogens. Rev. Infect. Dis. 12 (Suppl. 1):S73–S79.
8. Bryan, F. L. 1995. Hazard analyses of street foods and considerations for food safety. Dairy Food Environ. Sanit. 15 (2):64–69.
9. Bryan, F. L., I. Fukunaga, S. Tsutsumi, L. Miyshiro, D. Kagawa, B. Sakai, H. Matsuura, and M. Okamura. 1991. Hazard analysis of Japanese boxed lunches (bento). J. Environ. Health 54:29–32.
10. Bryan, F. L., M. Jermini, R. Schmitt, E. N. Chilufya, M. Mwanza, A. Matoba, E. Mfume, and H. Chibiya. 1997. Hazards associated with holding and reheating foods at vending sites in a small town in Zambia. J. Food Prot. 60:391–398.
11. Bryan, F. L., P. Teufel, S. Riaz, S. Roohi, F. Qadar, and Z. Malik. 1992. Hazards and critical control points of street-vended chat, a regionally popular food in Pakistan. J. Food Prot. 55:708–713.
12. Bryan, F. L., P. Teufel, S. Riaz, S. Roohi, F. Qadar, and Z. Malik. 1992. Hazards and critical control points of vending operations at a railway station and a bus station in Pakistan. J. Food Prot. 55:534–541.
13. Bryan, F. L., P. Teufel, S. Roohi, F. Qadar, S. Riaz, and Z. Malik. 1992. Hazards and critical control points of food preparation and storage in homes in a village and town in Pakistan. J. Food Prot. 55:714–721.
14. Bryan, F. L., S. C. Michanie, P. Alvarez, and A. Paniagua. 1988. Critical control points of street-vended foods in the Dominican Republic. J. Food Prot. 51:373–383.
15. Centers for Disease Control and Prevention. 1992. Cholera associated with an international airline flight, 1992. MMWR 41:134–135.
16. Centers for Disease Control and Prevention. 1993. Food and Water and Traveler's Di-

arrhea, Report 220004. U.S. Public Health Service, Department of Health and Human Services, Washington, DC.

17. Cerqueira, A. M. F., A. Tibana, and B. E. C. Guth. 1997. High occurrence of shiga-like toxin-producing strains among diarrheagenic *Escherichia coli* isolated from raw beef products in Rio de Janeiro City, Brazil. J. Food Prot. 60:177–180.
18. Cesarman, G., E. Cesarman, and G. Lagos. 1993. Prevention and treatment of traveler's diarrhea. N. Engl. J. Med. 329: 1584 (letter).
19. Communicable Disease Report (CDR) Weekly—UK. 1995. Scombrotoxic fish poisoning and imported tuna. 5 (29):135.
20. Communicable Disease Report (CDR) Weekly—UK. 1997. *E. coli* and hemolytic uremic syndrome—Canary Islands origin. 7(15):April 11.
21. DuPont, H. L., and C. D. Ericsson. 1993. Prevention and treatment of traveler's diarrhea. N. Engl. J. Med. 328:1821–1827.
22. El-Sherbeeny, M. R., M. F. Saddik, H. E. Aly, and F. L. Bryan. 1985. Microbiological profile and storage temperatures of Egyptian rice dishes. J. Food Prot. 48:39–43.
23. Foster, G. M., and F. K. Käferstein. 1985. Food safety and the behavioral sciences. Soc. Sci. Med. 21:1273–1277.
24. Gorbach, S. L., C. J. Carpenter, R. Grayson, J. D. Gryboski, R. L. Guerin, T. R. Hendrix, R. B. Hornick, J. S. Marr, J. Morris, B. F. Polk, S. R. Waters, G. W. Williams, and M. S. Wolfe. 1986. Consensus Development Conference Statement on Travelers' Diarrhea. Rev. Infect. Dis. 8 (suppl. 2):S227–S233.
25. Guerrant, R. L., J. M. Hughes, N. L. Lima, and J. Crane. 1990. Diarrhea in developed and developing countries: Magnitude, special settings, and etiologies. Rev. Infect. Dis. 12 (suppl. 1):S41–S50.
26. Hoge, C. W., D. R. Shim, P, Escheverria, R. Rajah, J. E. Herrmann, and J. H. Cross. 1996. Epidemiology of diarrhea among expatriate residents living in a highly endemic environment. JAMA 275:533–538.
27. Jermini, M., F. L., Bryan, R. Schmitt, C. Mwandwe, J. Mwenya, M. H. Zyuulu, E. N. Chilufya, A. Matoba, A. T. Hakalima, and M. Mwanza. 1997. Hazards and critical control points of food vending operations in a city in Zambia. J. Food Prot. 60: 288–299.
28. Käferstein, F. K. 1985. The global problem of foodborne infections and intoxications. Zbl. Bakt. Hyg. I. Abt. Orig. B180:335–342.
29. Käferstein, F., and Y. Motarjemi. 1991. Foodborne diseases as related to travellers. A public health challenge. Catering Health 2:41–52.
30. Kean, B. H. 1986. Travelers' diarrhea: An overview. J. Infect. Dis. 8 (suppl. 2): S111–S116.
31. Koo, D., K. Maloney, and R. Tauxe. 1996. Epidemiology of diarrheal disease outbreaks on cruise ships, 1986–1993. JAMA 275:545–547.
32. Lammerding, A. M., R. J. Irwin, C. A. Muckle, L. A. Elliot, J. E. Harris, and J. J. Kolar. 1995. Updates on recent food safety issues—Vaccines for enteric pathogens. Food Saf. Issues Update 9.
33. Lederberg, J. 1996. Infectious disease—A threat to global health and security. JAMA 276:417–419.
34. Liang, Y. M., and X. S. Yuan. 1991. Investigation of bacterial contamination of street-vended foods. Dairy Food Environ. Sanit. 11:725–727.

35. Manhart, M. D. 1990. In vitro antimicrobial activity of bismuth subsalicylate and other bismuth salt. Rev. Infect. Dis. 12 (suppl. 1):S11–S15.
36. Minnesota Department of Health. 1996. *Salmonella enteriditis* subtype 11 associated with travel to Mexico. Dis. Control New. 24(3):22.
37. Motarjemi, Y., F. Kaferstein, G. Moy, and F. Quevedo. 1991. The rationale for the education of foodhandlers. Regional Conference of Food Safety and Tourism for Africa and the Mediterranean, Tunis, Tunisia. World Health Organization, Geneva.
38. Murray, B. E. 1986. Resistance of *Shigella, Salmonella,* and other selected enteric pathogens to antimicrobial agents. Rev. Inf. Disease 2 (suppl. 2):S172–S181.
39. Okhuysen, P. C., and C. D. Ericsson. 1992. Travelers' diarrhea. Prevention and treatment. Med. Clin. North Am. 76:1357–1373.
40. Osterholm, M. T., C. W. Hedberg, and K. L. MacDonald. 1993. Prevention and treatment of traveler's diarrhea. N. Engl. J. Med. 329:1584 (letter).
41. Pierce, N. F., and N. Hirschhorn. 1977. Oral fluid—A simple weapon against dehydration in diarrhea: How it works and how to use it. WHO Chron. 31:87.
42. Poland, D. M., and O. P. Snyder. 1994. Hazards in the Food System. Hospitality Institute of Technology and Management, St. Paul, MN.
43. Prescott, L. M., J. P. Harley, and D. A. Klein. 1996. Microbiology, 3rd ed. W. C. Brown–Times Mirror Higher Education Group, Dubuque, IA, pp. 587–588.
44. Quevedo, F., F. Käferstein, G. Moy, and Y. Motarjemi. 1991. Food safety in the tourism sector. Regional Conference for Africa and the Mediterranean on Food Safety and Tourism. Tunis, Tunisia. World Health Organization. Geneva.
45. Ryder, R. W., J. G. Wells, and E. J. Gangarosa. 1977. A study of travelers' diarrhea in foreign visitors to the United States. J. Infect. Dis. 136:605–607.
46. Sack, R. B. 1990. Travelers' diarrhea: Microbiological bases for prevention and treatment. Rev. Inf. Dis. 12 (suppl. 1):S59–S63.
47. Schmitt, R., F. L. Bryan, M. Jermini, E. N. Chilufya, A. T. Hakalima, M. Zyuulu, E. Mfume, C. Mwandwe, E. Mullungushi, and D. Lubasi. 1997. Hazards and critical control points of food preparation in homes in which person had diarrhea in Zambia. J. Food Prot. 60:161–171.
48. Steffen, R. 1990. Worldwide efficacy of bismuth subsalicylate in the treatment of travelers' diarrhea. Rev. Inf. Dis. 12 (suppl. 1):S80–S86.
49. Steffen, R., R. Heusser, and H. L. DuPont. 1986. Prevention of travelers' diarrhea by nonantibiotic drugs. Rev. Infect. Dis. 8 (suppl. 2):S151–S158.
50. Todd, E. C. D. 1994. Surveillance of foodborne disease, in Y. H. Hui, J. R. Gorham, K. D. Murell, and D. O. Cliver (eds.), Foodborne Disease Handbook, Vol. 1, Diseases Caused by Bacteria. Marcel Dekker, New York, pp. 461–536.
51. Todd, E. C. D. 1996. Worldwide surveillance of foodborne disease: The need to improve. J. Food Prot. 59:82–92.
52. Torres-Vitella, M. R., A. Castilla, G. Finne, M. O. Rodriguez-Garcia, N. E. Martinez-Gonzales, and V. Navarro-Hidalgo. 1997. Incidence of *Vibria cholerae* in fresh fish and ceviche in Guadalajara, Mexico. J. Food Prot. 60:237–241.
53. WHO. 1989. Health Surveillance and Management Procedures for Food-Handling Personnel. World Health Organization, Geneva.
54. WHO. 1991. A Guide on Safe Food for Travelers. World Health Organization, Geneva.

55. WHO. 1996. *Vibrio cholerae* O139 Bengal infections among tourists to southeast Asia: An intercontinental foodborne outbreak. WHO Newsl. 48:3.
56. WHO. 1994. *Shigella* from imported iceberg lettuce. WHO Newl. 42:2.
57. Wittner, M., and H. B. Tanowitz. 1992. Intestinal parasites in returned travelers. Med. Clin. North Am. 76:1433–1448.

16

The Microbiological Safety of Bottled Waters

Donald W. Warburton
Health Canada, Ottawa, Ontario, Canada

I. INTRODUCTION

Bottled water is any potable water that is manufactured, distributed, or offered for sale that is sealed in food-grade bottles or other containers and is intended for human consumption (44,130). Thus, bottled water may be any potable water in any sealed container, ranging from 325 mL cans to 2.0 L bottles, and from 4.0 L jugs to 18.0 L carboys. Smaller and larger sizes are available in some localities. The approved supply for bottled water may be springs, municipal systems, or other sources (Table 1), and subsequently the water may be subjected to a number of treatments (distillation, carbonation, ozonation, and/or filtration, etc.) either in series or as a single treatment. The overall treatment of the source water is dependent on the quality of source water, the type of bottled water being manufactured, and where it is being manufactured (Table 1).

In some developing countries, the main source of safe drinking water may, in fact, be bottled water, since limited drinking waters are contaminated by industrial and human pollution. In Haiti, for example, the main bottled water is water demineralized via reverse osmosis (D. Warburton, Health Canada, unpublished data). In addition, in parts of China health officials have indicated that bottled water produced by distillation is the main source of drinking water (E. Todd, Health Canada, personal communication).

In Europe, bottled water labeled as "natural" may be given only a minimal treatment (Table 1), while in North America all bottled water may contain safe and suitable antimicrobial agents, such as ozone. Generally, most bottled waters do not

Table 1 Definitions of Approved Source Waters and Different Types of Bottled Water

Approved source	Approved source means the source of water whether it be from a spring, artesian well, drilled well, public or community water system, or any other source that has been inspected and the water analyzed, and found to be of a safe and sanitary quality (i.e., is potable) with or without treatment.
Spring and mineral water	
Spring water	Bottled potable water derived from an approved underground source (bore holes or springs that originate from a geological and physically protected underground water source and not from a public community water supply) that contains less than 500 mg/L total dissolved solids.[a] Spring water may be treated to remove unwanted chemical and microbiological components but may not be labelled as "natural" (see below).
Natural spring water	Same as "spring water," and in Europe must meet the collection requirements of "natural mineral water" (as below) without any treatment to remove bacteriological components.
Mineral water	Bottled potable water obtained from an approved underground source (bore holes or springs that originate from a geological and physically protected underground water source and not from a public community water supply) that contains not less than 500 mg/L of total dissolved solids.[a] In Europe, mineral water may be treated to remove unwanted chemical and microbiological components but may not be labelled as "natural" (see below).
Natural mineral water	Natural mineral water is mineral water (as defined above), but must meet the following conditions: it is collected under conditions that guarantee the original bacteriological purity; it is bottled close to the point of emergence of the source with particular hygienic precautions; it is not subjected to any treatments (other than removal of unstable constituents by decantation and/or filtration with the aid of aeration) that modify its essential mineral constituents; and cannot be shipped in bulk. A "naturally carbonated natural mineral water" is a natural mineral water which, after acceptable treatment, replacement of gas, and packaging, has the same content of gas as the source. A "noncarbonated natural mineral water" is a natural mineral water which, after acceptable treatment and packaging, does not contain free carbon dioxide in excess of the amount necessary to keep the hydrogen carbonate salts present in the water dissolved. A "decarbonated natural mineral water" is a natural mineral water which, after acceptable treatment and packaging, does not have the same carbon dioxide content as at emergence. A "carbonated natural mineral water" is a natural mineral water which, after acceptable treatment and packaging, has been made effervescent by the addition of carbon dioxide from another origin.
Other bottled water	
Artesian water	Bottled water from a well tapping a confined aquifer in which the water flows freely at the ground surface without pumping. It has been proposed that the collection of the water can be enhanced with the assistance of external pressure so long as such measures do not alter the physical properties, composition, and quality of the water.

Table 1 Continued

Other bottled water (*continued*)	
Bottled water	Water that is placed in a sealed container or package and is offered for sale for human consumption or other consumer uses.
Carbonated or sparkling water	Bottled water containing carbon dioxide. "Natural" carbonated, "naturally carbonated," or "sparkling" mineral or spring water (as above) refers to the water containing carbon dioxide and may have carbon dioxide added to it, provided that (1) the carbon dioxide added originates from the decarbonation of the water upon its emergence from the underground source; and (2) the carbon dioxide is not added to a level greater than was naturally occurring underground. "Carbonated" refers to any water where carbonation has been added at a quantity greater than originally in the source water or is from a process other than the decarbonation of the source water (L. Carpigna, Canadian Food Inspection Agency, personal communication).
Distilled water	Bottled water that has been produced by a process of distillation and has an electrical conductivity of not more than 10 μS/cm and total dissolved solids of less than 10 mg/L.
Drinking water	Bottled water obtained from an approved source that has undergone special treatment or that has undergone minimum treatment consisting of filtration [activated carbon and (or) particulate] and ozonation or equivalent disinfection process.
Deionized water	Bottled water that has been produced through a deionization process to reduce the total dissolved solids concentration to less than 10 mg/L.
Fluoridated water	Bottled water containing added fluoride in such an amount that the total concentration of added and naturally occurring fluoride does not exceed 1 mg/L.
Glacial water	Bottled water from a source that is direct from a glacier (or possibly from an iceberg). Glacial water shall meet the requirements of natural water.
Natural water	Bottled water (such as spring, mineral, artesian or well water) obtained from an approved source that is from an underground formation and not derived from a municipal or public water supply system. This water has undergone no treatment other than physical filtration, iron removal, and that has not had any significant change occur in the total concentration of the major ions in comparison with the concentrations occurring in the approved source water.
Purified water	Bottled water produced by distillation, deionization, reverse osmosis, or other suitable process that contains not more than 10 mg/L of total dissolved solids. Water that meets this definition and is vaporized, then condensed, may be labeled distilled water.
Well water	Bottled water from a hole bored, drilled, or otherwise constructed in the ground, which taps the water of an aquifer. Well water shall meet the requirements of natural water.

[a]The "cutoff" value is 250 mg/L in the United States.
Source: Ref. 130.

have added ingredients unless they are specialty products such as fluoridated water or those that have added flavors, such as fruit juices and simulated flavors.

Like other foods, bottled water must be processed, packaged, shipped, and stored in a safe and sanitary manner, and they must be truthfully and accurately labeled (79). Bottled waters are generally not sterile products and can contain naturally occurring bacteria, as well as those introduced during manufacturing or during consumer handling. While most bottled waters are safe consumable products, some products, along with some questionable manufacturing and consumer practices, may cause concern for certain segments of the population. This chapter provides an overview of the microbiological safety of bottled water.

A. History

In Europe, spring and mineral waters have been consumed for centuries by local inhabitants, immigrants, and invaders. No doubt, connoisseurs of these waters transported them in containers as they traveled away from the source waters. Inevitably, someone considered marketing the waters so that they would be available some distance from the source. For example, Evian Natural Mineral Water* (37) began its present-day history in 1789 when a French marquis began bottling the "miraculous" waters on his estate. Similarly, Perrier water (97) was first bottled as Perrier Sparkling Mineral Water in 1863. Since that time, and especially in the last decade, bottled water plants have come into existence the world over.

B. Outbreaks

Except for an isolated incident in which bottled water was reported to be the source of the causative agent in a cholera outbreak (14, 15), there have been no major waterborne outbreaks involving this product. However, bottled water has been occasionally associated with the diarrheal condition known as "traveler's disease" (46, 100).

During the cholera epidemic of 1974 in Portugal (14, 15), bottled mineral water was identified as one of the vehicles of transmission of *Vibrio cholerae*. While the source water from two springs was shown via microbiological analysis to be contaminated, epidemiological evidence implicated only the noncarbonated, variety of bottled mineral water, not the carbonated types. The source water, a limestone aquifer, was believed to be contaminated by raw river water and/or sewage from nearby villages, and may have seeped through fissures and channels in the limestone without being purified through natural filtration process through rock and soil layers. The basic pH of the spring water and noncarbonated bottled water would have allowed *V. cholerae* to survive, while in the more acidic car-

*Mention of specific brand names does not constitute endorsement by the author or Health Canada.

bonated products this bacterium would have died within 24 hours. From the information provided (14, 15), it is doubtful if the manufacturing plants tested the bottled waters for bacteria as part of their quality control/assurance program. Subsequent to this outbreak, Portuguese health officials insisted that a deep well be drilled away from the river, at a higher altitude and that ultraviolet (UV) water purification equipment be installed (15). The bottlers also had to demonstrate that the well water and finished products were free of pathogenic bacteria.

In general, other information concerning waterborne illness associated with bottled water is nonexistent. This lack of information reflects both the overall high quality of most products and the possible failure of epidemiologists and physicians to inquire about consumption of bottled water when investigating enteric illnesses. Another factor may be that many of the bacteria found in bottled water can cause mild, uncomplicated, and self-limiting gastroenteritis. Such illnesses are not likely to be treated by physicians, nor to be reported, and are even less likely to have the causative agent determined. Also in investigations of gastroenteritis linked to bottled water, the recovery of pathogens from water is usually suboptimal (127, 128) because of the viable but nonculturable nature of the stressed microorganisms. Despite its long safety record, bottled water has the potential to be the source of major water- and food-poisoning outbreaks. It is imperative that these products be manufactured under good manufacturing practice (GMP), using hazard analysis–critical control point (HACCP) systems to manage microbiological hazards.

II. SAFETY ISSUES FOR THE MANUFACTURER

A. Microbiological Contamination of Water Sources

The risk of contamination of bottled water by pathogenic bacteria and parasites is high, inasmuch as both can be found in the source water and can contaminate water during processing. The initial microbiological content of bottled water is dependent on the type of water, the initial source, its location relative to the surface and sources of contamination, the surrounding bedrock and soil constituents, its oxygen and mineral content, and the water flow.

Many aquifers, down to depths of 4 km, contain a variety of microorganisms (Table 2), at concentrations up to 10^5–10^7 cfu/g of sediment or per milliliter of water (8, 9, 56, 76, 136). In contrast, other studies report that the initial indigenous microbial level of spring or mineral water (both the source water and immediately after bottling) may be less than 10^2 cfu/mL (10, 91, 111) but can be up to 10^3 cfu/mL (57). These variations in bacterial levels are related to the type of rock stratum, the age of the water, the initial source of water, the temperature, and the oxygen and nutrient contents (62). Some types of bottled water use potable municipal water as the source, therefore the initial microbial levels are usually very low. Bacteria indigenous to municipal-treated and other source water may be innocuous or opportunistic pathogens [Table 2 (125)].

Table 2 Bacteria Found at Various Stages in the Production of Bottled Water and Their Medical Significance. Based on Refs. 126 and 130

Genus	Ground and spring water[a]	Municipal treated water[b]	ACRO filters and water softeners[c]	Bottled water[d]	Gastro-enteritis[e]
Achromobacter	+	+	+	+	?
Acidovorax	+	ND	ND	ND	N
Acinetobacter	+	+	+	+	S
Actinobacillus	+	ND	ND	ND	P
Aerobacter	–	–	+	ND	?
Aeromonas	+	+	+	+	P
Alcaligenes	+	+	+	+	S
Bacillus	+	+	+	+	P
Burkholderia	+	ND	ND	ND	?
Cellulomonas	+	ND	ND	ND	N
Chromobacterium	–	+	+	+	S
Citrobacter	+	+	+	+	S
Clostridium	+	+	+	ND	P
Comamonas	+	ND	ND	ND	N
Corynebacterium	+	+	+	+	S
Desulfobacter	+	ND	ND	ND	N
Enterobacter	+	+	+	+	S
Escherichia	+	+	+	+	P
Eucarya	+	ND	ND	ND	?
Flavimonas	+	ND	ND	ND	S
Flavobacterium	+	+	+	+	S
Hafnia	–	+	+	+	S
Hydrogenophaga	+	ND	ND	ND	N
Janthinobacterium	+	ND	ND	ND	N
Klebsiella	+	+	+	+	S
Kluyvera	–	–	+	ND	S
Legionella	–	+	+	ND	S
Methylobacterium	+	ND	ND	ND	N
Micrococcus	+	+	+	+	S
Moraxella	+	+	+	+	S
Mycobacterium	–	+	+	+	S
Nocardia	+	ND	ND	ND	S
Oerskovia	+	ND	ND	ND	?
Phyllobacterium	+	ND	ND	ND	N
Proteobacteria	+	ND	ND	ND	?
Proteus	+	+	–	ND	P
Pseudomonas	+	+	+	+	P
Salmonella	–	+	–	ND	P
Serratia	+	+	+	+	?
Shigella	–	+	–	+	P
Staphylococcus	+	+	+	+	P
Stenotrophomonas	+	ND	ND	ND	?
Streptococcus	+	+	+	+	P
Variovorax	+	ND	ND	ND	N
Weeksiella	+	ND	ND	ND	S
Xanthomonas	+	ND	ND	ND	S
Yersinia	+	+	–	+	P

[a]Groundwater including aquifers, springs, and underground sources (42, 43, 59, 68, 131).
[b]From Refs. 126 and 130.
[c]AC, activated carbon filters; RO, reverse-osmosis filters (126, 130).
[d]From Refs. 108, 126, and 130.
[e]From Refs. 67 and 83.
Key: +, isolated; –, not isolated; ND, not detected (i.e., probably not specifically tested for); P, pathogen causing gastroenteritis; S, secondary pathogen—not known if it causes gastroenteritis; ?, not known; N, not pathogenic.

In addition to indigenous bacteria from the natural water source (Table 1), bottled water may contain bacteria that enter as contaminants, including a wide range of indigenous saprophytic species (Table 2) as well as human pathogenic contaminants (10) and parasites (30, 107). Fecal contamination presents the most common, most widespread, and greatest source of risk to public health associated with all types of drinking water.

Concern has been expressed about the presence of more recently identified pathogenic agents in water, which can be generally classified within four broad groups: bacteria, viruses, protozoans, and helminths (53). These include some gram-negative bacteria, enteric viruses, and intestinal parasites, particularly *Giardia* and *Cryptosporidium* spp.

Pristine water resources, both surface waters and groundwater, are becoming more scarce because of global increases in population and the active intervention of humans in the environment (53). The microbial quality of groundwater is often superior to that of surface waters because of an effective barrier of soil on top of impervious rock strata that cap the aquifer. As a consequence, quality is uniformly excellent, with little influence from climatic changes and stormwater migrations through shallow depths of soil (53). However, groundwater is not immune to pollution either. There is growing evidence that soil barriers do not always assure groundwater quality protection (54).

Rock strata composed of limestone is very porous, often with holes and caverns through which surface water passes without the effective entrapment of microorganisms (54). This was the case in the 1974 cholera epidemic in Portugal (14, 15). In other situations, excessive surface application of minimally treated wastewaters may result in inundation of the natural soil barrier. Once the aquifer has become contaminated, restoration of water purity is very slow, even with an intervention such as pumping the water to a treatment site and then returning it to the aquifer. Much of the groundwater contamination is found in shallow wells (<33 m deep). In these wells, the source water is influenced by surface water runoff that percolates through the soil, since there is no protective bedrock perched on top to seal off the source from surface contaminants. The end result is that the quality of the water from these sources varies considerably (53).

In general, waterborne disease outbreaks are a result of either poor protection of the source water (groundwater, well, etc.) or inadequate treatment. Microbial contamination of groundwater has been responsible for large outbreaks of waterborne diseases, particularly gastroenteritis (27). Use of contaminated, untreated groundwater caused 35% of waterborne disease outbreaks during a 30-year period in the United States (86). Flooding sewage and stormwater can contaminate both surface waters and groundwater systems, and many private wells may become unsafe because the soil barrier is breached. Over the years, a variety of waterborne disease outbreaks have been attributed to contaminated aquifers or poorly protected well sites that have been contaminated by pathogens (53). Water (from springs, wells, municipal supplies, and other sources) has been associated with outbreaks or cases of foodborne illness due to enteric pathogens such as hepatitis A virus, *Aeromonas, Campylobacter, Escherichia, Pseudomonas, Shigella, Vibrio, Yersinia,* and *Giardia* species, as well as other parasites, viruses and bacteria (10, 30, 53, 65, 69, 70, 71, 78, 86, 107, 118, 125, 126, 132, 133). From 1920 to 1983, contaminated untreated groundwater in 661 public water systems accounted for 43.2% of water system deficiencies resulting in 82,528 reported cases of illness in the United States (54). Similarly, poor sanitation facilities led to the appearance of *Vibrio cholera* O1, biotype El Tor, in groundwater supplies and surface waters during the 1991 cholera outbreak in Central and South America (53).

Many intermittent contamination problems are caused by the overflow or seepage of sewage, surface water runoff, and periodic flooding over the well field. Just as bacterial populations may colonize an aquifer by filtration from above, they can do so by lateral migration (56), which can be over a considerable distance: 2–3 km in unconfined sandy aquifers (56) and over several hundred meters in granular aquifers (63). Bacteria may be retained by soil and prevented from reaching groundwater, whereas viruses can pass through the soil and reach the groundwater (13). Viruses have been isolated from groundwater flowing beneath landfill and treatment sites for domestic and agriculture sewage (75).

The appearance of coliform and pathogenic bacteria in water supply wells has confirmed that some contaminating (nonindigeneous) bacteria may survive for considerable lengths of time in groundwater (62). In general, bacteria and viruses persist longer in groundwater than in surface water (13). Recent studies of the survival of genetically engineered bacteria in aquifers have shown that these bacteria can survive from one to several months (and up to one year), depending on the bacteria, soil conditions (including chemical contaminants, pH, nutrients, and moisture), and competition with indigenous bacteria (62). *Salmonella typhimurium, E. coli, Enterococcus faecalis,* and poliovirus (type 1) survived more than 15 days in groundwater stored at 22°C (9) while total coliforms, fecal coliforms and fecal streptococci have been reported to survive more than 70 days in contaminated wells (12).

B. Production of Bottled Water

Only approved source water (Table 1) that has been inspected, analyzed, and found to be safe, sanitary, or potable (by federal, provincial, regional, or other health authorities) should be used for bottling. This will ensure that the water entering the bottling plant has low bacterial counts and is free from pathogens. Apart from the water source, the most likely sources of contamination are equipment used to pump or transport the water from its source to the bottling location, equipment used in processing or bottling the water, and the bottles and caps themselves. In an unsanitary manufacturing plant, contaminated bottles and caps can contain bacterial populations of up to 10^7 cfu/cm^2 (125). Heavy bacterial growth was obtained from some bottled water samples when the paper inside the cap was removed and analyzed (109). Bacteria such as *P. aeruginosa* can colonize equipment. Exposure of the water to air and contact with humans, insects, and other animals during bottling are potential sources of contamination (41).

During production and distribution of all types of bottled water, the cleanliness or bacterial content of the source water can change dramatically at each step of the manufacturing process, as well as during storage in bottles in the processing plant and on retailers' shelves (115). These changes can be brought about by contamination and by growth of indigenous microorganisms previously stressed, dormant, or starved (115) as a result of an altered environment. Contamination can occur at any time during processing owing to failure of hygienic or good manufacturing practices (85). Researchers (73, 112, 125) have found coliforms, *Staphylococcus epidermidis, S. humanis, S. aureus,* and *Enterococcus* spp. in bottles of water.

At the bottling plant, further microbiological contamination of the water may occur from equipment used in the processing or bottling process (such as deionizing columns and filters) and environmental sources [Table 2 (10, 41, 124, 125)]. A common water treatment system for bottled water (not including "natural" water) may include filtration, demineralization, and purification (Table 3). Microorganisms present in the source water are trapped in the ion-exchange and filtration units and may proliferate and contaminate the final product if these units are poorly maintained [Table 2 (66, 104, 114, 119, 122, 133)]. The microbial quality of water produced by these devices is extremely variable, and is a function of the contamination and nutrient content of the source water, the service life of the filter, the water temperature, the frequency of static water conditions, (51), and, maintenance of the filters. Water from reverse-osmosis filtration units can contain 10^4–10^7 cfu/mL (103), while activated carbon filtration units can release 10^2–10^5 cfu/mL (51, 122).

In 1991 Payment et al. (104) demonstrated a correlation between bacteria released from reverse-osmosis filtration units and gastrointestinal symptoms. Although this study used domestic point-of-use units, the bacteria released are common to all filtration units, including those used in the manufacture of bottled water (Table 2); therefore, the potential for disease exists.

Treatment of the source waters is dependent on the quality of the source water and the type of bottled water being manufactured. A minimal treatment for source waters must include filtration for protozoan parasite cyst removal, together with disinfection to eliminate pathogenic bacteria and viruses (51), although this may not be necessary to treat water from protected and approved sources where contamination is nonexistent. A filtration system with chemical pretreatment (including settling and precipitation of macro- or microparticulates) to entrap parasites appears to be the most logical, efficient, and cost-effective means of removal,

Table 3 A General Review of Bottled Water Treatment Systems Allowed in North America

Process and purpose	Hazards and degree of concern[a]
Source water collection	Biological, chemical, and physical
Approved sources:	CP
Artesian wells	
Glacial melt	
Municipal water	
Springs, etc.	
Aeration (removes volatile organics)	
Filtration	Biological and chemical
Activated carbon filter (removes solids, odors, organics, and bacteria)	CCP
Sand filter (removes coarse solids)	
Manganese filter (removes sulfur, iron, and solids)	
Demineralization or purification	Biological and chemical
Water softeners (remove total dissolved solids, coarse solids, and minerals)	CCP
Deionizer (removes dissolved minerals)	
Distiller (removes dissolved minerals)	
Reverse-osmosis filtration (removes dissolved minerals and coarse solids)	
Cation, anion, or mixed-bed filters (remove minerals)	
Mineral adjustment	
Mineral mix added [improves taste, mineral composition, and chemical (e.g., fluoride) composition]	
Purification and disinfection	Biological
Ozonation (0.4–0.6 ppm; kills bacteria)	CCP
UV irradiation (> 16,000 $\mu W/s/cm^2$ at 254 μm; kills bacteria)	
Filtration (< 1–5 μm; removes bacteria and parasites (for removal of the oocytes of some parasites the filter must be 1 μm)	
Carbonation (lowers pH and kills bacteria)	
Handling of final product	Biological and physical
Container filling	CCP
Capping	
Coding	
Distribution	

[a]CP, control point, CCP, critical control point.
Source: Ref. 130.

provided it is properly operated (88,105). This filtration process must be carefully managed because improper filter backwashing and instability of the filter media can provide opportunities for elution of the cysts into the final product (51). Subsequent disinfection must be sufficient to kill the cysts and, in some cases, the contact time may have to be hours rather than minutes (70).

Little is known at this time about the specific opportunistic organisms of concern, their virulence, and most effective way to control them (54). Control measures that should be introduced include careful surveillance of the safety of the source, be it a spring, well or supply of piped drinking water, and precautions to avoid contamination of the raw material by piping, containers, closures, closing machinery, packaging material, operators, production lines, and the factory environment (95, 115).

C. Disinfection/Sterilization of Water

"Natural" bottled waters (Table 1) come from an "approved" protected and sanitary underground source, where contamination from bacteria, parasites, and viruses has been shown to be negligible or nonexistent owing to the nature of the surrounding rock formations. By standards and definitions [Table 1 (22, 23)], these bottled waters must contain the indigenous innocuous microflora and therefore cannot be subjected to any process to remove them (4, 22, 23, 74).

To control the bacterial content, i.e., aerobic colony count (ACC), and the normal proliferation of indigenous and contaminating bacteria in bottled water (not those sold as "natural" products), manufacturers of bottled water should add a final disinfection step to their manufacturing process. Disinfection can be the most cost-effective public health insurance available. This treatment must reduce all bacteria, including pathogens, by at least 4 logs (50), assuming that the initial potable water was of good microbial quality.

Systems used within the bottled water plant to reduce microbial contamination can include carbonation, ozonation, and ultrafiltration (using a 1 μm "Absolute" filter), as well as other treatments listed in Table 3. In a bottled water manufacturing plant, the need for water treatment, including a disinfection system, is related to the type of bottled water and the target population.

Bacteriological quality is determined by the presence or absence of carbon dioxide ("sparkling" vs "still"), since carbonated products are less likely to be microbiological hazards because of the lower pH (95). In sparkling or carbonated water the carbon dioxide exerts a pronounced rapid bactericidal effect, while in still or flat water the bacteria can proliferate unless disinfection has been carried out during bottling. Flat water consistently has higher bacterial counts than carbonated water (108). The pH may be significantly lowered by carbonation of the bottled water, which can affect the bacterial growth or survival time in these products. This is not unexpected, owing to the documented antimicrobial action of carbon dioxide (72). In one study, noncarbonated mineral water had a pH of 6.2–6.5 (22), while the pH of carbonated mineral water was 4.9–5.1 (17). Warburton and Harrison (1997, unpublished data) have found the following pH ranges: noncarbonated mineral water, 7.8–8.23; carbonated mineral water, 3.34–7.0; spring water, 7.95–8.19; and other bottled waters 5.7–7.1. Enteropathogenic bacteria declined more rapidly in carbonated water because of the lower pH and the bactericidal effect of the CO_2 (17).

Carbonated products sampled in Canadian studies had rejection rates due to ACC of less than 4%, compared with 61% for other types of nontreated bottled water (125). Similarly, Ruskin et al. (110) found the lowest ACC to be in sparkling or carbonated products sampled in the U.S. Virgin Islands. Hunter (72) reports that in several studies only 3–8% of carbonated waters studied had ACCs exceeding 100 cfu/mL. In contrast, 52–72% of the still waters tested had ACCs exceeding 1000 cfu/mL (72).

There is no published literature on flavored bottled waters. However, most sold locally have been found to be high in carbonation and would have the bactericidal effects of low pH and high CO_2. In-house analysis of these products has shown ACC levels below 1 cfu/100 mL water (Warburton and Harrison, unpublished data).

Ozonation has also been shown to reduce microbial counts substantially, to retard bacterial growth of survivors, and to disinfect and even sterilize the product (125). Ozonation of some finished products is allowed by the bottled water associations, and the usage is dependent on the quality of the source water and the type of bottled water it will be marketed as; that is, "natural" bottled water cannot be ozonated, while other bottled waters can [Table 1 (130)].

Refrigerated storage will minimize bacterial multiplication in bottled water, and the consumer should be advised to keep bottled water refrigerated after opening (2, 55, 110, 125), especially products that may be consumed directly from the bottle and stored between uses. This will help deter growth of contaminants that are introduced from air entering the bottle upon opening and from oral contact. Water dispensers (coolers) that utilize the larger bottles should be equipped with refrigeration units.

Refrigerated storage of still waters can substantially delay their colonization by psychrotrophs. Secondary growth of most enteric pathogens is impossible at this temperature, while growth of psychrotrophic bacteria such as *Yersinia enterocolitica* and *A. hydrophila,* though possible, is markedly delayed (95). Studies by Rosenberg and coworkers indicated that bottled waters stored at room temperature showed a consistent and considerable increase in the total numbers of bacteria, while those kept at refrigeration temperatures were consistently lower than those products stored at ambient temperatures (61, 108, 109). However, with some bacteria the opposite may be true. For example, bottled water contaminated with

Listeria spp. will have higher levels in samples stored at 4°C than those stored at 22°C (J. Farber, Health Canada, personal communication).

D. The Effect of Storage on the Microbiology of Bottled Water

1. Indigenous Bacteria

Concern has arisen about the rapid colonization of bottled still waters by psychrotrophic, partly oligotrophic gram-negative rod-shaped bacteria during storage at ambient temperatures. Apparently, the original bacterial population, occurring in freshly drawn mineral waters, bears a rather specific character. The psychrotrophic oligotrophic association that, in course, will colonize still bottled waters is considered to be harmless.

Indigenous bacteria remain at low numbers while the water is in its natural environment but commence rapid growth after the water has been bottled (91, 125). Some bottled mineral waters do not support bacterial growth (130), which may reflect their high mineral content and the resulting high osmotic pressure. However, prolonged storage of most waters at room and refrigeration temperatures allows multiplication of both indigenous and contaminating bacteria to over 10^4 or 10^5 cfu/mL (10, 32, 38, 55, 91, 111, 125). During storage at ambient temperatures in plastic and glass containers, growth commences 1–2 days after bottling and may result in populations of 10^5–10^6 cfu/mL (99, 125), with counts occasionally exceeding 10^7 cfu/mL. These results are consistent with other studies, where bacterial populations in commercial samples (10^3–10^6 cfu/mL) were higher than those found in water collected from the spring (10^2–10^3 cfu/mL) (57). The reasons for this altered growth include increased oxygenation of the water during the bottling process, the increased surface area provided by the bottle, trace amounts of nutrients arising from the bottle (111), and the increased temperature of the bottled water compared with the source water.

In a 1987–1988 Canadian survey of bottled water (125), 82 lots of bottled water were obtained from 40 manufacturers across Canada. Approximately half of the bottles (410 sample units) were analyzed within 24 hours of bottling, while the other half (395 sample units) were stored at room temperature and analyzed after 30 days. In the first analyses, 91% of the bottled water had ACCs below 10^4 cfu/mL, while only sample units of distilled and glacial waters had ACCs over 10^5 cfu/mL. Every sample unit of drinking, deionized, unfiltered, mineral and spring waters had an ACC below 100 cfu/mL. After the 30 days of storage, the number of sample units of distilled, demineralized, treated, and bottled waters with ACCs between 10^2 and 10^6 cfu/mL had increased by 30%. Only a few sample units of distilled water had very high ACCs (> 10^6 cfu/mL). No other indicator bacteria (e.g., coliforms, *Aeromonas* spp., *Pseudomonas* spp.) were detected throughout this study.

The potential negative effects of the normal microbial proliferation must be taken into account when health agencies are monitoring products for ACC or setting standards. The nutrient level of some types of bottled still water allows the development of oligotrophic gram-negative rods, including pseudomonads, flavobacters, aeromonads, *Klebsiella,* and *Citrobacter* spp. (95). Secondary growth may occur as some species die, autolyse, and provide nutrients for the growth of other species, such as heterotrophic enteric pathogens (93, 111). When bacteria lyse, nutrients are provided for secondary (cryptic, cannibalistic, or cadaveric) growth of nutritionally more fastidious taxa. Hence, the potential resurgence of significant levels of thermotrophic Enterobacteriaceae, *Pseudomonas aeruginosa,* and a few other organisms of health significance present initially at very low numbers (< 1 cfu/L), must be monitored (95). Even small numbers of potentially harmful bacteria can survive for long periods in noncarbonated water, and than multiply, feeding on the products of lysis or on the metabolites of autotrophs.

Survival curves with a saw-toothed appearance are often typical of bacterial growth in water (58), and this appearance is probably indicative of a repeating cycle of bacterial multiplication, followed by die-off due to a shortage of nutrients, then cell lysis, subsequent release of nutrients, and another round of multiplication [cryptic or cadaveric growth (93)], although in some cases the overall bacterial population is slowly decreasing over time (130). Long-term changes in the major heterotrophic bacterial populations of bottled water occur which may reflect changes in nutrient availability, a decrease in viability of some strains, competition, antagonism, synergism, or other factors not yet understood (91, 127, 128).

Bacterial multiplication was more rapid, and bacterial numbers consistently higher, in bottled water stored in plastic containers than in those stored in glass (32, 108). A British study also found significantly lower colony counts from mineral and other bottled waters stored in glass than in plastic bottles (40). These results probably can be explained by noting that chemical components or substances can migrate or leach from the container walls into the water. These may include oxygen and other gases diffusing through the plastic containers into the water. Fewtrell et al. (40) found that the color of the container appeared to affect the ACC levels, with generally higher counts (37°C) obtained from both colored glass and colored plastic, compared to their clear counterparts. At 22°C, there was a significant difference only between colored and clear plastic bottles. However, the reason for this effect was not determined.

A variety of other Canadian surveys and routine monitoring of spring, mineral, and bottled waters (125) showed a wide range of ACC in these products. Generally, spring and mineral waters had lower ACC levels than did other bottled water. Of the varieties of bottled water tested, distilled water tended to have higher ACCs and also contained coliforms and *P. aeruginosa* (125). In contrast, glacial and deionized waters had lower ACCs than did some water classified as "bottled or purified" waters.

Similarly, HPB Regional Labs surveyed 3460 samples of bottled water during 1992–1997. Of these, 2466 samples (71.3%) contained ACC levels below 10^2 cfu/mL, 805 samples (23.3%) had 10^2–10^4 cfu/mL, while 189 samples (5.5%) exceeded 10^4 cfu/mL. In total, 994 samples (28.7%) had ACC levels greater than 10^2 cfu/mL. Coliforms, fecal coliforms, *A. hydrophila,* and *P. aeruginosa* were recovered from 3.7, 2.1, 0.63, and 1.2% of the samples, respectively (131).

2. Inoculation Studies

Inoculation studies have been conducted to determine the effects of postprocessing contamination due to poor GMP/HACCP, and storage on the survival of contaminating bacteria and subsequently the quality of bottled water (Table 4). When low levels of bacteria (≤ 100 cfu/mL) were inoculated into mineral and spring water (simulating low contamination), their numbers increased to 10^3–10^6/mL within one week (130,131). For instance, *A. hydrophila, C. freundii, E. coli O157,* and *P. aeruginosa* reached levels of exceeding 10^3 cfu/mL within 2 days, while *Salmonella typhimurium, S. senftenberg,* and *Serratia marcescens* multiplied more slowly, taking 5–6 days to reach these levels. It was determined that pathogens can proliferate to potentially infectious levels in bottled water even when the initial contamination levels are low.

Other inoculation studies were done using high levels of bacteria to simulate gross contamination of the source waters and final products by sewage or surface waters, as well as extremely poor GMPs in the manufacturing plant (127–130). When initially present at high levels, *A. hydrophila* could survive 60 days or longer in water stored at 22°C. The presence of other bacteria, such as *E. coli, Lactobacillus,* and *Flavobacterium,* and especially *P. aeruginosa,* enhanced the survival of *A. hydrophila* (127).

Warburton et al. (128) studied the survival of *P. aeruginosa* and *Salmonella* spp. after coinoculation into a variety of bottled waters. After inoculation into the water, the nutrient levels were sufficient to allow some growth (approximately 1 log) and subsequent survival of both *P. aeruginosa* and *Salmonella* for up to 60 days. After this time, the bacteria quickly decreased in numbers, probably because of nutrient depletion. In other trials, *P. aeruginosa* survived as the sole contaminant in distilled water for approximately 128 days (128). Alone, *Salmonella* survived in distilled water for 100 to over 158 days, while together the bacteria survived longer (> 145 days). As with *A. hydrophila* (127, 128), there was also a synergistic effect between *P. aeruginosa* and *Salmonella* spp.

The type of water (and thus the nutrient content) can influence the survival of bacteria [Table 4 (128, 130)]. It is well known that enteric bacteria become stressed by nutrient limitations in aquatic systems, especially where organic ma-

Table 4 The Survival of Bacteria in Bottled Water

Bacteria	Water type	Bottle type[a]	Inoculation level ($\log_{10}$ cfu/mL)	Survival (days)
A. hydrophila	Distilled	P	4.85–5.70	37–63
	Mineral		4.00	70
	Mineral	P	Natural contamination	69
Clostridium perfringens (vegetative cells)	Spring	P	3.78–5.70	7–14
	Mineral	P	5.60–5.65	7–35
E. cloacae	Mineral	P	4.75–5.43	> 20
		G	4.75	> 20
	Tap	G	4.75	> 20
E. coli	Mineral		4.00	42
	Mineral	P	4.35	≥ 20
		G	4.50	> 20
	Tap	G	4.35	> 20
E. coli O157	Distilled	G	5.42	> 309
	Mineral	P	4.16–6.54	> 44–> 309
		P	1.0	22–36
	River	P	4.45–5.75	24 to 44
	Spring	P	2.04–4.20	> 20–94
		G	3.24–4.20	> 48–> 102
	Tap	G	4.72	> 303
	Well	P	3.48	58–65
K. pneumoniae	Mineral	P	4.80–5.20	> 20
		G	5.35	> 20
	Tap	G	5.20	> 20
Pseudomonas spp.	Mineral			> 140
P. aeruginosa	Mineral		4.00	70
	Distilled		5.45–6.15	128–145
	Spring		6.08–6.43	> 137
	Mineral		1.99–6.45	> 135–> 365
	Mineral	G	5.40	> 20
		P	5.38–5.80	> 20
	Tap	G	5.50	> 20
	Tap		5.30	> 136
	Well		5.85–6.48	> 96–144
Salmonella senftenberg	Mineral		6.15	> 162
	Distilled		5.95–6.15	> 158
	Tap		5.00	> 136
	Well		4.60	8–12
S. typhimurium	Mineral		4.00	70
Salmonella spp.	Distilled		5.38–6.38	67–> 169
	Spring		6.15–6.40	> 137–> 162
	Mineral		6.30	> 135
	Tap		5.30	> 138
	Well		6.00	63–> 138

[a]G, glass bottle; P, plastic bottle.
Source: Ref. 130.

terial is persistently low (90). Reduction of bacterial size, use of stored nutrient reserves, and decreased physiological activity are known responses to nutrient limitations that appear to extend the viability of bacteria with time and to perpetuate a maintenance level of metabolism (90). Generally, those waters have a high content of minerals (including trace nutrients) and support bacterial growth for long periods of time, as opposed to distilled water (130).

Burge and Hunter (17) found that *S. typhimurium* and *P. aeruginosa* persisted in bottled mineral water for at least 70 days at 23–25°C. In other studies, *Pseudomonas* spp. was able to survive more than 140 days in mineral water bottled in both plastic and glass bottles (137) stored at 22°C. In contrast, Gonzalez et al. (58) found that *P. aeruginosa* not only survived but multiplied, reaching values greater than 10^4 bacteria/mL and were detectable for longer than one year in mineral water stored at 20, 30, and 37°C. At the conclusion of their studies, these authors (58) called *P. aeruginosa* a "permanent contaminant" because of its ability to survive "ad infinitum."

3. Biofilm Formation

In bottled water manufacturing plants, biofilms can play an important role in the survival of contaminating bacteria in pipes, filters, water reservoirs, and in the finished products. In a bottle containing liquid, the proportion of bacteria adhering to surfaces is usually much greater than in the liquid itself (25,87). Indeed, the majority of bacteria in aquatic systems grow within an exopolysaccharide-enclosed matrix (25,87). Contamination of water from biofilms adherent to container walls may occur as cells are gradually released, or as large fragments of biofilm become detached from the walls.

Hamilton and Rosenberg (61), using both electron microscopy and staining procedures, have demonstrated that biofilm proliferation on the walls of PVC water containers occurs as a function of storage time. Long-term survival of bacteria in bottled water may be due in part to biofilm formation on the walls of the container (121, 129). Heterotrophs, *A. hydrophila,* and *E. coli* were able to survive in biofilms at levels at or exceeding 3.1 $\log_{10}$ cfu/cm^2 for over 21 days at room temperature (121). There was a fluctuation in the biofilm and "planktonic" (free-floating) counts (121), which is a reflection of the continual incorporation of cells into the biofilm and sloughing off of cells into the liquid, as well as multiplication/die-off of the bacteria. *Aeromonas, E. coli* and other pathogens can become incorporated into biofilm communities, thus enhancing their viability (121).

Several authors have found higher counts and faster growth in bottled water stored in plastic or PVC bottles than in glass bottles (10,32,99). However, this may be dependent on the storage temperature (57). At 22°C, bacterial growth was more rapid and reached higher numbers in mineral water stored in plastic bottles, while at 37°C growth was higher in glass bottles (57). In contrast, De Felip *et al.* (31) found higher bacterial populations in mineral water stored at 20 and 37°C in PVC, followed by Tetrabrik-containers and then glass bottles (in that order). The inner surfaces of plastic bottles promote adhesion and colonization because they are rougher than glass inner surfaces (13). Also, it is thought that residual cleaning agents in glass bottles (more so in recycled bottles) may inhibit bacterial growth (10, 13). The ability to adhere to plastic walls allows bacteria to utilize the few chemical components and organic sources that migrate to solid surfaces, as well as the fatty acids that may be liberated by lipase activity from the plasticizers (32, 60, 111). Diffusion of gases through the walls of the plastic containers also aids bacterial growth in these bottles.

The colonization of surfaces enhances the bacteria's capacity to resist disinfection and starvation (92). For example, *P. aeruginosa* survived better in water bottled in PVC than glass, which may reflect the capacity of this bacterium to colonize this type of material (92). In contrast, the type of container did not influence the survival of Enterobacteriaceae (92).

E. Reusable Bottles and Safety

Reusable or refillable bottles may be manufactured out of glass, polyethylene terephthalate (PET), polycarbonate (PC), polypropylene (PP), polyvinyl chloride (PVC), or high density polyethylene (HDPE) (33). There is a trend for the increased use of refillable plastic bottles in the food industry. PET refillables, widespread over the world, are mainly used for carbonated soft drinks and mineral water. Since 1990, other refillable plastic bottles, such as polycarbonate, have been introduced for fresh milk, juices, and other beverages (33). The invention of new types of refillable plastic bottle served to combine all the advantages of the glass bottle and the one-way plastic bottle: their infragility, light weight, recyclability, and processibility.

However, studies have shown that certain chemical substances, such as pesticides, household chemicals, solvents, and automotive products, are absorbed by plastic when the bottles are misused by the consumer (39, 47–49, 80). Part of the chemical is retained by the plastic even after normal caustic washing and can migrate into the contents of a refilled bottle (33). The effects of exposure, storage time, pH, carbonation, temperature, repeated rewashing, and dilution strength on the absorption/desorption behavior depend on the individual chemical and for the most part are not fully predictable (39). In a study involving 62 chemical substances, including automotive, garden, and household products; all commercially available and candidates for inappropriate storage in PET refillable bottles, most were shown to migrate into the refilled beverage at a level that would pose a public health concern. However, only the pesticide parathion migrated at a level requiring a more detailed hazard assessment, which indicated no health hazard to adults or children from reuse of most plastic bottles (39).

Good manufacturing practices, including appropriate electronic and visual in-line detection systems, are required to eliminate stained bottles, bottles with solid residues, and bottles containing volatile contaminants, to minimize to an acceptable level the possibility of reusing misused bottles. The likelihood that a misused bottle will be returned to the marketplace is greatly minimized by the quality assurance practices in place at the beverage bottling plants. Pesticides and other chemicals that might not be normally detected by in-line detection systems are frequently dissolved in an organic solvent that can be detected by photoionization and other in-line detection devices (39). To decrease the risk of contamination, filling lines for returnable plastic bottles are equipped with a "sniffer" devices, checking each container before it is filled (33). Also, water-soluble polar chemicals (e.g., paraquat) are less likely to be absorbed into PET (39). From a chemical point of view, PET refillable bottles have a history of safe use in many countries, as millions of bottles have been used without any public health issues: this supports the continued use of these products (39).

However, from a bacteriological point of view the situation is different. Only a limited number of studies have been performed on the contamination risk by microorganisms when classical caustic cleaning is applied to rinse returnable containers. For example, Devlieghere and Huyghebaert (33) found that *B. cereus* showed less resistance against caustic cleaning than did *S. aureus*. It was concluded that *S. aureus* was able to survive the cleaning because of its secretion of exopolysaccharides. These authors found that under optimal conditions, the following classification could be made, in decreasing order of microbial rinsability of the examined materials: glass > polyethylene terephthalate (PET) > polycarbonate (PC) > polypropylene (PP) = polyvinyl chloride (PVC) > high density polyethylene (HDPE). Even at optimal rinsing conditions, it was not possible to totally remove all bacteria from the sides of the containers (33). It was possible to reduce the numbers of microorganisms by approximately $\log_{10}$ 1.25–3.0 (33). These findings support the recommendation that bottled water be disinfected by ozonation. Further investigation of the influence of material-related parameters (surface structure and tensions, etc.) and age (i.e., number of recycling) might be of interest to further understand the mechanism of removing attached microorganisms from surfaces (33).

III. MICROBIOLOGICAL TESTING AND CRITERIA FOR BOTTLED WATER

The examination of bottled water for indicators and some pathogens is essential after bottling (82). Rather than examining water for the presence of all possible pathogens using a variety of technically difficult, insensitive, time-consuming, and expensive tests, it is better to use relatively simple and more rapid procedures such as the detection and enumeration of indicator organisms (115, 118). Various indicators of fecal contamination have been proposed, including, ACC, total coliform, fecal coliform, *E. coli,* fecal streptococci (enterococci), acid-fast bacteria, and sulfite-reducing clostridia (53). Table 7 lists a variety of these indicator organisms, as suggested by the FAO/WHO Codex Alimentarius Commission (21, 24), the European Economic Community (57, 115), Health Canada (126), and the U.S. Food and Drug Administration (FDA) (26). The presence of fecal indicators (some of which may also be pathogenic), such as *E. coli* and other coliform, streptococci, and clostridia, suggests that the water is potentially dangerous and was manufactured under poor GMP. One study found that the presence of indicator bacteria, even at very low levels (< 10 cfu/mL), was associated with acute gastrointestinal disease (138). In contrast, the absence of fecal indicators is taken to denote the absence of bacterial pathogens, and to constitute an indication that the product is safe. The rationale for this is that pathogens are greatly outnumbered in feces by normal intestinal microorganisms (115).

In many instances, the presence of bacterial indicators suggests the occurrence of pathogenic organisms; however, a lack of indicators may not ensure the absence of certain pathogens (35). Of the indicator systems currently in use, the fecal coliform group satisfies requirements such as positive correlation with fecal contamination, simplicity of test performance, and minimal cost to a monitoring program. The chances of detecting *Salmonella* spp, increase in direct relation to the amount of fecal contamination present and the fecal coliform density (53). However, the dependence on traditional indicators such as coliform can be misleading, particularly when viral and/or protozoal contamination is suspected (45).

While bacteria that are unable to grow at 37°C do not constitute a health risk, any that can multiply at this temperature and possess virulence factors can cause disease in healthy as well as immunocompromised individuals. Bacterial regrowth is thus a matter of concern for public health officials, who must assess the quality of these waters. Regardless of their pathogenic potential, these heterotrophic bacteria are a nuisance because they interfere with the detection of indicator and pathogenic bacteria. There are two opposing views on the potential health effects of heterotrophic bacteria in drinking water: one advocates that they are without significance; the other suggests that some of them can cause disease if allowed to grow to significant numbers (101).

IV. SAFETY ISSUES FOR THE CONSUMER

A. Perceptions

Worldwide, the bottled water industry is experiencing an annual growth rate of 25% or more (18, 130, 135). The increase in the popularity of bottled water can be attributed to a number of factors (11, 16, 18, 77, 79, 108, 113, 126):

1. The public's unquestioning impression that bottled water is of good

quality because it is usually derived from a spring or underground source (1,55,118) that is pure, impeccably clean, scrupulously "protected," and "unharmed by man." Some consumers turn to bottled water because they wish to control their intake of additives (chlorine, fluorine, etc.) that are present in tap water. Some regular users consider bottled water purer than piped water from a bacteriological perspective (95).

2. Natural source waters are believed to have beneficial medicinal and therapeutic properties (41, 57, 89). While this has never been clinically demonstrated, it is nonetheless implied by the impressive labels used on certain brands (95). For example, the Italian brand Lynx states that "San Fermo Springs mineral water can aid digestion, is diuretic and can aid in the elimination of uric acid." Some health food stores also carry carbonated spring water to which aloe juice, fruit juices, and other herbal infusions (or a combination thereof) are added, and labels also state that these beverages can aid digestion.

3. Aesthetic considerations due to taste, color, and odor problems. Some people turn to bottled water because their tap water has an objectional taste or odor due, for example, to sulfur or chlorine content. A slight reddish or brown color in the water may indicate the presence of iron, manganese, or humic materials (residue from leaves, sediment, and other organic debris). Many consumers prefer bottled water because it has not been treated or polluted by chemicals, as well as for its pure taste, clarity, and in some instances effervescence (95).

4. The popularity of flavored water as a replacement for soft drinks and alcoholic beverages. Flavored carbonated water that may contain fruit juices is becoming increasingly popular. Strong promotional advertising by the bottled water industry has supported the impression that such products are favored by trendy baby-boomers and young professionals.

5. The health image. While bottled water manufacturers are not allowed to make health claims, the popular image of bottled water is often linked with exercise and youth. Increased consumer concern and awareness about health and fitness, the desire to replace soft drinks (containing unwanted sugar or sugar substitutes), and the changes in social attitudes toward alcohol consumption have also contributed to the increase in bottled water sales. It is estimated that 44% of consumers drink bottled water out of concern for their health (11).

6. Sodium and other dietary restrictions. In areas where drinking water must be softened, people on low sodium diets may turn to bottled water because softened water is high in sodium content. Persons who must restrict chlorine and fluorine intake because of allergies and/or other medical problems also may use bottled water.

7. Perceived or real problems with tap water. Problems with water in specific municipalities (especially highly publicized cases) can lead to a distrust of all municipal water, and in some localities 50% of the public does not trust tap water (11). In some areas consumers have legitimate concerns. Depending on the area,

private wells and smaller, rural municipalities may experience increasing contamination from such sources as agricultural pollutants, including pesticides, fertilizers, and animal wastes. Larger, urban areas can be affected by a multitude of pollution sources.

Published reports about waterborne disease outbreaks (7, 19, 53, 117, 123) including the 1993 *Cryptosporidium* episode in Milwaukee, and the 1989–1990 *E. coli* O157:H7 outbreak in Cabool, Missouri, as well as concern over other pollutants in water (77), have reduced the public's confidence in the municipal water systems. Payment et al. (105) measured the level of gastrointestinal illness related to the consumption of "acceptable" tap water (i.e., met current water quality criteria and was free of any detectable indicator bacteria and enteric viruses) prepared from raw contaminated surface waters and estimated that 35% of the reported gastrointestinal illnesses among the tap water drinkers was water-related and preventable. With the aging and deterioration of water and sewage treatment plants, the increase in usage of these systems due to population increases and pressures, and the general lack of government funds for improvement of these facilities (51, 52), people are using bottled water as a safety measure.

8. Personal taste. Many people feel that they can afford to switch to bottled water even if their concerns about tap water are not health-based. For instance, people who dislike the taste or odor of chlorine can choose bottled water that is treated with tasteless, odorless ozone (16).

9. The traveling experience. Many people rely on bottled water while traveling and visiting other countries. Having used bottled water while on these excursions and enjoyed the experience (i.e., taste and quality), they continue this "habit" upon their return home.

A summary of the public perceptions concerning bottled water compared with tap water can be seen in Table 5.

B. Susceptible Populations

Concerns about the safety of their drinking water have led many susceptible populations to consume large quantities of bottled water. Susceptible individuals—those who are particularly prone to infections by bacteria, viruses, and parasites—include "YOPIs" (the young, the old, the pregnant, and the immunocompromised) (10), such as the newborn, infants, cancer patients, AIDS and HIV-infected patients, and individuals with kidney disorders (28, 53, 54, 120).

These populations are particularly susceptible to infection by bacterial contaminants and parasites, and because of demographic changes (e.g., population aging), more people are at greater risk of contracting foodborne illness (98, 120). Susceptible populations have been warned to avoid drinking natural mineral waters (115).

Table 5 Summary of a Consumer Poll Showing Comparative Assessments of Tap and Bottled Water as Related to Specific Concerns

	Persons polled (%)[a]		
Concern	Tap water	Bottled water	Neither/both the same
Safer	28	57	14
Better quality	27	61	10
Better taste	30	58	10
Fewer chemicals	20	71	7
Fewer additives	26	67	6
Better quality control	33	59	7
More government regulations	59	29	8
More bacteria	68	18	9
More natural	33	60	6
More expensive	3	94	1
Contains chlorine	88	6	4
More minerals	41	51	6
More consistent taste	29	63	6
Contains fluorine	80	12	6

[a]600 persons were polled, however 100% of those polled did not answer each question.
Source: Ref. 11.

C. Potential Problems with Bottled Water

"Natural" bottled waters (Table 1) are sold with the understanding and, in Europe, the legal requirement, that they have not been subjected to any treatment that would remove natural indigenous bacteria (4, 5, 22, 23), which are believed to have medicinal and therapeutic qualities (50, 69, 105). Despite the wide use of mineral waters, it has never been proven that the ingested levels of indigenous microorganisms in bottled water have an adverse effect on health (32, 111). Yet, the susceptible populations have been warned to avoid drinking natural mineral waters (115). In contrast, it is known that bottled water has been given to hospital patients when there was uncertainty about the purity of tap water (29).

The original bacterial population (which is psychrotrophic and oligotrophic in nature) found in freshly drawn mineral water will colonize still bottled waters and is believed to be harmless (34, 95). Bottled water not only is consumed directly but is used to prepare foods and formula, and in humidifiers. Leclerc (81) reconstituted powdered infant formulas with still mineral water originating from an *extremely well-operated bottling plant* instead of the standard procedure of using boiled water. He did not observe clinical symptoms or any significant modifications of the intestinal or respiratory colonization of infants as a result of this substitution. However, most consumers will not know the manufacturing conditions of the bottled water at the manufacturing plant and should not make any assumptions that could endanger their infants, babies, or other susceptible individuals. Labels on products marketed for infants should indicate that the water is or is not sterile, and that the product should be used as directed by a physician or according to infant formula preparation instructions (79).

Despite the foregoing observations and studies, much controversy surrounds the question of the potential pathogenicity of indigenous microorganisms in mineral waters. Many of these microorganisms may not be pathogenic under normal conditions but have been responsible for infections in special circumstances, such as in hospitalized patients debilitated by illness or treatments that undermine the body's defences. Some authors advise that such patients should avoid drinking natural mineral waters (115, 126), while others recommend these products (96).

The transmission of bacterial (opportunistic) pathogens through water includes direct exposure through ingestion or contact with the contaminated water, and indirect exposure through ingestion of foods or products prepared with contaminated water, and exposure to aerosols (134).

Bottled water should be of high microbiological quality, especially if intended for use by vulnerable populations. When taken between meals, small amounts of water ($\leq$ 50 mL) can pass immediately through the stomach into the intestine and subsequently cause disease (94). Microorganisms (including pathogens) in this water may escape the bactericidal effect of gastric juices in the stomach, reaching the intestine in the same numbers as initially ingested. The minimum infective dose of pathogens may be quite low when ingested between meals. This may be especially true with the susceptible person, as the intragastric bactericidal mechanisms may be impaired to the extent that every microorganism ingested may reach the intestinal tract (93).

Bottled waters may be used in rather large quantities by debilitated and susceptible consumers (95), and the overall consumption of bottled water by such populations will probably increase as these populations increase (1). Contaminated water supplies might be a more significant source of infection than foods, especially for infants (116). Water consumption is encouraged for certain segments of the population; for instance, the following are recommendations offered by physicians, weight loss clinics, and bottled water manufacturers and associations: "drinking eight glasses of water a day will help you control your food intake during weight loss" and "drinking a minimum of 10 glasses of water a day is sufficient for both mother and baby during pregnancy" (20).

Other bacteria may be introduced into bottled water during manufacture (see Sec. II. B) and may have the potential for causing disease [Table 2 (104, 119)]. Many of these microorganisms, such as *Acinetobacter, Flavobacterium, Pseudomonas,* and other genera [Table 2 (60, 64, 102, 106, 125)], are considered secondary or opportunistic pathogens and may cause disease if present in high numbers

(119). These bacteria can proliferate to high numbers when bottled water is stored at room temperature (127–130) (see Sec. II. D).

Opportunistic waterborne pathogens deserve considerable attention. Unlike primary pathogens, which can affect persons of all ages, these organisms are particularly invasive to susceptible individuals (53, 54) and may cause illness with insidious onset after long incubation periods in immunocompromised persons (85). In other situations, contaminated water supplies often cause intestinal problems that go unreported if the problems are short-lived and are not associated with the ingestion of water (3). The occurrence of "minor" symptoms will usually not warrant epidemiological determination of their source in the absence of a large-scale occurrence (108). Whether individuals who consume contaminated water suffer adverse health effects depends on the bacteria involved, the number of bacteria ingested, and the individual's general health and resistance to that particular organism (51). There is also the possibility of secondary infection when compromised individuals on antibiotic therapy ingest water containing antibiotic-resistant organisms, such as *Pseudomonas* and *Acinetobacter* spp. (64, 108).

D. Consumer Use; Protection of Susceptible Populations

As a result of demographic changes, today more people than even are at greater risk of contracting a foodborne illness (120). Without appropriate intervention, the incidence of waterborne disease/epidemics/pandemics is also expected to increase dramatically (45). Since this statement refers to the use of source waters, one could, by extrapolation, include bottled water (and other drinking water) under this prediction. Adequate disinfection can be the most cost-effective public health insurance available, provided the treatment reduces all pathogenic bacteria by at least 4 logs (50).

For this reason, Health Canada recommends that only disinfected bottled water (i.e., carbonated, ozonated, UV-treated, etc.) should be used by susceptible populations (126). Similarly, a workshop on waterborne cryptosporidiosis (6) and the National Association of People with AIDS (NAPWA) (96) recommends that susceptible populations use bottled water that is manufactured from a protected source (Table 1) or from other source water that has been treated by distillation, reverse osmosis, or 1 μm "Absolute" filtration for the removal of protozoan cysts. Water from *protected* underground sources is unlikely to contain oocytes (such contamination usually results from intermittent mixing of the source water with surface water through fissures, etc.) when it has been consistently free of coliform bacteria (6, 28).

In contrast, the U.K. Expert Advisory Group on AIDS (130) recommends that AIDS patients with advanced symptoms boil their water (from any source) before drinking it, as a measure to prevent waterborne cryptosporidiosis. Persons with asymptomatic HIV infection should be apprised of the risks of parasite infection associated with the consumption of water, so they can make an informed decision about boiling water. Physicians treating other immunocompromised patients should consider advising their patients of the risks associated with parasites in water and the need to boil water, the patients' risk of contracting a waterborne disease, and the possible severity of an infection.

E. Contamination of Water Coolers

Contamination of water coolers is usually not related to GMPs in the bottled water plant, but to the use of water coolers as the dispensers for the water (36, 84). In a 1992 study in Quebec City, up to 36% of the water from water coolers was found to be contaminated with total coliforms, fecal coliforms, fecal streptococci, *Staphylococcus aureus,* or *Aeromonas* spp. (84).

The microbial content of the bottled water supply (usually 18 L) for use in the water cooler can be controlled during production. However, the environment to which the water bottles and dispensers are exposed cannot be strictly controlled, since there exists the possibility of contamination from the air and during use (36). The greatest potential for microbial contamination of bottled water occurs (a) during bottle changing of water dispensers (36), (b) from contamination of the water in the dispenser with skin bacteria from the handling of the external bottle surfaces (36), (c) from exposure of the reservoir to the air, and (d) from contamination via the metal shaft used to puncture the caps to allow the water to drain into the closed reservoir (36). Any bacterial contaminants, if present on the bottle cap or the recessed plastic channel used to guide the neck of the bottle onto the metal puncturing shaft, will be smeared across the shaft, or migrate in any residual water, when empty bottles are pulled out during bottle changes (36).

Regular cleaning of water coolers and disinfection with hypochlorite should limit contamination (84). In the 1992 study in Quebec City, only 44% of people possessing domestic water coolers and 36% of personnel responsible for workplace water coolers had been informed of the necessity of cleaning and disinfecting the equipment, let alone how to do it (84). Similarly, in a 1993–1994 survey of 48 Calgary schools using water coolers, 80–90% of individuals never received any information on cleaning and maintenance procedures (130). It is imperative that vendors and suppliers of water dispensers impress their clients the need for and methods of regular maintenance of their equipment (84, 130).

F. In-Store Manufactured Bottled Water

In-store manufactured bottled water generally is found in areas where the local water supply, including municipal water, is exceptionally hard (i.e., contains high levels of calcium, iron, sulfur, or other minerals, or high levels of sediments or contaminating organics). Grocery, department, and drug stores may have units that use reverse-osmosis filters, activated carbon filters, and sometimes steam distilling units to filter and purify municipal water. The clients usually bring in their own bottles and containers and fill them up at the unit, paying only for the water. In Canada, a large proportion of in-store manufacturing units are located on the prairies for the forementioned reasons. Alberta Public Health inspectors (F. Yan, Lakeland Regional Health Authority, personal communication) have indicated that such units can produce waters with excessive aerobic bacteria (ACC) levels ($\leq$ 1000 cfu/mL), in violation of municipal, provincial, and federal standards for drinking water.

In 1996 samples from 10 in-store bottled water manufacturers were sampled in the east central Alberta area (F. Yan, Lakeland Regional Health Authority, personal communication). Of these 10 manufacturers, seven had distillers, seven had activated carbon filters, two had water softeners, three had reverse-osmosis filters, eight had ultraviolet light treatment, and two used ozonation for disinfection. In this monitoring program, 136 samples were analyzed for ACC, coliforms, and fecal coliforms. For ACC, 70 (51.5%), 22 (16.2%), 13 (9.6%), and 31 (22.8%) of the 136 samples contained <10, 10^1–10^2, 10^2–10^3, and $>10^3$ cfu/mL, respectively. The latter 31 samples had all been UV-treated. None of the samples contained coliforms or fecal coliforms. This would indicate that the water was contaminated via the spigot or from air entering the holding tank, from dirty activated carbon filters, or from contaminated UV light chambers. Alternatively, the flow of the water through the UV light chamber may have been too fast, and therefore the contact time with the UV light too short.

The problems with this type of operation are as follows: (a) the faucets on these units are not usually flushed by the client before filling the bottle (Health Canada recommends that all taps be flushed after periods of stagnation and before obtaining drinking water); (b) the client may contaminate the faucet by touching and handling the faucets and other parts of the unit; (c) the client may use contaminated bottles for filling with the water, thus incurring exposure to bacteria through ingestion of the water or through foods in which the water is used; and (d) the manufacturing units may not be properly maintained by in-store staff or by the equipment's owner.

V. RECOMMENDATIONS FOR THE CONSUMER

A. What Should the Consumer Consider When Purchasing and Using Bottled Water?

1. Buy from a reputable firm. If concerned about chemical and bacterial content, contact the firm, which should provide consumers with analytical printouts. Most such vendors can be contacted via the phone numbers and addresses on the labels, or via Internet accounts.

2. Buy only bottles whose seals are unbroken. Examine the outside and interior of each bottle before purchase. Do not purchase any bottles bearing visible

extraneous material. Report any tampering or extraneous material to the store manager and health officials.

3. Do not be fooled by impressive labels or the history of the company. Examine the bottle and label for date of manufacturing or manufacturing code, "Best before" date, chemical analysis, treatment (e.g., ozonated, etc.), company contact number, location, and type of source water.

4. Keep the bottle clean, and preferably refrigerated (the water and inside cap and liner can support bacterial growth). Clean the outside of the bottle cap and neck before and after each use.

5. Do not refill old bottles without cleansing with hot soapy water and disinfecting with chlorinated water (see Table 6). It is preferable by far to buy newly manufactured bottled water.

6. Do not share bottles (i.e., do not have more than one person drink directly from a bottle). Dispense the water into clean cups or glasses if more than one person is to use the bottle.

7. When traveling, avoid bottled water unless it is carbonated or otherwise suitably disinfected; carbonation is easiest to recognize. Buy only sealed products that have been made by a reputable manufacturer. Wipe off the bottle or can top before drinking or pouring.

8. When refrigeration is impractical, store the bottled water in a cool, clean environment away from sources of heat and sunlight. Although manufacturers give bottled water a "best before" date or shelf life of 2 years, Health Canada suggests replacement after one year.

9. Clean water coolers regularly. See Table 6.

10. Buy products that have no spill caps. That is, water should not spill and air should not enter the bottled water when the bottle in the cooler is replaced. Use water coolers that filter the air as it enters the bottle when the water level lowers.

11. Use water dispensers that are actually coolers and can keep the water refrigerated. Those equipped with heating units should heat the water to boiling to ensure destruction of any bacteria.

12. If you are buying water marketed for a baby or an infant, check the label to determine whether the water is sterile. If not, consult your physician or use according to infant formula preparation instructions. If you are a member of any susceptible population, buy disinfected (ozonated) bottled water; and if you are particularly concerned, boil it before use.

VI. RECOMMENDATIONS

Better education and communication among bottled water associations, governments, health agencies, the media, the public (especially susceptible populations), and retail outlets is essential. The topics discussed should include the following.

Table 6 Maintaining the Cleanliness of a Water Cooler

Cleaning your water cooler
Unplug cord from electrical outlet of cooler.
Remove empty bottle.
Drain water from stainless steel reservoir(s) through faucet(s).
Prepare a disinfecting solution by (a) adding one tablespoon (15 mL) household bleach to one Imperial gallon (4.5 L) of water solution. This solution should not contain less than 100 ppm available chlorine. Heat the solution to a temperature of not less than 113°F (45°C) and pour into reservoir; or (b) some companies suggest using one part vinegar to three parts water solution to clean the reservoir of scale before cleaning with bleach. Check your manual. *Note:* Other disinfecting solutions may be suitable. Please check with your water cooler supplier.
Wash reservoir thoroughly with bleach solution and let stand for not less than 2 minutes (to be effective) and not more than 5 minutes (to prevent corrosion).
Drain bleach solution from reservoir through faucet(s).
Rinse reservoir thoroughly with clean tap water, draining water through faucets to remove traces of the bleach solution. *Note:* Clean your bottled water cooler with every bottle change.
Drip tray (located under faucets)
Lift off drip tray.
Remove the screen and wash both tray and screen in mild detergent. It is recommended to wipe the tray with the bleach solution to disinfect it.
Rinse well in clean water and replace on cooler.
Replacing bottle
Wash hands with soap and warm water before handling. If you choose to use clean protective gloves (e.g., latex), discard or disinfect after each use and prior to reuse. *Note:* Protective gloves should never replace proper hand washing and hygiene.
Wipe the top and neck of the new bottle with a paper towel dipped in household bleach solution [1 tablespoon (15 mL) of bleach, 1 gallon (4.5 L) of water]. Rubbing alcohol may also be used, but must be completely evaporated before placing the bottle in the cooler.
Remove cap from new bottle or follow instructions when using bottles with the non-spill caps.
Place new bottle on cooler.

Source: Ref. 130.

1. The types of bottled water and bacterial content. This information should include statements indicating that bottled water (especially "natural" waters) is not a sterile product and contains a variety of bacteria, the vast majority of which are innocuous. Explanations about production under GMPs, including statements about testing for pathogens and indicator bacteria would also be helpful.

2. The handling of products both by the consumer and the retail outlet. Most persons have the perception or assume that bottled water is sterile and do not

understand that microorganisms can survive and even grow in bottled water. Knowing this should affect the manner in which consumers (and especially susceptible populations) handle the product upon purchase. Proper handling includes refrigeration, proper preparation of infant formulas and other foods, and regular cleaning of water coolers and dispensers. Retail outlets must be encouraged to rotate their bottled water stocks on a regular basis so the consumer is purchasing the freshest product possible.

3. Provision of communiqués. Bottled water associations, governments, and health agencies should make available information letters or communiqués on bottled waters to parallel those available for the safe use and handling of other foods and beverages.

4. Labeling. Finished products should be labeled with information concerning processing treatments the water underwent. This is especially important for susceptible populations who are examining labells for statements concerning ozonation and microfiltration. Labels should also contain the bottled water associations' logos; a statement referring to membership in these trade associations would enable the consumer to purchase products produced under their strict and sanitary codes of practices. Each container should be marked with the bottling date, date of expiration, and lot number to assist the retailer and consumer in determining freshness (55, 110). These would aid in the follow-up of consumer complaints and the removal of contaminated product from retail shelves. Each container should be supplied with a protective seal to reduce the chances of contamination (2) and tampering.

Implementation of new standards and guidelines, as suggested in Table 7 as well as GMP codes, is necessary to aid government agencies, food testing laboratories, and bottled water manufacturers in providing the public with the best product possible. It is not surprising that both bottled water and municipal waters contain the same microorganisms (Table 1), since both can originate from the same sources. These bacterial genera are found in products from a variety of countries, and many are a cause for concern owing to their potential pathogenicity. When one considers that the target population may include pregnant women, the infirm, the immunocompromised, the elderly, and other susceptible people, bottled water must be shown to be safe. However, bottled water of excellent quality can be produced by following GMP/HACCP programs and by testing the source waters, inline samples, and finished products for a variety of indicator organisms (Table 7), while using the ACC as an indicator of GMP within 24 hours of production. New standards should include mandatory sterilization or disinfection of final products when bottled water is not manufactured from "approved sources." This would also include a mandatory filtration step or system to entrap parasites and their cysts. The definition of "approved source" needs to well defined and stated in regulations. Bacteriological analysis of "approved sources" must include all the indicator microorganisms listed in Table 7. There should also be cooperation between

Table 7 Present Regulations, Proposed Regulations, and Microbiological Limits for Bottled Water at the Source and at Other Critical Control Points

	Variables (see Key)			
Country and microorganism	n	c	m	M
Canada[a] (126)				
All bottled water				
ACC	5	2	10^2/mL	10^4/mL
Coliforms	10 (5)	1	0/100 mL	10/100 mL
A. hydrophila	5	0	0/100 mL	
P. aeruginosa	5	0	0/100 mL	
Fecal streptococci[b]	5	0	0/100 mL	
Spore-forming, sulfite-reducing anaerobes (clostridia)[b]	5	0	0/100 mL	
E. coli[b]	5	0	0/100 mL	
Parasites	5	0	0/100 mL	
Codex Alimentarius Commission (21–24)				
Natural mineral water				
Coliforms	5	1	0/250 mL	2/250 mL
E. coli	5	0	0/250 mL	
Fecal streptococci	5	1	0/250 mL	2/250 mL
Spore-forming sulfite-reducing anaerobes (clostridia)	5	1	0/250 mL	2/250 mL
P. aeruginosa	5	0	0/250 mL	
Bottled drinking water[c]				
No bacterial standards specified				
European Community (4, 5)				
Natural mineral water				
ACC within 12 h after bottling	?[d]	0	10^2/mL[e]	
		0	20/mL[f]	
ACC at source	?	0	20/mL[e]	
		0	5/mL[f]	
Coliforms	?	0	0/250 mL	
E. coli	?	0	0/250 mL	
Fecal streptococci	?	0	0/250 mL	
P. aeruginosa	?	0	0/250 mL	
spore-forming sulfite-reducing anaerobes (clostridia)	?	0	0/50 mL	
Parasites	?	0		
Bottled drinking waters				
ACC within 12 h of bottling	?	0	20/mL	100/mL
ACC within 12 h of bottling	?	0	5/mL	20/mL
Total and fecal coliforms, fecal streptococci, sulfite-reducing clostridia[g]	?	0	0/100 mL	
United States (FDA) (26, 44)				
Coliforms by MPN	10	1	2.2/100 mL	9.2/100 mL
Coliforms by membrane	10	1	1/100 mL	4/100 mL

[a]ACC standards should be applied to all bottled water within 24 hours of packaging. The number of sample units analyzed for coliforms is now 10, but will be changed to 5. *A. hydrophila* and *P. aeruginosa* regulations are in the process of being implemented through legislation.
[b]These guidelines are based on those of Codex (36) for the source water and critical control points.
[c]Bottled drinking waters (other than natural mineral water) must meet public drinking water standards if obtained from a public water system or must be of safe and suitable quality (i.e., potable) using multiple barrier treatments (filtration, disinfection, etc.). Regulations for specific bacteria are not given.
[d]The number of sample units is not specified.
[e]Determined at 20–22°C in 72 hours on agar–agar or agar–gelatin.
[f]Determined at 37°C in 24 hours on agar–agar.
[g]Other bacterial parameters 0/100 mL by membrane filter or Most Probable Number (MPN) < 1. Clostridia MPN ≤ 1/20 mL.
Key: n, number of sample units, usually selected at random, from a lot and examined in order to satisfy the requirements of a plan; c, maximum allowable number of marginally acceptable sample units per lot (i.e., the number of sample units allowed with counts between m and M.); m, maximum number of microorganisms per unit (mL) that is of no concern or is an acceptable level; M, number of bacteria per unit that indicate a potential health hazard, imminent spoilage, or gross insanitation, and if exceeded in any one sample unit causes rejection of the lot in question.
Source: Ref. 130.

governments, both federal and state or provincial, to standardize bottle water regulations. There should also be cooperation between governments, both federal and state or provincial, to standardize bottle water regulations.

More extensive surveillance of the bottled water industry is recommended, especially with respect to manufacturers who are not members of such bottled water associations as the International Bottled Water Association (IBWA) and the Canadian Bottled Water Association (CBWA). Public health officials should be encouraged to inquire about ingestion of bottled water when investigating waterborne and foodborne illness. More surveillance by government laboratories is often impractical because of limited resources. The standards established by these organizations for the production of bottled water require that all plants be inspected annually by independent agencies, in addition to government inspections. These organizations welcome additional and more stringent regulations (135). However, not all bottled water manufacturers belong to these organizations, and thus not all bottled water is produced in compliance with their code of practices (88).

However, based on available information, bottled water sold worldwide by members of the bottled water associations has generally been found to be of good microbiological quality and is not considered to pose any microbiological threat. Bottled water associations such as the IBWA and the CBWA issue sound guidance to members on GMPs for the production of bottled water. All bottlers should be encouraged to join such associations.

REFERENCES

1. Anonymous. 1990. Foodborne illness among elderly, AIDS patients seem likely to grow. Food Chem. News. Feb. 26, 1990, p. 25.
2. Anonymous. 1990. EPA researcher raises bottled water storage issues. Food Chem. News. Sept. 24, 1990, pp. 23–26.
3. Anonymous. 1991. Microbiological contamination found in 31% of bottled water. Food Chem. News, April 2, 1991, pp. 64–67.
4. Anonymous. 1980. Council directive of 15 July 1980 on the approximation of the member states relating to the exploitation and marketing of natural mineral waters. Off. J. Eur. Community 30, L229:1–10.
5. Anonymous. 1995. Proposal for a council directive concerning the quality of water intended for human consumption. Off. J. Eur. Community 38:5–24.
6. Anonymous. 1995. Assessing the public health threat associated with waterborne cryptosporidiosis: Report of a workshop. MMWR RR-6:1–20.
7. Bai, M. 1995. How safe is our water? New York Jan. 16, 1995, pp. 24–31.
8. Balkwill, D. L. 1988. Distribution and characterization of bacteria in deep aquifers, in Summary Report of the Conference on Mobility of Colloidal Particles in the Subsurface: Chemistry and Hydrology of Colloid–Aquifer Interactions, Monteo, NC, Oct. 4–6, 1988. U.S. Department of Energy, DOE/ER-0425.
9. Balkwill, D. L., J. K. Fredrickson, and J. M. Thomas. 1989. Vertical and horizontal

variations in the physiological diversity of the aerobic chemoheterotrophic bacterial microflora in deep southeast coastal plain subsurface sediments. Appl. Environ. Microbiol. 55:1058–1065.

10. Bischofberger, T., S. K. Cha, R. Schmitt, B. König, and W. Schmidt-Lorenz. 1990. The bacterial flora of non-carbonated, natural mineral water from springs to reservoir and glass and plastic bottles. Int. J. Food Microbiol. 11:51–72.
11. Bishop, J. N. 1992. Drinking water: Facts and issues in the 1990s. WaterPower Fall 1992:7–8, 10–13.
12. Bitton, G., S. R. Farrah, R. H. Ruskin, J. Butner, and Y. J. Chou. 1983. Survival of pathogenic and indicator organisms in ground water. Ground Water 21:405–410.
13. Bitton, G., and R. W. Harvey. 1992. Transport of pathogens through soils and aquifers, in R. Mitchell (ed.), Environmental Microbiology, Wiley-Liss, New York, pp. 103–124.
14. Blake, P. A., M. L. Rosenberg, J. B. Costa, P. S. Ferreira, C. L. Guimaraes, and E. J. Gangarosa. 1977. Cholera in Portugal, 1974. I. Modes of transmission. Am. J. Epidemiol. 105:337–343.
15. Blake, P. A., M. L. Rosenberg, J. Florencia, J. B. Costa, L. D. P. Quintino, and E. J. Gangarosa. 1977. Cholera in Portugal, 1974. II. Transmission by bottled mineral water. Am. J. Epidemiol. 105:344–348.
16. Bromberg, M., and D. Paterson. 1994. Bottled Water—Crystal clear choice or cloudy dilemma? University of Illinois at Urbana-Champaign. pp. 17. *http://hermes.ecn.purdue.edu:8001/cgi/conwqtest?/lw-19.il.ascii*
17. Burge, S. H., and P. R. Hunter. 1990. The survival of enteropathogenic bacteria in bottled mineral water. Riv. Ital. Ig. 50:401–406.
18. Cahill, J. 1994. Bottled water—do the public know what they are getting? Commun. Dis. Environ. Health Scotland. 28(94/51):3–8.
19. Carpenter, B., S. J. Hedges, C. Crabb, M. Reilly, and M. C. Bounds, 1991. Is your water safe? The dangerous state of drinking water in America. U.S. News World Rep. July 29, 1991, pp. 47–55.
20. Chandler, L. 1994. The essence of life . . . Water. WaterPower Spring 1994:8,10.
21. Codex Alimentarius Commission. 1986. Draft code of hygienic practice for collecting, processing and marketing of natural mineral waters, and methods of analysis for natural mineral waters. Report of the Fifteenth Session of the Coordinating Committee for Europe, Thun, Switzerland, June 16–20, 1986. Annex VI.
22. Codex Alimentarius Commission. 1993. Conversion of the Codex European regional standard for natural mineral waters to a world-wide Codex standard (step 3 of the Codex procedure). CL 1993/4-NMW.
23. Codex Alimentarius Commission. 1994. Natural Mineral Waters Codex Standard 108-1981. Codex Alimentarius 11:64–87.
24. Codex Alimentarius Commission. 1996. Proposed draft recommended international code of hygienic practice for the manufacture and marketing of bottled drinking waters (other than natural mineral water). Codex Standard CL 1996/13-FH, July.
25. Costerton, J. W., Z. Lewandowski, D. De Beer, D. Caldwell, D. Korber, and G. James. 1994. Biofilms, the customized microniche. J. Bacteriol. 176:2137–2142.
26. Cowman, S., and R. Kelsey. 1992. Bottled water, in Compendium of Methods for the Microbiological Examination of Foods, 3rd ed. American Public Health Association, Washington, DC, pp. 1031–1036.

27. Craun, G. F. 1979. Waterborne diseases: A status report emphasizing outbreaks in groundwater systems. Ground Water 17:185–191.
28. Craun, G. F., and W. Jakubowski. 1986. Status of waterborne giardiasis outbreaks and monitoring methods, in: C. L. Tate Jr. (ed.), Proceedings of an International Symposium on Water-Related Health Issues. American Water Resources Association. Bethesda, MD, pp. 167–174.
29. DeSha, C. M., and M. Hammett. 1977. Isolation of *Aeromonas hydrophila* from a natural mineral spring. Dev. Ind. Microbiol. 18:633–636.
30. Dive, D., J. P. Picard, and H. Leclerc. 1979. Amoebae in the water supply: An epidemiological study. Ann. Microbiol. (Inst. Pasteur) 130A:487–498.
31. De Felip, G., L. Toti, and P. Iannicelli. 1976. Osservazioni compartative sull' andamento dell flora microbica delle acque minerale naturale confezionate in vetro "PVC" a "Tetrabrik." Ann. Ist Super. Sanit. 12:203–209.
32. Del Vecchio, V., and M. Fischetti. 1972. Andamento nel tempo della flora saprofita presente in acque minerali: Confronto fra contenitori di vetro e contenitori di plastica. Nuovi Ann. Ig. Microbiol. 23:257–277.
33. Devlieghere, F., and A. Huyghebaert. 1997. Removal of microorganisms from polymer bottles by caustic washing. Leben. Wissen. Technol. 30:62–69.
34. Ducluzeau, R., S. Dufresne, and J. M. Bochand. 1976. Inoculation of the digestive tract of axenic mice autochthonous bacteria of mineral water. Eur. J. Appl. Microbiol. 2:127–134.
35. Dutka, B. J. 1973. Coliforms are an inadequate index of water quality. J. Environ. Health. 36:39–46.
36. Eckner, K. F. 1992. Comparison of resistance to microbial contamination of conventional and modified water dispensers. J. Food Prot. 55:627–631.
37. Evian. 1997. Evian Natural Mineral Water Home Page. *http://www.webevian.com/home.html*
38. Ferreira, A.-C., P. V. Morais, and M. S. Da Costa. 1994. Alterations in total bacteria, iodonitrophenylterazolium (INT)-positive bacteria, and heterotrophic plate counts of bottled mineral water. Can. J. Microbiol. 40:72–77.
39. Feron, V. J., J. Jetten, N. De Kruijf, and F. Van Den Berg. 1994. Polyethethylene terephthalate bottles (PRBs): A health and safety assessment. Food Additives Contamin. 11:571–594.
40. Fewtrell, L., D. Kay, A. Godfree, and G. O'Neill. 1996. Microbiological Quality of Bottled Water. Centre for Research into Environment and Health. University of Leeds, U.K. (Also cited in Food Saf. Inf. Bull. 78:October 1996).
41. Fleet, G. H., and F. Mann. 1986. Microbiology of natural mineral water: An overview with data on Australian waters. Food Technol. Aust. 38:106–110.
42. Fries, M. R., L. J. Forney, and J. M. Tiedje. 1997. Phenol- and toluene-degrading microbial populations from an aquifer in which successful trichlorethene cometabolism occurred. Appl. Environ. Microbiol. 63:1523–1530.
43. Fry, N. K., J. K. Frederickson, S. Fishbain, M. Wagner, and D. A. Stahl. 1997. Population structure of microbial communities associated with two deep, anaerobic, alkaline aquifers. Appl. Environ. Microbiol. 63:1498–1504.
44. Food and Drug Administration. 1995. Beverages: Bottled water; Final Rule. Fed. Regis. 21 CFR Part 103 et al. 60:57075–57130.
45. Ford, T. E., and R. R. Colwell. 1996. A Global Decline in Microbiological Safety

of Water: A Call for Action. American Academy of Microbiology Washington, D C, 40 pp.
46. Gangarosa, E. J., M. A. Kendrick, M. S. Lowenstein, M. H. Merson, and J. W. Mosley. 1980. Global travel and travelers' health. Aviat. Space Environ. Med. 51:265–270.
47. Gasaway, J. M. 1978. Significance of abuse chemical contamination of returnable dairy containers: Pesticide storage and detector evaluation. J. Food Prot. 41:851–862.
48. Gasaway, J. M. 1978. Significance of abuse chemical contamination of returnable dairy containers: Sensory and extraction studies. J. Food Prot. 41:863–876.
49. Gasaway, J. M. 1978. Significance of abuse chemical contamination of returnable dairy containers: Hazard assessment. J. Food Prot. 41:965–973.
50. Geldreich, E. E. 1986. Potable water: New directions in microbial regulations. ASM News 52:530–534.
51. Geldreich, E. E. 1989. Drinking water microbiology—New directions toward water quality enhancement. Int. J. Food Microbiol. 9:295–312.
52. Geldreich, E. E. 1990. Microbiological quality of source waters for water supply, in: G. A. McFeters (ed.), Drinking Water Microbiology. Springer-Verlag, New York, pp 1–31.
53. Geldreich, E. E. 1996. Pathogenic agents in freshwater resources. Hydrol. Process. 10:315–333.
54. Geldreich, E. E. 1997. Reinventing microbial regulations for safer water supplies, in: D. Kay and C. Fricker (eds.), Coliforms and *E. coli*. Royal Society of Chemistry, Cambridge, pp. 218–231.
55. Geldreich, E. E., H. D. Nash, D. J. Reasoner, and R. H. Taylor. 1975. The necessity of controlling bacterial populations in potable waters-bottled water and emergency water supplies. J. Am. Water Works Assoc. 67:117–124.
56. Ghiorse, W. C., and J. T. Wilson. 1988. Microbial ecology of the terrestrial subsurface. Adv. Appl. Microbiol. 33:107–172.
57. Gonzalez, C., C. Guttierrez, and T. Grande. 1987. Bacterial flora in bottled uncarbonated mineral drinking water. Can. J. Microbiol. 33:1120–1125.
58. Gonzalez, C., C. Ramirez, and M. Pereda. 1987. Multiplication and survival of *Pseudomonas aeruginosa* in uncarbonated natural mineral water. Microbiol. Aliment. Nutrit. 4:111–115.
59. Grant, M. A. 1997. Comparison of three methods for the analysis of coliforms in bottled water. J. Food Sci. Submitted for publication.
60. Guerzoni, M. E., R. Lanciotti, M. Sinigaglia, and F. Gardini. 1994. Analysis of the interaction between autochthonous bacteria and packaging material in PVC-bottled mineral water. Microbiol. Res. 149:115–122.
61. Hamilton, N., and F. A. Rosenberg. 1991. Examination of bottled drinking water containers for biofilm formation. Proceedings of a Water Quality Technology Conference, San Diego, CA, Part 2, pp. 1473–1490.
62. Harvey, R. W. 1993. Fate and transport of bacteria injected into aquifers. Curr. Opin. Biotechnol. 4:312–317.
63. Harvey, R. W., N. E. Kinner, D. MacDonald, D. W. Metge, and A. Bunn. 1993. Role of physical heterogeneity in the interpretation of small-scale laboratory and field observations of bacteria, microbial-sized microsphere, and bromide transport through aquifer sediments. Water Resour. Res. 29:2713–2721.

64. Hernandez-Duquino, H., and F. A. Rosenberg. 1987. Antibiotic-resistant *Pseudomonas* in bottled drinking water. Can. J. Microbiol. 33:286–289.
65. Hoadley, A. W. 1977. Potential health hazards associated with *Pseudomonas aeruginosa,* in A. W. Hoadley and B. J. Dutka (eds.), Bacterial Indicators/Health Hazards Associated with Water. American Society for Testing and Materials, Philadelphia, pp. 60–114.
66. Hoeschler F. A. 1980. Water treatment apparatus with means for automatic disenfection thereof. Off. Gaz. U.S. Patent Off. 999:680.
67. Holt, J. G., N. R. Krieg, P. H. A. Sneath, J. T. Staley, and S. T. Williams. 1994. Bergey's Manual of Determinative Bacteriology, 9th ed. Williams & Wilkins, Baltimore.
68. Holtzman, A. E., T. W. Aronson, N. Glover, S. Froman, G. N. Stelma, Jr., S. N. Sebata, M. G. Boian, T. T. Tran, and O. G. W., Berlin. 1997. Examination of bottled water for nontuberculous *Mycobacterium.* J. Food Prot. 60:185–187.
69. HPB (Health Protection Branch). 1981. Food-borne and water-borne disease in Canada. Annual Summary 1977. Health and Welfare Canada, Ottawa, Ont.
70. HPB. 1986. Food-borne and water-borne disease in Canada. Annual Summary 1980–82. Health and Welfare Canada, Ottawa, Ont.
71. HPB. 1988. Food-borne and water-borne disease in Canada. Annual Summary 1983–84. Health and Welfare Canada, Ottawa, Ont.
72. Hunter, P. R. 1993. The microbiology of bottled natural mineral waters. J. Appl. Bacteriol. 74:345–352.
73. Hunter, P. R., and S. H. Burge. 1987. The bacteriological quality of bottled natural mineral waters. Epidemiol. Infect. 99:439–443.
74. International Bottled Water Association. 1995. Model Bottled Water Regulation. IBWA, Alexandria, VA.
75. Keswick, B. H., and C. P. Gerba. 1980. Viruses in ground water. Environ. Sci. Technol. 14:1290–1297.
76. Kolbel-Boelke, J., E.-M. Anders, and A. Nehrkorn. 1988. Microbial communities in the saturated groundwater environment. II. Diversity of bacterial communities in a Pleistocene sand aquifer and their in vitro activities. Microb. Ecol. 16:31–48.
77. Kotz, D. 1995. How safe is your water? Good Housekeeping November, pp. 128–130, 213–214.
78. Krovacek, K., M. Peterz, A. Faris, and I. Mansson. 1989. Enterotoxigenicity and drug sensitivity of *Aeromonas hydrophila* isolated from well water in Sweden: A case study. Int. J. Food Microbiol. 8:149–154.
79. Lambert, V. 1993. Bottled water—New trends, new rules. FDA Consumer June, pp. 8–11.
80. Landsberg, J. D., F. W. Bodyfelt, and M. E. Morgan. 1977. Retention of chemical contaminants by glass, polyethelene, and polycarbonate multiuse milk containers. J. Food Prot. 40:772–777.
81. Leclerc, H. 1980. Bacteriological characterization of mineral water produced at Evian, in De L'Eau Minérale d'Evian. S. A. des Eaux d'Evian, Evian, France, pp. 24–27.
82. Leclerc, H., D. A. A. Mossel, and C. Savage. 1985. Monitoring non-carbonated ("still") mineral waters for aerobic colonization. Int. J. Food Microbiol. 2:341–347.

83. Lennette, E. H., A. Balows, W. J. Hausler, and J. P. Truant. (eds.). 1980. Manual of Clinical Microbiology, 3rd ed. American Society for Microbiology, Washington, DC, 1044 pp.
84. Levesque, B., P. Simard, D. Gauvin, S. Gingras, E. Dewailly, and R. Letarte. 1994. Comparison of the microbiological quality of water coolers and that of municipal water systems. Appl. Environ. Microbiol. 60:1174–1178.
85. Levine, W. C., and G. F. Craun. 1990. Waterborne disease outbreaks, 1986–1988. MMWR 39:1–13.
86. Lippy, E. C., and S. C. Waltrip. 1984. Waterborne disease outbreaks, 1946–1980: A thirty-five-year perspective. J. Am. Water Works Assoc. 76:60–67.
87. Lock, M. A., R. R. Wallace, J. W. Costerton, R. M. Ventullo, and S. E. Charlton. 1984. River epilithon: Towards a structural–functional model. Oikos 44:10–12.
88. Logsdon, G. S. 1988. Comparison of some filtration processes appropriate for *Giardia* cyst removal, in: P. M. Wallis and B. R. Hammond (eds.), Advances in *Giardia* Research. University of Calgary Press, Calgary, Alb; Canada, pp. 95–102.
89. Mavridou, A., M. Papapetropoulou, P. Boufa, M. Lambriri, and J. A. Papadakis. 1994. Microbiological quality of bottled water in Greece. Lett. Appl. Microbiol. 19:213–216.
90. McFeters, G. A. 1989. Detection and significance of injured indicator and pathogenic bacteria in water, in: B. Ray (ed.), Injured Index and Pathogenic Bacteria: Occurrence and Detection in Foods, Water and Feeds. CRC Press, Boca Raton, FL, pp. 179–210.
91. Morais, P. V., and M. S. Da Costa. 1990. Alterations in the major heterotrophic bacterial populations isolated from a still bottled mineral water. J. Appl. Bacteriol. 69:750–757.
92. Moreira, L., P. Agostinho, P. V. Morais, and M. S. Da Costa. 1994. Survival of allocthonous bacteria in still mineral water bottled in polyvinyl chloride (PVC) and glass. J. Appl. Bacteriol., 77:334–339.
93. Mossel, D. A. A. 1976. Various taxo- and ecogroups of bacteria as index organisms for the enteric contamination of bottled waters: Their significance and enumeration. Ann. Ist Super. Sanit. 12:177–190.
94. Mossel, D. A. A., and H. Y. Oei. 1975. Person-to-person transmission of enteric bacterial infection. Lancet i:751.
95. Mossel, D. A. A., J. E. L. Corry, C. B. Struijk, and R. M. Baird (eds.). 1995. Bottled Waters, in: Essentials of the Microbiology of Foods. A Textbook for Advanced Studies. John Wiley & Sons, Toronto, Ont., Canada, pp. 391–394.
96. National Association of People with AIDS. 1995. Should I Be Concerned About the Water I Drink? NAPWA, Washington, DC.
97. Nestlé. 1997. Nestlé's home page. *http://www.nestle.com/html/h1.html*
98. Nunes, K. 1997. Foodborne illness. Meat Poult. 2:22–23.
99. Oger, C., J. F. Hernandez. J. M. Delattre, A. H. Delabroise, and S. Krupsky. 1987. Study by epifluorescence of the fate of total bacterial flora in a bottled mineral water. Water Res. 21:469–474.
100. Pavia, A. T. 1987. Travel to the Soviet Union: Is diarrhea a risk? JAMA 258:1661.
101. Payment, P. 1996. Should we regulate the bacterial heterotrophic plate count (HPC)

in drinking water? in: Planning for Tomorrow. Proceedings of the Sixth National Conference on Drinking Water, Victoria, BC, Canada, Oct. 16–18, 1994. American Water Works Association, Denver.

102. Payment, P., F. Gramade, and G. Paquette. 1988. Microbiological and virological analysis of water from two water filtration plants and their distribution systems. Can. J. Microbiol. 34:1304–1309.
103. Payment, P., F. Gamache, and G. Paquette. 1989. Comparison of microbiological data from two water filtration plants and their distribution system. Wat Sci. Technol. 21:287–289.
104. Payment, P., E. Franco, L. Richardson, and J. Siemiatycki. 1991. Gastrointestinal health effects associated with consumption of drinking water produced by point-of-use domestic reverse-osmosis filtration units. Appl. Environ. Microbiol. 57:945–948.
105. Payment, P., L. Richardson, J. Siemiatycki, R. Dewar, M. Edwardes, and E. Franco. 1991. A randomized trial to evaluate the risk of gastrointestinal disease due to the consumption of drinking water meeting current microbiological standards. Am. J. Public Health 81:703–708.
106. Pollack, M. 1983. The role of exotoxin A in *Pseudomonas* disease and immunity. Rev. Infect. Dis. 5:5979–5984.
107. Riveria, F., M. Glavan, E. Robles, P. Leal, L. Gonzalez, and A. M. Lacey. 1981. Bottled mineral waters polluted by protozoa in Mexico. J. Protozool. 28:54–56.
108. Rosenberg, F. A. 1990. The bacterial flora of natural mineral waters and potential problems associated with its ingestion. Riv. Ital. Ig. 50:301–310.
109. Rosenberg, F. A., H. Zakaria, and M. Rose. 1989. Effect of storage on the microbiology of bottled water. Abstracts of the ASM Annual Meeting, 1989. Abstr. Q-229, p. 368.
110. Ruskin, R. H., J. H. Krishna, and G. A. Beretta. 1991. Microbiological quality of selected bottled water brands in the U.S. Virgin Islands. Abstracts of the 91st General Meeting of the American Society for Microbiology, May 5–9, 1991, Dallas, TX, Q-246, p. 317.
111. Schmidt-Lorenz, W. 1976. Microbiological characteristics of natural mineral water. Ann. Ist. Super. Sanit. 12:93–112.
112. Sekla, L., J. Drummond, D. Milley, W. Stackiw, D. Sargeant, J. Drew, and J. Sisler. 1990. Are the alternatives to municipal waters truly safer? Can. Dis. Wkly. Rep. 16:223–226.
113. Slater, D. 1988. Bottled waters the beverage of the future. Dairy Food Environ. Sanit. 8:303–304.
114. Stamm, J. M., W. E. Engelhard, and J. E. Parsons. 1969. Microbiological study of water-softener resins. Appl. Microbiol. 18:376–386.
115. Stickler, D. 1989. The microbiology of bottled natural mineral waters. J. Soc. Health., 109:118–124.
116. Stiles, M. E. 1989. Less recognized or presumptive foodborne pathogenic bacteria, in: M. P. Doyle (ed.), Foodborne Bacterial Pathogens. Marcel Dekker, New York, pp. 689–692.
117. Terry, S. 1993. Troubled water. New York Times Magazine Sept. 26, pp. 1–46, 62–63.

118. Tobin, R. S. 1989. Microbiology of potable water in Canada: An overview of the Health and Welfare Canada program. Toxicity Assess. 4:257–270.
119. Tobin, R. S., D. K. Smith, and J. A. Lindsay. 1981. Effects of activated carbon and bacteriostatic filters on microbiological quality of drinking water. Appl. Environ. Microbiol. 41:646–651.
120. U.S. General Accounting Office. 1996. Food Safety. Information on Foodborne Illnesses. Washington, DC. GAO/RCED-96-96. 31 pp.
121. Walker, J. T., C. W. Mackerness, J. Rogers, and C. W. Keevil. 1995. Heterogeneous mosaic biofilm—A haven for waterborne pathogens, in: H. M. Lappinscott and J. W. Costerton (eds.), Microbial Biofilms. Cambridge University Press, Cambridge.
122. Wallis, C., C. H. Stagg, and J. L. Melnick. 1974. The hazards of incorporating charcoal filters into domestic water systems. Water Res. 8:111–113.
123. Wallis, P. M., S. L. Erlandsen, J. L. Isaac-Renton, M. E. Olson, W. J. Robertson, and H. Van Kuelen. 1996. Prevalence of *Giardia* cysts and *Cryptosporidium* oocysts and characterization of *Giardia* spp. isolated from drinking water in Canada. Appl. Environ. Microbiol. 62:2789–2797.
124. Warburton, D. W., P. I. Peterkin, K. F. Weiss, and M. A. Johnston. 1986. Microbiological quality of bottled water sold in Canada. Can. J. Microbiol. 32:891–893.
125. Warburton, D. W., K. L. Dodds, R. Burke, M. A. Johnston, and P. J. Laffey. 1992. A review of the microbiological quality of bottled water sold in Canada between 1981 and 1989. Can. J. Microbiol. 38:12–19.
126. Warburton, D. W. 1993. A review of the microbiological quality of bottled water sold in Canada. Part 2. The need for more stringent standards and regulations. Can. J. Microbiol. 39:158–168.
127. Warburton, D. W., J. K. McCormick, and B. Bowen. 1994. Survival and recovery of *Aeromonas hydrophila* in water: Development of methodology for testing bottled water in Canada. Can. J. Microbiol. 40:145–148.
128. Warburton, D. W., B. Bowen, and A. Konkle. 1994. The survival and recovery of *Psedomonas aeruginosa* and its effect upon salmonellae in water: Methodology to test bottled water in Canada. Can. J. Microbiol. 40:987–992.
129. Warburton, D. W., J. W. Austin, B. H. Harrison, and G. Sanders. 1998. Survival and recovery of *Escherichia coli* O157 in inoculated bottled water. J. Food Prot. 61(8): 948–952.
130. Warburton, D. W., and J. W. Austin. 1997. Bottled Water, in: B. M. Lund, A. C. Baird-Parker, G. W. Gould (eds.), Microbiology of Food. Chapman & Hall. London.
131. Warburton, D. W., B. Harrison, C. Crawford, C. Fox, L. Gour, R. Foster, and P. Krol. 1998. A further review of the microbiological quality of bottled water sold in Canada; 1992–1997 survey results. Int. J. Food Microbiol. 39:221–226.
132. Weber, G., H.-P. Werner, and H. Matschnigg. 1971. Death cases in newborns caused by *Pseudomonas aeruginosa* contaminated drinking water. Zbl. Bakt., I. Abt. Orig. 216:210–214.
133. WHO. 1984. Guidelines for Drinking-Water Quality, Vol. 1, Recommendation. World Health Organization, Geneva.
134. WHO. 1984. Guidelines for Drinking-Water Quality, Vol. 2, Health Criteria and Other Supporting Information. World Health Organization, Geneva.
135. Wilson, C. 1991. Hitting the bottle. Food Canada. 51:14–17.

136. Wilson, J. T., J. F. McNabb, D. L. Balkwill, and W. C. Ghiorse. 1983. Enumeration and characterization of bacteria indigenous to a shallow water table aquifer. Ground Water 21:134–142.
137. Yurdusev, N., and R. Ducluzeau. 1985. Qualitative and quantitative development of the bacterial flora of Vittel mineral water in glass or plastic bottles. Sci. Aliment. 5:231–238.
138. Zmirou, D., J. P. Ferley, J. F. Collin, M. Charrel, and J. Berlin. 1987. A follow-up study of gastrointestinal diseases related to bacteriological substandard drinking water. Am. J. Public Health 77:582–584.

17

The Use of the Internet for Food Safety Information and Education

Jeffrey M. Farber
Health Canada, Ottawa, Ontario, Canada

Don Schaffner
Rutgers University, New Brunswick, New Jersey

I. INTRODUCTION

There are numerous published reviews, books, educational pamphlets, videotapes, and software programs dealing with various aspects of food safety. In many cases, there is much duplication of material, and attempts should be made to coordinate efforts in this area. This is already happening in Canada and in the United States, where national food safety education programs are being developed. This chapter points out good sources of food safety education material focusing on the newest source of material presently available, the Internet.

With the advent of the Internet, the world has grown much smaller for those involved in food safety. Numerous Internet news servers and electronic discussion groups have emerged to give everyone who is interested a daily "blow by blow" description of what is happening around the world. Although this has dramatically speeded up the rate at which we hear food safety information, the downside is that we have become inundated with information. In addition, there is a lot of good and not so good information out there. So while we are getting "food safety breaking stories" much faster these days, overall these stories are less reliable. That is, there can be problems in authenticating what you see on the Internet.

Before describing some of the best Internet sites to access for food safety information, we present a short description of Internet nomenclature for the benefit of those not familiar with the terminology of cyberspace.

The Internet is an extensive worldwide system of interlinked but independent computer networks now being used for many purposes, but originally created by the United States Department of Defense to protect the country's computer networks in the event of a war. The day-to-day functions of the Internet are carried out by 13 computers located in the United States called root servers, which translate addresses such as *http://www.usda.gov/fsis/* into the numeric code used by the system's basic software. The World Wide Web and electronic mail (e-mail) are the two most important technologies on the Internet. The World Wide Web (more commonly known as WWW or the Web), is the fastest growing part of the Internet and provides an easy entry point to it. Web pages can display text, video, sound, pictures and animated graphics. Originally developed at CERN, the European Laboratory for Particle Physics, the Web was created to provide a simple system for scientists to exchange all types of data, such as text, graphics, or illustrations. One common feature of Web pages and really the key to the WWW is the use of hypertext, which is a way of creating and showing text in a way that lets it be linked in multiple ways and available at various levels of detail. Hypertext allows one to "jump" from one place to the next through a series of "links" created by someone. Once you have called up a certain home page, all you have to do is use your mouse to move the arrow on your screen over a word that is usually underlined and in color, until you see the arrow change to a little hand or other similar icon. At this point, you know you are over a hypertext link, and upon clicking once with your left mouse button on this word(s), you will be automatically transferred to another screen containing information related to that link. A link can literally connect a chapter on foodborne pathogens saved on a computer in the U.S. Department of Agriculture (USDA) in Washington, with a chapter on waterborne pathogens on a computer in Auckland, New Zealand. Thus, just by clicking on a word, you can move from one page of information to another; clicking on a link brings up a new screen containing information related to the previous link.

The behind-the-scenes work of the Web is done by means of an Internet protocol called HTTP or HyperText Transport Protocol. HTTP is very efficient because it does not have any "search" functions to slow it down. One simply clicks on a word and another screen of information appears, from anywhere in the world. The language that HTTP "speaks" best is called HTML (HyperText Markup Language), which is a simpler subset of another powerful language (SGML: Standardized Generalized Markup Language), which is used for "tagging" documents for electronic format and structural uses. Any computer on the Internet that speaks HTTP is referred to as a "Web server," and any computer that can access that server is a "client."

Getting onto the Web requires a software tool called a browser or client. To access a browser you can either have one installed on the computer you are using to access the Internet, or telnet (i.e., transfer) to another computer on the Internet that offers public access to its WWW browser. If your computer is connected directly to the Internet through a LAN or dial-up (SLIP/PPP) connection, you can choose any one of the many graphical browsers currently on the market such as Netscape or Microsoft Explorer. In addition, you may be required to use another type of software if you sign up with one of the many major commercial online services. To start accessing some of these food safety sites, open your favorite browser, go to "File," and then "Open Location," and then type in one of the uniform resource locators (URLs) given in this chapter. URLs are "addresses" that specify the location of something on the Internet. An example of a URL in a standard form is:

http://www.usda.gov/

The part of the URL before the colon indicates the access method for the type of resource you want to retrieve. In the example above, the http at the beginning of the address tells you and the Internet that you are looking for a WWW server. Many URLs for Web pages include www after the *http://.* (Other addresses may start with file, telnet, news, ftp, or gopher, which are used for accessing or transfer of specific files, transferring to login screens on remote computers, or reaching newsgroups). The part of the URL following the double slash specifies a machine name or site, in the case above, the U.S. Department of Agriculture. The last part of the address will tell you something about the location or type of site you are in. In the example given, *.gov* stands for government. When *.com* is used at the end, this signifies that the addressee is a company (commercial), not an institution or government agency. Other common endings, also referred to as domains, are names of countries: *.ca* (Canada), *.uk* (United Kingdom), *.us* (United States). Currently there are more than 200 country domains. Other common domain names designate universities (*.edu*), nonprofit organizations (*.org*), networks (*.net*), and military (*.mil*) entities.

One should be aware that Web sites constantly change. If a URL does not work, check each character to make sure that you have typed in the address correctly. You can also try deleting some of the filename and/or directories, and then track down the information you are looking for by following the links. For example, suppose that the following URL does not work: *http://bigcompany.com/food/safety.html.* You can then try *http://bigcompany.com/food/,* or just *http://bigcompany.com.* The sections that follow list Web sites available by food commodity, area, company, and so on.

II. WEB SITES

A. Non Profit Organizations

1. Government/Federal Institutions/Organizations

http://www.cdc.gov/ This is the official home page of the U.S. Centers for Disease Control and Prevention (CDC) based in Atlanta. It is an excellent home page that contains general information about CDC, as well as a great deal of information about the latest news, health information, travelers' health, publications [e.g., the journals *Morbidity and Mortality Weekly Reports (MMWR)* and *Emerging Infectious Diseases*], products, data and statistics, training and employment, and funding.

http://www.fda.gov/ This is the official home page of the U.S. Food and Drug Administration, another excellent site for all information related to food safety. The home page will provide you with a link to the food area (*http://vm.cfsan.fda.gov/list.html*). The food area of the FDA Web site covers topics such as programs, special interest areas (includes the prime connection system), ideas on how to interact with the FDA Center for Food Safety and Applied Nutrition, as well as other sources of food safety information. Some of the headings listed under "Programs" include biotechnology, cosmetics, dietary supplements, food additives and premarket approval, the very popular bad bug book (Foodborne Pathogenic Microorganisms and Natural Toxins), food labeling and nutrition, imports, exports, inspections, recalls and HACCP, pesticides and chemical contaminants, and seafood. There are even food safety messages that move across the bottom of the screen as you view it.

http://www.usda.gov/ This is the official home page of the U.S. Department of Agriculture. From the home page, you can click on "agencies" to get you to a listing of the Agricultural Research Service (ARS. *http://www.ars.usda.gov/*) and the Food Safety and Inspection Service (FSIS: *http://www.usda.gov/agency/fsis/homepage.htm*), both excellent sites. Included in the FSIS home page are topics such as innovations ("What's New?"), the agency's mission and activities, publications, consumer education and information, organization/program areas, and HACCP implementation.

http://www.cfia-acia.agr.ca/ The official home page of the Canadian Food Inspection Agency provides direct links to such diverse topics and areas as acts and regulations, agricultural areas, animal health and disease surveillance, feeds, plants and plant health, seeds, biotechnology in agriculture (including its regulation), inspection manuals and codes of practice, a meat hygiene manual of procedures, guides to food labeling, news about food recalls, food and seafood safety fact sheets, risk assessments, and related sites.

http://www.hc-sc.gc.ca/ This is the Health Canada home page. Click on the word "Health" to get a listing of all the branches related to science. Of interest to food safety professionals will be the Laboratory Centre for Disease Control home page (*http://www.hc-sc.gc.ca/hpb/lcdc/hp eng.html*), Canada's equivalent to the CDC, and the Food Program, which is responsible for health and safety as related to food consumption. In the LCDC home page, under the "Disease Prevention and Control Guidelines," one can find topics such as biosafety, HIV/AIDS, infection control, perinatal health, travel health/quarantine, vaccines, and vaccine-preventable diseases. A Food Program home page (*http://www.hc-sc.gc.ca/food-aliment*) has now been set up.

http://www.foodsafety.org/ This Web site is the home of the U.S. National Food Safety Database, a project that catalogs many food safety–related training and educational materials available today. A quick glance at the home page will reveal topics such as consumer-related food safety materials, industry-related food safety materials, educator/trainer-related materials, and a food safety quiz. In the section on consumer-related food safety materials, one finds topics such as additives and residues in food; canning, freezing, and drying; consumer perceptions of food risks; disaster preparation and prevention; foodborne illnesses; general food safety information; glossary of food terms; informational resources; labeling; and a big section on food irradiation. There is also specific information available for meat and poultry, produce, and seafood processors.

http://www.nal.usda.gov/fnic/foodborne/foodborn.htm This site, which is supported by the U.S. Department of Agriculture Food Safety and Inspection Service and the Food and Drug Administration Center for Food Safety and Applied Nutrition, is the USDA/FDA Foodborne Illness Education Information Center. Among many other things, there is a USDA/FDA Foodborne Illness Educational Materials Database that contains descriptions and ordering information for consumer educational materials and food worker training resources, including games and teaching guides for elementary and secondary school education, training materials for the management and workers of retail food markets, food service establishments, and institutions; audiovisuals materials; and posters. There is also an interactive electronic discussion group called "Foodsafe," which is intended as a communication tool to link professionals interested in food safety issues. You can subscribe online to a mailing list, which directs messages to be delivered to your Internet e-mail address. Be careful, this is a "high traffic" list, and if you subscribe you may get 30 or more messages a day! If you do subscribe to this list, also be sure to save the information on how to unsubscribe. There are also foodborne illness education information center publications and various links to FDA and USDA food safety sites. This is an excellent site.

http://www.phls.co.uk/ The official home page of the U.K. Public Health Laboratory Service provides links to the Communicable Disease Surveillance Centre (CDSC) home page, the Central Public Health Laboratory home page, and access to the popular Communicable Disease Report, or CDR, which is CDSC's main vehicle for published information. It also provides an alphabetized list of links to other sites.

http://www.nfpa-food.org/ This is the home page of the National Food Processors Association (NFPA), which is the largest U.S. food trade association and the scientific voice for the U.S. food processing industry on scientific and public policy issues. A good link entitled "Food Science" provides fact sheets on selected areas of food safety [e.g., juice and *E. coli,* NFPA opinions and positions reflected in editorials published in *Food Quality* magazine, technical assistance (for members only), and updates on the status of NFPA research projects].

http://www.sfpa.sk.ca/ This is the home page of the Saskatchewan Food Processors Association. One can locate Saskatchewan food processors currently involved in the production of specific food products. As well, one can find links to Saskatchewan food processing company sites, as well as to leading Canadian and U.S. food industry Web sites.

http://foodnet.fic.ca/alliance/cfta.html This is the official home page of the Canadian Food Trade Alliance, which includes some of the key national and provincial food industry associations. On the home page, one can find links to the Food Institute of Canada, Canada Pork International, Food Beverage Canada, Further Poultry Processors Association, Canadian Poultry & Egg Processors Council, Canadian Council of Grocery Distributors, Saskatchewan Food Processors Association, and the Manitoba Food Processors Association.

2. Other Government Sites of Interest

http://www.epa.gov/ The home page of the U.S. Environmental Protection Agency.

http://www.cfis.agr.ca/ The home page of the Canadian Food Inspection Service.

http://www.econ.ag.gov/ The home page of the Economic Research Service of the USDA—the official source for economic analysis and information on agriculture, food, natural resources, and rural America.

http://www.fedstats.gov/ A searchable site from the more than 70 agencies in the U.S. federal government that produce statistics of interest to the public. The Federal Interagency Council on Statistical Policy maintains this site.

http://waffle.nal.usda.gov/agdb/ The home page of the Agriculture Network Information Center. The AgDB is a database directory of quality agriculture-related databases, datasets, and information systems.

http://www.nih.gov/ The home page of the U.S. National Institutes of Health.

http://www.access.gpo.gov/su_docs/aces/aces140.html/ The online version of the U.S. Federal Register.

http://europa.eu.int The home page of the European Union (EU).

http://www.fsai.ie The home page of the Food Safety Authority of Ireland. Quite good!

3. Food Trade and Consumer Organizations

http://www.ilsi.org/ This is the official home page of the the International Life Sciences Institute (ILSI), which is a nonprofit, worldwide foundation established in 1978 to advance the understanding of scientific issues relating to nutrition, food safety, toxicology, and the environment. ILSI is headquartered in Washington, D.C., and has branches worldwide. ILSI funds research in several areas and also organizes symposia on cutting edge topics.

http://www.apha.org/ The American Public Health Association (APHA), the oldest and largest organization of public health professionals in the world, is concerned with a wide range of issues affecting personal and environmental health. APHA is highly visible through its scientific and practice programs, publications, annual meeting, and advocacy efforts. Familiar publications include the *American Journal of Public Health,* the very popular food microbiology text *Compendium of Methods for the Microbiological Examination of Foods,* and the association's *Control of Communicable Diseases Manual.*

http://www.campden.co.uk/ This is the official home page for the Campden and Chorleywood Food Research Association (CCFRA), which is the largest independent membership-based organization in the United Kingdom, carrying out research and development for the food and drinks industry worldwide. Research at CCFRA covers the whole spectrum of the food industry from food safety through product development to quality management. On the home page, one can find links to information about CCFRA, research, training, contract services, publications, and software as well as links to hot sites. Access to certain information such as the full-length research publications is available to members only.

http://www.netins.net/showcase/cast/ This is the official home page of the Council for Agricultural Science and Technology (CAST), in Ames, Iowa, whose mission is to identify food and fiber, environmental, and other agricultural issues and to interpret related scientific research information for legislators, regulators, and the media for use in public policy decision making. CAST is known as the science source for food, agricultural, and environmental issues. In 1994, it published a very well-known report on foodborne pathogens, which is now being updated.

http://www.fao.org/ This is the home page of FAO, the Food and Agriculture Organization of the United Nations, which was founded in October 1945 with a mandate to raise the levels of nutrition and standards of living, to improve agricultural productivity, and to better the condition of rural populations. FAO is the largest independent agency within the United Nations. On the home page, you will find interesting topics such as statistical databases, nutrition, economics, fisheries, sustainable development, partnership programs and technical cooperation. In addition, one can find news and highlights and in-depth highly focused articles. Also included on the home page are links to Codex Alimentarius (an international food standards "organization") and FAO/WHO food standards.

http://www.who.org/ The home page of the World Health Organization provides topic links to information sources, WHO reports, general information about WHO, and links to other UN Web sites. One of the best sites of interest for food safety professionals is under "Information Sources," where there are links to

"Health Topics" (listing of health topics and associated WHO programs), current information such as press releases and fact sheets, documents published by the organization, and health-related World Wide Web links. Areas covered as health topics include communicable/infectious/tropical diseases, vaccine-preventable and noncommunicable diseases, environment and lifestyle, family and reproductive health, and health policies and systems. See the topic *Environment and Lifestyle* to find information on food safety (http://www.who.ch/fsf/). Included in this interesting site are links to information on safe food for travelers, publications and documents on food safety, as well as links to many other sites on food safety.

http://www.asmusa.org/ This is the home page of the American Society for Microbiology (ASM), the oldest and largest single life science society, which has over 40,000 members located throughout the world. ASM represents 24 disciplines of microbiological specialization, plus a division for microbiology educators. This site contains a wealth of information and is a must for any microbiologist. Besides all the general information, you can search for members' addresses and phone numbers, see the very popular journal *ASM News* online, register for the Annual Meeting, and give feedback on previous meetings. For educators, there is online and e-mail access to *The Focus on Microbiology Education,* a newsletter published three times a year.

http://nrcbsa.bio.nrc.ca/~csm/ The home page for the Canadian Society of Microbiologists publishes information about the organization, its members, upcoming events, membership requirements, publications, and the annual conference.

http://home.vicnet.net.au/~asm/welcom2.htm The home page of the Australian Society for Microbiology has links to information about the society and to publications, conferences, meetings, seminars, and general sites of interest.

http://www.ift.org/ This is the official home page of the well-known Institute of Food Technologists (IFT), which is a nonprofit scientific society with 28,000 members working in food science, food technology, and related professions in industry, academia, and government. The mission of the organization is to support improvement of the food supply and its use through science, technology, and education. IFT has affiliations with food science and technology associations throughout the world, and it publishes the popular journal *Food Technology,* as well as the many position papers (Scientific Status Summaries). The home page makes available information about all the meetings, workshops, and publications of the society.

http://www.easynet.co.uk/ifst/ This is the home page of the Institute of Food Science & Technology (IFST). Although based in the United Kingdom, IFST has members throughout the world and is keen to strengthen these links. IFST is an independent incorporated professional qualifying body for food scientists and technologists. Its main modus operandi is to serve the public interest, both nationally and internationally, by furthering the application of science and technology to all aspects of the supply of safe, wholesome, nutritious, and attractive food. There is a lot of very good information at this site, such as IFST's Position Statements on hot food-related topics (like BSE, BST, irradiation, *Cryptosporidium,* and many others). One can also access links to numerous useful external food-related resources on the World Wide Web, order IFST publications, and so on.

http://www.maff.gov.uk/ This is the home page of the U.K. Ministry of Agriculture, Fisheries, and Food (MAFF).

http://www.cma.ca/ This is the home page of the Canadian Medical Association.

http://www.pasteur.fr/welcome-uk.html This is the home page of the Institut Pasteur in France.

http://www.iamfes.org/ This is the home page of the International Association of Milk, Food and Environmental Sanitarians. IAMFES publishes two scientific journals, *Dairy, Food and Environmental Sanitation* and *Journal of Food Protection.* One can view the table of contents as well as the abstracts from the *Journal of Food Protection,* obtain information about the video lending library, and link to other food safety sites.

4. Universities

http://www.mit.edu:8001/people/cdemello/univ.html This one site is all you really need to access college and university home pages from around the world. Its alphabetical listing has over 3000 entries. A great site.

B. Food Industry

1. Food Processors

Many familiar food companies have home pages on the Web. Just enter the name of a company in the field. As long as you are searching the Web and the URL ends in *.com,* you only have to type the name you are looking for in the location or file open box.

http://www.dole.com/ The home page for Dole Foods comes complete with information on the company as well as job openings. The "5 a day" site (*http://www.dole5aday.com*), which features Bobby the banana moving on a skateboard, as well as a nutrition center, is really fun.

http://www.mmm.com/ This is the home page for 3M. A good search engine is available by which you can search by concept or keywords.

http://www.kraftfoods.com/ This is a really fun site that brings you into Kraft's interactive kitchen. This site is full of information, recipes, nutrition information, new products, and so on. There is also information on Kraft Canada (*http://www.kraftfoods.com/canada/*) and the Oscar Mayer virtual lunchbox tour (*http://www.oscar-mayer.com/*). The food experts at the Kraft Creative Kitchens have also put together a month of different meal ideas, for those of us who have trouble thinking of what to make for supper each night!

http://www.unilever.com/ Home page for Unilever, a well-known manufacturer of products such as tea, frozen foods, and dairy products. It also has links to DiverseyLever, Unilever's global cleaning and hygiene business.

http://www.nestle.com/ This is the home page for the Nestlé food company. A good search engine is available by which you can search by concept or keywords.

2. Foodservice

http://www.mcdonalds.com/ Home page of the famous McDonald's chain of restaurants.

http://www.burgerking.com/ Home page for the Burger King chain of restaurants.

http://www.wendys.com/ Home page for the Wendy's fast-food chain.

http://www.pepsi.com/ Home page for the soft drink giant.

http://www.jackinthebox.com/ Home page for the Jack in the Box restaurants, with a separate link to food safety.

3. Food Industry Support and Service Companies

http://www.qualiconweb.com/ This is the home page of Qualicon, Inc., the DuPont subsidiary dedicated to improving commercial diagnostics in food, pharmaceutical, and personal care products. You may be familiar with the RiboPrinter Microbial Characterization and the BAX Pathogen Detection Systems, which are the first commercial products to use the polymerase chain reaction (PCR) for the detection of pathogenic bacteria in food and environmental samples.

http://www3.vicam.com/VICAM/ VICAM supplies many kits for the food industry, such as ListerTest and OchraTest, as well as *Salmonella* testing and tests for other microorganisms.

http://www.biomerieux-vitek.com/ The home page of bioMerieux-Vitek, a supplier of automated microbiology and immunoassay systems for clinical microbiology and industrial quality control labs, offers topics such as "The Wide World of Bugs," "Clinical Laboratory Systems," and "bioMerieux Vitek in Space." BioMerieux makes the popular automated VIDAS ELISA machine.

http://www.silliker.com/ This is the home page of Silliker Labs, which provides lab services and technical support to the Food Industry. Silliker's resource section provides a good collection of technical and regulatory information, white papers, articles, and related links. You can also view the latest issue of *SCOPE,* a quarterly technical bulletin. There are also very good related links in "food science," an HACCP question and answer guide.

http://www.abcr.com/ This is the home page of ABC Research, which provides a wide range of analytical and technical services to the food industry. There are links to areas such as microbiology, applied biotechnology (ELISA tests), HACCP, and sanitation.

http://www.atcc.org/ This is the home page for the American Type Culture Collection, a repository for thousands of reference strains of microorganisms. ATCC also provides biological products, technical services, and educational programs to organizations around the world.

http://www.dynal.no/ This is the home page for Dynal, the company that specializes in the use of biomagnetic separation for separating foodborne pathogens from broth or foods.

http://www.horizonpress.com/gateway/commercial.html Click on this site for a comprehensive listing of all general lab suppliers such as Biorad, Fisher, Beckman, and Millipore.

C. Publishing and Conferences

1. Journals—Scientific

http://www.cdc.gov/ncidod/eid/index.htm This is the homepage for *Emerging Infectious Diseases,* a published four times a year by the National Center for Infectious Diseases, Centers for Disease Control and Prevention (CDC), in Atlanta. This journal provides the latest research and information on emerging infectious diseases.

http://www.fas.org/promed/ This is the homepage for ProMED. ProMED is really not a journal, but rather an electronic information resource. ProMED is a project of the Federation of American Scientists put together to promote the establishment of a global program for monitoring emerging diseases. See the ProMed entry in Sec. III. A if you want to subscribe to the ProMED newsgroup.

http://www.foodincanada.com/ Home page for the magazine *Food in Canada.*

http://www.sciencenews.org/ This is the home page for *Science News,* and a good general information on science-related issues. Has links to "Science Safari in Cyberspace," leading to a site called "Cells Alive!" This latter site (*http://www.cellsalive.com/index.htm*) contains time-lapse movies showing, for example, how *E. coli,* given a suitable environment for growth, can divide and form a colony of hundreds of bacteria in just a few hours.

http://www.sciencemag.org/ This is the home page for *Science,* the popular global research weekly, published by the American Association for the Advancement of Science. "Science Online" is published with the assistance of Standford University's High Wire Press.

http://194.216.217.166/bmj/ Home page of the *British Medical Journal* (*BMJ*).

http://highwire.stanford.edu/ This site from High Wire Press, Stanford University Libraries, is an excellent site that provides links to many journals. Some of the journals include the *Journal of Bacteriology* (*http://asmusa.edoc.com/jb/*), *Journal of Clinical Microbiology* (*http://asmusa.edoc.com/jcm/*), *Clinical Microbiology Reviews* (*http://asmusa.edoc.com/cmr/*), *Infection and Immunity* (*http://asmusa.edoc.com/iai/*), *Applied and Environmental Microbiology* (*http://asmusa.edoc.com/aem/*), and the *Annual Review* series (*http://www.AnnualReviews.org/ari/*).

http://www.hbuk.co.uk/ap/journals/fd.htm Home page for *Food Microbiology.* Includes information on the editorial board, instructions for authors, specifications for electronic submissions, and current table of contents.

http://www.cisti.nrc.ca/cisti/journals/microep.html Home page for the *Canadian Journal of Microbiology.*

http://www.ama-assn.org/public/journals/jama/jamahome.htm The home page for the *Journal of the American Medical Association.* It includes a current issue table of contents, past issues of *JAMA,* book and journal reviews, and instructions for authors.

http://www.cdc.gov/epo/mmwr/mmwr.html/ Home page for the popular *Morbidity and Mortality Weekly Report* from the CDC. This site contains a valuable searchable index of *MMWR* publications from 1993 through the present. It also includes interactive *MMWR* morbidity and mortality tables, bulletins from around the world, and case definitions for infectious conditions under public health surveillance.

http://www.journals.uchicago.edu/JID/home.html Home page for the *Journal of Infectious Diseases.*

http://www.thelancet.com/ Home page for the British Journal *The Lancet* (interactive format).

http://www.nature.com/ Home page for *Nature* magazine.

http://asmusa.edoc.com/jcm/ Home page for the *Journal of Clinical Microbiology.*

http://www.nejm.org/ Home page for the *New England Journal of Medicine* online. Has a search function. You can receive each week's table of contents by e-mail.

http://www.elsevier.nl:80/inca/publications/store/5/0/6/0/3/4/ Home page for the *Journal of Microbiological Methods.*

http://www.newscientist.com/ Home page for *New Scientist* magazine.

http://www.blackwell-science.com/~cgilib/jnlpage.bin?Journal=LAM&File=LAM&Page=aims/ Home page for *Letters in Applied Microbiology.*

http://www.library.tudelft.nl/gids/eng/infotype/ejournal.htm A great site for finding all about electronic journals. Contains a full listing of hundreds of freely accessible e-journals.

http://www.mco.edu/lib/instr/libinsta.html This home page contains a list of links to Web sites that provide instructions to authors for over 2000 journals in the health sciences. A great site for scientists who publish a lot.

2. Book Publishers

http://www.dekker.com/homepage/home top.htm Home page of Marcel Dekker, Inc.

http://www.foodsci.com/ Home page of Chapman and Hall—Food Science and Technology Resource Centre.

http://www.blacksci.co.uk/default.htm Home page of Blackwell Science.

http://www.elsevier.com/ Home page of Elsevier Science.

http://www.nap.edu/ Home page of National Academy Press.

http://www.ietc.ca/polysci/index.htm Home page of PolyScience Publications Inc. The "Food Technology" section includes information on the *Compendium of Analytical Methods,* a manual that provides a ready reference of the methods used by the Health Protection Branch (HPB) of Health Canada. In addition, information is available on *Foodborne and Waterborne Disease in Canada,* a publication that describes all the foodborne incidents occurring in Canada each year from information compiled in HPB by the Foodborne Disease Reporting Centre. See also the HPB "Food Program" home page.

3. Conferences

One can find a wide range of conferences listed on the Web. In many cases, one can register for the conference directly online. Some other general examples include the following.

http://www.iefp.org/ Register for the International Exposition for Food Processors.

http://www.grc.uri.edu/ All the information you need on the Gordon Research Conferences can be found at this site. Information on future meetings is also available.

http://www.isopol.com/ This is the official site for information on the international *Listeria* meetings, which are held every 3 to 4 years.

D. Commodity-Specific Information

1. Dairy

http://www.dairyinfo.agr.ca/ This is the official home page of the Canadian Dairy Information Centre, which is a joint venture between the Dairy Farmers of Canada (DFC), National Dairy Council of Canada (NDCC), Canadian Dairy Commission (CDC), and Agriculture & Agri-Food Canada (AAFC). Clicking on the home page will lead you to information on the highlights of the Canadian dairy industry, dairy market information, FIL-IDF Canada (International Dairy Federation), and Canadian Dairy Extension Committee (CDEC). Other pertinent sites cover issues related to programs and policies, regulations, trade agreements, dairy sites around the world (e.g., eastern and western Europe, Australia, France, New Zealand, Scotland, and the United States), and many statistical sites. This site holds a wealth of information for the dairy enthusiast.

http://www.adsa.uiuc.edu/ The home page of the American Dairy Science Association provides information about the association, its annual meeting, and the popular *Journal of Dairy Science.* It also contains good links to other sites of interest to dairy enthusiasts.

2. Meat

http://www.hooked.net/users/nma/ This is the home page for the National Meat Association (NMA), which is a nonprofit trade association for meat packers and processors, as well as equipment manufacturers and suppliers who provide services to the meat industry. It contains excellent links to many meat-related sites on the Web, such as magazines, companies, and government associations.

http://www.ncanet.org/ This is the home page for the National Cattlemen's Beef Association, the marketing organization and trade association for America's cattle farmers and ranchers.

http://www.abs.sdstate.edu/flcs/foodsafety/menulist/doc/meatsafe.htm This is the home page of the SDSU Department of Animal and Range Sciences, Cooperative Extension Service, South Dakota State University–U.S. Department of Agriculture, which includes links to meat safety.

http://ifse.tamu.edu/alliance/foodsafety.html This is the home page for the International Meat & Poultry HACCP Alliance. At the home page site, food safety information is organized under three main headings: HACCP and food safety, hazards, and a reference corner.

http://www.nzmeat.co.nz/ This is the home page of the New Zealand Meat Board, which is an organization funded entirely by meat producers (i.e., not by government). There are search functions at this site that one can use to search topics related to meat.

http://www.csiro.au/index.html Australia's Commonwealth Scientific and Industrial Research Organisation (CSIRO) is an independent statutory authority constituted and operating under the provisions of the Science and Industry Research Act, 1949. CSIRO's current structure was established by the Science and Industry Research Amendment Act, 1986. CSIRO is actually an agency in the Industry, Science and Tourism government portfolio under the Minister for Industry, Science and Tourism. The CSIRO research areas of interest would be found on the home page under the heading "Agribusiness," where one finds links to topics such as field crops, food processing, meat, dairy, and aquaculture.

http://www.mtgplace.com/ This site serves as an information provider for the red meat and poultry industry. Users must register to access sites.

http://cmsa.ca/ This is the home page of the Canadian Meat Science Association (CMSA). Some of the goals of this society are to promote the application of science and technology to the production, processing, packaging, distribution, preparation, evaluation, and utilization of all meat and meat products, as well as to promote the coordination of educational, research, development and service activities in meat science and related areas. Links on the home page will take you to material covering, among other things, meat-oriented Web sites and an HACCP toolbox.

http://www.cmc-cvc.com/ This is the home page of the Canadian Meat Council, which represents the meat packing/processing industry in Canada.

3. Produce

http://www.thepacker.com/ This is the home page for "The Packer," a site that gives readers access to up-to-date and late-breaking news on all issues affecting the produce industry.

http://www.fresh-cuts.org/ This is the home page of the International Fresh-Cut Produce Association, one of the leading U.S. produce associations, which has over 500 member companies. IFPA represents commercial producers, suppliers, and distributors of fresh-cut produce, as well as affiliated companies such as equipment and packaging manufacturers. IFPA is very active in the food safety arena.

http://www.pma.com/ This is the home page for the Produce Marketing Association, a not-for-profit trade association that serves about 2500 members who market fresh fruits, vegetables, and floral products worldwide.

E. Food-Safety-Related Topics

1. Biotechnology

http://www.wi.mit.edu/bio/biopage2.html This site is compliments of the Whitehead Institute, a nonprofit, independent basic research and teaching institution that specializes in programs in cancer and AIDS research, developmental biology, structural biology, infectious disease, and genetics. The site is excellent and contains loads of information on molecular database searching; that is, topic headings include submitting sequences to databases, text-based searching of databases, sequence/pattern searching via forms, protein structures, and links to lists with many sequence analysis tools.

http://www.ncbi.nlm.nih.gov/ This is the home page for the National Center for Biotechnology Information of the U.S. National Institutes of Health. This is an excellent site for molecular biologists.

http://www.horizonpress.com/gateway/micro.html This home page is also listed below, but is worth mentioning twice. An extremely comprehensive resource for all matters related to molecular biology/microbiology.

2. Food Safety

http://www.fst.vt.edu/haccp97/ This site contains the document entitled "Hazard Analysis and Critical Control Point (HACCP) Principles and Application Guidelines," which was adopted on August 14, 1997, by the National Advisory Committee on Microbiological Criteria for Foods. You must have Adobe Acrobat Reader, version 3.0 or later, to view this document.

http://ificinfo.health.org/ This is the home page for the International Food Information Council (IFIC) Foundation, a nonprofit organization based in Washington, D.C., whose role is to provide sound, scientific information on food safety and nutrition to journalists, health professionals, educators, government officials, and consumers. The home page contains links to information on food safety and nutrition, information for reporters and educators, search sites, and so on. The section on food safety and nutrition is comprehensive and covers many areas, such as food safety, biotechnology, food allergies, food labeling, international food regulation, agriculture, and food production.

http://www.foodinstitute.com/ Home page of the Food Institute, a nonprofit information and reporting association, with members in more than 2200 companies in 50 states and in over 40 foreign countries. This institute is involved with the entire food distribution system, with members including growers, food processors, wholesalers, supermarket chains, independent retailers, food industry suppliers, and foodservice distributors, among others. The institute publishes *Food Safety Issues,* a monthly journal providing fresh insights on food safety issues. To get full benefit from this site, it is necessary to become a member of the institute.

http://www.ces.ncsu.edu/depts/foodsci/agentinfo/index.html This home page is courtesy of the Cooperative Extension Service at North Carolina State University, whose goals are to promote food safety education via the Internet. It is not intended to repeat information that is available elsewhere on the Internet. This very bright and colorful home page has links to major food groups such as dairy, meat, fruits and vegetables, eggs, fish and seafood, as well as organisms of concern. A wealth of information!

3. Public Health/Health

http://www.healthfinder.gov/ This is the home page of Healthfinder, a gateway consumer health and human services information Web site from the U.S. government. The site provides links to select online publications, clearinghouses, databases, Web sites, and support and self-help groups, as well as government agencies and not-for-profit organizations. This is a very good site.

http://www.healthworks.co.uk/ This site contains information on health and medical information for health professionals. U.S. and U.K. residents can access Medline for free. Medline access for Canadians can be done through *http://www.medscape.com.*

http://www.medicinalfoodnews.com/ The home page of HealthGate Data Corp is full of general health information. One can search by subject as well as by using Medline and six other databases.

http://navigator.tufts.edu/ This is the very popular home site of the Nutrition Navigator, compliments of Tufts University. This is an excellent site for nutrition information.

http://www.cyinfo.com/ This is the home page for GIDEON, a medical epidemiology software package. GIDEON (**g**lobal **i**nfectious **d**iseases and **e**pidem**io**logy **n**etwork) is designed to diagnose all the world's infectious diseases (including foodborne ones) based on the symptoms, signs, and laboratory findings that are entered for a patient. While the GIDEON software has a very primitive user interface (it is currently a DOS application that runs under Windows), its cumbersome nature is more than offset by the depth and breadth of the information it provides.

http://www.magnet.state.ma.us/dph/ This is the home page for the Massachusetts Department of Public Health (DPH). At this site, one can find a very good manual on foodborne illness investigation.

http://www.nsf.org/ NSF International is an independent, not-for-profit organization that is committed to public health safety and protection of the environment by developing standards, by providing education, and by providing third-party conformity assessment services. NSF is widely recognized for its scientific and technical expertise in the health and environmental sciences. WHO recently named NSF International a Collaborating Centre for Food Safety.

http://www.eurekalert.org/ EurekAlert! is a comprehensive news server for the latest research advances in science, medicine, and engineering. It is produced by the American Association for the Advancement of Science (AAAS), with technical support provided by Stanford University. This is a very comprehensive service.

http://channels.reed-elsevier.com/sciencertw/elsevierscience/docs/toolbar.asp?_FileName=welcome.htm Another all-science news home page (news and views, hot topics, daily updates, etc.) from Elsevier which is quite good.

http://fester.his.path.cam.ac.uk/phealth/phweb.html The University of Cambridge public health page. A very good site with links to Web sites in public health, epidemiology, biostatistics, and evidence-based health.

http://www.hc-sc.gc.ca/hpb/lcdc/biosafty/msds/index.html This home page of the Office of Biosafety in Health Canada gives information on all Material Safety Data Sheets (MSDS) regulated under Workplace Hazardous Materials Information System (WHMIS) legislation. These MSDS are very useful for personnel working in the microbiology or food microbiology lab as they give quick safety reference material relating to infectious microorganisms. The MSDS are organized to contain health hazard information such as infectious dose, viability, decontamination, medical information, laboratory hazard, recommended precautions, handling information, and spill procedures. There is an alphabetical listing at the top of the page, along with a line listing of microorganisms. A real handy site!

http://hippo.findlaw.com/ This is the home page for "Health Hippo," a good U.S. health policy site. At this site you will find a good collection of policy and regulatory materials related to health care.

http://pages.prodigy.net/pdeziel/ An infectious disease specialty hot site that contains links to topics such as infectious disease journals, online images, medical and microbiology Web sites, travel medicine Web sites, and so on. Great for the infectious disease enthusiast!

4. Microbiology

http://hna.ffh.vic.gov.au/vidrl/Links_page.html#FIDEP An excellent site on Internet microbiology and infectious disease (ID) resources. Contains headings such as culture and biologicals collections, emerging and environmental infectious diseases, regional microbiology/epidemiology reports, foodborne infectious diseases and food poisoning, hepatitis, HIV/AIDS, hospital infection control, Internet/Web resources, journals with important micro/ID papers, molecular microbiology of ID, public health sites, searchable medical/ID databases, travel health, universities and research institutes, miscellaneous microbiology and virology sites, and general WWW/Internet search engines.

http://www.gen.emory.edu/MEDWEB/keyword/infectious_diseases.html This is another very comprehensive site dealing with all areas related to medicine and infectious diseases. For the generalist, there is a wealth of information here related to such wide-ranging topics as epidemiology, genetics and molecular biology, hepatitis, immunology, journal clubs, lists of Internet resources, microbiology and virology, nutrition, and parasitology.

http://www.horizonpress.com/gateway/micro.html This excellent site, aptly named the "Microbiology Jump Station," contains a very comprehensive collection of links for all microbiologists. Some of the topics include gene cloning and analysis, an introduction to molecular biology, an introduction to the Internet for the molecular biologist, jump sites and directories to other sites of interest to the microbiologist, journal links to a wide range of life science journals and books, all types of online information resources, bugs in the news, virtual lectures in microbiology, microbiological institutes and organizations, and microbiology societies.

http://www.exp.ie/flair.html This is the home page for FLAIR-FLOW EUROPE, a specialized dissemination project for the results emerging from EU-sponsored food research and technical development projects. At this Web site, one can find various summaries of EU-funded food R&D projects together with pointers to sources of more detailed information.

5. Risk Assessment

http://www.nal.usda.gov/fnic/foodborne/risk.htm This site contains a nice up to-date listing of references on food safety risk assessment, management, and communication, compiled by the USDA/FDA Foodborne Illness Education Information Center.

http://www.fao.org/waicent/faoinfo/economic/esn/risk/riskcont.htm This FAO site contains all the documents from a report on "Risk Management and Food Safety," a result of a joint FAO/WHO consultation held in Rome, January 27–31, 1997.

http://www.riskworld.com/ The home page for Risk World covers all the news and views on risk assessment and risk management. It contains a wealth of information for risk assessors. On the home page one can find links to topics such as information on newsgroups, software, jobs, news articles, courses, and workshops, all related to risk assessment. There is a very comprehensive listing of risk-related World Wide Web sites. This is an excellent site.

6. Microscopy

http://www.MME-Microscopy.com/education/ This site is the home page for the Microscopy/Microscopy Education (MME), North America's first national consortium of microscopy experts specializing in customized on-site courses, covering all areas of microscopy.

III. OTHER ELECTRONIC FORUMS AND SEARCH TOOLS

A. Mailing Lists (Listservers)

E-mail mailing lists (listservers or listservs in Internet jargon) provide electronic discussion forums that aims to facilitate information sharing, (e.g., workshops, seminars, conferences, and new research) and to promote links, collaborative working, joint problem-solving, and mutual support. One way to see if there is a listserver out there discussing a topic of interest to you is to check out the Liszt mailing list directory web site: *http://www.liszt.com/*. This Web site indexes thousand of discussion groups on many, many topics.

Listservers work automatically using ordinary e-mail. One joins these Internet mailing lists by sending a "subscribe" message, plus the name of the list, as well as your first and last name to the listserver address. For example, to subscribe to the National Food Safety Educators' Net (EdNet), send the message.

"SUBSCRIBE EDNET-L firstname lastname" to *listserv@foodsafety.goy.*

You will remain a "subscriber" until you send an "unsubscribe" message to the listserver address. Be sure to save the "unsubscribe" information you receive when you first subscribe to a list. This is especially important for some lists, which can generate so many messages that the novice user is easily overwhelmed.

One can also post a message or a response to a previous posting by sending an e-mail to the list address, which is not the same as the listserver address. Listed below are some of the key mailing lists in the food safety arena.

Bovine Spongiform Encephalopathy This list is the definitive place to go to chat with international experts on BSE, and to learn about cutting-edge debate in this rapidly expanding field of science.

Listserver address: *listserv@rz.uni-karlsruhe.de*

FOOD-LAW This list is hosted by the University of Minnesota, and typically has fewer than 10 messages a day. This is a good place to go to get those tough food law questions answered. The focus is primarily U.S. and European food law.

Listserver address: *listserv@tc.umn.edu*

FOODSAFE An excellent general international discussion forum on a wide range of food safety matters. Be forewarned, this list can generate 30 messages a day, or more.

Listserver address: *majordomo@nal.usda.gov*

FNSPEC This food and nutrition specialist list is one of the earliest food-related discussion lists on the Internet. Many food-safety-related messages seem to have shifted to FOODSAFE, but this list is still very useful for finding answers to consumer-related food safety and nutrition questions.

Listserver address: *listserv@ecn.purdue.edu*

FSNET-L More a distribution list than a discussion list; daily messages contain updates on food safety topics appearing in the news media and scientific literature. If you need to keep up with the food safety headlines of the day without spending your whole day to do it, FSNET is the list to be on. Sister lists (AGNET and AnimalNET) contain similar summaries for agricultural and animal-related information.

Listserver address: *listserv@listserv.uoguelph.ca*

IFT The purpose of this list is to instantaneously connect competent food scientists to fast-breaking food issues today. This is a closed list, and to join, a current subscriber must sponsor you. For more details, see: http://www.fst.ohio-state.edu/Lee/list-ift.htm

Listserver address:*listserver@lists.acs.ohio-state.edu*

PROMED-MAIL This moderated list deals with reporting and monitoring of all emerging diseases, including foodborne disease. Not every message posted here will be of interest to food scientists, but some will be very useful.

Listserver address: *majordomo@usa.healthnet.org*

RISKANAL This is a discussion list focused on risk analysis, which is broadly defined to include risk assessment, risk characterization, risk communication, risk management, and policy relating to risk, in the context of risks of concern to individuals, to public and private sector organizations, and to society as a local, regional, national, or global level.

Listserver address: *listserv@listserv.pnl.gov*

SEAFOOD The primary purpose of this Internet-based seafood network is to facilitate information exchange about the Hazard Analysis and Critical Control Point (HACCP) system of food safety control in the seafood industry.

Listserver address: *listproc@ucdavis.edu*

B. Newsgroups

Newsgroups are similar to e-mail lists but require specialized software beyond a simple e-mail program. Newsgroup software is now included with many Internet browsers like Netscape or the Internet Explorer. As with e-mail lists, the people participating in the newsgroup usually have a common goal or interest that attracts them to join the group. There are thousands of such groups, with one specifically dedicated to food science: *sci.bio.food-science.*

One of the best ways to search for newsgroups you may be interested in is to access the Deja News Web site located at *http://www.dejanews.com.* Besides the powerful newsgroup search feature, one can also use Deja News to browse groups by category and post messages. The "Power Search" form is a handy feature that will allow you to control which database you search and how your results are displayed.

C. Searching Tools

It is beyond the scope of this chapter to discuss searching techniques on the Internet. Suffice it to say that many powerful search engines can be found on the Internet and that searching has become a way of life for many scientists and food industry professionals. A handy listing of the most commonly used home pages for searching is provided. Readers who are interested in learning more about searching techniques and tools on the Internet can consult recent books on the subject (see References the end of this chapter).

http://www.altavista.digital.com/
http://www.dejanews.com/
http://www.metacrawler.com/
http://www.lycos.com/

http://www.northernlight.com/
http://www.isleuth.com/
http://www.dogpile.com/
http://www.yahoo.com/
http://www.excite.com/
http://www.hotbot.com/
http://www.infoseek.com/
http://www.healthatoz.com/ (specialized search engine for health and medicine)
http://www.google.com/
http://www.directhit.com/
http://www.mamma.com/
http://www.goto.com/

REFERENCES

A. General

Baxevanis, A. D., and B. F. Francis Oueliette. 1998. Internet basics for biologists, in: Current Protocols in Molecular Biology. John Wiley & Sons, New York.
Kiley, R. 1996. Medical Information on the Internet. A Guide for Health Professionals. Churchill & Livingstone, New York.
O'Donnell, K., and L. Winger. 1998. The Internet for Scientists. Harwood Academic Publishers, Australia.
Peruski L. F. Jr., and A. H. Peruski. 1997. The Internet and the New Biology: Tools for Genomic and Molecular Research. American Society for Microbiology, Washington, DC.
Swindell, R., R. Miller, and G.S.A. Myers (eds.). 1996. Internet for the Molecular Biologist, Portland, OR.

B. Specific

1. Anonymous. 1997. Research information on the Internet. AARN News Lett. 53:9.
2. Baxevanis, A. D. 1998. The Internet and the biologist. Methods Biochem Anal. 39: 1–15.
3. Butler, D. 1999. The writing is on the Web for science journals in print. Nature 397: 195–200.
4. East, S. P. 1998. Combing the Web. Chem. Biol. 5:R337.
5. Glowniak, J. 1998. History, structure, and function of the Internet. Semin. Nucl. Med.. 28:135–144.
6. Harmsen, D., J. Rothganger, C. Singer, J. Albert, and M. Frosch. 1999. Intuitive hypertext-based molecular identification of micro-organisms. Lancet. 353:291.
7. Houston, J. D., and D. C. Fiore. 1998. Online medical surveys: Using the Internet as a research tool. MD Comput. 15:116–120.

8. Johnson, W. 1998. Serving up food safety. Occup. Health Saf. 67:174–176.
9. Kushi, L. H., J. Finnegan, B. Martinson, J. Rightmyer, C. Vachon, and L. Yochum. 1997. Epidemiology and the Internet. Epidemiology 8:689–690.
10. McLeod, S. D. 1998. The quality of medical information on the Internet. A new public health concern. Arch. Ophthalmol. 116:1663–1665.
11. Recipon, H., and W. Makalowski. 1997. The biologist and the World Wide Web: An overview of the search engines technology, current status and future perspectives. Curr. Opin. Biotechnol. 8:115–118.
12. Rothman, K. J., C. I. Cann, and A. M. Walker. 1997. Epidemiology and the Internet. Epidemiology 8:123–125.
13. Simmler, M. C., and P. Dessen. 1998. The Internet for the medical and scientific community. Mol. Hum. Reprod. 4:725–730.

Index